Predictive Modeling
with SAS® Enterprise Miner™
Practical Solutions for Business Applications
Second Edition

Kattamuri S. Sarma, PhD

support.sas.com/bookstore

The correct bibliographic citation for this manual is as follows: Sarma, Kattamuri S., PhD. 2013. *Predictive Modeling with SAS® Enterprise Miner™: Practical Solutions for Business Applications, Second Edition*. Cary, NC: SAS Institute Inc.

Predictive Modeling with SAS® Enterprise Miner™: Practical Solutions for Business Applications, Second Edition

Copyright © 2013, SAS Institute Inc., Cary, NC, USA

ISBN 978-1-60764-767-6

SAS Institute Inc., SAS Campus Drive, Cary, North Carolina 27513-2414.

December 2013

SAS provides a complete selection of books and electronic products to help customers use SAS® software to its fullest potential. For more information about our offerings, visit **support.sas.com/bookstore** or call 1-800-727-3228.

SAS® and all other SAS Institute Inc. product or service names are registered trademarks or trademarks of SAS Institute Inc. in the USA and other countries. ® indicates USA registration.

Other brand and product names are trademarks of their respective companies.

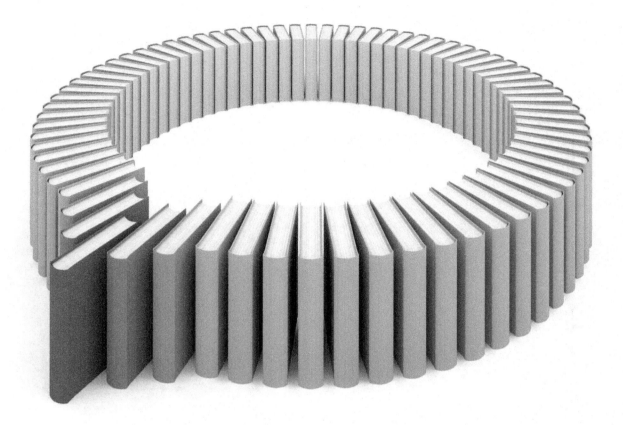

Gain Greater Insight into Your SAS® Software with SAS Books.

Discover all that you need on your journey to knowledge and empowerment.

 support.sas.com/bookstore
for additional books and resources.

Contents

Preface

In order to make effective use of the tools provided in SAS Enterprise Miner, you need to be able to do more than just set the software in motion. While it is essential to know the mechanics of how to use each tool (or *node*, as Enterprise Miner tools are called), you should also understand the methodology behind each one, be able to interpret the output each produces, and know the multitude of options they have. Although this book will appeal to beginners because of its step-by-step, screen-by-screen introduction to the basic tasks that can be accomplished with Enterprise Miner, it also provides the depth and background needed to master many of the more complex but rewarding concepts contained within this versatile product.

The book begins by introducing the basics of creating a project, manipulating data sources, choosing the right property values for each node, and navigating through different results windows. It then demonstrates various pre-processing tools required for building predictive models before treating the three main predictive modeling tools: **Decision Tree**, **Neural Network**, and **Regression**. These are addressed in considerable detail, with numerous examples of practical business applications that are illustrated with tables, charts, displays, equations, and even manual calculations that let you see the essence of what Enterprise Miner is doing as it estimates or optimizes a given model. By the time you finish with this book, Enterprise Miner will no longer be a "black box": you will have an in-depth understanding of the product's inner workings. This book strives to show the link between the output generated by Enterprise Miner and the statistical theory behind the business analysis that requires Enterprise Miner. I also examine the SAS code generated by each node and show the correspondence between the theory and the results produced by Enterprise Miner. In many places, however, I give intuitive explanations of the way that various nodes such as **Decision Tree**, **Neural Network**, **Regression**, and **Variable Selection** operate and how different options such as Model Selection Criteria and Model Assessment are implemented. These explanations are intended not to replicate the exact steps that SAS uses internally to make these computations, but to give a good practical sense of how these tools work. Overall, I believe this approach will help you use the tools in Enterprise Miner with greater comprehension and confidence.

Several examples of business questions drawn from the insurance and banking industries that are based on simulated, but realistic, data are used to illustrate the Enterprise Miner tools. However, the procedures discussed are relevant for any industry. I also include tables and graphs from the output data sets created by various nodes to show how to make custom tables from the results produced by Enterprise Miner. In the end, you should have gained enough understanding of Enterprise Miner to become comfortable and innovative in adapting the applications discussed here to solve your own business problems.

Chapter Details

Chapter 1 discusses research strategy. This includes general issues such as defining the target population, defining the target (or dependent) variable, collecting data, cleaning the data, and selecting an appropriate model.

Chapter 2 shows how to open Enterprise Miner, start a new project, and create data sources. It shows various components of the Enterprise Miner window and shows how to create a process flow diagram. In this chapter, I use example data sets to demonstrate in detail how to use the **Input Data**, **Data Partition**, **Filter**, **File Import**, **Time Series**, **Merge**, **Append**, **StatExplore**, **MultiPlot**, **Graph Explore**, **Variable Clustering**, **Cluster**, **Variable Selection**, **Drop**, **Replacement**, **Impute**, **Interactive Binning**, **Principal Components**, **Transform Variables**, and **SAS Code** nodes. I also discuss the output and SAS code generated by some of these nodes. I manually compute certain statistics, such as Cramer's V, and compare the results with those produced by **StatExplore**. Finally, I explain the details of how to compute Eigenvalues, Eigenvectors, and Principal Components.

Chapter 3 covers the **Variable Selection** and **Transform Variables** nodes in detail. When using the **Variable Selection** node you have many options, depending on the type of target and the measurement scale of the inputs. To help clarify the concepts, I illustrate each situation with a separate data set. This chapter also shows how to make variable selection using the **Variable Clustering** and **Decision Tree** nodes.

Chapter 4 discusses decision trees and regression trees. First, I present the general tree methodology and, using a simple example, I manually work through the sequence of steps— growing the tree, classifying the nodes, and pruning—which is performed by the **Decision Tree** node. I then show how decision tree models are built for predicting response and risk by presenting two examples based on a hypothetical auto insurance company. The first model predicts the probability of response to a mail order campaign. The second model predicts risk as measured by claim frequency, and since claim frequency is measured as a continuous variable, the model built in this case is a regression tree. A detailed discussion of the SAS code generated by the **Decision Tree** node is included at the end of the chapter. This chapter shows how to develop decision trees interactively.

Chapter 5 provides an introduction to neural networks. Here I try to demystify the neural networks methodology by giving an intuitive explanation using simple algebra. I show how to configure neural networks to be consistent with economic and statistical theory and how to interpret the results correctly. The neural network architecture—input layer, hidden layers, and output layer—is illustrated algebraically with numerical examples. Although the formulas presented here may look complex, they do not require a high-level knowledge of mathematics, and patience in working through them will be rewarded with a thorough understanding of neural networks and their applications.

In this chapter, I first intuitively discuss the iterative processes of estimating the model using the training data set as well as selecting the optimal weights for the model using the validation data set. Next, explicit numerical examples are given to clarify each step. As in Chapter 4, two models are developed using the hypothetical insurance data: a response model with a binary target, and a risk model with (in this case) an ordinal target representing accident frequency. Line by line, I examine the SAS code generated by each node and show the correspondence between the theory and the results produced by Enterprise Miner.

The calculations behind the Receiver Operating Characteristic (ROC) Charts are illustrated using the results produced by the **Model Comparison** node.

Alternative specifications of the Neural Networks including Multi Layer Perceptron (MLP) Neural Network, Radial Basis Function neural networks and various built-in architectures of the Neural Network node are illustrated through mathematical representation and SAS code generated by the Neural Network node.

AutoNeural, DMNeural and Dmine Regression nodes are illustrated and the models developed by **DMNeural, AutoNeural and Dmine Regression** nodes are compared.

Chapter 6 demonstrates how to develop logistic regression models for targets with different measurement scales: binary, categorical with more than two categories, ordinal, and continuous (interval-scaled). Using an example data set with a binary target, I demonstrate various model selection criteria and model selection methods. I also present business applications from the banking industry involving two predictive models: one with a binary target, and one with a continuous target. The model with a binary target predicts the probability of response to a mail campaign while the model with a continuous target predicts the increase in deposits that is due to an interest rate increase. This chapter also shows how to calculate the lift and capture rates of the models when the target is continuous.

In Chapter 7, I compare the results of three modeling tools—**Decision Tree**, **Neural Network**, and **Regression**—that were presented in earlier chapters. For this purpose I develop two predictive models and then take turns applying the three modeling tools to each model. The first model has a binary target and predicts the probability of customer attrition for a fictitious bank. The second model has an ordinal target, which is a discrete version of a continuous variable, and predicts risk (as measured by loss frequency) for a fictitious auto insurance company. This chapter also provides a method of computing the lift and capture rates of these models using the expected value of the target variable.

In this chapter the methods of boosting and combining predictive models is illustrated using the **Gradient Boosting** and **Ensemble** nodes. The predictive performance of these two methods are compared.

Chapter 8 shows how to calculate profitability for each of the ten deciles created when a data set of prospective customers is scored using the output of the modeling process. It then shows how to use these profitability estimates to address questions such as how to choose an optimum cut-off point for a mailing campaign. Here my objective is to introduce the notion of the marginal cost and marginal revenue associated with risk and response and to show how they can be used to make rational quantitative decisions in the marketing sphere.

Chapter 9 gives an introduction to predictive modeling using unstructured textual data. Quantifying and textual and put it into a spread sheet or SAS table form is an important pre-requisite for developing predictive models with textual data. Quantifying textual data involves several steps: These are parsing the documents, filtering, and reducing decreasing the dimension reduction. Dimension reduction is done by Singular Value Decomposition (SVD). I first illustrated the quantification of textual data, Boolean Retrieval method and dimension reduction using SVD method using a simplified example. Then I showed how to use Text Parsing, Text Filter, Text Topic and Text Cluster nodes. Then I illustrated how to use the output data set produced by the Text Topic node for estimating a logistic regression equation. Using a simple example I demonstrated the Expectation-Maximization (EM) Clustering. I have explained the Hierarchical clustering method with simple algebra.

Exercises are included at the end of Chapters 2 -7 and 9.

How to Use the Book

- To get the most out of this book, open Enterprise Miner and follow the sequence of tasks performed in each chapter, using either the data sets stored on the CD included with this book or, even better, your own data sets.
- Work through the manual calculations as well as the mathematical derivations presented in the book to get an in-depth understanding of the logic behind different models.
- To learn predictive modeling, read the general explanation and the theory, and then follow the steps given in the book to develop models using either the data sets provided on the CD or your own data sets. Try variations of what is done in the book to strengthen your understanding of the topics covered.
- If you already know Enterprise Miner and want to get a good understanding of decision trees and neural networks, focus on the examples and detailed derivations given in Chapters 4, 5, and 6. These derivations are not as complex as they appear to be.

Prerequisites

- Elementary algebra and basic training (equivalent to one to two semesters of course work) in statistics covering inference, hypothesis testing, probability, and regression
- Familiarity with measurement scales of variables—continuous, categorical, ordinal, etc.
- Experience with Base SAS software and some understanding of simple SAS macros and macro variables

About This Book

Purpose

The purpose of this book is to demonstrate how to develop predictive models quickly and effectively using SAS Enterprise Miner, to demonstrate how to use Enterprise Miner tools for accomplishing various tasks involved in developing predictive models, and to provide an in-depth explanation of the theory and computations of each tool.

Is This Book for You?

If you are a graduate student, researcher, or statistician interested in predictive modeling, a data mining expert who wants to learn SAS Enterprise Miner, or a business analyst looking for an introduction to predictive modeling using SAS Enterprise Miner, you'll be able to efficiently and skillfully develop predictive models using the theory and examples presented in this book.

Prerequisites

You will need a knowledge of basic statistics that includes an understanding of regression analysis, statistical hypothesis testing, analysis of variance knowledge of linear algebra, and a good grasp of the business problem being analyzed.

What's New in This Edition

This second edition features expanded coverage of the SAS Enterprise Miner nodes, now including File Import, Time Series, Variable Clustering, Cluster, Interactive Binning, Principal Components, AutoNeural, DMNeural, Dmine Regression, Gradient Boosting, Ensemble, and Text Mining.

About the Examples

Software Used to Develop the Book's Content

SAS Enterprise Miner 12.1 and SAS Text Miner.

The data and programs used in this book will be available from the author's page at:
http://support.sas.com/publishing/authors

Example Code and Data

You can access the example code and data for this book by linking to its author page at:
http://support.sas.com/publishing/authors. Select the name of the author. Then, look for the cover thumbnail of this book, and select Example Code and Data to display the SAS programs that are included in this book.

For an alphabetical listing of all books for which example code and data is available, see
http://support.sas.com/bookcode. Select a title to display the book's example code.

If you are unable to access the code through the website, send an e-mail to saspress@sas.com.

Output and Graphics Used in This Book

The output and graphs are from the Results windows of the Enterprise Miner. Additional graphics are generated using SAS/Graph. Most of the tables are from the results of various nodes. Additional tables are created using SAS code in the SAS code node. Exercises are added at the end of chapters 2 through 7 and 9.

Exercise Solutions

The exercises given are for reinforcing the methods used. For most exercises, the solutions will be obvious when the steps given are followed. For other exercises, answers will be posted on author's page.

Additional Resources

SAS offers you a rich variety of resources to help build your SAS skills and explore and apply the full power of SAS software. Whether you are in a professional or academic setting, we have learning products that can help you maximize your investment in SAS.

Bookstore	http://support.sas.com/bookstore/
Training	http://support.sas.com/training/
Certification	http://support.sas.com/certify/
SAS Global Academic Program	http://support.sas.com/learn/ap/
SAS OnDemand	http://support.sas.com/learn/ondemand/
Support	http://support.sas.com/techsup/
Training and Bookstore	http://support.sas.com/learn/
Community	http://support.sas.com/community/

Keep in Touch

We look forward to hearing from you. We invite questions, comments, and concerns. If you want to contact us about a specific book, please include the book title in your correspondence.

To Contact the Author through SAS Press

By e-mail: saspress@sas.com

Via the Web: http://support.sas.com/author_feedback

SAS Books

For a complete list of books available through SAS, visit http://support.sas.com/bookstore.

Phone: 1-800-727-3228

Fax: 1-919-677-8166

E-mail: sasbook@sas.com

SAS Book Report

Receive up-to-date information about all new SAS publications via e-mail by subscribing to the SAS Book Report monthly eNewsletter. Visit http://support.sas.com/sbr.

Publish with SAS

SAS is recruiting authors! Are you interested in writing a book? Visit http://support.sas.com/saspress for more information.

About The Author

 Kattamuri S. Sarma, PhD, is an economist and statistician with 30 years of experience in American business, including stints with IBM and AT&T. He is the founder and president of Ecostat Research Corp., a consulting firm specializing in predictive modeling and forecasting. Over the years, Dr. Sarma has developed predictive models for the banking, insurance, telecommunication, and technology industries. He has been a SAS user since 1992, and he has extensive experience with multivariate statistical methods, econometrics, decision trees, and data mining with neural networks. The author of numerous professional papers and publications, Dr. Sarma is a SAS Certified Professional and a SAS Alliance Partner. He received his bachelor's degree in mathematics and his master's degree in economic statistics from universities in India. Dr. Sarma received his PhD in economics from the University of Pennsylvania, where he worked under the supervision of Nobel Laureate Lawrence R. Klein.

Learn more about this author by visiting his author page at http://support.sas.com/publishing/authors/sarma.html. There you can download free book excerpts, access example code and data, read the latest reviews, get updates, and more.

About The Author

Acknowledgments

At the beginning of this project, I had many discussions with my son Ravi Sarma about practical aspects of modeling techniques, especially neural networks and decision trees. Ravi, who was then a graduate student at MIT working on a neural networks project, read the manuscript at several stages and contributed editorial assistance, in-depth questions, and insightful comments that greatly improved the text. Ravi has been a source of encouragement and intellectual support during the entire project.

I am grateful to Linda Kellogg for helping me in the customized installation of various versions of SAS Enterprise Miner and for coming to my rescue whenever I faced a technical question. The second edition of this book would not have been possible without Linda's generous help.

Many thanks go to other SAS technical support personnel who thoroughly answered my questions about different nodes of the SAS Enterprise Miner.

In addition, I would like to thank Joan Keyser for her outstanding editorial work. I would like to thank Goutam Chakraborty who, in addition to reviewing the book, provided me very positive suggestions for making improvements. I would like to thank Russell Albright for generously reviewing the complex topic of text mining.

I would like to thank Julie Platt and Stacey Hamilton for their solid support and extraordinary patience when I was revising the book several times with various versions of the SAS Enterprise Miner.

I would like to thank my wife, Lokamatha, for putting up with me while I spent many weekends and holidays working on this book.

xxii

Chapter 1: Research Strategy

1.1 Introduction

This chapter discusses the planning and organization of a predictive modeling project. Planning involves tasks such as these:

- defining and measuring the target variable in accordance with the business question
- collecting the data
- comparing the distributions of key variables between the modeling data set and the target population to verify that the sample adequately represents the target population
- defining sampling weights if necessary
- performing data-cleaning tasks that need to be done prior to launching SAS Enterprise Miner

Alternative strategies for developing predictive models using SAS Enterprise Miner are discussed at the end of this chapter.

1.2 Measurement Scales for Variables

Because many of the steps above involve a discussion of the data and types of variables in our data sets, I will first define the measurement scales for variables that are used in this book. In general, I have tried to follow the definitions given by Alan Agresti

- A *categorical variable* is one for which the measurement scale consists of a set of categories.

- Categorical variables for which levels (categories) do not have a natural ordering are called *nominal*.
- Categorical variables that do have a natural ordering of their levels are called *ordinal*.
- An *interval variable* is one that has numerical distances between any two levels of the scale.[1]

According to the above definitions, the variables INCOME and AGE in Tables 1.1 to 1.5 and BAL_AFTER in Table 1.3 are interval-scaled variables. Because the variable RESP in Table 1.1 is categorical and has only two levels, it is called a binary variable. The variable LOSSFRQ in Table 1.2 is ordinal. (In SAS Enterprise Miner you can change its measurement scale to interval, but I have left it as ordinal.) The variables PRIORPR and NEXTPR in Table 1.5 are nominal.

Interval-scaled variables are sometimes called *continuous*. Continuous variables are treated as interval variables. Therefore I use the terms *interval-scaled* and *continuous* interchangeably.

I also use the terms ordered polychotomous variables and ordinal variables interchangeably. Similarly, I use the terms unordered polychotomous variables and nominal variables interchangeably.

1.3 Defining the Target

The first step in any data mining project is to define and measure the target variable to be predicted by the model that emerges from your analysis of the data. This section presents examples of this step applied to five different business questions.

1.3.1 Predicting Response to Direct Mail

In this example, a hypothetical auto insurance company wants to acquire customers through direct mail. The company wants to minimize mailing costs by targeting only the most responsive customers. Therefore, the company decides to use a response model. The target variable for this model is RESP, and it is binary, taking the value of 1 for response and 0 for no response.

Table 1.1 shows a simplified version of a data set used for modeling the binary target response (RESP).

Table 1.1

CUSTOMER	AGE	INCOME	STATUS	PC	NC	RESP
1	25	$45,000	S	1	1	0
2	45	$61,000	MC	1	2	1
3	54	.	MC	1	3	0
4	32	$24,000	MNC	0	4	0
5	43	$31,000	MC	0	5	0
6	56	$23,456	MC	1	6	1
7	78	.	W	0	7	0
8	6	$100,256	D	1	1	0
9	26	$345,678	MNC	1	2	1
10	32	$100,211	S	0	3	0
11	51	$21,312	MC	1	4	0
12	31	$83,456		0	5	1
13	23	$24,234	MNC	1	1	0
14	47	$43,566	MC	0	3	0
15	77	$12,002	MC	1	4	1
16	83	$32,454	W	1	5	0
17	25	$61,345	S	0	6	0
18	32	$76,123	MC	1	7	0
19	52	$25,324		1	8	0
20	32	$31,886	MNC	0	1	0
21	23	$78,345	S	1	8	0
22	80	$61,234	MNC	1	2	0
23	123	$76,876	S	1	4	0
24	45	$24,002		3	5	0

In Table 1.1 the variables AGE, INCOME, STATUS, PC, and NC are input variables (or explanatory variables). AGE and INCOME are numeric and, although they could theoretically be considered continuous, it is simply more practical to treat them as interval-scaled variables.

The variable STATUS is categorical and nominal-scaled. The categories of this variable are S if the customer is single and never married, MC if married with children, MNC if married without children, W if widowed, and D if divorced.

The variable PC is numeric and binary. It indicates whether the customers own a personal computer, taking the value 1 if they do and 0 if not. The variable NC represents the number of credit cards the customers own. You can decide whether this variable is ordinal or interval scaled.

The target variable is RESP and takes the value 1 if the customer responded, for example, to a mailing campaign, and 0 otherwise. A binary target can be either numeric or character; I could have recorded a response as Y instead of 1, and a non- response as N instead of 0, with virtually no implications for the form of the final equation.

Note that there are some extreme values in the table. For example, one customer's age is recorded as 6. This is obviously a recording error, and the age should be corrected to show the actual value, if possible. Income has missing values that are shown as dots, while the nominal variable STATUS has missing values that are represented by blanks. The **Impute** node of SAS Enterprise Miner can be used to impute such missing values. See Chapters 2, 6, and 7 for details.

1.3.2 Predicting Risk in the Auto Insurance Industry

The auto insurance company wants to examine its customer data and classify its customers into different risk groups. The objective is to align the premiums it is charging with the risk rates of its customers. If high-risk customers are charged low premiums, the loss ratios will be too high and the company will be driven out of business. If low-risk customers are charged disproportionately high rates, then the company will lose customers to its competitors. By accurately assessing the risk profiles of its customers, the company hopes to set customers' insurance premiums at an optimum level consistent with risk. A risk model is needed to assign a risk score to each existing customer.

In a risk model, *loss frequency* can be used as the target variable. Loss frequency is calculated as the number of losses due to accidents per *car-year*, where car-year is equal to the time since the auto insurance policy went into effect, expressed in years, multiplied by the number of cars covered by the policy. Loss frequency can be treated as either a continuous (interval-scaled) variable or a discrete (ordinal) variable that classifies each customer's losses into a limited number of bins. (See Chapters 5 and 7 for details about bins.) For purposes of illustration, I model loss frequency as a continuous variable in Chapter 4 and as a discrete ordinal variable in Chapters 5 and 7. The loss frequency considered here is the loss arising from an accident in which the customer was "at fault," so it could also be referred to as "at-fault accident frequency". I use *loss frequency, claim frequency,* and *accident frequency* interchangeably.

Table 1.2 shows what the modeling data set might look like for developing a model with loss frequency as an ordinal target.

Table 1.2

CUSTOMER	AGE	INCOME	NPRVIO	LOSSFRQ
1	25	$45,000	0	0
2	45	$61,000	1	1
3	54	.	2	0
4	32	$24,000	3	3
5	43	$31,000	4	0
6	56	$23,456	0	0
7	78	.	1	2
8	6	$100,256	3	2
9	26	$345,678	4	1
10	32	$100,211	5	3
11	51	$21,312	3	2
12	31	.	1	1
13	23	$24,234	0	0
14	47	$43,566	1	0
15	77	$12,002	0	0
16	83	$32,454	0	2
17	25	$61,345	1	1
18	32	$76,123	1	0
19	52	$25,324	3	1
20	32	$31,886	1	0
21	23	$78,345	3	3
22	80	$61,234	2	0
23	123	$76,876	2	0
24	45	$24,002	1	1

The target variable is LOSSFRQ, which represents the accidents per car-year incurred by a customer over a period of time. This variable is discussed in more detail in subsequent chapters in this book. For now it is sufficient to note that it is an ordinal variable that takes on values of 0, 1, 2, and 3. The input variables are AGE, INCOME, and NPRVIO. The variable NPRVIO represents the number of previous violations a customer had before he purchased the insurance policy.

1.3.3 Predicting Rate Sensitivity of Bank Deposit Products

In order to assess customers' sensitivity to an increase in the interest rate on a savings account, a bank may conduct price tests. Suppose one such test involves offering a higher rate for a fixed period of time, called the *promotion window*.

In order to assess customer sensitivity to a rate increase, it is possible to fit three types of models to the data generated by the experiment:

- a response model to predict the probability of response
- a short-term demand model to predict the expected change in deposits during the promotion period
- a long-term demand model to predict the increase in the level of deposits beyond the promotion period

The target variable for the response model is binary: response or no response. The target variable for the short-term demand model is the increase in savings deposits during the promotion period net[2] of any concomitant declines in other accounts. The target variable for the long-term demand model is the amount of the increase remaining in customers' bank accounts after the promotion period. In the case of this model, the promotion window for analysis has to be clearly defined, and only customer transactions that have occurred prior to the promotion window should be included as inputs in the modeling sample.

Table 1.3 shows what the data set looks like for modeling a continuous target.

Table 1.3

CUSTOMER	AGE	INCOME	B_JAN	B_FEB	B_MAR	B_APR	BAL_AFTER
1	25	$45,000	$4,000	$4,230	$4,400	$4,900	$5,900
2	45	$61,000	$5,000	$4,000	$3,000	$0	$2,000
3	54	.	$1,200	$1,100	$3,000	$100	$200
4	32	$24,000	$5,234	$345	$5,678	$78	$878
5	43	$31,000	$4,000	$4,232	$4,100	$4,700	$4,950
6	56	$23,456	$2,000	$4,000	$3,000	$20	$1,000
7	78	.	$1,200	$1,100	$3,000	$100	$1,300
8	6	$100,256	$5,234	$345	$5,678	$78	$1,088
9	26	$345,678	$3,435	$4,674	$678	$80,000	$80,000
10	32	$100,211	$787	$4,230	$4,400	$4,900	$5,900
11	51	$21,312	$8,750	$7,800	$3,456	$50	$10,000
12	31	.	$5,000	$4,000	$3,000	$100	$4,000
13	23	$24,234	$4,000	$4,230	$4,400	$4,376	$5,900
14	47	$43,566	$4,674	$678	$800	$7,890	$8,890
15	77	$12,002	$5,234	$345	$5,678	$78	$1,078
16	83	$32,454	$4,000	$4,230	$4,400	$4,900	$5,900
17	25	$61,345	$2,000	$4,000	$3,000	$120	$1,000
18	32	$76,123	$1,200	$1,100	$3,000	$100	$1,100
19	52	$25,324	$5,234	$345	$5,678	$78	$1,078
20	32	$31,886	$3,435	$4,674	$678	$8,000	$9,000
21	23	$78,345	$787	$4,230	$4,400	$4,900	$5,900
22	80	$61,234	$8,780	$7,800	$3,456	$0	$100
23	123	$76,876	$5,000	$4,000	$3,000	$250	$1,034
24	45	$24,002	$4,300	$4,200	$4,400	$4,900	$7,245

The data set shown in Table 1.3 represents an attempt by a hypothetical bank to induce its customers to increase their savings deposits by increasing the interest paid to them by a predetermined number of basis points. This increased interest rate was offered (let us assume) in May 2006. Customer deposits were then recorded at the end of May 2006 and stored in the data set shown in Table 1.3 under the variable name BAL_AFTER. The bank would like to know what type of customer is likely to increase her savings balances the most in response to a future incentive of the same amount. The target variable for this is the dollar amount of change in balances from a point before the promotion period to a point after the promotion period. The target variable is continuous. The inputs, or explanatory variables, are AGE, INCOME, B_JAN, B_FEB, B_MAR, and B_APR. The variables B_JAN, B_FEB, B_MAR, and B_APR refer to customers' balances in all their accounts at the end of January, February, March, and April of 2006, respectively.

1.3.4 Predicting Customer Attrition

In banking, attrition may mean a customer closing a savings account, a checking account, or an investment account. In a model to predict attrition, the target variable can be either binary or continuous. For example, if a bank wants to identify customers who are likely to terminate their accounts at *any* time within a pre-defined interval of time in the future, it is possible to model attrition as a binary target. However, if the bank is interested in predicting the *specific* time at which the customer is likely to "attrit," then it is better to model attrition as a continuous target—time to attrition.

In this example, attrition is modeled as a binary target. When you model attrition using a binary target, you must define a performance window during which you observe the occurrence or non-occurrence of the event. If a customer attrited during the performance window, the record shows 1 for the event and 0 otherwise.

Any customer transactions (deposits, withdrawals, and transfers of funds) that are used as inputs for developing the model should take place during the period prior to the performance window. The *inputs window* during which the transactions are observed, the *performance window* during which the event is observed, and the *operational lag*, which is the time delay in acquiring the inputs, are discussed in detail in Chapter 7 where an attrition model is developed.

Table 1.4 shows what the data set looks like for modeling customer attrition.

Table 1.4

CUSTOMER	AGE	INCOME	B_JAN	B_FEB	B_MAR	B_APR	ATTR
1	25	$45,000	$4,000	$4,230	$4,400	$4,900	0
2	45	$61,000	$5,000	$4,000	$3,000	$0	1
3	54	.	$1,200	$1,100	$3,000	$100	0
4	32	$24,000	$5,234	$345	$5,678	$78	0
5	43	$31,000	$4,000	$4,232	$4,100	$4,700	0
6	56	$23,456	$2,000	$4,000	$3,000	$20	1
7	78	.	$1,200	$1,100	$3,000	$100	0
8	6	$100,256	$5,234	$345	$5,678	$78	0
9	26	$345,678	$3,435	$4,674	$678	$80,000	1
10	32	$100,211	$787	$4,230	$4,400	$4,900	0
11	51	$21,312	$8,750	$7,800	$3,456	$50	1
12	31	.	$5,000	$4,000	$3,000	$100	1
13	23	$24,234	$4,000	$4,230	$4,400	$4,376	0
14	47	$43,566	$4,674	$678	$800	$7,890	0
15	77	$12,002	$5,234	$345	$5,678	$78	1
16	83	$32,454	$4,000	$4,230	$4,400	$4,900	0
17	25	$61,345	$2,000	$4,000	$3,000	$120	0
18	32	$76,123	$1,200	$1,100	$3,000	$100	0
19	52	$25,324	$5,234	$345	$5,678	$78	0
20	32	$31,886	$3,435	$4,674	$678	$8,000	0
21	23	$78,345	$787	$4,230	$4,400	$4,900	0
22	80	$61,234	$8,780	$7,800	$3,456	$0	0
23	123	$76,876	$5,000	$4,000	$3,000	$250	0
24	45	$24,002	$4,300	$4,200	$4,400	$4,900	0

In the data set shown in Table 1.4, the variable ATTR represents the customer attrition observed during the performance window, consisting of the months of June, July, and August of 2006. The target variable takes the value of 1 if a customer attrits during the performance window and 0 otherwise. Table1.4 shows the input variables for the model. They are AGE, INCOME, B_JAN, B_FEB, B_MAR, and B_APR. The variables B_JAN, B_FEB, B_MAR, and B_APR refer to customers' balances for all of their accounts at the end of January, February, March, and April of 2006, respectively.

1.3.5 Predicting a Nominal Categorical (Unordered Polychotomous)Target

Assume that a hypothetical bank wants to predict, based on the products a customer currently owns and other characteristics, which product the customer is likely to purchase next. For example, a customer may currently have a savings account and a checking account, and the bank would like to know if the customer is likely to open an investment account or an IRA, or take out a mortgage. The target variable for this situation is nominal. Models with nominal targets are also used by market researchers who need to understand consumer preferences for different products or brands. Chapter 6 shows some examples of models with nominal targets.

Table 1.5 shows what a data set might look like for modeling a nominal categorical target.

Table 1.5

CUSTOMER	AGE	INCOME	PRIORPR	NEXTPR
1	25	$45,000	A	X
2	45	$61,000	B	Z
3	54	.	C	Y
4	32	$24,000	A	X
5	43	$31,000	B	Z
6	56	$23,456	C	Z
7	78	.	C	Z
8	6	$100,256	A	X
9	26	$345,678	AB	X
10	32	$100,211	CD	Z
11	51	$21,312	AC	Y
12	31	.	AB	X
13	23	$24,234	CD	Z
14	47	$43,566	D	Z
15	77	$12,002	E	Z
16	83	$32,454	A	X
17	25	$61,345	B	X
18	32	$76,123	A	Z
19	52	$25,324	A	Y
20	32	$31,886	C	X
21	23	$78,345	D	Z
22	80	$61,234	A	Z
23	123	$76,876	B	Z
24	45	$24,002	D	X

In Table 1.5, the input data includes the variable PRIORPR, which indicates the product or products owned by the customer of a hypothetical bank at the beginning of the performance window. The *performance window*, defined in the same way as in Section 1.3.4, is the time period during which a customer's purchases are observed. Given that a customer owned certain products at the beginning of the performance window, we observe the next product that the customer purchased during the performance window and indicate it by the variable NEXTPR.

For each customer, the value for the variable PRIORPR indicates the product that was owned by the customer at the beginning of the performance window. The letter A might stand for a savings account, B might stand for a certificate of deposit, etc. Similarly, the value for the variable NEXTPR indicates the first product purchased by a customer during the performance window. For example, if the customer owned product B at the beginning of the performance window and purchased products X and Z, in that order, during the performance window, then the variable NEXTPR takes the value X. If the customer purchased Z and X, in that order, the variable NEXTPR takes the value Z, and the variable PRIORPR takes the value B on the customer's record.

1.4 Sources of Modeling Data

There are two different scenarios by which data becomes available for modeling. For example, consider a marketing campaign. In the first scenario, the data is based on an experiment carried out by conducting a marketing campaign on a well-designed sample of customers drawn from the target population. In the second scenario, the data is a sample drawn from the results of a past marketing campaign and not from the target population. While the latter scenario is clearly less desirable, it is often necessary to make do with whatever data is available. In such cases, you can make some adjustments through observation weights to compensate for the lack of perfect compatibility between the modeling sample and the target population.

In either case, for modeling purposes, the file with the marketing campaign results is appended to data on customer characteristics and customer transactions. Although transaction data is not always available, these tend to be key drivers for predicting the attrition event.

1.4.1 Comparability between the Sample and Target Universe

Before launching a modeling project, you must verify that the sample is a good representation of the target universe. You can do this by comparing the distributions of some key variables in the sample and the target universe. For example, if the key characteristics are age and income, then you should compare the age and income distribution between the sample and the target universe.

1.4.2 Observation Weights

If the distributions of key characteristics in the sample and the target population are different, sometimes observation weights are used to correct for any bias. In order to detect the difference between the target population and the sample, you must have some prior knowledge of the target population. Assuming that age and income are the key characteristics, you can derive the weights as follows: Divide income into, let's say, four groups and age into, say, three groups. Suppose that the target universe has N_{ij} people in the ith age group and j^{th} income group, and assume that the sample has n_{ij} people in the same age-income group. In addition, suppose the total number of people in the target population is N, and the total number of people in the sample is n. In this case, the appropriate observation weight is $(N_{ij}/N)/(n_{ij}/n)$ for the individual in the ith age group and j^{th} income group in the sample. You should construct these observation weights and include them for each record in the modeling sample prior to launching SAS Enterprise Miner, in effect creating an additional variable in your data set. In SAS Enterprise Miner, you assign the role of **Frequency** to this variable in order for the modeling tools to consider these weights in estimating the models. This situation inevitably arises when you do not have a scientific sample drawn from the target population, which is very often the case.

However, another source of bias is often deliberately introduced. This bias is due to over-sampling of rare events. For example, in response modeling, if the response rate is very low, you must include all the responders available and only a random fraction of non-responders. The bias introduced by such over-sampling is corrected by adjusting the predicted probabilities with prior probabilities. These techniques are discussed in Section 4.8.2.

1.5 Pre-Processing the Data

Pre-processing has several purposes:

- eliminate obviously irrelevant data elements, e.g., name, social security number, street address, etc., that clearly have no effect on the target variable
- convert the data to an appropriate measurement scale, especially converting categorical (nominal scaled) data to interval scaled when appropriate
- eliminate variables with highly skewed distributions
- eliminate inputs which are really target variables disguised as inputs
- impute missing values

Although you can do many cleaning tasks within SAS Enterprise Miner, there are some that you should do prior to launching SAS Enterprise Miner.

1.5.1 Data Cleaning Before Launching SAS Enterprise Miner

Data vendors sometimes treat interval-scaled variables, such as birth date or income, as character variables. If a variable such as birth date is entered as a character variable, it is treated by SAS Enterprise Miner as a categorical variable with many categories. To avoid such a situation, it is better to derive a numeric variable from the character variable and then drop the original character variable from your data set.

Similarly, income is sometimes represented as a character variable. The character A may stand for $20K ($20,000), B for $30K, etc. To convert the income variable to an ordinal or interval scale, it is best to create a new version of the income variable in which all the values are numeric, and then eliminate the character version of income.

Another situation which requires data cleaning that cannot be done within SAS Enterprise Miner arises when the target variable is disguised as an input variable. For example, a financial institution wants to model customer attrition in its brokerage accounts. The model needs to predict the probability of attrition during a time interval of three months in the future. The institution decides to develop the model based on actual attrition during a performance window of three months. The objective is to predict attritions based on customers' demographic and income profiles, and balance activity in their brokerage accounts prior to the window. The binary target variable takes the value of 1 if the customer attrits and 0 otherwise. If a customer's balance in his brokerage account is 0 for two consecutive months, then he is considered an attritor, and the target value is set to 1. If the data set includes both the target variable (attrition/no attrition) and the balances during the performance window, then the account balances may be inadvertently treated as input variables. To prevent this, inputs which are really target variables disguised as input variables should be removed before launching SAS Enterprise Miner.

1.5.2 Data Cleaning After Launching SAS Enterprise Miner

Display 1.1 shows an example of a variable that is highly skewed. The variable is MS, which indicates the marital status of a customer. The variable RESP represents customer response to mail. It takes the value of 1 if a customer responds, and 0 otherwise. In this hypothetical sample, there are only 100 customers with marital status M (married), and 2900 with S (single). None of the married customers are responders. An unusual situation such as this may cause the marital status variable to play a much more significant role in the predictive model than is really warranted, because the model tends to infer that all the married customers were non-responders because they were married. The real reason there were no responders among them is simply that there were so few married customers in the sample.

Display 1.1

The SAS System			
The FREQ Procedure			
Frequency	Table of MS by Resp		
Percent		Resp	
Row Pct	MS(Marital		
Col Pct	Status)	0	1	Total
M	100	0	100
	3.33	0.00	3.33
	100.00	0.00	
	3.42	0.00	
S	2826	74	2900
	94.20	2.47	96.67
	97.45	2.55	
	96.58	100.00	
Total	2926	74	3000
	97.53	2.47	100.00

These kinds of variables can produce spurious results if used in the model. You can identify these variables using the **StatExplore** node, set their roles to Rejected in the **Input Data** node, and drop them from the table using the **Drop** node.

The **Filter** node can be used for eliminating observations with extreme values, although I do not recommend elimination of observations. Correcting them or capping them instead might be better, in order to avoid introducing any bias into the model parameters. The **Impute** node offers a variety of methods for imputing missing values. These nodes are discussed in the next chapter. Imputing missing values is necessary when you use **Regression** or **Neural Network** nodes.

1.6 Alternative Modeling Strategies

The choice of modeling strategy depends on the modeling tool and the number of inputs under consideration for modeling. Here are examples of two possible strategies when using the **Regression** node.

1.6.1 Regression with a Moderate Number of Input Variables

Pre-process the data:

- Eliminate obviously irrelevant variables.
- Convert nominal-scaled inputs with too many levels to numeric interval-scaled inputs, if appropriate.
- Create composite variables (such as average balance in a savings account during the six months prior to a promotion campaign) from the original variables if necessary. This can also be done with SAS Enterprise Miner using the **SAS Code** node.

Next, use SAS Enterprise Miner to perform these tasks:

- Impute missing values.
- Transform the input variables.
- Partition the modeling data set into train, validate, and test (when the available data is large enough) samples. Partitioning can be done prior to imputation and transformation, because SAS Enterprise Miner automatically applies these to all parts of the data.
- Run the **Regression** node with the Stepwise option.

1.6.2 Regression with a Large Number of Input Variables

Pre-process the data:

- Eliminate obviously irrelevant variables.
- Convert nominal-scaled inputs with too many levels to numeric interval-scaled inputs, if appropriate.
- Combine variables if necessary.

Next, use SAS Enterprise Miner to perform these tasks:

- Impute missing values.
- Make a preliminary variable selection. (Note: This step is not included in Section 1.6.1.)
- Group categorical variables (collapse levels).
- Transform interval-scaled inputs.
- Partition the data set into train, validate, and test samples.
- Run the **Regression** node with the Stepwise option.

The steps given in Sections 1.6.1 and 1.6.2 are only two of many possibilities. For example, one can use the **Decision Tree** node to make a variable selection and create dummy variables to then use in the **Regression** node.

1.7 Notes

1. Alan Agresti, *Categorical Data Analysis* (New York, NY: John Wiley & Sons, 1990), 2.
2. If a customer increased savings deposits by $100 but decreased checking deposits by $20, then the net increase is $80. Here, *net* means *excluding*.

Chapter 2: Getting Started with Predictive Modeling

2.1 Introduction

This chapter introduces you to SAS Enterprise Miner 12.1 and some of the preprocessing and data cleaning tools (nodes) needed for data mining and predictive modeling projects. SAS Enterprise Miner's modeling tools are not included in this chapter as they are covered extensively in Chapters 4, 5, and 6.

2.2 Opening SAS Enterprise Miner 12.1

To start SAS Enterprise Miner 12.1, click the SAS Enterprise Miner icon on your desktop.[1] If you have a Workstation configuration, the Welcome to Enterprise Miner window opens, as shown in Display 2.1.

Display 2.1

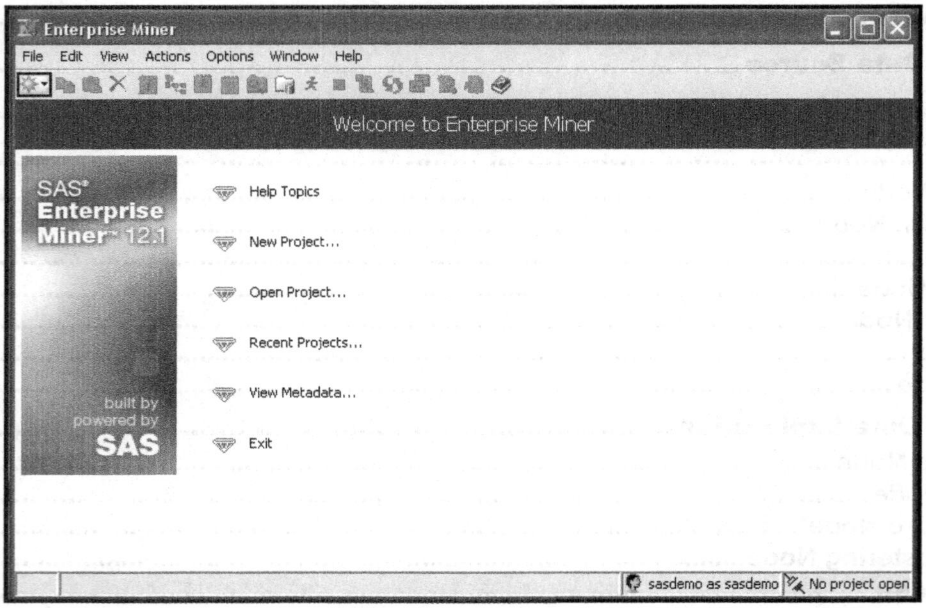

2.3 Creating a New Project in SAS Enterprise Miner 12.1

When you select **New Project** in the Enterprise Miner window, the Create New Project window opens.

In this window, enter the name of the project and the directory where you want to save the project. This example uses Chapter2 and C:\TheBook\EM12.1\EMProjects (the directory where the project will be stored). Click **Next**. A new window opens, which shows the **New Project Information**.

Display 2.2

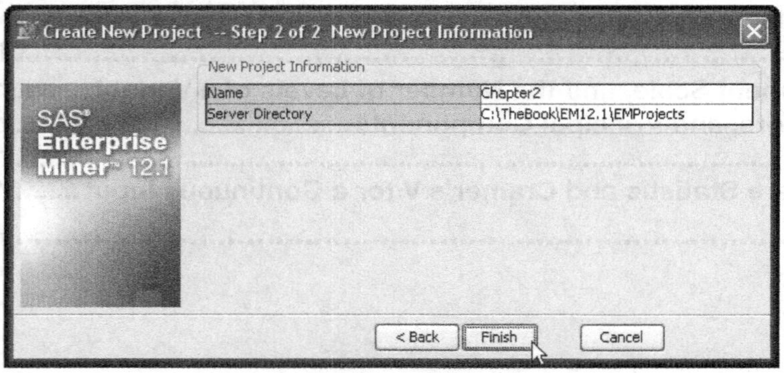

Click **Finish**. The new project is created, and the SAS Enterprise Miner 12.1 interface window opens, showing the new project.

2.4 The SAS Enterprise Miner Window

This is the window where you create the process flow diagram for your data mining project. The numbers in Display 2.3 correspond to the descriptions below the display.

Display 2.3

① **Menu bar**

② **Tools bar**: This contains the Enterprise Miner node (tool) icons. The icons displayed on the toolbar change according to the tab you select in the area indicated by ③.

③ **Node (Tool) group tabs**: These tabs are for selecting different groups of nodes. The toolbar ② changes according to the node group selected. If you select the **Sample** tab on this line, you will see the icons for **Append, Data Partition, File Import, Filter, Input Data, Merge, Sample**, and **Time Series** in ②. If you select the **Explore** tab, you will see the icons for **Association, Cluster, DMDB, Graph Explore, Market Basket, Multiplot, Path Analysis, SOM/Kohonen, StatExplore, Variable Clustering**, and **Variable Selection** in ②.

④ **Project Panel:** This is for viewing, creating, deleting, and modifying the **Data Sources, Diagrams**, and **Model Packages**. For example, if you want to create a data source (tell SAS Enterprise Miner where your data is and give information about the variables, etc.), you click **Data Sources** and proceed. For creating a new diagram, you right-click **Diagrams** and proceed. To open an existing diagram, double-click on the diagram you want.

⑤ **Properties Panel:** In this panel, you would see properties of **Project, Data Sources, Diagrams, Nodes**, and **Model Packages** by selecting them. In this example, the nodes are not yet created; hence, you do not see them in Display 2.3. You can view and edit the properties of any object selected. If you want to specify or change any options in a node such as **Decision Tree** or **Neural Network**, you must use the **Properties** panel.

⑥ **Help Panel:** This displays a description of the property that you select in the **Properties** panel.

⑦ **Status Bar:** This indicates the execution status of the SAS Enterprise Miner task.

⑧ **Toolbar Shortcut Buttons:** These are shortcut buttons for **Create Data Source**, **Create Diagram**, **Run**, etc. To display the text name of these buttons, position the mouse pointer over the button.

⑨ **Diagram Workspace:** This is used for building and running the process flow diagram for the project with various nodes (tools) of SAS Enterprise Miner.

Project Start Code

For any SAS Enterprise Miner Project, you must specify the directory where the data sets required for the project are located. Open the Enterprise Miner window and click ⬚ located in the value column of **Project Start Code** row in the Properties panel (see Display 2.3). A window opens where you can type the path of the library where your data sets for the project are located. The Project Start Code window is shown in Display 2.4.

Display 2.4

The data for this project is located in the folder C:\TheBook\EM12.1\Data\Chapter2. This is indicated by the libref TheBook. When you click **Run Now**, the library reference to the path is created. You can check whether the library is successfully created by opening the log window by clicking the **Log** tab.

2.5 Creating a SAS Data Source

You must create a data source before you start working on your SAS Enterprise Miner Project. After the data source is created, it contains all the information associated with your data—the directory path to the file that contains the data, the name of the data file, the names and measurement scales of the variables in the data set, and the cost and decision matrices and target profiles you specify. Profit matrix is also known as decision weights in SAS Enterprise Miner 12.1, which is used in decisions such as assigning a target class to an observation and assessing the models. This section shows how a data source is created, covering the essential steps. For additional capabilities and features of data source creation, refer to the Help menu in SAS Enterprise Miner. SAS Enterprise Miner saves all of this information, or *metadata*, as different data sets in a folder called **Data Sources** in the project directory.

To create a data source, click on the toolbar shortcut button or right-click **Data Sources** in the Project panel, as shown in Display 2.5.

Display 2.5

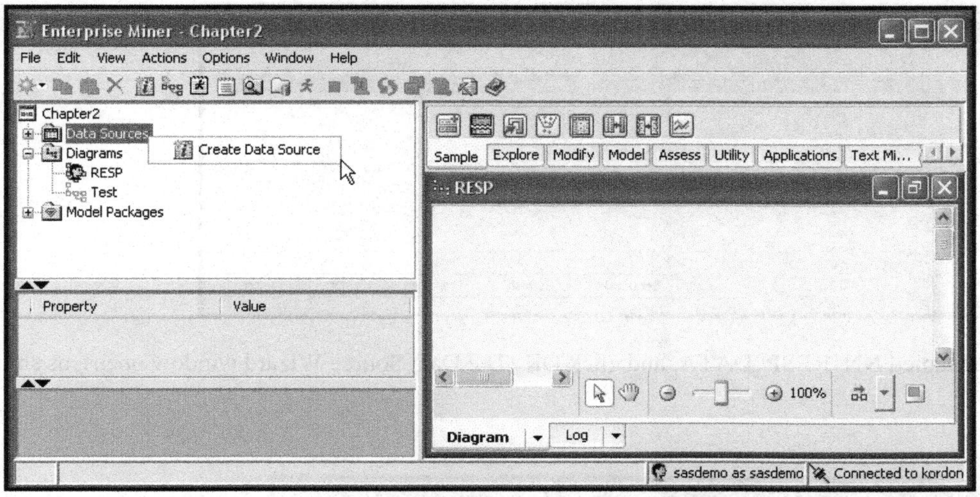

When you click **Create Data Source**, the Data Source Wizard window opens, and SAS Enterprise Miner prompts you to enter the data source.

If you are using a SAS data set in your project, use the default value SAS Table in the **Source** box and click **Next**. Then another window opens, prompting you to give the location of the SAS data set.

When you click **Browse**, a window opens that shows the list of library references. This window is shown in Display 2.6.

Display 2.6

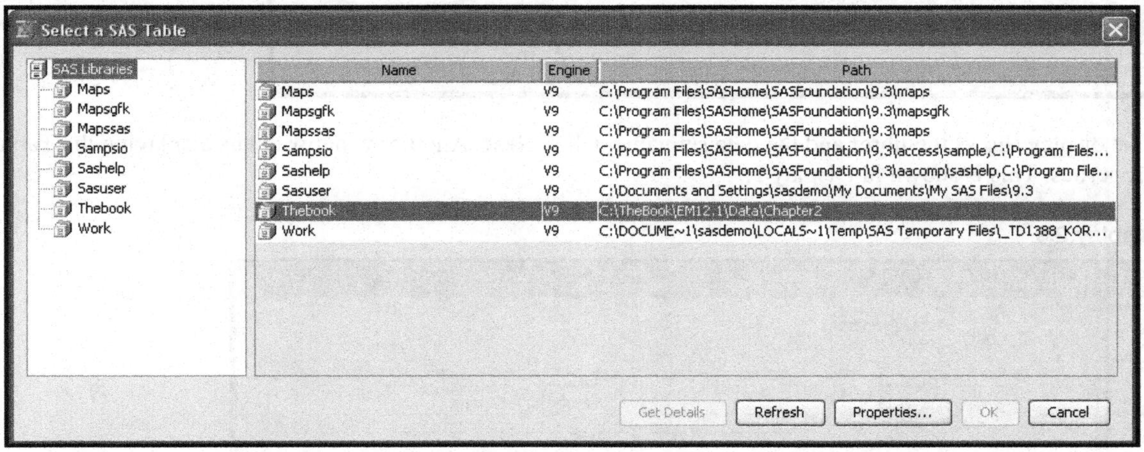

Since the data for this project is in the library Thebook, double-click on Thebook. The window opens with a list of all the data sets in that library. This window is shown in Display 2.7.

Display 2.7

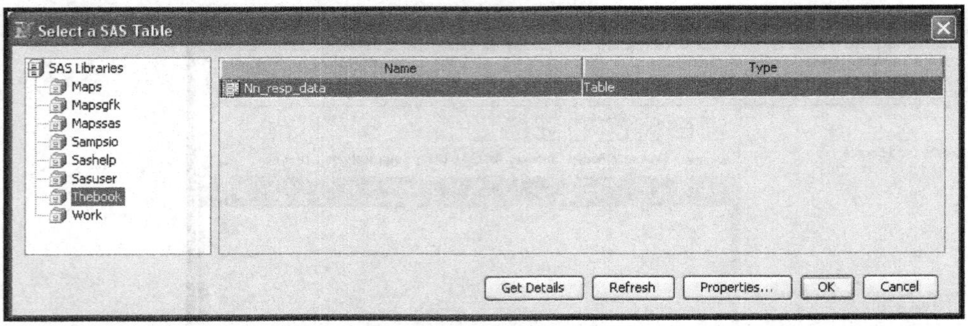

Select the data set named NN_RESP_DATA, and click OK. The Data Source Wizard window opens, as shown in Display 2.8.

Display 2.8

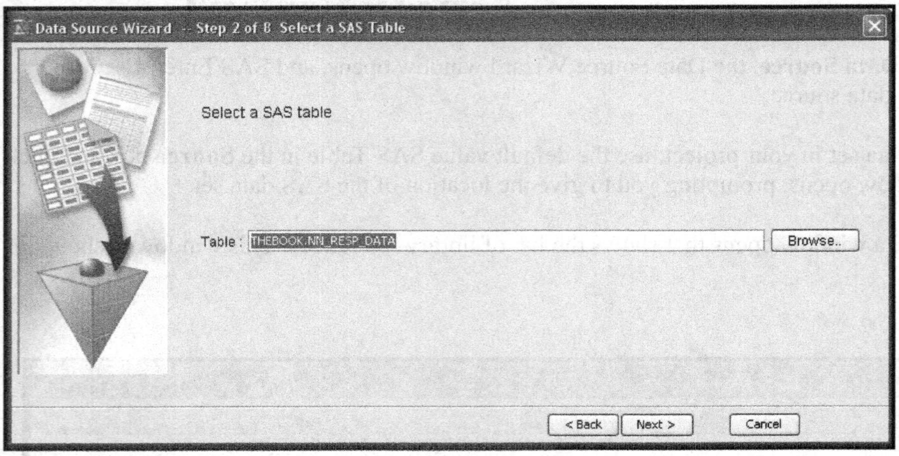

This display shows the libref and the data set name. Click **Next**. Another window opens displaying the Table Properties. This is shown in Display 2.9.

Display 2.9

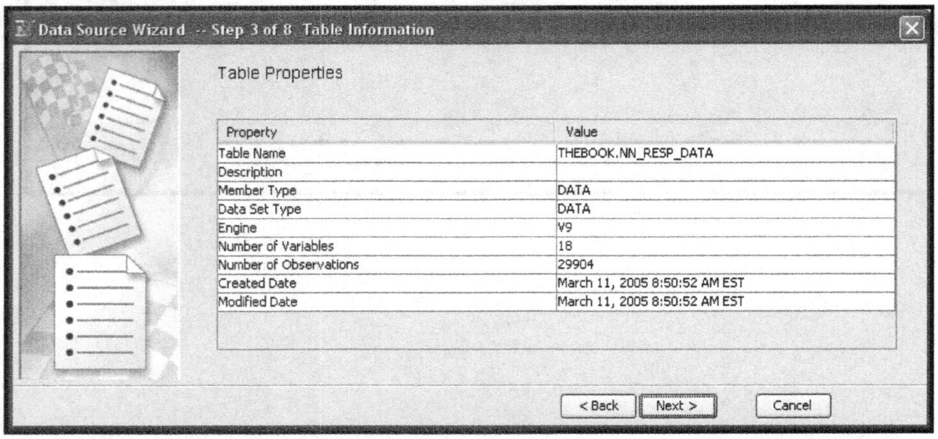

Property	Value
Table Name	THEBOOK.NN_RESP_DATA
Description	
Member Type	DATA
Data Set Type	DATA
Engine	V9
Number of Variables	18
Number of Observations	29904
Created Date	March 11, 2005 8:50:52 AM EST
Modified Date	March 11, 2005 8:50:52 AM EST

Click **Next** to show the Metadata Advisor Options. This is shown in Display 2.10.

Display 2.10

Use the Metadata Advisor Options window to define the metadata. Metadata is data about data sets. It specifies how each variable is used in the modeling process. The metadata contains information about the role of each variable, its measurement scale,[2] etc.

If you select the **Basic** option, the initial measurement scales and roles are based on the variable attributes. It means that if a variable is numeric, its measurement scale is designated to be interval, irrespective of how many distinct values the variable may have. For example, a numeric binary variable will also be initially given the interval scale. If your target variable is binary in numeric form, it will be treated as an interval-scaled variable, and it will be treated as such in the subsequent nodes. If the subsequent node is a **Regression** node, SAS Enterprise Miner automatically uses ordinary least squares regression, instead of the logistic regression, which is usually appropriate with a binary target variable.

With the **Basic** option, all character variables are assigned the measurement scale of nominal, and all numeric variables are assigned the measurement scale of interval.

If you select **Advanced**, SAS Enterprise Miner applies a bit more logic as it automatically sets the variable roles and measurement scales. If a variable is numeric and has more than 20 distinct values, SAS Enterprise Miner sets its measurement scale (level) to interval. In addition, if you select the **Advanced** option, you can customize the measurement scales. For example, by default the **Advanced** option sets the measurement scale of any numeric variable to nominal if it takes 20 or fewer unique values, but you can change this number by clicking **Customize** and setting the **Class Levels Count Threshold** property (See Display 2.11) to a number other than the default value of 20.

For example, consider a numeric variable such as X, where X may be the number of times a credit card holder was more than 60 days past due in payment in the last 24 months. In the modeling data set, X takes the values 0, 1, 2, 3, 4, and 5 only. With the **Advanced Advisor** option, SAS Enterprise Miner will assign the measurement scale of X to Nominal by default. But, if you change the **Class Levels Count Threshold** property from 20 to 3, 4, or 5, SAS Enterprise Miner will set the measurement scale of X to interval. A detailed discussion of the measurement scale assigned when you select the **Basic** Advanced Advisor Options with default values of the properties, and **Advanced** Advisor Options with customized properties is given later in this chapter.

Display 2.11 shows the default settings for the Advanced Advisor Options.

Display 2.11

Property	Value
Missing Percentage Threshold	50
Reject Vars with Excessive Missing Values	Yes
Class Levels Count Threshold	20
Detect Class Levels	Yes
Reject Levels Count Threshold	20
Reject Vars with Excessive Class Values	Yes
Database Pass-Through	Yes

Class Levels Count Threshold

If "Detect class levels"=Yes, interval variables with less than the number specified for this property will be marked as NOMINAL. The default value is 20.

One advantage of selecting the **Advanced** option is that SAS Enterprise Miner automatically sets the role of each unary variable to Rejected. If any of the settings are not appropriate, you can change them later in the window shown in Display 2.12.

In this example, the **Class Levels Count Threshold** property is changed to 10. I closed the Advanced Advisor Options window by clicking **OK,** and then I clicked **Next**. This opens the window shown in Display 2.12.

Display 2.12

Name	Role	Level	Report	Order	Drop	Lower Limit
AGE	Input	Interval	No		No	.
CRED	Input	Interval	No		No	.
DELINQ	Input	Nominal	No		No	.
DEPC	Input	Binary	No		No	.
EMP_STA	Input	Nominal	No		No	.
GENDER	Input	Binary	No		No	.
HEQ	Input	Nominal	No		No	.
INCOME	Input	Nominal	No		No	.
MFDU	Input	Binary	No		No	.
MILEAGE	Input	Interval	No		No	.
MOB	Input	Binary	No		No	.
MRTGI	Input	Nominal	No		No	.
MS	Input	Nominal	No		No	.
NUMTR	Input	Nominal	No		No	.
RESTYPE	Input	Nominal	No		No	.
RES_STA	Input	Binary	No		No	.
cuscode	Rejected	Nominal	No		No	.
resp	Target	Binary	No		No	.

This window shows the Variable List table with the variable names, model roles, and measurement levels of the variables in the data set. This example specifies the model role of the variable resp as the target.

If you check the box **Statistics** at the top of the Variable List table, the Advanced Option function of the Data Source Wizard calculates important statistics such as the number of levels, percent missing, minimum, maximum, mean, standard deviation, skewness, and kurtosis for each variable. If you check the **Basic** box, the

Variable List table also shows what type (character or numeric) each variable belongs. Display 2.13 shows a partial view of these additional statistics and variable types.

Display 2.13

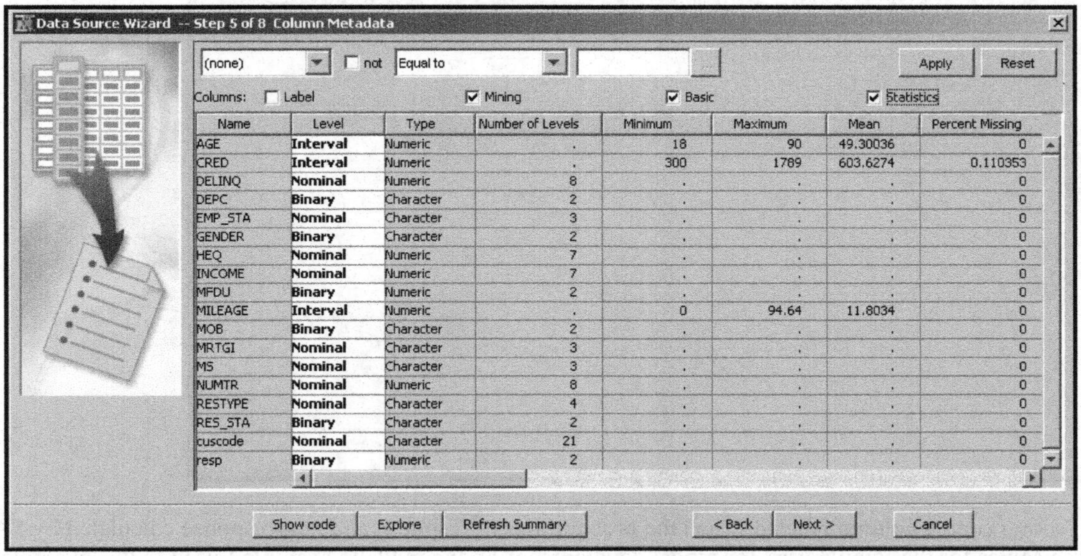

If the target is categorical, when you click **Next**, another window opens with the question "Do you want to build models based on the values of the decisions?"

If you are using a profit matrix (decision weights), cost variables, and posterior probabilities, select **Yes**, and click **Next** to enter these values (you can also enter or modify these matrices later). The window shown in Display 2.14 opens.

Display 2.14

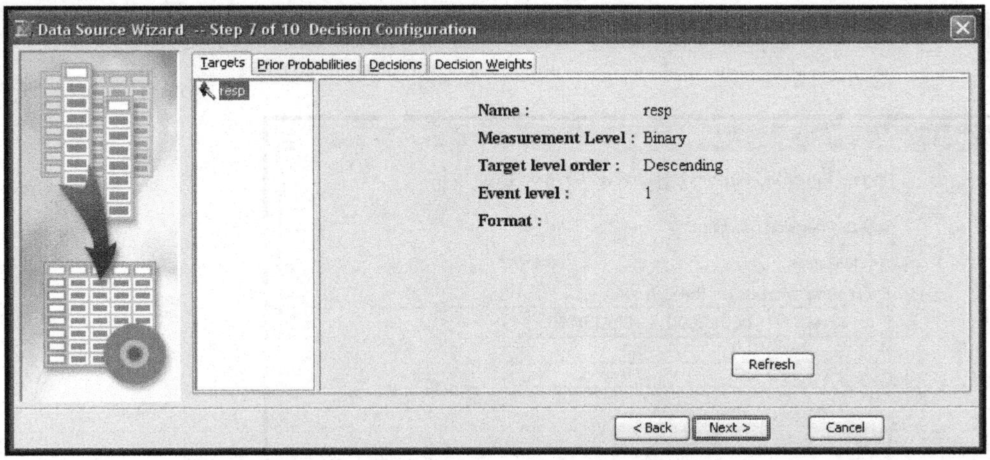

The **Targets** tab displays the name of the target variable and its measurement level. It also gives the target levels of interest. In this example, the variable resp is the target and is binary, which means that it has two levels: response, indicated by 1, and non-response, indicated by 0. The event of interest is response. That is, the model is set up to estimate the probability of response. If the target has more than two levels, this window will show all of its levels. (In later chapters, I model an ordinal target that has more than two levels, each level indicating frequency of losses or accidents, with 0 indicating no accidents, 1 indicating one accident, and 2 indicating four accidents, and so on.)

Display 2.15 shows the **Prior Probabilities** tab.

Display 2.15

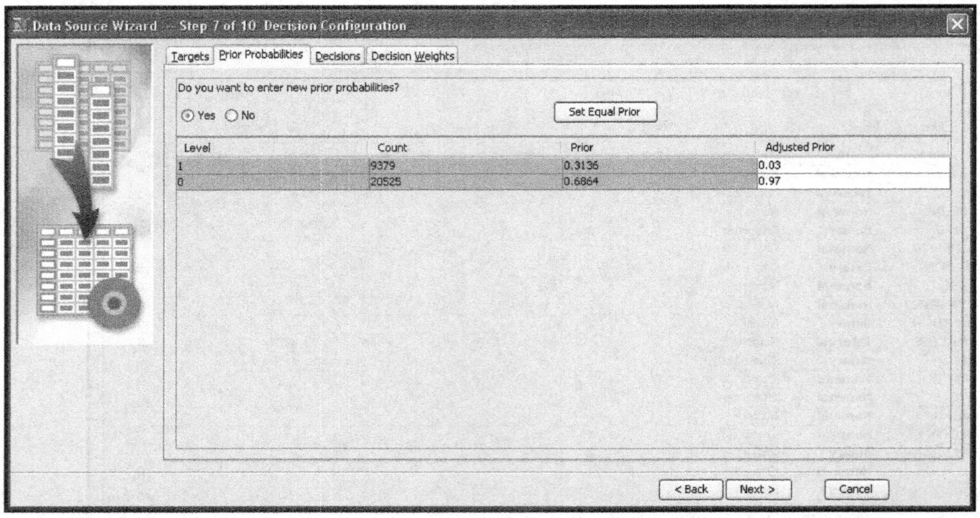

This tab shows (in the column labeled Prior) the probabilities of response and non-response calculated by SAS Enterprise Miner for the sample used for model development. In the modeling sample I used in this example, the responders are over-represented. In the sample, there are 31.36% responders and 68.64% non-responders, shown under the column "Prior." So the models developed from the modeling sample at hand will be biased unless a correction is made for the bias caused by over-representation of the responders. In the entire population, there are 3% responders and 97% non responders. These are the true *prior probabilities*. If you enter these true prior probabilities in the Adjusted Prior column as I have done, SAS Enterprise Miner will correct the models for the bias and produce unbiased predictions. To enter these adjusted prior probabilities, select **Yes** in response to the question **Do you want to enter new prior probabilities?** Then enter the probabilities that are calculated for the entire population (0.03 and 0.97 in my example).

To enter a profit matrix, click the **Decision Weights** tab, shown in Display 2.16.

Display 2.16

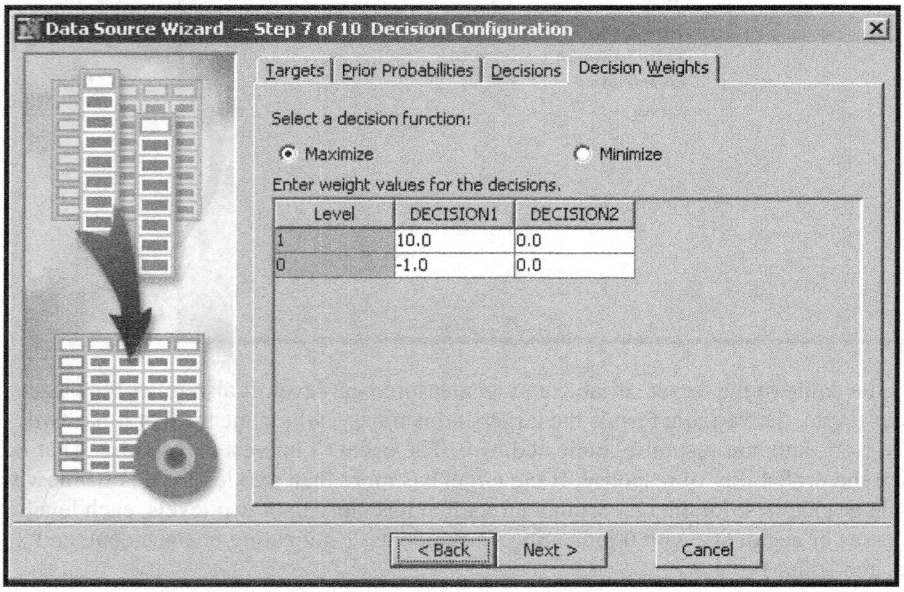

The columns of this matrix refer to the different decisions that need to be made based on the model's predictions. In this example, DECISION1 means classifying or labeling a customer as a responder, while DECISION2 means classifying a customer as a non-responder. The entries in the matrix indicate the profit or loss associated with a correct or incorrect assignment (decision), so the matrix in this example implies that if a customer is classified as a responder, and he is in fact a responder, then the profit is $10. If a customer is classified as a responder, but she is in fact a non-responder, then there will be a loss of $1. Other cells of the matrix can be interpreted similarly.

In developing predictive models, SAS Enterprise Miner assigns target levels to the records in a data set. In the case of a response model, assigning target levels to the records means classifying each customer as a responder or non- responder. In a latter step of any modeling project, SAS Enterprise Miner also compares different models, on the basis of a user-supplied criterion, to select the best model. In order to have SAS Enterprise Miner use the criterion of profit maximization when assigning target levels to the records in a data set and when choosing among competing models, select **Maximize** for the option **Select a decision function**. The values in the matrix shown here are arbitrary and given only for illustration.

Display 2.17 shows the window for cost variables, which you can open by clicking the **Decisions** tab.

Display 2.17

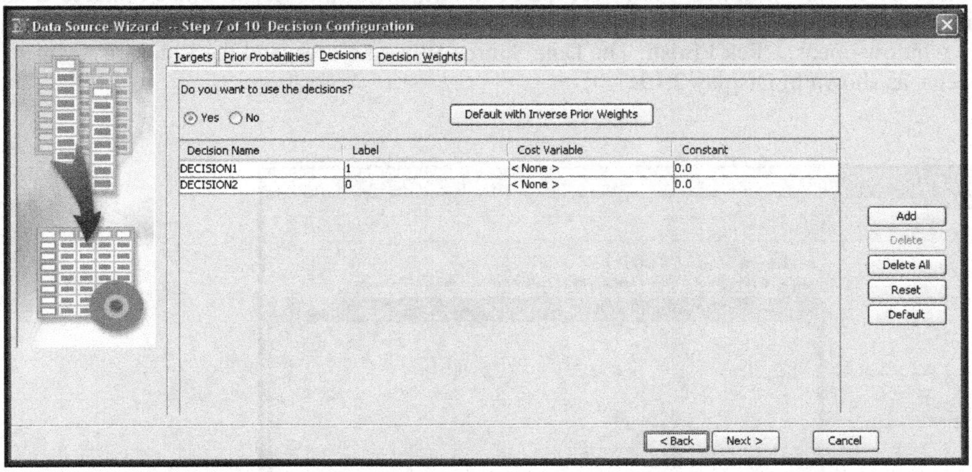

If you want to maximize profit instead of minimizing cost, then there is no need to enter cost variables. Costs are already taken into account in profits. Therefore, in this example, cost variables are not entered.

When you click **Next**, another window opens with the question "Do you wish to create a sample data set?" Since I want to use the entire data for this project, I chose **No**, and click **Next**. A window opens that shows the data set and its role, shown in Display 2.18. In this example, the data set NN_RESP_DATA is assigned the role Raw.

Diagram 2.18

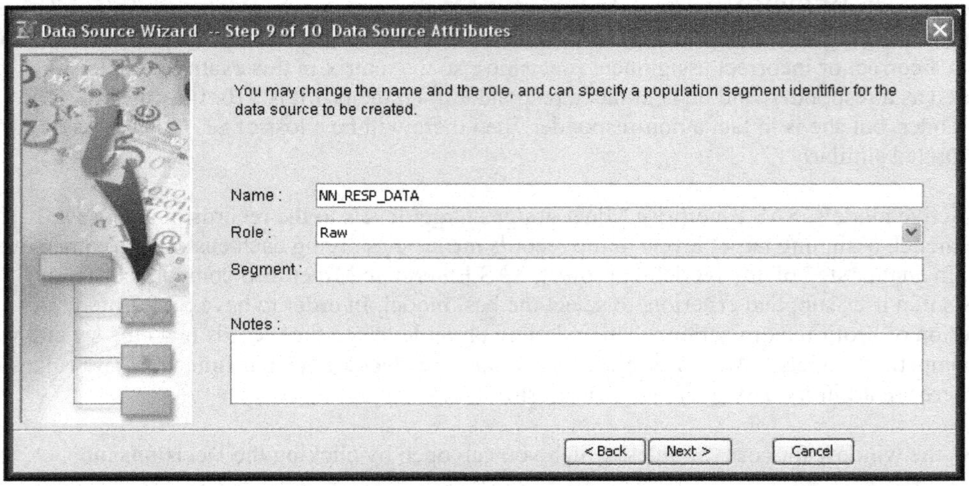

Other options for **Role** are **Train**, **Validate**, **Test**, **Score**, **Document**, and **Transaction**. Since I plan to create the **Train**, **Validate**, and **Test** data sets from the sample data set, I leave its role as **Raw**. When I click **Next**, the metadata summary window opens. Click **Finish**. The Data Source Wizard closes, and the Enterprise Miner Project window opens, as shown in Display 2.19.

Display 2.19

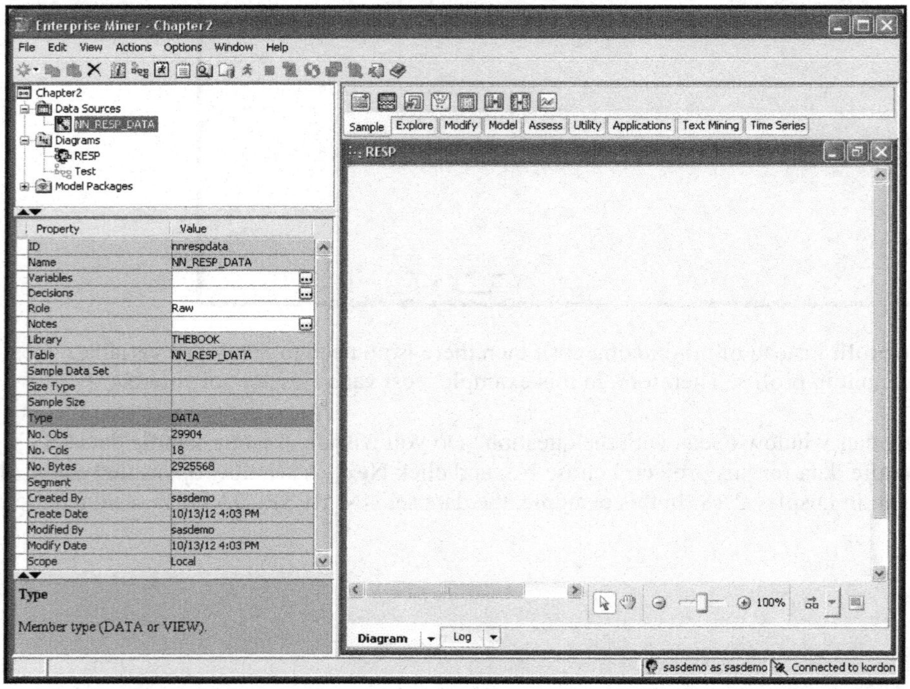

In order to see the properties of the data set, expand **Data Sources** in the project panel. Select the data set NN_RESP_DATA; the Properties panel shows the properties of the data source as shown in Display 2.19.

You can view and edit the properties. Open the variables table (shown in Display 2.20) by clicking [...] located on the right of the **Variables** property in the Value column.

Display 2.20

This variable table shows the name, role, measurement scale, etc., for each variable in the data set. You can change the role of any variable, change its measurement scale, or drop a variable from the data set. If you drop a variable from the data set, it will not be available in the next node.

By checking the "Basic" box located above the columns, you can see the variable type and length. By checking the "Statistics" box you can see statistics such as mean standard deviation etc. for interval scaled variables as shown in Display 2.20.

Note that the variable Income is numeric (see under the column Type) but its level (measurement scale) is set to Nominal because income has only 7 levels (unique values) in the sample. When you select the advanced advisor options in creating the data source, by default the measurement scale of a numeric variable is set to nominal if it has less than 20 unique values. See display 2.11, where you see that the class level count threshold is 20 by default. Although we changed this threshold to 10, the measurement scale of Income is still nominal since it has only 7 levels (less than 10).

2.6 Creating a Process Flow Diagram

To create a process flow diagram, right-click **Diagrams** in the project panel (shown in Display 2.19), and click **Create Diagram**. You are prompted to enter the name of the diagram in a text box labeled **Diagram Name**. After entering a name for your diagram, click **OK**. A blank workspace opens, as shown in Display 2.21, where you create your process flow diagram.

Display 2.21

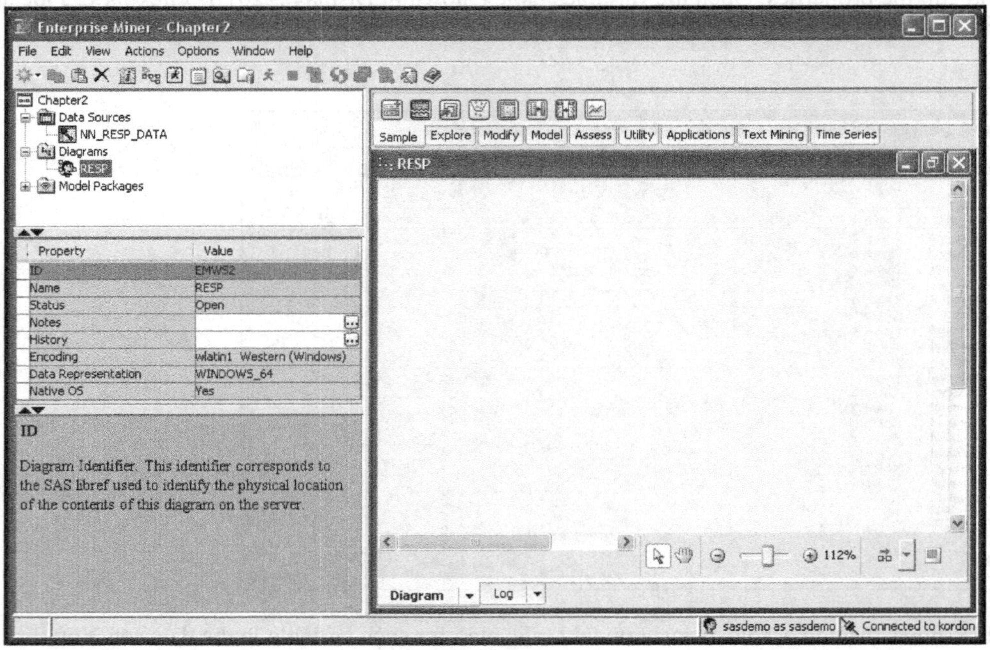

To create a process flow diagram drag and connect the nodes (tools) you need for your task. The following sections show some examples of how to use the nodes available in SAS Enterprise Miner.

2.7 Sample Nodes

If you open the **Sample** tab, the tool bar is populated with the icons for the following nodes: Append, Data Partition, File Import, Filter, Input Data, Merge, Sample, and Time Series. In this section, I provide an overview of some of these nodes, starting with the **Input Data** node.

2.7.1 Input Data Node

This is the first node in any diagram (unless you start with the **SAS Code** node). In this node, you specify the data set that you want to use in the diagram. You might have already created several data sources for this project, as discussed in Section 2.5. From these sources, you need to select one for this diagram. A data set can be assigned to an input in one of two ways:

- When you expand **Data Sources** in the project panel by clicking the + on the left of Data Sources, all the data sources appear. Then click on the icon to the left of the data set you want, and drag it to the Diagram Workspace. This creates the **Input Data** node with the desired data set assigned to it.

- Alternatively, first drag the **Input Data** node from the toolbar into the Diagram Workspace. Then set the **Data Source** property of the Input Data node to the name of the data set. To do this, select the

  Input Data node, then click ⬛ located to the right of the **Data Source** property as shown in Display 2.22. The Select Data Source window opens. Click on the data set you want to use in the diagram. Then click **OK**.

When you follow either procedure, the **Input Data** node is created as shown in Display 2.23.

Display 2.22

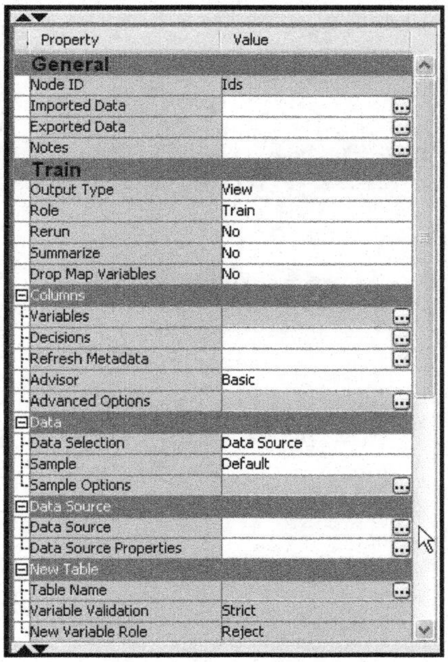

Property	Value
General	
Node ID	Ids
Imported Data	
Exported Data	
Notes	
Train	
Output Type	View
Role	Train
Rerun	No
Summarize	No
Drop Map Variables	No
⊟Columns	
Variables	
Decisions	
Refresh Metadata	
Advisor	Basic
Advanced Options	
⊟Data	
Data Selection	Data Source
Sample	Default
Sample Options	
⊟Data Source	
Data Source	
Data Source Properties	
⊟New Table	
Table Name	
Variable Validation	Strict
New Variable Role	Reject

Display 2.23

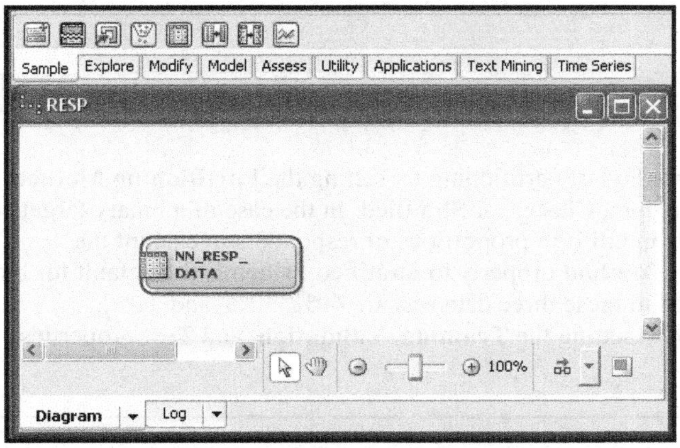

2.7.2 Data Partition Node

In developing predictive models, you must partition the sample into **Training**, **Validation**, and **Test**. The **Training** data is used for developing the model using tools such as **Regression**, **Decision Tree**, and **Neural Network**. During the training process, these tools generate a number of models. The Validation data set is used to evaluate these models, and then to select the best one. The process of selecting the best model is often referred to as *fine tuning*. The Test data set is used for an independent assessment of the selected model.

From the Tools bar, drag the **Data Partition** node into the Diagram Workspace, connect it to the **Input Data** node and select it, so that the Property panel shows the properties of the **Data Partition** node, as shown in Display 2.25.

The **Data Partition** node is shown in Display 2.24, and its Property panel is shown in Display 2.25.

Display 2.24

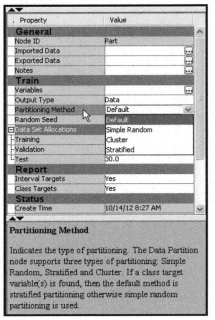

Display 2.25

Property	Value
General	
Node ID	Part
Imported Data	
Exported Data	
Notes	
Train	
Variables	
Output Type	Data
Partitioning Method	Default
Random Seed	Default
⊟ Data Set Allocations	Simple Random
├ Training	Cluster
├ Validation	Stratified
└ Test	30.0
Report	
Interval Targets	Yes
Class Targets	Yes
Status	
Create Time	10/14/12 8:27 AM

Partitioning Method

Indicates the type of partitioning. The Data Partition node supports three types of partitioning: Simple Random, Stratified and Cluster. If a class target variable(s) is found, then the default method is stratified partitioning otherwise simple random partitioning is used.

In the **Data Partition** node, you can specify the method of partitioning by setting the **Partitioning Method** property to one of four values: Default, Simple random, Cluster, or Stratified. In the case of a binary target such as response, the stratified sampling method results in uniform proportions of responders in each of the partitioned data sets. Hence, I set the **Partitioning Method** property to Stratified, which is the default for binary targets. The default proportion of records allocated to these three data sets are 40%, 30%, and 30%, respectively. You can change these proportions by resetting the **Training**, **Validation**, and **Test** properties under the **Data Set Allocations** property.

2.7.3 Filter Node

The **Filter** node can be used for eliminating observations with extreme values (outliers) in the variables.

You should not use this node routinely to eliminate outliers. While it may be reasonable to eliminate some outliers for very large data sets for predictive models, the outliers often have interesting information that leads to insights about the data and customer behavior.

Before using this node, you should first find out the source of extreme value. If the extreme value is due to an error, the error should be corrected. If there is no error, you can truncate the value so that the extreme value does not have an undue influence on the model.

Display 2.26 shows the flow diagram with the **Filter** node. The **Filter** node follows the **Data Partition** node. Alternatively, you can use the **Filter** node before the **Data Partition** node.

Display 2.26

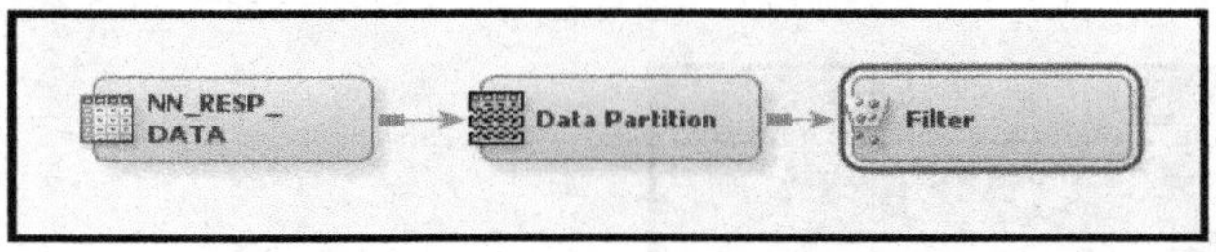

To use the **Filter** node, select it, and set the **Default Filtering Method** property for **Interval Variables** to one of the values, as shown in Display 2.27. (Different values are available for Class variables.)

Display 2.27

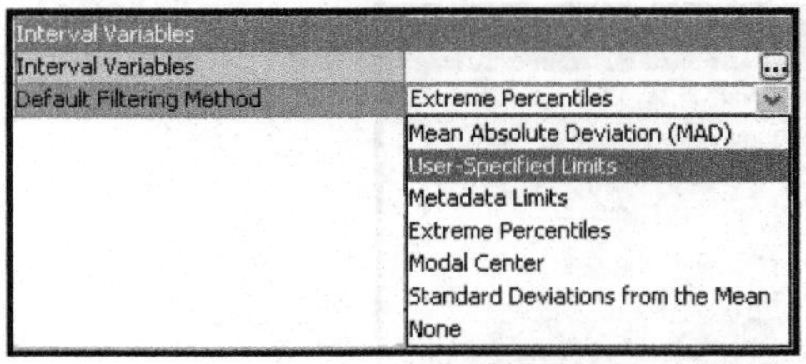

Display 2.28 shows the Properties panel of the **Filter** node.

Display 2.28

I set the **Tables to Filter** property to All Data Sets so that outliers are filtered in all three data sets—Training, Validation, and Test—and then ran the **Filter** node.

The Results window shows the number of observations eliminated due to outliers of the variables. Output 2.1 (from the output of the Results window) shows the number of records excluded from the Train, Validate, and Test data sets due to outliers.

Output 2.1

Number Of Observations			
Data Role	Filtered	Excluded	DATA
TRAIN	11557	403	11960
VALIDATE	8658	314	8972
TEST	8673	299	8972

The number of records exported to the next node is 11557 from the Train data set, 8658 records from the Validate data set and 8673 records from the Test data set. Displays 2.29 and 2.30 show the criteria used for filtering the observations.

Display 2.29

Variable	Role	Minimum	Maximum	Filter Method	Keep Missing Values	Label
AGE	INPUT	22	88	PERCENTS	Y	
CRED	INPUT	309	931	PERCENTS	Y	
MILEAGE	INPUT	0.151	20.77	PERCENTS	Y	

Limits for Interval Variables

Display 2.30

Variable	Role	Level	Train Count	Train Percent	Label	Filter Method
DELINQ	INPUT	5	5	0.041806		MINPCT
DELINQ	INPUT	6	12	0.100334		MINPCT
DELINQ	INPUT	7	12	0.100334		MINPCT
NUMTR	INPUT	5	8	0.06689		MINPCT
NUMTR	INPUT	6	18	0.150502		MINPCT
NUMTR	INPUT	7	13	0.108696		MINPCT

Excluded Class Values

To see the SAS code that is used to perform the filters, click **View→SAS Results→Flow Code**. The SAS code is shown in Display 2.31.

Display 2.31

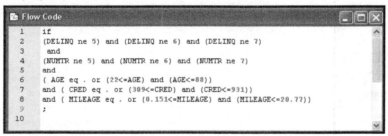

```
1   if
2   (DELINQ ne 5) and (DELINQ ne 6) and (DELINQ ne 7)
3    and
4   (NUMTR ne 5) and (NUMTR ne 6) and (NUMTR ne 7)
5   and
6   ( AGE eq . or (22<=AGE) and (AGE<=88))
7   and ( CRED eq . or (309<=CRED) and (CRED<=931))
8   and ( MILEAGE eq . or (0.151<=MILEAGE) and (MILEAGE<=20.77))
9    ;
10
```

Instead of using the default filtering method for all variables, you can specify different filtering methods to individual variables. Do this by opening the Variables window. To open the Variable window for interval variables, click ⊡ located to the right of the **Interval Variables** property. The Interactive Interval Filter window opens, as shown in Display 2.32.

Display 2.32

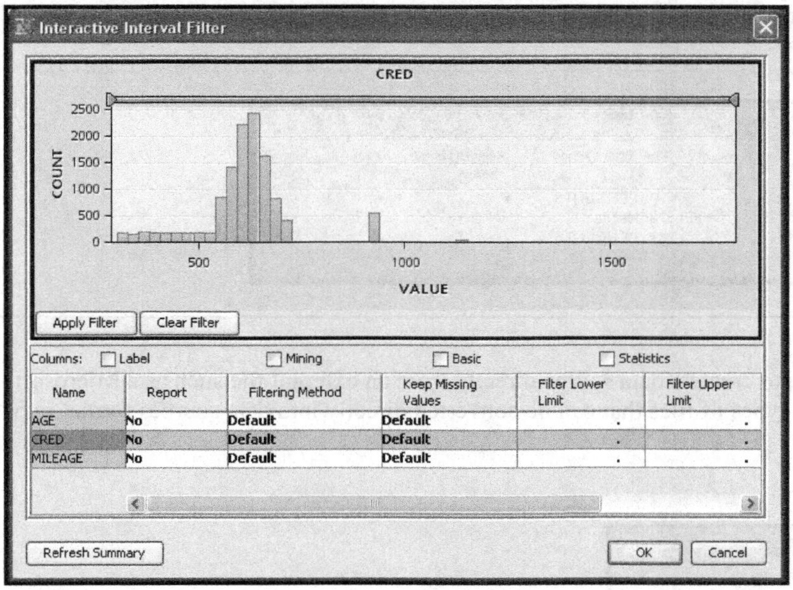

For example, if you want to change the filtering method for the variable CRED(stands for Credit Score) , select the row for the variable CRED as shown in Display 2.32 and click in the Filtering Method column corresponding to CRED. A drop-down menu of all the filtering methods available appears. You can then select the method you want.

You can also interactively set the limits for filtering out extreme values by sliding the handles that appear above the chart in Display 2.32. Let me illustrate this by manually setting the filtering limits for the variable CREDIT.

Display 2.32 shows that some customers have a credit rating of 1000 or above. In general the maximum credit rating is around 950, so a rating above this value is almost certainly erroneous. So I set the lower and upper limits for the CREDIT variable at around 301 and 950 respectively. I set these limits by sliding the handles located at the top of the graph to the desired limits and clicking **Apply Filter**, as shown in Display 2.33. Click **OK** to close the window.

Display 2.33

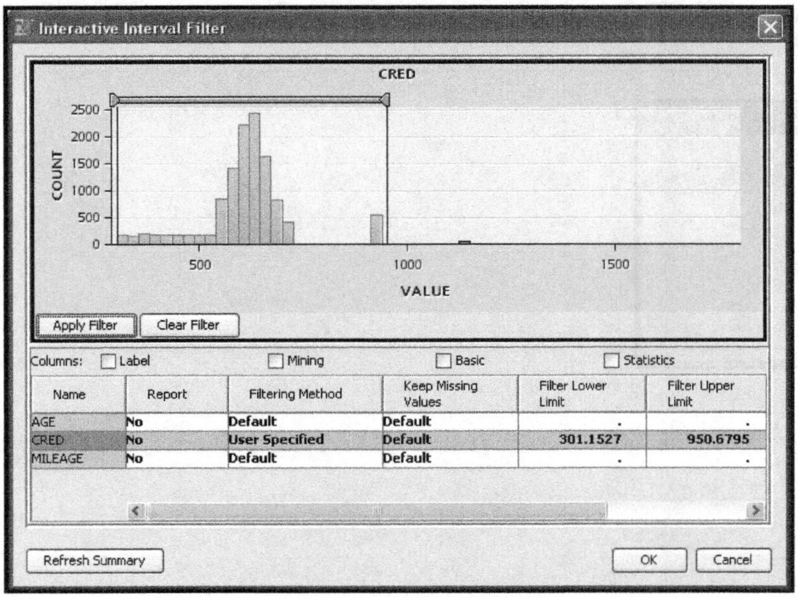

After running the **Filter** node and opening the Results window, you can see that the limits I set for the credit variable have been applied in filtering out the extreme values, as shown in Display 2.34.

Display 2.34

Variable	Role	Minimum	Maximum	Filter Method	Keep Missing Values	Label
AGE	INPUT	22	88	PERCENTS	Y	
CRED	INPUT	301.1527	950.6795	MANUAL	Y	
MILEAGE	INPUT	0.151	20.77	PERCENTS	Y	

2.7.4 File Import Node

The **File Import** node enables you to create a data source directly from an external file such as a Microsoft Excel file. Display 2.35 shows the types of files that can be converted directly into data sources format in SAS Enterprise Miner.

Display 2.35

```
File Import Types

Input Data Source                    Extension
=============================        =============================
dBASE 5.0, IV, III+, and III files   .dbf
Stata                                .dta
Microsoft Excel                      .xls, .xlsx
SAS JMP files                        .jmp
Paradox .DB files                    .db
SPSS                                 .sav
Lotus                                .wk1, .wk3, .wk4
Tab-Delimited values                 .txt
Comma-Separated values               .csv
Delimited values (user defined)      .dlm

                    OK
```

I will illustrate the **File Import** node by importing an Excel file. You can pass the imported data to any other node. I demonstrate this by connecting the **File Import** node to the **StatExplore** node.

To use the **File Import** node, first create a diagram in the current project by right-clicking **Diagrams** in the project panel, as shown in Display 2.36

Display 2.36

```
Enterprise Miner - Chapter2

File  Edit  View  Actions  Options  Window  Help

Chapter2
  Data Sources
  Diagram      Create Diagram
  Model P.        Import Diagram from XML...
```

Type a diagram name in the text box in the Create Diagram dialog box, and click **OK**. A blank workspace is created. Drag the **File Import** tool from the **Sample** tab, as shown in Display 2.37.

Display 2.37

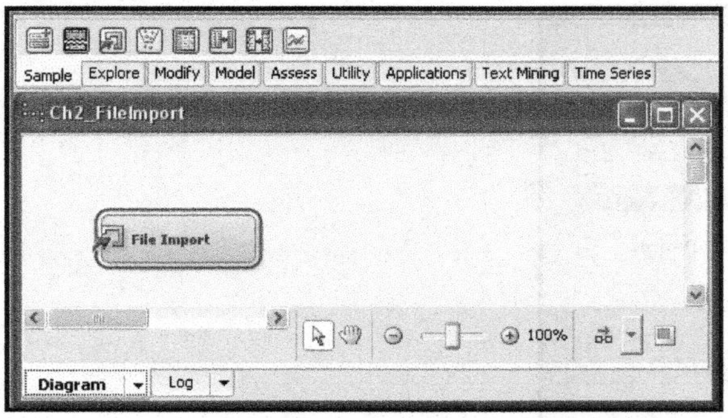

The Properties panel of the **File Import** node is shown in Display 2.38.

Display 2.38

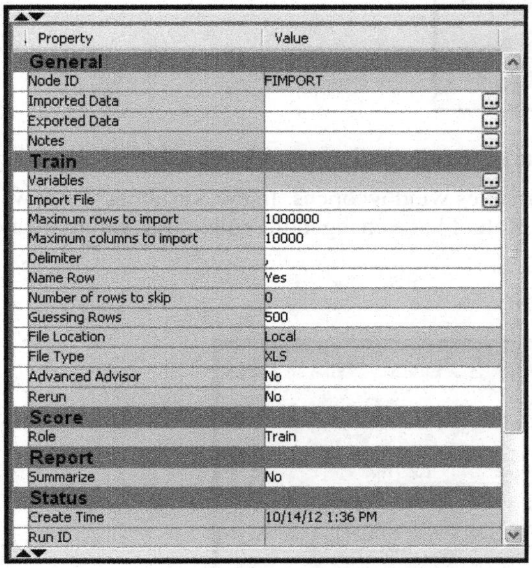

In order to configure the metadata, click ⬚ to the right of **Import File** property in the Properties panel (see Display 2.38). The File Import dialog box appears.

Since the Excel file I want to import is on my C drive, I select the **My Computer** radio button, and type the directory path, including the file name, in the File Import dialog box. Now I can preview my data in the Excel sheet by clicking the **Preview** button, or I can complete the file import task by clicking **OK**. I chose to click **OK**.

The imported Excel file is now shown in the value of the **Import** File property in the Properties panel, as shown in Display 2.39.

Display 2.39

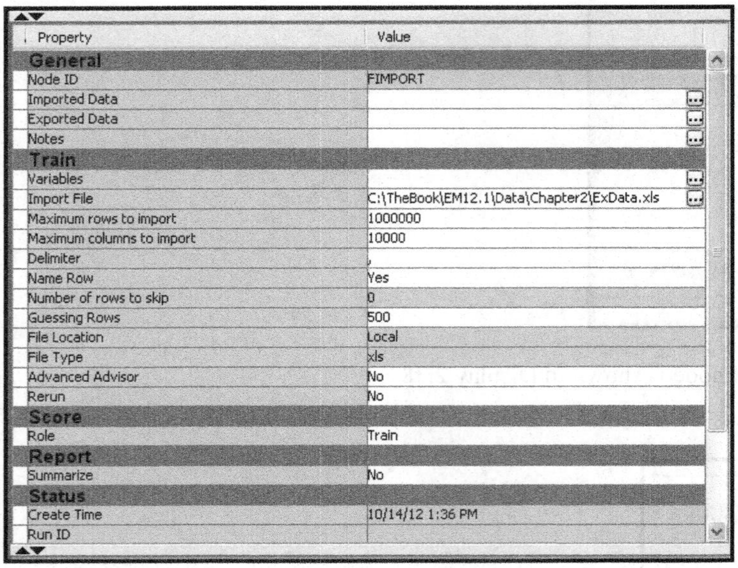

Next we have to assign the Metadata Information such as variable roles and measurement scales. To do this click [...] located to the right of **Variables** property. The Variables window opens. In the Variables window, you can change the role of a variable by clicking on the column Role. I have changed the role of the variable Sales to Target, as shown in Display 2.40.

Display 2.40

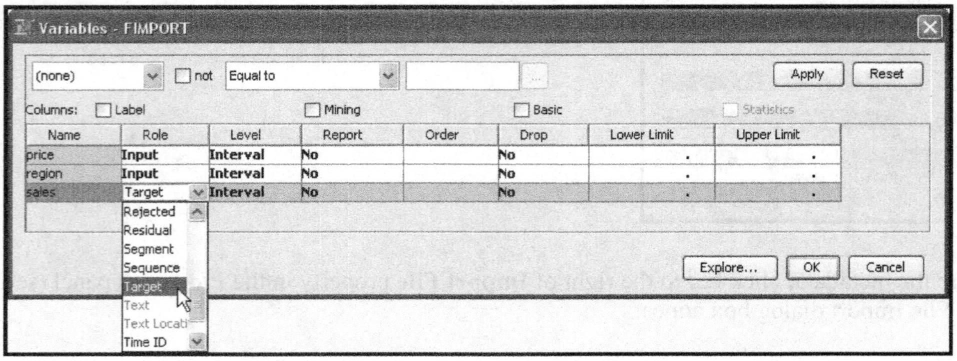

Click **OK**, and the data set is ready for use in the project. It can be passed to the next node. I have connected the **StatExplore** node to the **File Import** node in order to verify that the data can be used, as shown in Display 2.41.

Display 2.41

Display 2.42 shows the table that is passed from the **File Import** node to the **StatExplore** node.

Display 2.42

You can now run the **StatExplore** node. I successfully ran **StatExplore** node shown in Display 2.41. I discuss the results later in this chapter.

2.7.5 Time Series Node

Converting transactions data to time series

The **Time Series** node in SAS Enterprise Miner 12.1 can be used to condense transactional data to time series form, which is suitable for analyzing trends and seasonal variations in customer transactions. Both the transactional data and the time series data are time stamped. But the observations in transactional data may not occur at any particular frequency, whereas the observations in time series data pertain to consecutive time periods of a specific frequency, such as annual, quarterly, monthly, weekly or daily.

In order to introduce the reader to the **Time Series** node, I present an example of a transactional data set that shows the sales of two products (termed A and B) over a 60-month period by a hypothetical company. The company sells these products in 3 states—Connecticut, New Jersey, and New York. The sales occur in different weeks within each month. Display 2.43 shows a partial view of this transactions data.

Display 2.43

Transactions Data				
MNTH_YR	Sales	Product	Week	STATE
JAN2005	2121.29	A	1	CT
JAN2005	1590.97	A	2	CT
JAN2005	1121.29	B	2	CT
JAN2005	639.14	B	3	CT
JAN2005	530.32	A	4	CT
JAN2005	751.26	B	4	CT
FEB2005	2120.40	A	1	CT
FEB2005	1590.30	A	2	CT
FEB2005	1120.40	B	3	CT
FEB2005	638.63	B	3	CT
FEB2005	530.10	A	4	CT
FEB2005	750.67	B	4	CT
MAR2005	2123.33	A	1	CT
MAR2005	1592.50	A	2	CT
MAR2005	1123.33	B	3	CT
MAR2005	640.30	B	3	CT
MAR2005	530.83	A	4	CT
MAR2005	752.63	B	4	CT

Note that, in January 2005, customers purchased product A during the weeks 1, 2, and 4, and they purchased Product B during the weeks 2, 3 and 4. In February 2005, customers purchased product A during the weeks 1, 2 and 4, and they purchased B in weeks 3 and 4. If you view the entire data set, you will find that there are sales of both products in all 60 months (Jan 2005 through Dec 2009), but there may not be sales during every week of every month. No data was entered for weeks when there were no sales, hence there is no observation in the data set for the weeks during which no purchases were made. In general, in a transaction data set, an observation is recorded only when a transaction takes place.

In order to analyze the data for trends, or seasonal or cyclical factors, you have to convert the transactional data into weekly or monthly time series. Converting to weekly data entails entering zeroes for the weeks that had missing observations to represent no sales. In this example, we convert the transaction data set to a monthly time series.

Display 2.44 shows monthly sales of product A, and Display 2.45 shows monthly sales of product B in Connecticut. Time series of this type are used by the **Time Series** node for analyzing trends and seasonal factors.

Display 2.44

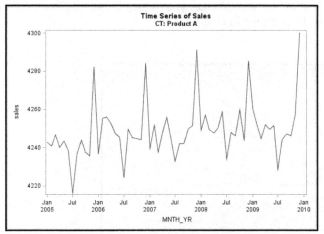

Display 2.45

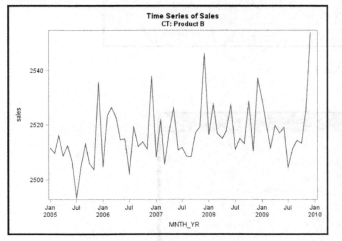

In order to perform an analysis on monthly time series derived from the transactions data shown in Display 2.43, you need to specify the time ID (month_yr in this example), and the cross ID variables (State and Product) and the target variable (Sales) in the **Time Series** node. The steps described in the following sections illustrate how to identify seasonal factors in the sales of products A and B (from the transactions data discussed earlier) using the **Time Series** node. You can use the **Time Series** node to find trend and cyclical factors in the sales also, but here I show only the seasonal decomposition of the sales.

First open SAS Enterprise Miner, create a project, and then create a data source for the transaction data. The first two steps are the same as described earlier, but creating a data source from transaction data is slightly different.

Creating a Data Source for the Transaction data

To create the data source, open an existing project. Right-click **Data Sources** in the Project panel and select **Create Data Source** (as shown earlier in Display 2.5 when we created a new SAS data source). The Data Source Wizard opens. For this example, use the default value of SAS Table for the **Source** field. Click **Next**.

For step 3, enter the name of the transactions data set (called TRANSACT in this example), and click **Next**. The Table Properties table opens, as shown in Display 2.46A.

In the "Source" text box type in the data type. Since the transactions data is a SAS data set in this example, I entered "SAS Table". By clicking on the "Next" button, the Wizard takes you to Step 2. In Step2 enter the name of the transaction data set (THEBOOK.TRANSACT) as shown in Display 2.46.

Display 2.46

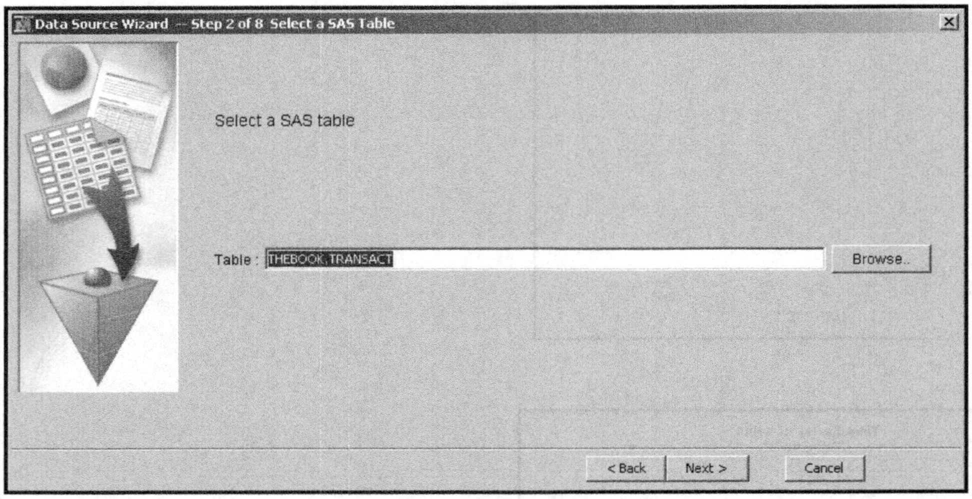

By clicking "Next" the Wizard takes you to Step 3.

Display 2.46A

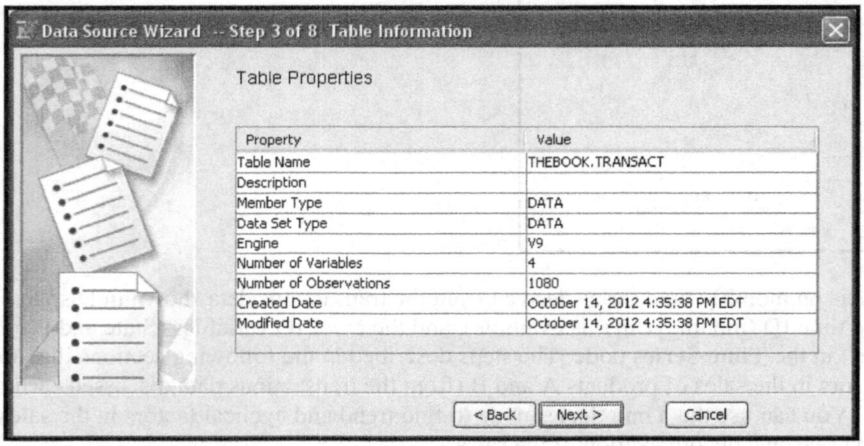

Click **Next** to move to step 4 of the Wizard. Select the **Advanced** option and click **Next**.

In step 5, names, roles, measurement levels, etc. of the variables in your data set are displayed.

Display 2.47

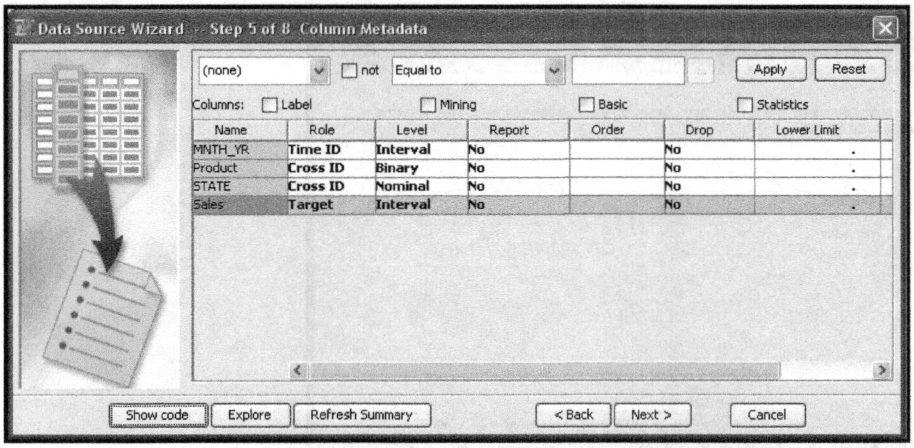

I changed the role of the variable Month_Yr to Time ID and its measurement level to Interval, the roles of the variables Product and State to Cross ID, and the role of Sales to Target. These changes result in the use of monthly time series of sales for analysis. Since there are three states and two products, six time series will be analyzed when I run the **Time Series** node.

Click **Next**. In step 6, the Wizard asks if you want to create a sample data set. I selected **No**, as shown in Display 2.48.

Display 2.48

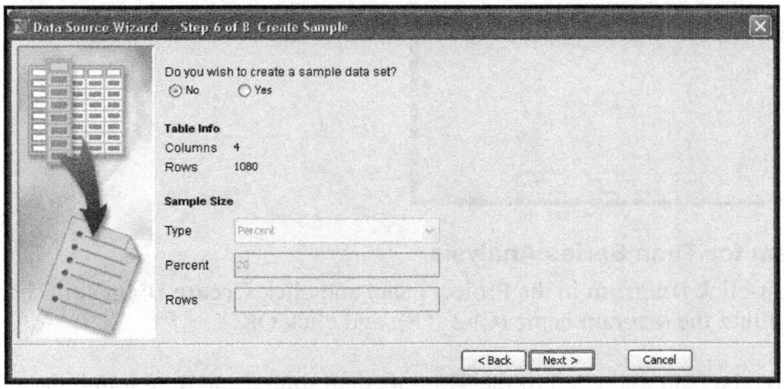

Click **Next**. In step 7, assign Transaction as the Role for the data source (shown in Display 2.49), and click **Next**.

Display 2.49

Display 2.50

Step 8 shows a summary of the metadata created. When you click **Finish**, the Project window opens.

Display 2.50

Creating a Process Flow Diagram for Time Series Analysis

To create a process flow diagram, right-click **Diagram** in the Project panel and click **Create Diagram**. The Create New Diagram window opens. Enter the diagram name (Ch2_TS), and click OK.

From the **Data Sources** folder, click on the icon to the left of the Transact data source and drag it into the Diagram Workspace. In the same way, drag the **Time Series** node from the **Sample** tab into the Diagram Workspace and connect it to the **Input Data** node, as shown in Display 2.51.

Display 2.51

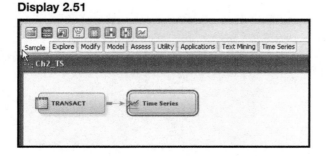

Analyzing Time Series: Seasonal Decomposition

By clicking the **Time Series** node in the Diagram Workspace (Display 2.51), the Properties panel of the **Time Series** node appears, as shown in Display 2.52.

Display 2.52

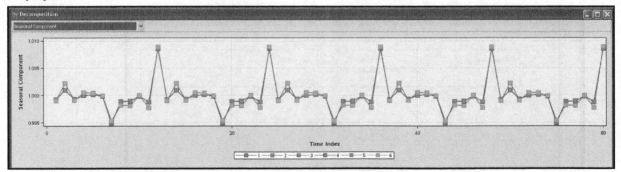

I have set the **Select an Analysis** property to Seasonal Decomposition since I am interested in getting the monthly seasonal factors for sales of the two products for the three states included in the data set. Then I ran the **Time Series** node and opened the Results window. In the Decomposition window, I selected the **Seasonal Component** graph. This graph in Display 2.53 shows the seasonal effects for the six monthly time series created by the **Time Series** node.

Display 2.53

Display 2.53 shows that there are seasonal factors for the months of July and December. You can view the seasonal factors of any individual time series by right-clicking on the graph area to open the Data Options dialog box, shown in Display 2.54.

Display 2.54

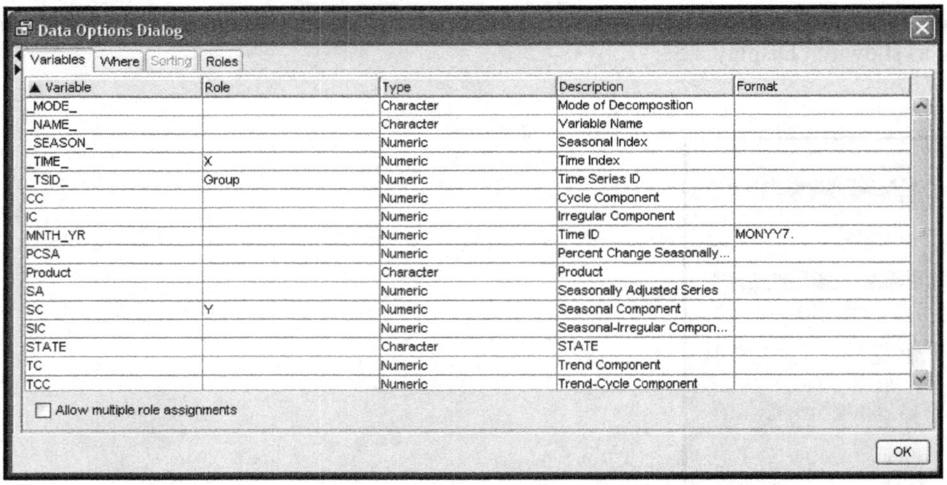

Data Options Dialog
Variables
Where
Sorting
Roles
▲ Variable | Role | Type | Description | Format
MODE | | Character | Mode of Decomposition |
NAME | | Character | Variable Name |
SEASON | | Numeric | Seasonal Index |
TIME | X | Numeric | Time Index |
TSID | Group | Numeric | Time Series ID |
CC | | Numeric | Cycle Component |
IC | | Numeric | Irregular Component |
MNTH_YR | | Numeric | Time ID | MONYY7.
PCSA | | Numeric | Percent Change Seasonally... |
Product | | Character | Product |
SA | | Numeric | Seasonally Adjusted Series |
SC | Y | Numeric | Seasonal Component |
SIC | | Numeric | Seasonal-Irregular Compon... |
STATE | | Character | STATE |
TC | | Numeric | Trend Component |
TCC | | Numeric | Trend-Cycle Component |
Allow multiple role assignments
OK

To see the seasonal factors in the sales for given product in a given state, click the **Where** tab. The window for making the selection of a time series opens, as shown in Display 2.55.

Display 2.55

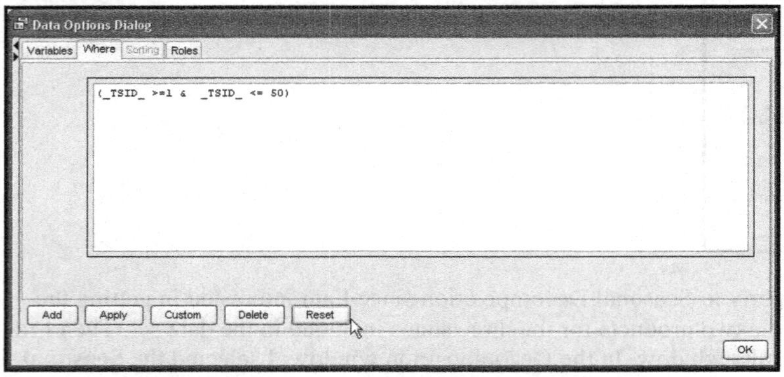

Data Options Dialog
Variables
Where
Sorting
Roles
(_TSID_ >=1 & _TSID_ <= 50)
Add
Apply
Custom
Delete
Reset
OK

Click **Reset**. A Data Options dialog box opens as shown in Display 2.56.

Display 2.56

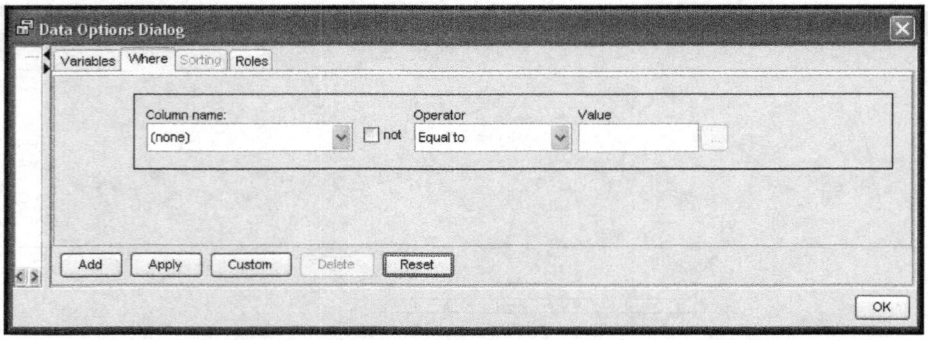

Data Options Dialog
Variables
Where
Sorting
Roles
Column name:
(none)
not
Operator
Equal to
Value
Add
Apply
Custom
Delete
Reset
OK

In the **Column name** box, select the variable name State. Enter CT in the **Value** field, and click **Add**. A new area appears, as shown in Display 2.57.

Display 2.57

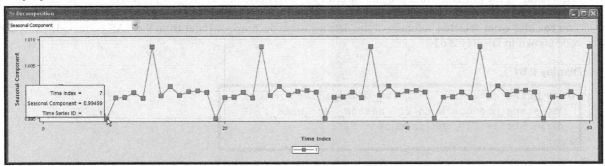

Select Product in the **Column name** box, and select the value A. Click **Apply** and **OK**. The seasonal component graph for Sales of Product A in the state of Connecticut appears, as shown in Display 2.58.

Display 2.58

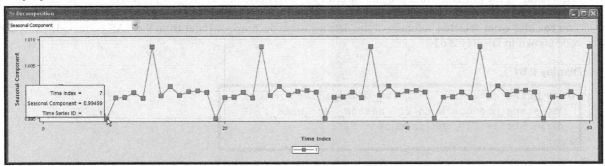

By hovering over any point on the graph, you can see the seasonal components for that month. As Display 2.58 shows, the seasonal component (factor) is 0.9949 for the month of July 2005. Since the component is below 1, it means that the sales were slightly 1% below normal. Similarly, you can see that the seasonal factors account for a slightly higher sales during the month of December.

To learn how to estimate the seasonal components of a time series, refer to:

- Dagum, E. B. (1980), The X-11-ARIMA Seasonal Adjustment Method, Statistics Canada.
- Dagum, E. B. (1983), The X-11-ARIMA Seasonal Adjustment Method, Technical Report 12-564E, Statistics Canada.
- Ladiray, D. and Quenneville, B. (2001), Seasonal Adjustment with the X-11 Method, New York: Springer-Verlag.

The example presented here is very simple. The full benefit of the **Time Series** node becomes clear when you use more complex data than that presented here.

Output Data Sets

To see a list of output data sets that are created by the **Time Series** node, click ![icon] located in the Value column of the **Exported Data** property of the Properties panel, as shown in Display 2.59. Display 2.60 shows a list of the output data sets.

Display 2.59

Property	Value
General	
Node ID	TIME
Imported Data	
Exported Data	
Notes	

Display 2.60

Port	Table	Role	Data Exists
TRAIN	EMWS4.TIME_TRAIN	Train	Yes
SEASON	EMWS4.TIME_SEASON	SEASON	No
TREND	EMWS4.TIME_TREND	TRANSPOSEDTREND	No
DECOMP	EMWS4.TIME_DECOMP	DECOMP	Yes
CORRSTAT	EMWS4.TIME_CORRSTAT	CORRSTAT	No

Exported Data - Time Series

Browse... Explore... Properties... OK

The seasonal decomposition data is saved as a data set with the name time_decomp. On my computer, this table is saved as C:\TheBook\EM12.1\EMProjects\Chapter2\Workspaces\EMWS4\time_decomp.sas7bdat. Chapter2 is the name of the project, and it is a sub-directory in C:\TheBook\EM12.1\EMProjects.

You can also print selected columns of the data set time_decomp.sas7bdat from the SAS code node using the code shown in Display 2.61.

Display 2.61

```
proc print data=&EM_LIB..time_decomp label noobs;
 var state product Mnth_Yr _season_ sc _name_;
 where state="CT" and product = "A";
run;
```

Partial output generated by this code is shown in Output 2.2.

Output 2.2

STATE	Product	Time ID	Seasonal Index	Seasonal Component	Variable Name
CT	A	JAN2005	1	0.99932	Sales
CT	A	FEB2005	2	1.00102	Sales
CT	A	MAR2005	3	0.99936	Sales
CT	A	APR2005	4	1.00006	Sales
CT	A	MAY2005	5	1.00015	Sales
CT	A	JUN2005	6	0.99988	Sales
CT	A	JUL2005	7	0.99499	Sales
CT	A	AUG2005	8	0.99894	Sales
CT	A	SEP2005	9	0.99899	Sales
CT	A	OCT2005	10	0.99985	Sales
CT	A	NOV2005	11	0.99880	Sales
CT	A	DEC2005	12	1.00865	Sales
CT	A	JAN2006	1	0.99932	Sales
CT	A	FEB2006	2	1.00102	Sales
CT	A	MAR2006	3	0.99936	Sales
CT	A	APR2006	4	1.00006	Sales
CT	A	MAY2006	5	1.00015	Sales
CT	A	JUN2006	6	0.99988	Sales
CT	A	JUL2006	7	0.99499	Sales
CT	A	AUG2006	8	0.99894	Sales
CT	A	SEP2006	9	0.99899	Sales
CT	A	OCT2006	10	0.99985	Sales
CT	A	NOV2006	11	0.99880	Sales
CT	A	DEC2006	12	1.00865	Sales
CT	A	JAN2007	1	0.99932	Sales
CT	A	FEB2007	2	1.00102	Sales

2.7.6 Merge Node

The **Merge** node can be used to combine different data sets within a SAS Enterprise Miner project. Occasionally you may need to combine the outputs generated by two or more nodes in the process flow. For example, you may want to test two different types of transformations of interval inputs together, where each type of transformation is generated by different instances of the **Transform Variables** node. To do this you can attach two **Transform Variables** nodes, as shown in Display 2.62. You can set the properties of the first **Transform Variables** node such that it applies a particular type of transformation to all interval inputs. You can set the properties of the second **Transform Variables** node to perform a different type of transformation on the same interval variables. Then, using the **Merge** node, you can combine the output data sets created by these two **Transform Variables** nodes. The resulting merged data set can be used in a **Regression** node to test which variables and transformations are the best for your purposes.

Display 2.62

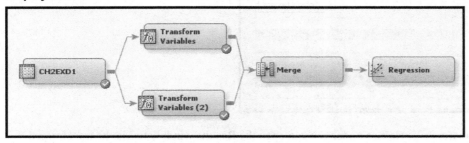

To make this example more concrete, I have generated a small data set for a hypothetical bank with two interval inputs and an interval target. The two inputs are: (1) interest rate differential (SPREAD) between the interest rate offered by the bank and the rate offered by its competitors, and (2) amount spent by the bank on advertisements (AdExp) to attract new customers and/or induce current customers to increase their savings balances. The target variable is the month-to-month change in the savings balance (DBAL) of each customer for each of a series of months, which is a continuous variable.

The details of different transformations and how to set the properties of the **Transform Variables** node to generate the desired transformations are discussed later in this chapter and also in Chapter 4. Here it is sufficient to know that the two **Transform Variables** nodes shown in Display 2.62 will each produce an output data set, and the **Merge** node merges these two output data sets into a single combined data set.

In the upper **Transform Variables** node, all interval inputs are transformed using the "*optimal binning*" method. (See Chapter 4 for more detail.) The optimal binning method creates a categorical variable from each continuous variable; the categories are the input ranges (or class intervals or bins). In order for all continuous and interval scaled inputs to be transformed by the Optimal Binning method, I set the **Interval Inputs** property in the Default Methods group to Optimal Binning, as shown in Display 2.63.

Display 2.63

Property	Value
General	
Node ID	Trans
Imported Data	
Exported Data	
Notes	
Train	
Variables	
Formulas	
Interactions	
SAS Code	
Default Methods	
Interval Inputs	Optimal Binning
Interval Targets	None
Class Inputs	None
Class Targets	None
Treat Missing as Level	No
Sample Properties	
Method	First N
Size	Default
Random Seed	12345
Optimal Binning	
Number of Bins	4

After running the **Transform Variables** node, you can open the Results window to see the transformed variables created, as shown in Display 2.64. The transformed variables are: OPT_AdExp and OPT_SPREAD.

Display 2.64

Results - Node: Transform Variables Diagram: Ch2_MergeNode

File   Edit   View   Window

Transformations Statistics

Source	Method	Variable Name	Formula	Number of Levels
Input	Original	AdExp		
Input	Original	SPREAD		
Output	Computed	OPT_AdExp	Optimal Binning(4)	2
Output	Computed	OPT_SPREAD	Optimal Binning(4)	4

You can view the output data set by clicking ⊡ located to the right of the Exported Data row in the Property table in Display 2.63. A partial view of the output data set created by the upper **Transform Variables** node is shown in Display 2.65.

Display 2.65

EMWS5.Trans_TRAIN

	AdExp	Month	SPREAD	DBAL	Transformed AdExp	Transformed SPREAD
1	0.0	1.0	0.06691475162375382	0.00692096224786376	01:low-0.9404285, MISSING	02:-0.058923-0.106362, MISSING
2	0.0	2.0	0.014998669661109076	-1.0664723322629457	01:low-0.9404285, MISSING	02:-0.058923-0.106362, MISSING
3	0.0	3.0	0.0076085131482510615	-1.1066431169034483	01:low-0.9404285, MISSING	02:-0.058923-0.106362, MISSING
4	0.0	4.0	0.0120922603910469938	-2.134382441750344	01:low-0.9404285, MISSING	02:-0.058923-0.106362, MISSING
5	0.0	5.0	0.028332158225225612	1.3533032273438241	01:low-0.9404285, MISSING	02:-0.058923-0.106362, MISSING
6	0.0	6.0	0.017225112195288883	-0.7538131423775758	01:low-0.9404285, MISSING	02:-0.058923-0.106362, MISSING
7	0.0	7.0	-0.004363195000600895	-0.5135291840501511	01:low-0.9404285, MISSING	02:-0.058923-0.106362, MISSING
8	0.0	8.0	-0.019779262240584794	-3.2131523715533437	01:low-0.9404285, MISSING	02:-0.058923-0.106362, MISSING
9	0.0	9.0	-0.05664722911127473	0.32422144871489733	01:low-0.9404285, MISSING	02:-0.058923-0.106362, MISSING
10	0.0	10.0	-0.0611979244295392	-2.0596819136393982	01:low-0.9404285, MISSING	01:low--0.058923

To generate a second set of transformations, click on the second (lower) **Transform Variables** node and set the **Interval Inputs** property in the Default Methods group to Exponential so that all the interval inputs are transformed using the exponential funtion. Display 2.66 shows the new variables created by the second **Transform Variables** node.

Display 2.66

Source	Method	Variable Name	Formula	Number of Levels
Input	Original	AdExp		
Input	Original	SPREAD		
Output	Computed	EXP_AdExp	exp(AdExp)	
Output	Computed	EXP_SPREAD	exp(SPREAD)	

Results - Node: Transform Variables (2) Diagram: Ch2_MergeNode
File Edit View Window
Transformations Statistics

The two **Transform Variables** nodes are then connected to the **Merge** node, as shown in Display 2.62. I have used the default properties of the **Merge** node.

After running the **Merge** node, click on it and click [...] located to the right of **Exported Data** in the Properties panel. Then, select the exported data set, as shown in Display 2.67.

Display 2.67

Exported Data - Merge

Port	Table	Role	Data Exists
TRAIN	EMWS5.Merge_TRAIN	Train	Yes
VALIDATE	EMWS5.Merge_VALIDATE	Validate	No
TEST	EMWS5.Merge_TEST	Test	No
SCORE	EMWS5.Merge_SCORE	Score	No

Browse... | Explore... | Properties... | OK

Click **Explore** to see the Sample Properties, Sample Statistics, and a view of the merged data set.

The next step is to connect a **Regression** node to the **Merge** node, then click Update Path. The variables exported to the **Regression** node are shown in Display 2.68.

Display 2.68

Variables - Reg

(none) | not | Equal to | | Apply | Reset

Columns: Label | Mining | ✓ Basic | Statistics

Name	Use	Report	Role	Level	Type	Format
DBAL	Yes	No	Target	Interval	Numeric	
EXP_AdExp	Default	No	Input	Interval	Numeric	
EXP_SPREAD	Default	No	Input	Interval	Numeric	
OPT_AdExp	Default	No	Input	Nominal	Character	
OPT_SPREAD	Default	No	Input	Nominal	Character	

Explore... | Update Path | OK | Cancel

Display 2.68 shows that the transformed variables created by the Optimal Binning method are nominal and those created by the second **Transform Variables** node are interval scaled. You can now run the **Regression** node and test all the transformations together and make a selection. Since we have not covered the **Regression** node, I have not run it here. But you can try.

2.7.7 Append Node

The **Append** node can be used to combine data sets created by different paths of a process flow in a SAS Enterprise Miner project. The way the **Append** node combines the data sets is similar to the way a SET statement in a SAS program stacks the data sets. This is different from the side-by-side combination that is done by the **Merge** node.

Display 2.69 shows an example in which the **Append** node is used.

Display 2.69

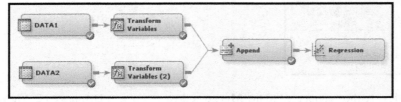

In Display 2.69 two data sources are used. The first data source is created by a data set called Data1, which contains data on Sales and Price in Region A at different points of time (months). The second data source is created from the data set Data2, which contains data on Sales and Price for Region B.

To illustrate the **Append** node, I have used two instances of the **Transform Variables** node. In both instances, the **Transform Variables** node makes a logarithmic transformation of the variables Sales and Price, creates data sets with transformed variables, and exports them to the **Append** node.

The output data sets produced by the two instances of the **Transform Variables** node are then combined by the **Append** node[3] and passed to the **Regression** node, where you can estimate the price elasticity of sales using the combined data set and the following specification of the demand equation:

$$\log\_ \textbf{sales} = \alpha + \beta \log\_ \textbf{price} \text{ or}$$

$$\textbf{Sales} = A \; \Pr \textit{ice}^{\beta}$$

$$\text{where } A = e^{\alpha}$$

In an equation of this form, β measures the price elasticity of demand;

in this example it is -1.1098

The first data set (Data1) has 100 observations and four columns (variables)—Price, Sales, Month, and Region. The second data set (Data2) also has 100 observations with the four columns Price, Sales, Month, and Region. The combined data set contains 200 observations.

The first instance of **Transform Variables** node creates new variables log_sales and log_Price from Data1, stores the transformed variables in a new data set, and exports it to the **Append** node. The data set exported by the first instance of the **Transform Variables** node has 100 observations.

The second instance of the **Transform Variables** node performs the same transformations done by the first instance, creates a new data set with the transformed variables, and exports it to the **Append** node. The data set exported by the second instance also has 100 observations.

The **Append** node creates a new data set by stacking the two data sets generated by the two instances of the **Transform Variables** node. Because of stacking (as opposed to side-by-side merging) the new data set has 200 observations. The data set exported to the **Regression** node has four variables, as shown in Display 2.70, and 200 observations shown in Output 2.3.

Display 2.70

```
┌─────────────────────────────────────────────────────────────────────────┐
│ ⊠ Variables - Reg                                                    [×]  │
├─────────────────────────────────────────────────────────────────────────┤
│                                                                           │
│  (none)          ▼   ☐ not  Equal to       ▼              ...  [ Apply ] [ Reset ] │
│                                                                           │
│  Columns:  ☐ Label       ☐ Mining       ☐ Basic       ☐ Statistics       │
│   ┌──────────┬──────┬────────┬──────────┬──────────┐                      │
│   │  Name    │ Use  │ Report │  Role    │  Level   │                      │
│   ├──────────┼──────┼────────┼──────────┼──────────┤                      │
│   │ LOG_Price│ Yes  │ No     │ Input    │ Interval │                      │
│   │ LOG_Sales│ Yes  │ No     │ Target   │ Interval │                      │
│   │ Region   │ No   │ No     │ Input    │ Nominal  │                      │
│   │ _NODEID_ │ No   │ No     │ Input    │ Nominal  │                      │
│   └──────────┴──────┴────────┴──────────┴──────────┘                      │
│                                                                           │
│                      [ Explore... ]  [ Update Path ]  [ OK ]  [ Cancel ]  │
└─────────────────────────────────────────────────────────────────────────┘
```

Output 2.3

```
The DMREG Procedure

               Model Information

Training Data Set         WORK.EM_DMREG.VIEW
DMDB Catalog              WORK.REG_DMDB
Target Variable           LOG_Sales (Transformed Sales)
Target Measurement Level  Interval
Error                     Normal
Link Function             Identity
Number of Model Parameters 2
Number of Observations    200


               Analysis of Variance

                        Sum of
Source          DF      Squares    Mean Square   F Value   Pr > F

Model            1    78.260914    78.260914    2821.80   <.0001
Error          198     5.491412     0.027734
Corrected Total 199    83.752326


          Model Fit Statistics

R-Square      0.9344    Adj R-Sq      0.9341
AIC        -715.0264    BIC        -712.9862
SBC        -708.4297    C(p)          2.0000


      Analysis of Maximum Likelihood Estimates

                         Standard
Parameter  DF   Estimate   Error    t Value   Pr > |t|

Intercept   1    5.4540    0.0363    150.06    <.0001
LOG_Price   1   -1.1098    0.0209    -53.12    <.0001
```

Display 2.71 shows the property settings of the **Append** node for the example presented here.

Display 2.71

Property	Value
General	
Node ID	APPEND
Imported Data	
Exported Data	
Notes	
Train	
Data Selector	
Output Type	Data
Action	By Role
Status	
Create Time	10/16/12 9:31 AM
Run ID	540d4393-564c-4e31-8a7c-08662c9a9
Last Error	
Last Status	Complete
Last Run Time	10/16/12 9:43 AM
Run Duration	0 Hr. 0 Min. 8.70 Sec.
Grid Host	
User-Added Node	No

Action

Indicates what append action should be performed. When set to "By Role", the node will combine data sets of the same role. When set to "Combine" the node will append data sets of role TRAIN, VALIDATE, and TEST.

This type of appending is useful in pooling the price and sales data for two regions and estimating a common equation. Here my intention is only to demonstrate how the **Append** node can be used for pooling the data sets for estimating a pooled regression and not to suggest or recommend pooling. Whether you pool depends on statistical and business considerations.

2.8 Tools for Initial Data Exploration

In this section I introduce the **StatExplore**, **MultiPlot**, **GraphExplore**, **Variable Clustering**, **Cluster** and **Variable Selection** nodes, which are useful in predictive modeling projects.

I will use two example data sets in demonstrating the use of **StatExplore**, **MultiPlot**, and **GraphExplore**. The first data set shows the response of a sample of customers to solicitation by a hypothetical auto insurance company. The data set consists of an indicator of response to the solicitation and several input variables that measure various characteristics of the customers who were approached by the insurance company. Based on the results from this sample, the insurance company wants to predict the probability of response based on customer's characteristics. Hence the target variable is the response indicator variable. It is a binary variable taking only two values, namely 0 and 1, where 1 represents response and 0 represents non-response. The actual development of predictive models is illustrated in detail in subsequent chapters, but here I provide an initial look into various tools of SAS Enterprise Miner that can be used for data exploration and discovery of important predictor variables in the data set.

The second data set used for illustration consists of month-to-month change in the savings balances of all customers (DBAL) of a bank, interest rate differential (Spread) between the interest rate offered by the bank and its competitors, and amount spent by the bank on advertisements to attract new customers and/or induce current customers to increase their savings balances.

The bank wants to predict the change in customer balances in response to change in the interest differential and the amount of advertising dollars spent. In this example, the target variable is change in the savings balances, which is a continuous variable.

Click the **Explore** tab located above the Diagram Workspace so that the data exploration tools appear on the tool bar. Drag the **Stat Explore**, **MultiPlot**, and **Graph Explore** nodes and connect them to the **Input Data Source** node, as shown in Display 2.72

Display 2.72

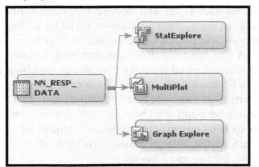

2.8.1 Stat Explore Node

Stat Explore Node: Binary Target (Response)

Select the **Stat Explore** node in the Diagram Workspace to see the properties of the **StatExplore** node in the Properties panel, shown in Display 2.73.

Display 2.73

Property	Value
General	
Node ID	Stat
Imported Data	
Exported Data	
Notes	
Train	
Variables	
Data	
Number of Observations	100000
Validation	No
Test	No
Standard Reports	
Interval Distributions	Yes
Class Distributions	Yes
Level Summary	Yes
Use Segment Variables	No
Cross-Tabulation	
Variable Selection	
Hide Rejected Variables	Yes
Number of Selected Variables	1000
Chi-Square Statistics	
Chi-Square	Yes
Interval Variables	Yes
Number of Bins	5
Correlation Statistics	
Correlations	Yes
Pearson Correlations	Yes
Spearman Correlations	No
Status	
Create Time	10/17/12 12:12 PM

Interval Variables

Generates Chi-Square statistics for interval variables by binning the variables.

If you set the **Chi-Square** property to Yes, a Chi-Square statistic is calculated and displayed for each variable. The Chi-Square statistic shows the strength of the relationship between the target variable (Response, in this example) and each categorical input variable. The appendix to this chapter shows how the Chi-Square statistic is computed.

In order to calculate Chi-Square statistics for continuous variables such as age and income, you have to first create categorical variables from them. Derivation of categorical variables from continuous variables is done by partitioning the ranges of continuous scaled variables into intervals. These intervals constitute different categories or levels of the newly derived categorical variables. A Chi-Square statistic can then be calculated to measure the strength of association between the categorical variables derived from the continuous variables and the target variable. The process of deriving categorical variables from continuous variables is called *binning*. If you want the **StatExplore** node to calculate the Chi-Square statistic for interval scaled variables, you must set the **Interval Variables** property to Yes, and you must also specify the number of bins into which you want the interval variables to be partitioned. To do this, set the **Number of Bins** property to the desired number of bins. The default value of the **Number of Bins** property is 5. For example, the interval scaled variable AGE is grouped to five bins, which are 18–32.4, 32.4-46.8, 46.8-61.2, 61.2-75.6, and 75.6-90.

When you run the **StatExplore** node and open the Results window, you see a Chi-Square plot, Variable Worth plot and an Output window. The Chi-Square plot shows the Chi-Square value of each categorical variable and binned variable paired with the target variable, as shown in Display 2.74. The plot shows the strength of relationship of each categorical or binned variable with the target variable.

Display 2.74

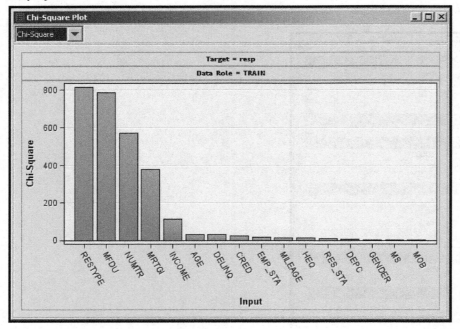

The results window also displays, in a separate panel, the worth of each input. The worth is calculated from the p-value corresponding to the calculated Chi-Square test statistic. The p-value corresponding to the calculated chi-square statistic is calculated as

$$P(\chi^2 \geq calculated \; chi-Square \; statistic) = p$$

Worth of the input is $-2\log(p)$.

The Variable Worth plot is shown in Display 2.75.

Display 2.75

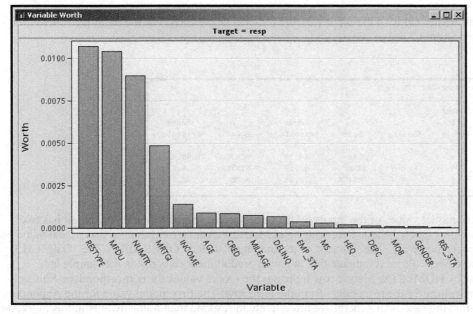

Both the Chi-Square plot and the Variable Worth plot show that the variable RESTYPE (the type of residence) is the most important variable since it has the highest Chi-Square value (Display 2.74) and also the highest worth (Display 2.75). Next in importance is MFDU (an indicator of multifamily dwelling unit). From the **StatExplore** node you can make a preliminary assessment of the importance of the variables.

An alternative measure of calculating the worth of an input, called *impurity reduction*, is discussed in the context of decision trees in Chapter 4. In that chapter, I discuss how impurity measures can be applied to calculate the worth of categorical inputs one at a time.

In addition to the Chi-Square statistic, you can display the Cramer's V for categorical and binned interval inputs in the Chi-Square Plot window. Select Cramer's V from the drop-down list, as shown in Display 2.76, in order to open the plot.

Display 2.76

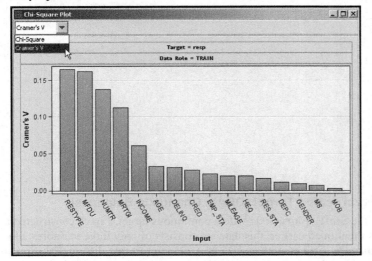

You can get the formulae for Chi-Square statistic and Cramer's V from the **Help** tab of SAS Enterprise Miner. The calculation of the Chi-Square statistic and Cramer's V are illustrated step-by-step in the appendix to this chapter.

The Output window shows the mode of the input variable for each target class. For the input RESTYPE, the modal values are shown Output 2.4.

Output 2.4

```
Data Role=TRAIN Variable Name=RESTYPE

                       Number
               Target    of                           Mode                    Mode2
Target         Level   Levels   Missing    Mode    Percentage    Mode2     Percentage

_OVERALL_                 4        0        HOME      54.73      RENTER       40.42
resp             0        4        0        HOME      60.19      RENTER       35.11
resp             1        4        0        RENTER    52.05      HOME         42.77
```

This output arranges the modal values of the inputs by the target levels. In this example, the target has two levels: 0 and 1. The columns labeled Mode Percentage and Mode2 Percentage exhibit the first modal value and the second modal value, respectively. The first row of the output is labeled _OVERALL_. The _OVERALL_ row values for Mode and Mode Percent indicate that the most predominant category in the sample is homeowners, indicated by HOME. The second row indicates the modal values for non-responders. Similarly, you can read the modal values for the responders from the third row. The first modal value for the responders is RENTER, suggesting that the renters in general are more likely to respond than home owners in this marketing campaign. These numbers can be verified by running PROC FREQ from the Program Editor, as shown in Display 2.77.

Display 2.77

```
proc freq data=TheBook.NN_RESP_DATA;
  table RESTYPE*RESP/nopercent norow missing;
run ;
```

The results of PROC FREQ are shown in Output 2.5.

Output 2.5

The FREQ Procedure			
Frequency Col Pct	Table of RESTYPE by resp		
		resp	
RESTYPE	0	1	Total
CONDO	380 1.85	186 1.98	566
COOP	585 2.85	300 3.20	885
HOME	12354 60.19	4011 42.77	16365
RENTER	7206 35.11	4882 52.05	12088
Total	20525	9379	29904

StatExplore Node: Continuous/Interval scaled Target (DBAL: Change in Balances)

In order to demonstrate how you can use the **StatExplore** node with a continuous target, I have constructed a small data set for a hypothetical bank. As mentioned earlier, this data set consists of only three variables: (1) month-to-month change in the savings balances of all customers (DBAL), (2) interest rate differential (Spread) between the interest rate offered by the bank and its competitors, and (3) amount spent by the bank on advertisements (AdExp) to attract new customers and/or induce current customers to increase their savings balances. This small data set is used for illustration, although in practice you can use the **StatExplore** node to explore much larger data sets consisting of hundreds of variables.

Display 2.78 shows the process flow diagram for this example. The process flow diagram is identical to the one shown for a binary target, except that the input data source is different.

Display 2.78

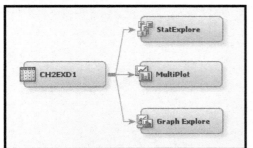

The property settings for the **StatExplore** node for this example are the same as those shown in Display 2.73, with the following exceptions: the **Interval Variables** property (in the Chi-Square Statistics group) is set to No and the **Correlations**, **Pearson Correlations** and **Spearman Correlations** properties are all set to Yes.

After we run the **Stat Explore** node, we get the correlation plot, which shows the Pearson correlation between the target variable and the interval scaled inputs. This plot is shown in Display 2.79.

Display 2.79

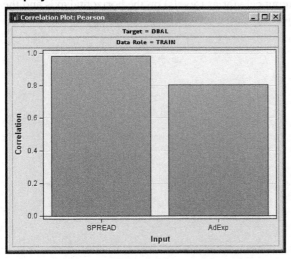

Both the SPREAD and Advertisement Expenditure (AdExp) are positively correlated with Changes in Balances (DBAL), although the correlation between AdExp and DBAL is lower than the correlation between the spread and DBAL.

Display 2.80 shows a comparison of the worth of the two inputs advertisement expenditure and interest rate differential (spread).

Display 2.80

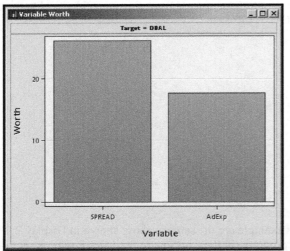

2.8.2 MultiPlot Node

MultiPlot Node: Binary Target (Response)

After you run the **MultiPlot** node, the results window shows plots of all inputs against the target variable. If an input is a categorical variable, then the plot shows the input categories (levels) on the horizontal axis and the percentage distribution of the target classes on the vertical axis, as shown in Display 2.81.

Display 2.81

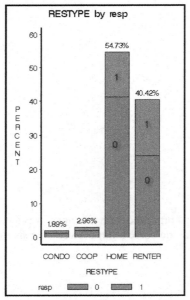

The variable RESTYPE is a categorical variable, along with categories CONDO, COOP, HOME, and RENTER. The categories refer to the type of residence the customer lives. The plot above shows that the percentage of responders (indicated by 1) among the renters is higher than among the home owners.

When the input is continuous, the distribution of responders and non-responders is given for different intervals of the input, as shown in Display 2.82. The midpoint of each interval is shown on the horizontal axis, and the distribution of target class (response and non-response) is shown on the vertical axis.

Display 2.82

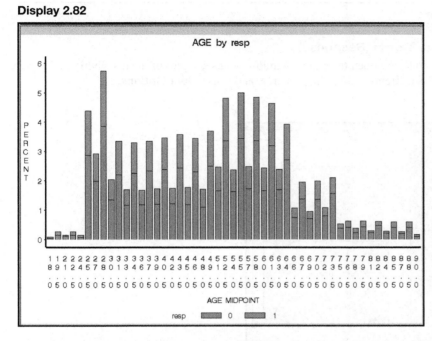

MultiPlot Node: Continuous/Interval scaled Target (DBAL: Change in Balances)

If you run the **MultiPlot** node in the Process Flow shown in Display 2.78 and open the Results window, you see a number of charts that show the relation between each input and the target variable. One such chart shows how the change in balances is related to interest rate differential (SPREAD), as shown in Display 2.83. The **MultiPlot** node shows the mean of target variable in different intervals of continuous inputs.

Display 2.83

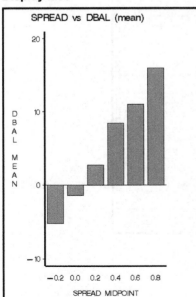

In Display 2.83, the continuous variable SPREAD is divided into six intervals and the midpoint of each interval is shown on the horizontal axis. The mean of the target variable in each interval is shown along the vertical axis. As expected, the chart shows that the larger the spread, the higher the increase in balances is.

2.8.3 Graph Explore Node

Graph Explore Node: Binary Target (Response)

If you run the **Graph Explore** node and open the results window, you see a plot of the distribution of the target variable, as shown in Display 2.84. Right-click in the plot area, and select **Data Options**.

Display 2.84

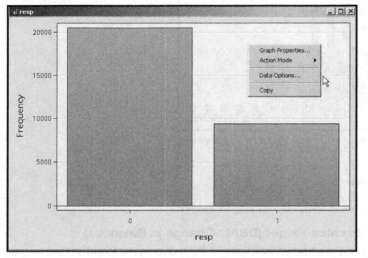

In the Data Options dialog box, shown in the Display 2.85, select the variables you want to plot and their roles.

Display 2.85

▲ Variable	Role	Type	Description	Format
DELINQ		Numeric	DELINQ	
DEPC		Character	DEPC	
EMP_STA		Character	EMP_STA	
GENDER	Category	Character	GENDER	
HEQ		Numeric	HEQ	
INCOME		Numeric	INCOME	
MFDU		Numeric	MFDU	
MILEAGE		Numeric	MILEAGE	
MOB		Character	MOB	
MRTGI		Character	MRTGI	
MS		Character	MS	
NUMTR		Numeric	NUMTR	
RES_STA		Character	RES_STA	
resp	Response	Numeric	resp	
RESTYPE		Character	RESTYPE	

Response statistic: Mean

☐ Allow multiple role assignments

OK

I assigned the role category to GENDER and the role Response to the target variable resp. I selected the response statistic to be the Mean and clicked **OK**. The result is shown in Display 2.86, which shows the Gender category on the horizontal axis and the mean of the response variable for each gender category on the vertical axis. The response rate for males is slightly higher than that for females.

Display 2.86

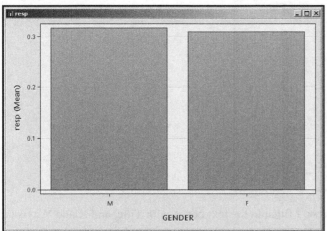

If you explore the **Graph Explore** node further, you will find there are many types of charts available.

Graph Explore Node: Continuous/Interval Scaled Target (DBAL: Change in Balances)

Run the **Graph Explore** node and open the results window. Click **View** on the menu bar, and select **Plot**. The Select a Chart Type dialog box opens, as shown in Display 2.87.

Display 2.87

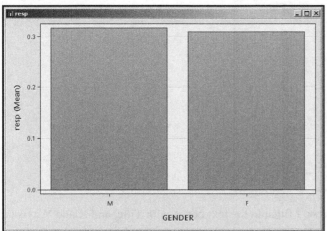

I selected the Scatter chart and clicked **Next**. Then I selected the roles of the variables SPREAD and DBAL, designating SPREAD to be the X variable and DBAL to be the Y variable, as shown in Display 2.88.

Display 2.88

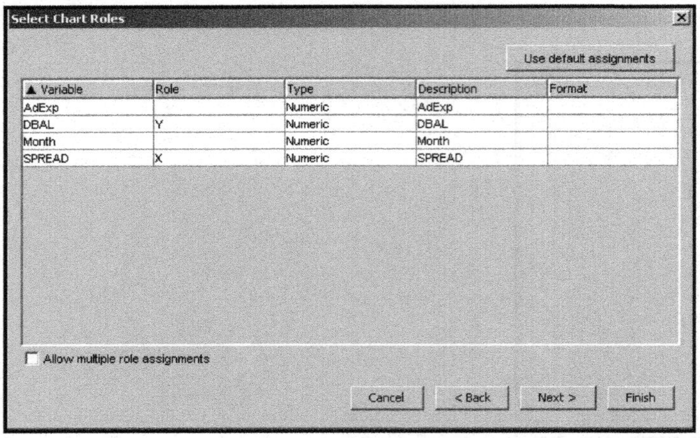

I clicked **Next** twice. In the Chart Titles dialog box, I filled in the text boxes with Title, and X and Y axis labels as shown in Display 2.89 and clicked **Next**.

Display 2.89

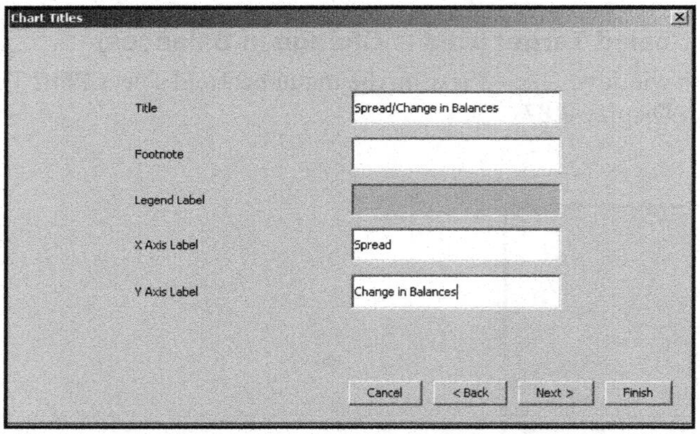

In the last dialog box, I clicked **Finish**, which resulted in the plot shown in Display 2.90.

Display 2.90

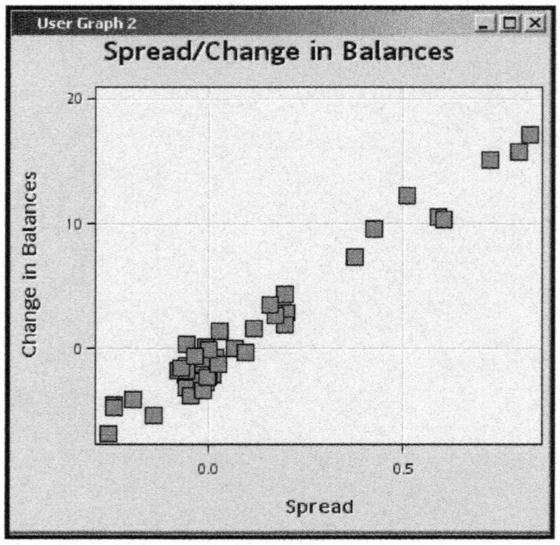

To change the marker size, right-click in the chart area and select **Graph Properties**. The properties window opens, shown in Display 2.91.

Display 2.91

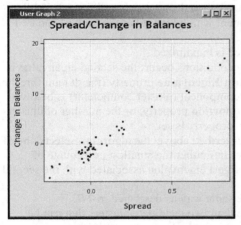

Clear the Autosize Markers check box, and slide the scale to the left until you see that the **Size** is set to 3. Click **Apply** and **OK**. These settings result in a scatter chart with smaller markers, as shown in Display 2.92.

Display 2.92

The scatter chart shows that there is direct relation between SPREAD and Change in Balances.

2.8.4 Variable Clustering Node

The **Variable Cluster** node divides the inputs (variables) in a predictive modeling data set into disjoint clusters or groups. Disjoint means that if an input is included in one cluster, it does not appear in any other cluster. The inputs included in a cluster are strongly inter-correlated, and the inputs included in any one cluster are not strongly correlated with the inputs in any other cluster.

If you estimate a predictive model by including only one variable from each cluster or a linear combination of all the variables in that cluster, you not only reduce the severity of collinearity to a great extent, you also have fewer variables to deal with in developing the predictive model.

In order to learn how to use the **Variable Clustering** node and interpret the results correctly, you must understand how the **Variable Clustering** node clusters (groups) variables.

The **Variable Clustering** algorithm is both divisive and iterative. The algorithm starts with all variables in one single cluster and successively divides it into smaller and smaller clusters. The splitting is binary in the sense that at each point in the process, a cluster is split into two sub-clusters or child clusters, provided certain criteria are met. The process of splitting can be described as follows:[4]

1. Initially there is a single cluster with all variables in the data set included in it.
2. Eigenvalues are extracted from the correlation (or covariance) matrix of the variables included in the cluster. The largest eigenvalue and the next largest eigenvalue are used for calculating the eigenvectors and creating cluster components. If you arrange the eigenvalues in descending order of magnitude, then the first eigenvalue is the largest and the second eigenvalue is the next largest.
3. Eigenvectors corresponding to the first two (the largest and the next largest) eigenvalues are calculated. We can call them first and second eigenvectors.
4. Perform an oblique rotation on the eigenvectors and calculate "rotated" components.
5. Cluster components (principal components), which are linear combinations of all the variables included in the cluster, are calculated from the first and second eigenvectors. A linear combination can be thought of as a weighted sum of the variables with the elements of a rotated eigenvector as weights. Two cluster components are created – one corresponding to the first eigenvalue and the other corresponding to the second eigenvalue. We can call these first and second cluster components or first and second principal components since they correspond to the first and second eigenvalues as described in 2 above.
6. If the criterion for splitting is met (as described in 8 below), then the cluster is split into two child clusters by assigning each variable to the cluster component with which it has the highest squared multiple correlation.
7. Reassign variables iteratively until the explained variance is maximized
8. Steps 6 and 7 are performed only if one of the following conditions occur: the second eigenvalue is larger than the threshold value specified by the **Maximum Eigenvalue** property (the default threshold value is 1), the variance explained by the first principal component (cluster component) is below a specified threshold value specified by the **Variation Proportion** property, or the number of clusters is smaller than the value to which the **Maximum Clusters** property is set.
   After the initial cluster is split into two child-clusters as described above, the algorithm selects one of the child-clusters for further splitting. The selected cluster has either the smallest percentage of variation explained by its first cluster component or the largest eigenvalue associated with the second cluster component.
9. The selected cluster is split into two child-clusters in the same way as described in 2-7.
10. At any point in the splitting process, there may be more than two candidate clusters that can be considered for splitting. The cluster that is selected from this set has either the smallest percentage of variation explained by its first cluster component or the largest eigenvalue associated with the second cluster component.
11. Steps 2 – 7 are repeated until no cluster is eligible for further splitting.
   When all the clusters meet the stopping criterion, the splitting stops.

The stopping criterion is met when at least one of the following is true: (a) The second eigenvalue in all the clusters is smaller than the threshold value specified by the **Maximum Eigenvalue** property; (b) The proportion of variance explained by the first principal component in all the clusters is above the threshold value specified by the **Variation Proportion** property; (c) the number of clusters is equal to the value set for the **Maximum Clusters** property.

To demonstrate the **Variable Cluster** node, I have created a small data set with 11 variables. Display 2.93 shows the correlation matrix of variables.

Display 2.93

Matrix of Pearson Correlation Coefficients											
Variable	VAR1	VAR2	VAR3	VAR4	VAR5	VAR6	VAR7	VAR8	VAR9	VAR10	VAR11
VAR1	1.00000	0.49186	0.16896	0.66478	-0.01287	-0.00963	-0.00459	-0.01073	0.00528	0.00252	0.00875
VAR2	0.49186	1.00000	-0.35089	0.19182	-0.00195	0.01000	0.02177	-0.00477	-0.00251	-0.00354	-0.00283
VAR3	0.16896	-0.35089	1.00000	0.32858	-0.05146	-0.03284	-0.03536	-0.03355	0.00868	0.01240	0.00504
VAR4	0.66478	0.19182	0.32858	1.00000	-0.02524	-0.03490	-0.01250	-0.01072	0.02125	0.02169	0.02640
VAR5	-0.01287	-0.00195	-0.05146	-0.02524	1.00000	0.44284	0.49929	0.55291	0.02035	0.03638	0.03159
VAR6	-0.00963	0.01000	-0.03284	-0.03490	0.44284	1.00000	0.61245	0.08093	0.00467	-0.00182	0.01018
VAR7	-0.00459	0.02177	-0.03536	-0.01250	0.49929	0.61245	1.00000	0.38055	0.01908	0.02492	0.04154
VAR8	-0.01073	-0.00477	-0.03355	-0.01072	0.55291	0.08093	0.38055	1.00000	0.00547	0.01614	0.02536
VAR9	0.00528	-0.00251	0.00868	0.02125	0.02035	0.00467	0.01908	0.00547	1.00000	0.90936	0.80759
VAR10	0.00252	-0.00354	0.01240	0.02169	0.03638	-0.00182	0.02492	0.01614	0.90936	1.00000	0.71193
VAR11	0.00875	-0.00283	0.00504	0.02640	0.03159	0.01018	0.04154	0.02536	0.80759	0.71193	1.00000

Display 2.94 shows the process flow diagram for clustering the variables in the data set.

Display 2.94

The properties of the **Variable Clustering** node are shown in Display 2.95.

Display 2.95

Property	Value
General	
Node ID	VarClus
Imported Data	
Exported Data	
Notes	
Train	
Variables	
Clustering Source	Correlation
Keeps Hierarchies	Yes
Includes Class Variables	No
Two Stage Clustering	Auto
⊟ Stopping Criteria	
⊢Maximum Clusters	Default
⊢Maximum Eigenvalue	Default
⊢Variation Proportion	Default
Print Option	Short
Suppress Sampling Warning	No
Score	
Variable Selection	Cluster Component
Interactive Selection	
Hides Rejected Variables	No
Status	
Create Time	10/26/12 7:17 AM

In this example, I set the **Clustering Source** property to Correlation. I have chosen to use the correlation matrix as the source for clustering, so the eigenvalues are calculated from the correlation matrix. From each eigenvalue, the algorithm calculates an eigenvector. The method of calculating eigenvalues and corresponding eigenvectors can be found in any linear algebra textbook or any multivariate statistics book such as "Applied Multivariate Statistical Analysis" by Richard A. Johnson and Dean W. Wichern.

In principle, if the number of variables in a cluster is k, you can extract k eigenvalues for that cluster of variables. But if the variables in the cluster are closely related to each other, you may find only a few eigenvalues larger than 1, and it may be more useful to group the closely related variables together in sub-clusters.

In our example, at the very beginning of the splitting process $k = 11$ since the very first cluster includes all the variables in the data set. But as the splitting process continues, the number of variables (k) in each resulting cluster will be less than 11.

Let the largest eigenvalue of the correlation matrix be represented by λ_1 and the next largest eigenvalue by λ_2. A cluster is not split if $\lambda_2 < \tau$, where τ is the value of the **Maximum Eigenvalue** property.

Calculation of Cluster Components

The eigenvalues of the correlation (or covariance) matrix and the corresponding eigenvectors are needed to calculate the cluster components. From the first eigenvalue λ_1, the algorithm calculates an eigenvector $e_1 = (e_{11}, e_{21}, e_{31}, ..., e_{k1})'$. From the second eigenvalue λ_2, it calculates the eigenvector $e_2 = (e_{12}, e_{22}, e_{32}, ..., e_{k2})'$. These eigenvectors e_1 and e_2 are normalized so that they are of unit length. The eigenvectors are then rotated. The rotation creates two new vectors from the two eigenvectors e_1 and e_2. the new vectors created by the rotation are $W_1 = (w_{11}, w_{21}, w_{31}, ..., w_{k1})'$ and $W_2 = (w_{12}, w_{22}, w_{32}, ..., w_{k2})'$.

The algorithm then calculates two new variables called cluster components from the original variables using the elements of W_1 and W_2 as the weights. The cluster components are linear combinations or weighted averages of the variables that have been chosen to be together in a cluster, so it is a way of combining the closely related variables in your data set into a single variable that can be used to represent those variables. The exact calculation of the cluster components is given by the following two equations:

The first cluster component is:

$$C_{i1} = \sum_{j=1}^{k} w_{j1} x_{ij}, \text{where}$$

i stands for the ith observation and

x_{ij} = the value of the jth input for the ith observation

the second cluster component is:

$$C_{i2} = \sum_{j=1}^{k} w_{j2} x_{ij}.$$

The cluster membership of a variable is unique if its weight in one cluster component is zero while its weight in the other cluster component is nonzero. Although the rotation takes us in this direction, it does not always result in a clear-cut assignment of variables to the clusters. In other words, it does not result in a variable having zero weight in one component and a nonzero weight in another component.

Assignment of variables to clusters

In order to achieve this uniqueness of membership (or non-overlapping of clusters), the algorithm compares the squared multiple correlations of each variable in the cluster with the two components C_{i1} and C_{i2}. For example, suppose $R^2(x_1, C_1) > R^2(x_1, C_2)$, where R^2 stands for squared multiple correlation. That is, $R^2(x_1, C_1)$ is the squared multiple correlation of a regression of x_1 on C_1, and $R^2(x_1, C_2)$ is the squared multiple correlation of a regression of x_1 on C_2. Then x_1 is put in child cluster 1; otherwise, it is assigned to cluster 2. Thus the variables are kept in non-overlapping or "disjoint" clusters.

The algorithm goes further by iteratively re-assigning the variables to different clusters in order to maximize the variance accounted for by the cluster components. You can request that the hierarchical structure developed earlier in the process not be destroyed during this iterative re-assignment process by setting the **KeepHierarchies** property to Yes (see Diagram 2.96).

After the iteration process is completed, the cluster components are recomputed for each cluster, and they are exported to the next node.

Proportion of variance explained by a cluster component

One of the criterion used in determining whether to split a cluster is based on the proportion of variance explained by the first cluster component (or first principal component) of the cluster. This proportion of variance explained by the first cluster component is the ratio:

$$\frac{\text{Variance of the first cluster component}}{\text{Total variance of all the variables included in the cluster}}$$

$$= \frac{V(C_1)}{\sum_{i=1}^{l} V(X_i)}, \text{ where } l \text{ is the number of variables included in the cluster.}$$

To get a general idea of how to calculate the proportion of variance explained by a principal component in terms of the eigenvalues, refer to "Applied Multivariate Statistical Analysis" by Richard A. Johnson and Dean W. Wichern. The exact calculation of the proportion of variance explained by the first cluster component by the **Variable Clustering** node in SAS Enterprise Miner may be different from the calculations shown in the book cited above.

Variable clustering using the example data set

I ran the **Variable Clustering** node shown in Display 2.94. The output in the Results window is reproduced in output 2.6.

Output 2.6

```
Variable Summary

          Measurement     Frequency
Role      Level           Count

INPUT     INTERVAL         11




Oblique Principal Component Cluster Analysis

Observations          1500     Proportion          0
Variables               11     Maxeigen            1


Clustering algorithm converged.


                    Cluster Summary for 1 Cluster

                        Cluster   Variation    Proportion     Second
Cluster    Members     Variation  Explained    Explained    Eigenvalue
-----------------------------------------------------------------------
   1          11          11       2.637892      0.2398       2.3078

Total variation explained = 2.637892 Proportion = 0.2398

Cluster 1 will be split because it has the largest second eigenvalue,2.307765,
which is greater than the MAXEIGEN=1 value.
```

Output 2.6 (cont'd)

```
Clustering algorithm converged.


                    Cluster Summary for 2 Clusters

                        Cluster   Variation    Proportion     Second
Cluster    Members     Variation  Explained    Explained    Eigenvalue
-----------------------------------------------------------------------
   1           3           3       2.621943      0.8740       0.3029
   2           8           8       2.319889      0.2900       1.9449

Total variation explained = 4.941832 Proportion = 0.4493


                           R-squared with
                         -------------------
2 Clusters                 Own      Next      1-R**2
Cluster    Variable      Cluster   Closest    Ratio
-----------------------------------------------------------
Cluster 1    VAR10       0.8783    0.0006     0.1217
             VAR11       0.8008    0.0011     0.1995
             VAR9        0.9429    0.0002     0.0672
-----------------------------------------------------------
Cluster 2    VAR1        0.0127    0.0000     0.9873
             VAR2        0.0001    0.0000     0.9999
             VAR3        0.0180    0.0001     0.9821
             VAR4        0.0194    0.0006     0.9812
             VAR5        0.6810    0.0010     0.3193
             VAR6        0.5011    0.0000     0.4989
             VAR7        0.6863    0.0009     0.3140
             VAR8        0.4013    0.0003     0.5989

Cluster 2 will be split because it has the largest second eigenvalue, 1.944901,
which is greater than the MAXEIGEN=1 value.
```

Output 2.6 (cont'd)

```
Clustering algorithm converged.


                     Cluster Summary for 3 Clusters

                       Cluster     Variation    Proportion      Second
  Cluster    Members   Variation   Explained    Explained    Eigenvalue
  -----------------------------------------------------------------------
      1         3          3        2.621943       0.8740        0.3029
      2         4          4        2.307878       0.5770        0.9805
      3         4          4        1.950298       0.4876        1.3867

Total variation explained = 6.880119 Proportion = 0.6255


                               R-squared with
  3 Clusters                  -------------------
                               Own       Next      1-R**2
  Cluster     Variable       Cluster   Closest     Ratio
  -----------------------------------------------------------
  Cluster 1   VAR10          0.8783    0.0007      0.1218
              VAR11          0.8008    0.0013      0.1995
              VAR9           0.9429    0.0003      0.0572
  -----------------------------------------------------------
  Cluster 2   VAR5           0.6907    0.0010      0.3096
              VAR6           0.5082    0.0004      0.4920
              VAR7           0.7006    0.0009      0.2996
              VAR8           0.4083    0.0003      0.5918
  -----------------------------------------------------------
  Cluster 3   VAR1           0.8578    0.0001      0.1422
              VAR2           0.3107    0.0001      0.6893
              VAR3           0.0634    0.0026      0.9390
              VAR4           0.7183    0.0007      0.2819

Cluster 3 will be split because it has the largest second eigenvalue, 1.386702,
which is greater than the MAXEIGEN=1 value.
```

Output 2.6 (cont'd)

```
Clustering algorithm converged.


                     Cluster Summary for 4 Clusters

                       Cluster     Variation    Proportion      Second
  Cluster    Members   Variation   Explained    Explained    Eigenvalue
  -----------------------------------------------------------------------
      1         3          3        2.621943       0.8740        0.3029
      2         4          4        2.307878       0.5770        0.9805
      3         2          2        1.664776       0.8324        0.3352
      4         2          2        1.350887       0.6754        0.6491

Total variation explained = 7.945484 Proportion = 0.7223


                               R-squared with
  4 Clusters                  -------------------
                               Own       Next      1-R**2
  Cluster     Variable       Cluster   Closest     Ratio
  -----------------------------------------------------------
  Cluster 1   VAR10          0.8783    0.0007      0.1218
              VAR11          0.8008    0.0013      0.1995
              VAR9           0.9429    0.0003      0.0572
  -----------------------------------------------------------
  Cluster 2   VAR5           0.6907    0.0010      0.3096
              VAR6           0.5082    0.0007      0.4921
              VAR7           0.7006    0.0012      0.2997
              VAR8           0.4083    0.0003      0.5919
  -----------------------------------------------------------
  Cluster 3   VAR1           0.8324    0.0386      0.1743
              VAR4           0.8324    0.0069      0.1688
  -----------------------------------------------------------
  Cluster 4   VAR2           0.6754    0.1404      0.3776
              VAR3           0.6754    0.0743      0.3506

No cluster meets the criterion for splitting.
```

As a way of reducing the number of inputs (dimension reduction), you can select one representative variable from each cluster. A representative variable of a cluster can be defined as the one with the lowest 1-R**2 Ratio. This ratio is calculated as:

$$1-R^2_{\ ratio} = \frac{1-R^2_{\ own\ cluster}}{1-R^2_{\ next\ closest\ cluster}}, \text{ where}$$

$$R^2_{\ own\ cluster} = \text{Correlation with own cluster and}$$

$$R^2_{\ next\ closest\ cluster} = \text{Correlation with next closest cluster.}$$

An alternative way of achieving dimension reduction is to replace the original variables with cluster components.

Display 2.96 shows the cluster plot.

Display 2.96

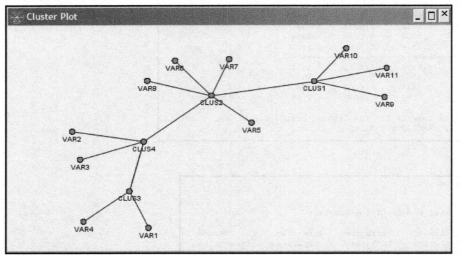

The variables exported to the next node are shown in Display 2.97.

Display 2.97

Name	Use	Report	Role	Level
Clus1	Default	No	Input	Interval
Clus2	Default	No	Input	Interval
Clus3	Default	No	Input	Interval
Clus4	Default	No	Input	Interval
VAR1	Default	No	Rejected	Interval
VAR10	Default	No	Rejected	Interval
VAR11	Default	No	Rejected	Interval
VAR2	Default	No	Rejected	Interval
VAR3	Default	No	Rejected	Interval
VAR4	Default	No	Rejected	Interval
VAR5	Default	No	Rejected	Interval
VAR6	Default	No	Rejected	Interval
VAR7	Default	No	Rejected	Interval
VAR8	Default	No	Rejected	Interval
VAR9	Default	No	Rejected	Interval

The 11 original variables and the 4 new variables (Clus1, Clus2, Clus3, and Clus4) are exported to the next node. You could develop a predictive model using only the four newly constructed variables Clus1, Clus2, Clus3, and Clus4 instead of the original 11 variables. Thus, the **Variable Clustering** node helps in reducing the inputs and reducing the danger of collinearity.

An alternative to using the constructed variables Clus1, Clus2, Clus3, and Clus4 is to select one input variable from each of the four clusters presented in Output 2.6. The best input variable to select from each cluster can be determined on the basis of the correlation of the input to the cluster component or its correlation with the target variable.

Note that I have not included a target variable in the example data set I used to demonstrate the **Variable Clustering** node. If you include another variable with the role of TARGET, then the **Variable Clustering** node excludes it from creating clusters. This becomes clear in Chapter 3 where I demonstrate how the **Variable Clustering** node is used for variable selection.

2.8.5 Cluster Node

The **Cluster** node can be used to create clusters of customers (observations) with similar characteristics. An examination of these clusters enables you to discover patterns in your data and also helps identify the inputs and input combinations that are potentially good predictors of the target variable. Cluster analysis is an unsupervised learning technique in the sense that it creates the clusters from the input variables alone, without reference to any target variable. Cluster analysis by itself does not tell you how the inputs are related to the target. Additional analysis is needed to find out the relationship of the clusters to the target variable. This sort of clustering can be used, for example, to segment your customers, without reference to any particular target variable, just to see if your customers (observations) naturally fall into different groups or clusters. Customers who end up in the same cluster are similar in some way, while customers in different clusters are relatively dissimilar.

In order to demonstrate the properties and the results of the **Cluster** node, I have used a data set of 5,000 credit card customers of a hypothetical bank. These customers were observed during an interval of 6 months, which is referred to as the performance window. During this performance window, some of the customers cancelled their credit cards and some retained them. I created an indicator variable, called Cancel, which takes the value 1 if the customer cancelled his credit card during the performance window and 0 if he did not. In addition, data was appended to each customer's record showing his demographic characteristics and other information relating to the customer's behavior prior to the time interval specified by the performance window. Some examples of customer behavior are: the number of credit card delinquencies, the number of purchases, and the amount of his/her credit card balance – all for the period prior to the performance window.

We use only the input variables to create the clusters, and we pass the cluster labels and cluster numbers created by the **Cluster** node to other nodes such as the **Regression** node for further analysis. The cluster labels or numbers can be used as nominal inputs in the **Regression** node. Display 2.98 shows the process flow for demonstrating the **Cluster** node.

Display 2.98

A **Regression** node is attached to the **Cluster** node to show how the variables created by the **Cluster** node are passed to the next node. Instead of the **Regression** node, you could use the **Stat Explore** or **Graph Explore** node to compare the profiles of the clusters.

Display 2.99 shows the Properties window for the **Cluster** node.

Display 2.99

Property	Value
General	
Node ID	Clus
Imported Data	
Exported Data	
Notes	
Train	
Variables	
Cluster Variable Role	Input
Internal Standardization	Standardization
⊟ Number of Clusters	
Specification Method	Automatic
Maximum Number of Clusters	10
⊟ Selection Criterion	
Clustering Method	Ward
Preliminary Maximum	50
Minimum	2
Final Maximum	5
CCC Cutoff	3
⊟ Encoding of Class Variables	
Ordinal Encoding	Rank
Nominal Encoding	GLM
⊟ Initial Cluster Seeds	
Seed Initialization Method	Default
Minimum Radius	0.0
Drift During Training	No
⊟ Training Options	
Use Defaults	Yes
Settings	
⊟ Missing Values	
Interval Variables	Default
Nominal Variables	Default
Ordinal Variables	Default
Scoring Imputation Method	None
Score	
Cluster Variable Role	Input
Hide Original Variables	Yes
Cluster Label Editor	
Report	
Cluster Graphs	Yes
Tree Profile	Yes
Distance Plot and Table	Yes
Status	
Create Time	10/27/12 5:03 PM

In the Train group I set the **Cluster Variable Role** to Input. After the cluster node creates the clusters, it creates a nominal variable called _Segment_, which takes the value 1, 2, 3, etc., showing the cluster to which an observation belongs. When you set the value of the **Cluster Variable Role** property to Input, the variable _segment_ is assigned the role of Input and passed to the next node. Another nominal variable called _Segment _Label_ takes the values Cluster1, Cluster2, Cluster3, etc., showing the label of the cluster. The variable _Segment_ can be used as a nominal input in the next node. Alternatively, you can set the value of the **Cluster Variable Role** property to Segment resulting in a segment variable with the model role of Segment. This can be used for by-group processing in the next node.

The settings in the Number of Clusters group and in the Selection Criterion cause the **Cluster** node to perform clustering in two steps. In the first step, it creates a number of clusters not exceeding 50 (as specified by the **Preliminary Maximum** property), and in the second step it reduces the number of clusters by combining clusters, subject to two conditions specified in the properties: (1) The number of clusters must be greater than 2 (as specified by the value of the **Minimum** property), and (2) the value of the Cubic Clustering Criterion (CCC) statistic is greater than or equal to the Cutoff, which is 3. The Ward method is used to determine the final maximum number of clusters.

I then ran the **Cluster** node and opened the results window. A partial list of the Mean Statistics by cluster is shown in Display 2.100

Display 2.100

Segment Id	Frequency of Cluster	Gross Margin	Annual Percentage Rate	Months on Book	Balance	Count of purchases in last 12 months	Unpaid Balance 4 Mos Pd label	Balance without delinquency
1	216	25.31455	13.02042	64.62037	1704.673	9.181319	0.476397	643.0701
2	1	32.8224	12.47497	17	1123.431	13.61607	1321.54	10.31091
3	3206	23.19642	13.93	53.60356	539.8907	6.320806	0.510535	446.9325
4	911	22.91063	13.58061	54.6674	6230.749	6.89414	0.48673	5975.812
5	666	21.23	13.61971	16.86036	295.4395	3.970195	1.323517	390.6896

Display 2.100 shows how the mean of the inputs differs from cluster to cluster. Cluster 2 (segment id =2) has only one observation. SAS Enterprise Miner created a separate cluster for this observation because it has some outliers in the inputs. The customer in cluster 2 generates the highest average gross margin (32.8) for the company. (Gross margin is revenue less cost of service.) The customer in cluster 2 differs from customers in other clusters in other ways: although she makes 13.6 purchases per year with an average daily balance of $1123.43, this customer has an unpaid balance of $1321.54 with four months past due, and an average balance of 10.31 without delinquency. While cluster 1 ranks highest in average gross margin (a measure of customer profitability), Clusters 3 ranks highest in total profits (gross margins) earned by the company since it contains the largest number of customers (3206). Cluster analysis can often give important insights into customer behavior as it did in this example.

To illustrate how the output of the **Cluster** node is exported to the subsequent nodes in the process flow, I attached a **Regression** node to the **Cluster** node. The variables passed to the **Regression** node are shown in Display 2.101.

Display 2.101

Name	Use	Report	Role	Level
APR	Default	No	Input	Interval
BAL	Default	No	Input	Interval
CANCEL	Yes	No	Target	Binary
Distance	Default	No	Rejected	Interval
High_Credit	Default	No	Input	Interval
Margin	Default	No	Input	Interval
MaritalStatus	Default	No	Input	Nominal
MultLang	Default	No	Input	Nominal
Pct_unpaid_bal_4MPD	Default	No	Input	Interval
Purch_count_12M	Default	No	Input	Interval
TENURE	Default	No	Input	Interval
SEGMENT	Default	No	Input	Nominal
_SEGMENT_LABEL_	Default	No	Rejected	Nominal
creditline	Default	No	Input	Interval
gendern	Default	No	Input	Nominal
good_bal	Default	No	Input	Interval
num_tran_bal_GT_0	Default	No	Input	Interval
occupation	Default	No	Input	Nominal
purch_amount	Default	No	Input	Interval

Since I set the value of the **Cluster Variable Role** property to Input, the variable _SEGMENT_ is assigned the role of Input by the **Cluster** node and is passed on to the **Regression** node as a nominal variable.

You can use this nominal variable as an input in the regression. Alternatively, if I had selected Segment as the value for the **Cluster Variable Role** property, then the **Cluster** node would have assigned the role of Segment to the variable _SEGMENT_, in which case it could be used for by-group processing.

2.8.6 Variable Selection Node

The **Variable Selection** node can be used for variable selection for predictive modeling. There are number of alternative methods with various options for selecting variables. The methods of variable selection depend on the measurement scales of inputs and the targets.

These options are discussed in detail in Chapter 3. In this chapter, I present a brief review of the techniques used by the **Variable Selection** node for different types of targets, and I show how to set the properties of the **Variable Selection** node for choosing an appropriate technique.

There are two basic techniques used by the **Variable Selection** node. They are the R-Square selection method and the Chi-Square selection method. Both these techniques select variables based on the strength of their relationship with the target variable. For interval targets, only the R-Square selection method is available. For binary targets both the R-Square and Chi-Square selection methods are available.

2.8.6.1 R-Square Selection Method

To use the R-Square selection method, you must set the **Target Model** property of the **Variable Selection** node to R-Square, as shown in the Display 2.102.

Display 2.102

Property	Value
General	
Node ID	Varsel
Imported Data	
Exported Data	
Notes	
Train	
Variables	
Max Class Level	100
Max Missing Percentage	50
Target Model	R-Square
Manual Selector	
Rejects Unused Input	Yes
⊟ Bypass Options	
Variable	None
Role	Input
⊟ Chi-Square Options	
Number of Bins	50
Maximum Pass Number	6
Minimum Chi-Square	3.84
⊟ R-Square Options	
Maximum Variable Number	3000
Minimum R-Square	0.0050
Stop R-Square	1.0E-5
Use AOV16 Variables	Yes
Use Group Variables	Yes
Use Interactions	No
Use SPD Engine Library	Yes
Print Option	Default
Score	
Hides Rejected Variables	Yes
Hides Unused Variables	Yes
Status	
Create Time	10/28/12 8:49 AM
Run ID	

When the R-Square selection method is used, the variable selection is done in two steps. In Step 1, the **Variable Selection** node computes an R-Square value (the squared correlation coefficient) with the target for each variable, and then assigns the Rejected role to those variables that have a value less than the minimum R-square. The default minimum R-square cut-off is set to 0.005. You can change the value of the minimum R-square by setting the **Minimum R-Square** property to a value other than the default.

The R-square value (or the squared correlation coefficient) is the proportion of variation in the target variable explained by a single input variable, ignoring the effect of other input variables.

In Step 2, the **Variable Selection** node performs a forward stepwise regression to evaluate the variables chosen in the first step. Those variables that have a stepwise R-square improvement less than the cut-off criterion have the role of rejected. The default cut-off for R-square improvement is set to 0.0005. You can change this value by setting the **Stop R-Square** property to a different value.

For interval variables, R-square is calculated directly by means of a linear regression of the target variable on the interval variable, assessing only the linear relation between the interval variable and the target.

To detect non-linear relations, the **Variable Selection** node creates binned variables from each interval variable. The binned variables are called AOV16 variables. Each AOV16 variable has a maximum of 16 intervals of equal width. The AOV16 variable is treated as a class variable. A one-way analysis of variance is performed to calculate the R-Square between an AOV16 variable and the target. These AOV16 variables are included in Step 1 and Step 2 of the selection process.

Some of the interval variables may be selected both in their original form, as well as in their binned (AOV16) form. If you set the **Use AOV16 Variables** property to Yes, then the AOV16 variables are passed to the next node. For variable selection, the number of categories of a nominal categorical variable can be reduced by combining categories that have similar distribution of the target levels. To use this option, set the **Use Group Variables** property to Yes.

2.8.6.2 Chi-Square Selection Method

When you select the Chi-Square selection method, the **Variable Selection** node creates a tree based on Chi-square maximization. The **Variable Selection** node first bins interval variables, and uses the binned variable rather than the original inputs in building the tree. The default number of bins is 50, but you can change the number by setting the **Number of Bins** property to a different value. Any split with Chi-square below the specified threshold is rejected. The default value for the Chi-square threshold is 3.84. You can change this value by setting the **Minimum Chi-Square** property to the desired level. The inputs that give the best splits are included in the final tree, and passed to the next node with the role of Input. All other inputs are given the role of Rejected.

2.8.6.3 Variable Selection Node: An Example with R-Square Selection

I use a data set that consists of 17,880 observations but only nine inputs and a binary target in order to show the results produced by different settings of the properties of the **Variable Selection** node. The flow diagram for this demonstration is shown in Display 2.103.

Display 2.103

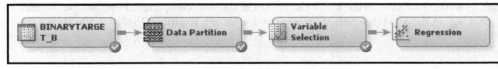

Display 2.104 shows the Variables window of the **Input Data** node. The data set is partitioned such that 50% is allocated to Training, 30% to Validation, and 20% to Test.

Display 2.104

Name	Role	Level	Type
VAR1N	Input	Interval	Numeric
VAR2O	Input	Nominal	Numeric
VAR3N	Input	Interval	Numeric
VAR4C	Input	Nominal	Character
VAR5O	Input	Nominal	Numeric
VAR6C	Input	Nominal	Character
VAR7C	Input	Nominal	Character
VAR8O	Input	Nominal	Numeric
credscore	Input	Interval	Numeric
resp	Target	Binary	Numeric

Since the target variable is binary in this example, I could use any of the options available for the **Target Model** property (see Display 2.102). I chose the R-Square selection method.

Next, I select the threshold values for the **Minimum R-Square** and **Stop R-Square** properties. The **Minimum R-Square** property is used in Step 1 and the **Stop R-Square** property is used in Step 2. In this example, the **Minimum R-Square** property is set to the default value, and the **Stop R-Square** property to 0.00001. The **Stop R-Square** property is deliberately set at a very low level so that most of the variables selected in the first step will also be selected in the second step. In other words, I effectively by-passed the second step so that all or most of the variables selected in Step 1 are passed to the **Decision Tree** node or **Regression** node that follows. I did this because both the **Decision Tree** and **Regression** nodes make their own selection of variables, and I wanted the second step of variable selection to be done in these nodes rather than in the **Variable Selection** node.

Display 2.105

R-Square Options	
Maximum Variable Number	3000
Minimum R-Square	0.0050
Stop R-Square	1.0E-5
Use AOV16 Variables	Yes
Use Group Variables	Yes
Use Interactions	No
Use SPD Engine Library	Yes
Print Option	Default

From Display 2.105, you can see that I set the **Use AOV16 Variables** property to Yes. If I chose No, the **Variable Selection** node (i.e., the underlying PROC DMINE) would still create the AOV16 variables and include them in Step 1, but not in Step 2. In addition, if you select No, they will not be passed to the next node.

When you set the **Use Group Variables** property to Yes, the categories of the class variables are combined, and new variables with the prefix of G_ are created. (This is explained in more detail in the next chapter.)

With these settings, I ran the **Variable Selection** node. After the run was completed, I opened the Results window by right-clicking on the **Variable Selection** node and selecting **Results**. Display 2.106 shows the results.

Display 2.106

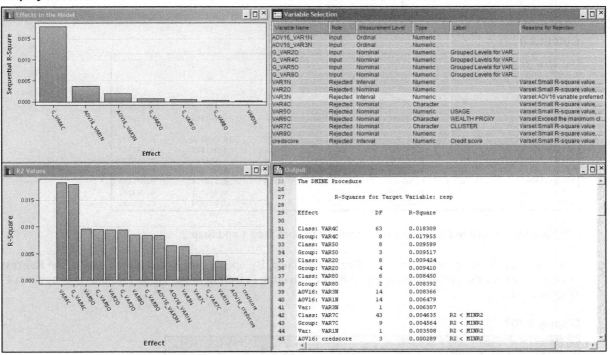

The bar chart in the top left quadrant of Display 2.106 shows the variables selected in Step 2. The bottom left quadrant shows a histogram of the R-Square values computed in Step 1. The table in the top right quadrant shows the variables selected in Step 2 and passed to the next node, along with their assigned roles. The output pane in the bottom right quadrant of the Results window shows the variables selected in Step 1. These are shown in Output 2.7.

Output 2.7

```
The DMINE Procedure

          R-Squares for Target Variable: resp

Effect               DF      R-Square

Class: VAR4C         63      0.018309
Group: VAR4C          8      0.017955
Class: VAR5O          8      0.009589
Group: VAR5O          3      0.009517
Class: VAR2O          8      0.009424
Group: VAR2O          4      0.009410
Class: VAR8O          6      0.008450
Group: VAR8O          2      0.008392
AOV16: VAR3N         14      0.008366
AOV16: VAR1N         14      0.006479
Var:   VAR3N          1      0.006307
Class: VAR7C         43      0.004635   R2 < MINR2
Group: VAR7C          9      0.004564   R2 < MINR2
Var:   VAR1N          1      0.003508   R2 < MINR2
AOV16: credscore      3      0.000289   R2 < MINR2
Var:   credscore      1      0.000162   R2 < MINR2
```

Scrolling down further in the output pane, you can see the variables selected in Step 2. These are shown in Output 2.8.

Output 2.8

```
The DMINE Procedure


                          Effects Chosen for Target: resp


                                                            Sum of      Error Mean
Effect              DF       R-Square      F Value   p-Value    Squares          Square

Group: VAR4C        8        0.017955     20.411192   <.0001   14.446732       0.088473
AOV16: VAR1N        14       0.003767      2.452744   0.0019    3.031124       0.088272
AOV16: VAR3N        14       0.001933      1.259276   0.2244    1.555591       0.088236
Group: VAR5O        3        0.000520      1.581224   0.1916    0.418482       0.088219
Group: VAR2O        4        0.000767      1.748304   0.1364    0.616726       0.088189
Group: VAR8O        2        0.000299      1.363888   0.2557    0.240541       0.088182
Var:   VAR3N        1        0.000112      1.018060   0.3130    0.089774       0.088182
```

The variables shown in Output 2.8 were selected in both Step 1 and Step 2.

The next node in the process flow is the **Regression** node. By opening the Variables window of the **Regression** node, you can see the variables passed to it by the **Variable Selection** node. These variables are shown in Display 2.107.

Display 2.107

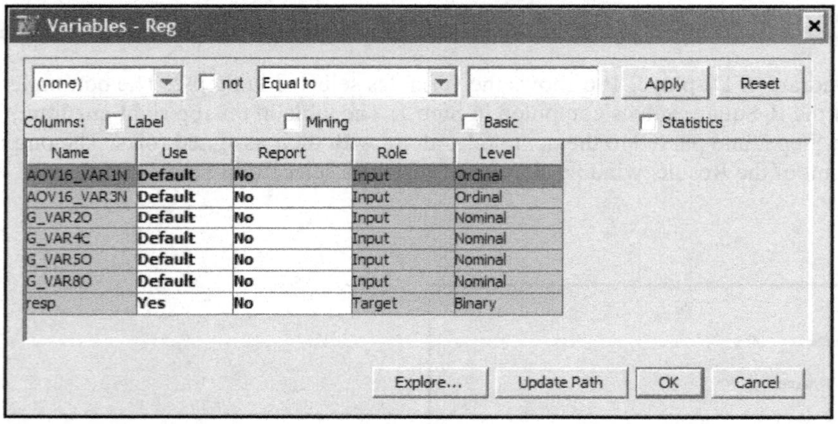

In output 2.7 and output 2.8, you see that VAR1N is rejected in Step 1, but AOV16_VAR1N is selected in both Step 1 and Step 2, indicting a non-linear relationship between VAR1N and the target.

The definitions of the AOV16 variables and the grouped class variables (G variables) are included in the SAS code generated by the **Variable Selection** node. You can access this code by right-clicking on the Variable Selection node and clicking **Results → View → Scoring → SAS Code**. Display 2.108 shows a segment of the SAS code.

Display 2.108

```
/*----AOV16_VAR1N begin----*/
if ( missing( VAR1N ) ) then do;
  substr(_WARN_, 1, 1) = 'M';
  AOV16_VAR1N = 3;
end;
else
if (VAR1N < 5.8125) then AOV16_VAR1N = 1;
else
if (VAR1N < 10.625) then AOV16_VAR1N = 2;
else
if (VAR1N < 15.4375) then AOV16_VAR1N = 3;
else
if (VAR1N < 20.25) then AOV16_VAR1N = 4;
else
if (VAR1N < 25.0625) then AOV16_VAR1N = 5;
else
if (VAR1N < 29.875) then AOV16_VAR1N = 6;
else
if (VAR1N < 34.6875) then AOV16_VAR1N = 7;
else
if (VAR1N < 39.5) then AOV16_VAR1N = 8;
else
if (VAR1N < 44.3125) then AOV16_VAR1N = 9;
else
if (VAR1N < 49.125) then AOV16_VAR1N = 10;
else
if (VAR1N < 53.9375) then AOV16_VAR1N = 11;
else
if (VAR1N < 58.75) then AOV16_VAR1N = 12;
else
if (VAR1N < 63.5625) then AOV16_VAR1N = 13;
else
if (VAR1N < 68.375) then AOV16_VAR1N = 14;
else
if (VAR1N < 73.1875) then AOV16_VAR1N = 15;
else AOV16_VAR1N = 16;
/*----AOV16_VAR1N end----*/
```

Display 2.108 (*continued*)

```
/*----G_VAR50 begin----*/
length _NORM12 $ 12;
_NORM12 = put( VAR50 , BEST12. );
%DMNORMIP( _NORM12 )
drop _NORM12;
select(_NORM12);
  when(".          " ) G_VAR50 = 3;
  when("0          " ) G_VAR50 = 3;
  when("1          " ) G_VAR50 = 2;
  when("2          " ) G_VAR50 = 1;
  when("3          " ) G_VAR50 = 0;
  when("4          " ) G_VAR50 = 1;
  when("5          " ) G_VAR50 = 1;
  when("6          " ) G_VAR50 = 3;
  when("8          " ) G_VAR50 = 3;
  otherwise substr(_WARN_, 2, 1) = 'U';
end;
label G_VAR50="Grouped Levels for VAR50";
/*----G_VAR50 end----*/
```

2.8.6.4 Variable Selection Node: An Example with Chi-Square Selection

To perform variable selection using the Chi-square criterion, set the **Target Model** property to Chi-Square. SAS Enterprise Miner constructs a CHAID type of tree, and the variables used in the tree become the selected variables. The relative importance of the variables is shown in Display 2.109.

Display 2.109

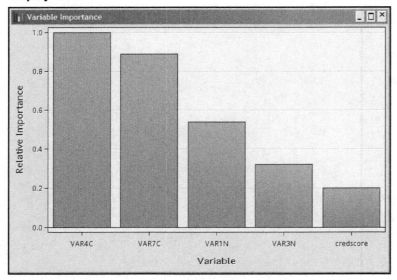

Display 2.110 shows the variables selected by Chi-square selection process.

Display 2.110

Variable Name	Role	Measurement Level	Type	Label	Reasons for Rejection
VAR1N	Input	Interval	Numeric		
VAR2O	Rejected	Nominal	Numeric		Varsel:Small Chi-square value
VAR3N	Input	Interval	Numeric		
VAR4C	Input	Nominal	Character		
VAR5O	Rejected	Nominal	Numeric	USAGE	Varsel:Small Chi-square value
VAR6C	Rejected	Nominal	Character	WEALTH PROXY	Varsel:Exceed the maximum class level of 100
VAR7C	Input	Nominal	Character	CLUSTER	
VAR8O	Rejected	Nominal	Numeric		Varsel:Small Chi-square value
credscore	Input	Interval	Numeric	Credit score	

2.8.6.5 Saving the SAS Code Generated by the Variable Selection Node

If the **Target Model** property is set to R-Square, the **Use AOV16 Variables** property is set to Yes, and the **Use Group Variables** property is set to Yes, then the **Variable Selection** node generates SAS code with definitions ofAOV16 and Group variables. Right-click the Variable Selection node and then click **Results** to open the Results window. Click **View → Scoring →SAS Code** in the Results window in order to view the SAS code. Save the SAS code by clicking **File →Save** as and giving a name to the file where the SAS code is being saved and pointing to the directory where the file is being saved.

2.8.6.6 The Procedures behind the Variable Selection Node

If you set the **Target Model** property to R-Square and run the **Variable Selection** node, SAS Enterprise Miner uses PROC DMINE in variable selection. Prior to running this procedure, it creates a data mining database (DMDB) and a data mining database catalog (DMDBCAT) using PROC DMDB.

When you set the **Target Model** property to Chi-Square, the **Variable Selection** node uses PROC DMSPLIT and PROC SPLIT for selecting variables and for assessing the variable importance.

To see the syntax of these procedures, open the Results window, click **View** on the menu bar, and click SAS Results→ **Log**.

Display 2.111 shows the log from PROC DMDB.

Display 2.111

```
Run proc dmdb with the specified maxlevel criterion.
 *---------------------------------------------------------* ;
 * EM: DMDBClass Macro ;
 *---------------------------------------------------------* ;
 %macro DMDBClass;
     VAR2O(ASC) VAR4C(ASC) VAR5O(ASC) VAR6C(ASC) VAR7C(ASC) VAR8O(ASC) resp(DESC)
 %mend DMDBClass;
 *---------------------------------------------------------* ;
 * EM: DMDBVar Macro ;
 *---------------------------------------------------------* ;
 %macro DMDBVar;
     VAR1N VAR3N credscore
 %mend DMDBVar;
 *---------------------------------------------------------*;
 * EM: Create DMDB;
 *---------------------------------------------------------*;
 libname _spdslib SPDE "C:\DOCUME~1\DELLUS~1\LOCALS~1\Temp\SAS Temporary Files\_TD4204_DH084NF1_";
 Libref _SPDSLIB was successfully assigned as follows:
 Engine:        SPDE
 Physical Name: C:\DOCUME~1\DELLUS~1\LOCALS~1\Temp\SAS Temporary Files\_TD4204_DH084NF1_\
 proc dmdb batch data=EMWS3.Part_TRAIN
 dmdbcat=WORK.EM_DMDB
 maxlevel = 101
 out=_spdslib.EM_DMDB
 ;
 class %DMDBClass;
 var %DMDBVar;
 target
 resp
 ;
 run;
```

Display 2.112 shows the log from PROC DMINE.

Display 2.112

```
 * Varsel: Input Variables Macro ;
 *---------------------------------------------------------* ;
 %macro INPUTS;
     VAR1N VAR2O VAR3N VAR4C VAR5O VAR7C VAR8O CREDSCORE
 %mend INPUTS;
 proc dmine data=_spdslib.EM_DMDB dmdbcat=WORK.EM_DMDB
 minr2=0.005 maxrows=3000 stopr2=0.00001 NOINTER USEGROUPS OUTGROUP=EMWS3.Varsel_OUTGROUP
 outest=EMWS3.Varsel_OUTESTDMINE outeffect=EMWS3.Varsel_OUTEFFECT outrsquare =EMWS3.Varsel_OUTRSQUARE
 NOMONITOR
 PSHORT
 ;
 var %INPUTS;
 target resp;
 code file="C:\TheBook\EM12.1\EmProjects\Chapter2\Workspaces\EMWS3\Varsel\EMFLOWSCORE.sas";
 code file="C:\TheBook\EM12.1\EmProjects\Chapter2\Workspaces\EMWS3\Varsel\EMPUBLISHSCORE.sas";
 run;
```

2.9 Tools for Data Modification

2.9.1 Drop Node

The **Drop** node can be used for dropping variables from the data set or metadata. The **Drop** node can be used to drop variables from a data set created by the predecessor node. For example, suppose I used the **StatExplore** node to examine the data from the BinaryTarget_B and the exploratory analysis led me to believe that the variable VAR6C is not needed in my project. So I used the **Drop** node to drop the variable VAR6C from the data set. Display 2.113 shows the process flow diagram I used for performing this task.

Display 2.113

You can drop any irrelevant variable by setting the **Drop from Tables** property to Yes. This setting drops the selected variables from the table that is exported to the next node.

To select the variables to be dropped, click , located to the right of **Variables** property of the **Drop** node. If you want to drop any variable, click in the Drop column in the variable and select Yes, as shown in Display 2.114.

Display 2.114

Name	Drop	Role	Level
VAR1N	Default	Input	Interval
VAR2O	Default	Input	Nominal
VAR3N	Default	Input	Interval
VAR4C	Default	Input	Nominal
VAR5O	Default	Input	Nominal
VAR6C	Yes	Input	Nominal
VAR7C	Default	Input	Nominal
VAR8O	Default	Input	Nominal

Click **OK** and run the **Drop** node. You can open the Exported Data set and verify that the variable VAR6C is dropped from the Exported Data.

2.9.2 Replacement Node

In contrast to the **Filter** node discussed in Section 2.7.3, the **Replacement** node can be used to filter out extreme values in a variable *without losing any observations*. The **Replacement** node can also be used to change the distribution of any variable in the sample, if it is appropriate to do so. If the variable has either a bi-modal or extremely skewed distribution, you can use the **Replacement** node to make its distribution more centralized or symmetric.

In this section I use the **Replacement** node to eliminate certain implausible values that some of the variables have in the modeling data set. This process is also called filtering, but it is filtering without loss of observations.

The data set used in this illustration has 5000 observations and 17 variables, including the target variable.

To use the **Replacement** node to filter (clean up) the data, I create the process flow diagram shown in Display 2.115.

Display 2.115

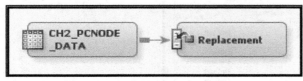

There are a number of default methods available of filtering. A description of these methods can be found in the online help. You can choose default method by setting the **Default Limits Method** property to one of the values shown in Display 2.116.

Display 2.116

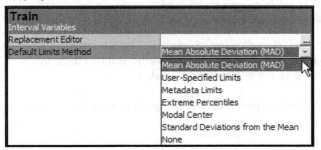

Specify the cut-off values for the selected method by setting the **Cutoff Values** property, as shown in Display 2.117.

Display 2.117

Property	Value
MAD	9.0
Percentiles for Extreme Percentiles	0.5
Modal Center	9.0
Standard Deviation	3.0

MAD

Specifies the number of deviations from the median to be used as cutoff value. That is, values that are that many mean absolute deviations away from the median will be used as the limit values. When set to User-Specified the values specified using the Interval Editor are used. When set to Missing, blanks or missing values are used as the replacement values.

Run the **Replacement** node and click **OK**.

The next example demonstrates how to use the Replacement Editor to access the interval variable table, and then enter the cut-off values either manually or by using movable sliding line.

I set the **Default Limits Method** property to User-Specified Limits.

Next, I open the Replacement Editor window by clicking ⊡ to the right of the **Replacement Editor** property under the Interval Variables group. The Interactive Replacement Interval Filter window, showing a list of all the interval variables, opens as shown in Display 2.118.

Display 2.118

I select the variable APR (Annual Percentage Rate) and click Generate Summary. The Summary Table dialog box opens with the prompt "Summary statistics have not yet been calculated. Would you like to do this now?" I click **Yes**. Then a chart opens in the Interactive Replacement Interval Filter window.

By moving the reference line above the chart and clicking **Apply Filter**, I changed the Replacement Lower Limit to 3.56115, as shown in Display 2.119.

Display 2.119

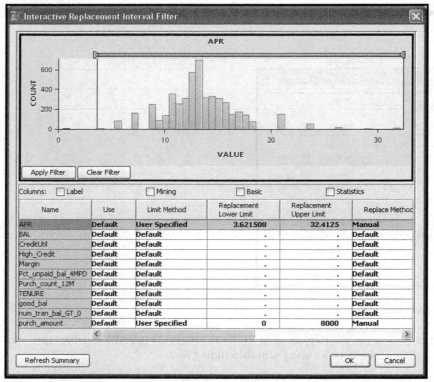

I changed the lower cut-off value for the variable purch_amount to 0 by clicking on the variable purch_amount, typing 0 in the column Replacement Lower Limit, and clicking **Apply Filter**.

I then clicked **Refresh Summary**. After the summary table refreshed, I clicked **OK**, and the Interactive Replacement Interval Filter window closed. I ran the **Replacement** node and opened the results window. Display 2.120 shows the cut-off values I specified for the variables APR and purch_amount.

Display 2.120

Interval Variables									
Variable	Replace Variable	Limits Method	Lower limit	Upper Limit	Label	Replacement Method	Lower Replacement Value	Upper Replacement Value
APR	REP_APR	MANUAL	3.56115	32.4125	Annual Percentage Rate	MANUAL	3.56115	32.4125
purch_amount	REP_purch_amount	MANUAL	0	8000	Spending	MANUAL	0	8000

2.9.3 Impute Node

The **Impute** node is used for imputing missing values of inputs. The icon of this node, along with other icons of the Modify group, appears on the toolbar when you select the **Modify** tab of the SAS Enterprise Miner window (Display 2.3).

Display 2.121 shows the **Impute** node selected in a process flow diagram.

Display 2.121

The **Impute** node has a number of techniques for imputation. To select a particular technique, you must first select the **Impute** node on the Diagram Workspace and set the **Default Input Method** property to the desired technique. For interval variables, the options available are Mean, Maximum, Minimum, Median, Mid-Range, Distribution, Tree, Tree Surrogate, Mid-Minimum Space, Tukey's Biweight, Huber, Andrew's Wave, Default Constant Value, and None. For class variables, the options are Count, Default Constant Value, Distribution, Tree, Tree Surrogate, and None.

You can select an imputation technique that is different from the default technique for any variable by opening the variables window. Click ⬚ located in the Value column on the **Variables** property in the **Impute** node's Properties panel. After the Variables window opens, you can select a method from the method column correponding to the variable you are interested in imputing by a particular method of your choice.

2.9.4 Interactive Binning Node

I will demonstrate the **Interactive Binning** node using a data set for a hypothetical bank. There is a binary target variable that takes the value 1 if the customer cancelled her account during a certain time interval called the performance window. In addition to the target variable, the data set has 15 inputs, some of which are interval-scaled and some of which are nominal variables, which capture the demographic characteristics and customer behavior of each customer prior to the performance window.

Binning helps in uncovering complex non-linear relationships between the inputs and the target variable in a modeling data set.

Display 2.122 shows the process flow diagram used for demonstrating the **Interactive Binning** node.

Display 2.122

Display 2.123 shows the properties of the **Interactive Binning** node.

Display 2.123

Property	Value
General	
Node ID	BINNING
Imported Data	
Exported Data	
Notes	
Train	
Variables	
Interactive Binning	
Treat Missing as Level	Yes
Use Frozen Groupings	No
Interval Variable Options	
Apply Level Rule	No
Method	Quantile
Number of Groups	4
Class Variable Options	
Group Rare Levels	Yes
Cutoff Value Percentage	0.5
Import/Export Options	
Import Grouping Data	No
Import Data Set	
Score	
Group Level	Nominal
Variable Selection Method	Gini Statistic
Gini Cutoff	20.0
Report	
Create Grouping Data	Yes
Create Method	Overwrite
Number of Variables	10
Status	
Create Time	11/1/12 2:03 PM

If you run the **Interactive Binning** node using these property settings, each interval-scaled variable will first be "binned" into four bins (groups). Binning is a method for converting an interval-scaled variable, with its infinite range of possible numerical values, into a categorical variable, which takes on a limited number of values, corresponding to the classes into which the observations may fall. Binning involves grouping the observations by sorting them into bins or buckets corresponding to more or less narrow ranges of the values that the original interval-scaled variable takes on. Each of these bins becomes one of the classes of the resulting categorical variable. Since I set the **Method** property in the Interval Options property group to Quantile, the ranges of values assigned to each bin are defined such that each interval[5] contains an approximately equal number of observations. With the settings as shown in Display 2.123, for example, the interval-scaled variable APR is divided into four intervals, namely, values less than 11.88, values from 11.88 to 13.62, values from 13.62 to 15.52, and values 15.52 and above. In addition, a separate bin is created for missing values, since I set the **Treat Missing as Level** property to Yes.

For nominal variables, the bin is defined as a category. For example, the input "occupation" has 19 levels. So initially each level is treated as a bin. An additional bin is assigned to the missing category.

After the bins are formed, the Gini statistic is computed for each input. If the Gini statistic for a variable has a value above the minimum cut-off specified by the **Gini Cutoff** property, the variable is assigned the role of "input"; otherwise, it is designated as "rejected."

Display 2.124 shows the output statistics from results window after I ran the **Interactive Binning** node with the properties specified in Display 2.123.

Display 2.124

Variable	Gini Statistic	Level for Interactive	Calculated Role	New Role	Level	Label	Gini Ordering
MultLang	60.82	NOMINAL	Input	Default	NOMINAL	Speaks multiple languages	1
occupation	56.473	NOMINAL	Input	Default	NOMINAL	Occupation Code	2
good_bal	44.853	INTERVAL	Input	Default	INTERVAL	Balance without delinque...	3
APR	20.901	INTERVAL	Input	Default	INTERVAL	Annual Percentage Rate	4
purch_amount	19.889	INTERVAL	Rejected	Default	INTERVAL	Spending	5
TENURE	18.631	INTERVAL	Rejected	Default	INTERVAL	Months on Book	6
BAL	17.679	INTERVAL	Rejected	Default	INTERVAL	Balance	7
High_Credit	9.441	INTERVAL	Rejected	Default	INTERVAL	Highest amount of credit	8
Purch_count_12M	7.035	INTERVAL	Rejected	Default	INTERVAL	Count of purchases in las...	9
num_tran_bal_GT_0	5.483	INTERVAL	Rejected	Default	INTERVAL	#Transactions with Bal >0...	10
creditline	4.955	INTERVAL	Rejected	Default	INTERVAL	creditline	11
Margin	4.093	INTERVAL	Rejected	Default	INTERVAL	Gross Margin	12
gendern	4.009	NOMINAL	Rejected	Default	NOMINAL	Gender Code: Numeric	13
MaritalStatus	2.952	NOMINAL	Rejected	Default	NOMINAL	Marital Status, numeric	14
Pct_unpaid_bal_4MPD	2.858	INTERVAL	Rejected	Default	INTERVAL	Unpaid Balance 4 Mos Pd...	15

You can see that all but four variables are rejected on the basis of the minimum cut-off value of 20, which is set by the **Gini-Cutoff** property.

Splitting a Group (Bin)

Using the variable TENURE from our example, this example shows how you can interactively change the split points or bins. Display 2.124 shows that the variable TENURE is rejected because it has a Gini coefficient of 18.631, which is below the minimum cut-off value of 20. By judiciously changing the split points of the bins, we may be able to improve the Gini statistic of this variable. To start the process, click ⊡ located to the right of the **Interactive Binning** property as shown in Display 2.123.

The Interactive Binning window opens, as shown in Display 2.125.

Display 2.125

Variable	Label	Level	Calculated ...	New Role	Original Gini	Gini Statistic
MultLang	Speaks multiple languages	NOMINAL	Input	Default	60.82	60.82
occupation	Occupation Code	NOMINAL	Input	Default	56.473	56.473
good_bal	Balance without delinquency	INTERVAL	Input	Default	44.853	44.853
APR	Annual Percentage Rate	INTERVAL	Input	Default	20.901	20.901
purch_amount	Spending	INTERVAL	Rejected	Default	19.889	19.889
TENURE	Months on Book	INTERVAL	Rejected	Default	18.631	18.631
BAL	Balance	INTERVAL	Rejected	Default	17.679	17.679
High_Credit	Highest amount of credit	INTERVAL	Rejected	Default	9.441	9.441
Purch_count_12M	Count of purchases in last 12 months	INTERVAL	Rejected	Default	7.035	7.035
num_tran_bal_GT_0	#Transactions with Bal >0 within 0-3 Months	INTERVAL	Rejected	Default	5.483	5.483
creditline	creditline	INTERVAL	Rejected	Default	4.955	4.955
Margin	Gross Margin	INTERVAL	Rejected	Default	4.093	4.093
gendern	Gender Code: Numeric	NOMINAL	Rejected	Default	4.009	4.009
MaritalStatus	Marital Status, numeric	NOMINAL	Rejected	Default	2.952	2.952
Pct_unpaid_bal_4MPD	Unpaid Balance 4 Mos Pd label	INTERVAL	Rejected	Default	2.858	2.858

Because I am interested in re-defining the bins for the TENURE variable, I selected TENURE from the variable list by clicking **Select** and then clicking the **Groupings** tab. The details of the bins of the TENURE variable are shown in display 2.126.

Display 2.126

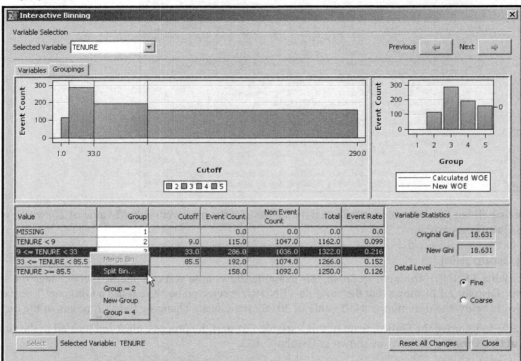

I right-click on Group 3 and select **Split Bin**. A text box opens, prompting me for a new cutoff value. I enter the new cut-off value (24) and click **OK**. Display 2.127 shows the new bins that were created.

Display 2.127

Value	Group	Cutoff	Event Count	Non Event Count	Total	Event Rate
MISSING	1		0.0	0.0	0.0	0.0
TENURE < 9	2	9.0	115.0	1047.0	1162.0	0.099
9 <= TENURE < 24	3	24.0	196.0	962.0	1158.0	0.169
24 <= TENURE < 33	3	33.0	90.0	74.0	164.0	0.549
33 <= TENURE < 85.5	4	85.5	192.0	1074.0	1266.0	0.152
85.5 <= TENURE	5		158.0	1092.0	1250.0	0.126

The original group 3 is split into two groups, but the same Group number is assigned to both of the newly created bins. To change this, right-click on the row corresponding to any of the newly constructed groups and select New Group, as shown in Display 2.127. As a result, we get a new definition of the bins, as shown in Display 2.128. Now the Gini value is 20.890.

Display 2.128

Value	Gr...	Cutoff	Event Count	Non Event Count	Total	Event Rate	Variable Statistics
MISSING	1		0.0	0.0	0.0	0.0	Original Gini 18.631
TENURE < 9	2	9.0	115.0	1047.0	1162.0	0.099	New Gini 20.890
9 <= TENURE < 24	3	24.0	196.0	962.0	1158.0	0.169	
24 <= TENURE < 33	4	33.0	90.0	74.0	164.0	0.549	Detail Level
33 <= TENURE < 85.5	5	85.5	192.0	1074.0	1266.0	0.152	⦿ Fine
85.5 <= TENURE	6		158.0	1092.0	1250.0	0.126	○ Coarse

Select   Selected Variable: TENURE   Reset All Changes   Close

To save the changes, click **Close**, and then click Yes in the Save Changes dialog box.

You can open the Results window and view the Output Variables table. Display 2.129 shows the output variables from the Results window.

Display 2.129

Variable	Gini Statistic	Level for Interactive	New Role	Calculated Role	Level	Label	Gini Ordering
MultLang	60.82	NOMINAL	Default	Input	NOMINAL	Speaks mult...	1
occupation	56.473	NOMINAL	Default	Input	NOMINAL	Occupation ...	2
good_bal	44.853	INTERVAL	Default	Input	INTERVAL	Balance wit...	3
APR	20.901	INTERVAL	Default	Input	INTERVAL	Annual Perc...	4
TENURE	20.89	INTERVAL	Default	Input	INTERVAL	Months on B...	5
purch_amo...	19.889	INTERVAL	Default	Rejected	INTERVAL	Spending	6
BAL	17.679	INTERVAL	Default	Rejected	INTERVAL	Balance	7
High_Credit	9.441	INTERVAL	Default	Rejected	INTERVAL	Highest am...	8
Purch_coun...	7.035	INTERVAL	Default	Rejected	INTERVAL	Count of pur...	9
num_tran_b...	5.483	INTERVAL	Default	Rejected	INTERVAL	#Transactio...	10
creditline	4.955	INTERVAL	Default	Rejected	INTERVAL	creditline	11
Margin	4.093	INTERVAL	Default	Rejected	INTERVAL	Gross Margin	12
gendern	4.009	NOMINAL	Default	Rejected	NOMINAL	Gender Cod...	13
MaritalStatus	2.952	NOMINAL	Default	Rejected	NOMINAL	Marital Statu...	14
Pct_unpaid...	2.858	INTERVAL	Default	Rejected	INTERVAL	Unpaid Bala...	15

The Gini Statistic for TENURE is 20.89 after we interactively created new bins. The role of the variable TENURE changed from Rejected to Input after the split.

Display 2.128 shows that the newly created bin consisting of customers with tenure between 24 and 33 months has an event (cancellation) rate of 54.9%—far above the overall average rate of 15.02%. Thus the splitting has highlighted the fact that a customer is most vulnerable after two years of being with the company. But since this newly formed group has only 164 customers, we should investigate this phenomenon further with a larger sample.

The variables that are passed to the node following the **Interactive Binning** node, and their roles, are shown in Display 2.130.

Display 2.130

Name	Use /	Report	Role	Level
TENURE	Default	No	Rejected	Interval
Purch_count_12M	Default	No	Rejected	Interval
creditline	Default	No	Rejected	Interval
MultLang	Default	No	Rejected	Nominal
MaritalStatus	Default	No	Rejected	Nominal
Pct_unpaid_bal_4MPD	Default	No	Rejected	Interval
occupation	Default	No	Rejected	Nominal
num_tran_bal_GT_0	Default	No	Rejected	Interval
purch_amount	Default	No	Rejected	Interval
gendern	Default	No	Rejected	Nominal
good_bal	Default	No	Rejected	Interval
GRP_APR	Default	No	Input	Nominal
CANCEL	Default	No	Target	Binary
GRP_MultLang	Default	No	Input	Nominal
APR	Default	No	Rejected	Interval
BAL	Default	No	Rejected	Interval
High_Credit	Default	No	Rejected	Interval
GRP_occupation	Default	No	Input	Nominal
Margin	Default	No	Rejected	Interval
GRP_TENURE	Default	No	Input	Nominal
GRP_good_bal	Default	No	Input	Nominal

Combine Groups

Now I will show how to combine groups using the **Interactive Binning** node. We will use the variable OCCUPATION for this illustration.

As before, launch the interactive binning process by clicking ▢, located to the right of the **Interactive Binning** property. Then, in the Interactive Binning window, select the variable occupation from the Variables tab, as shown in Display 2.131.

Display 2.131

Variable	Label	Level	Calculated Role	New Role	Original Gini	Gini Statistic
MultLang	Speaks multiple lang...	NOMINAL	Input	Default	60.82	60.82
occupation	Occupation Code	NOMINAL	Input	Default	56.473	56.473
good_bal	Balance without deli...	INTERVAL	Input	Default	44.853	44.853
APR	Annual Percentage ...	INTERVAL	Input	Default	20.901	20.901
TENURE	Months on Book	INTERVAL	Input	Default	20.89	20.89
purch_amount	Spending	INTERVAL	Rejected	Default	19.889	19.889
BAL	Balance	INTERVAL	Rejected	Default	17.679	17.679
High_Credit	Highest amount of c...	INTERVAL	Rejected	Default	9.441	9.441
Purch_count_12M	Count of purchases i...	INTERVAL	Rejected	Default	7.035	7.035
num_tran_bal_GT_0	#Transactions with ...	INTERVAL	Rejected	Default	5.483	5.483
creditline	creditline	INTERVAL	Rejected	Default	4.955	4.955
Margin	Gross Margin	INTERVAL	Rejected	Default	4.093	4.093
gendern	Gender Code: Numeric	NOMINAL	Rejected	Default	4.009	4.009
MaritalStatus	Marital Status, numeric	NOMINAL	Rejected	Default	2.952	2.952
Pct_unpaid_bal_4MPD	Unpaid Balance 4 Mo...	INTERVAL	Rejected	Default	2.858	2.858

Click the **Groupings** tab, and select the **Coarse** Detail Level, as shown in Display 2.132. In the initial run, occupation groups 11, 12, 13, 20, 23, 51, 52, 61, 62, and 80 are combined into Group 11 (Bin 11).

Display 2.132

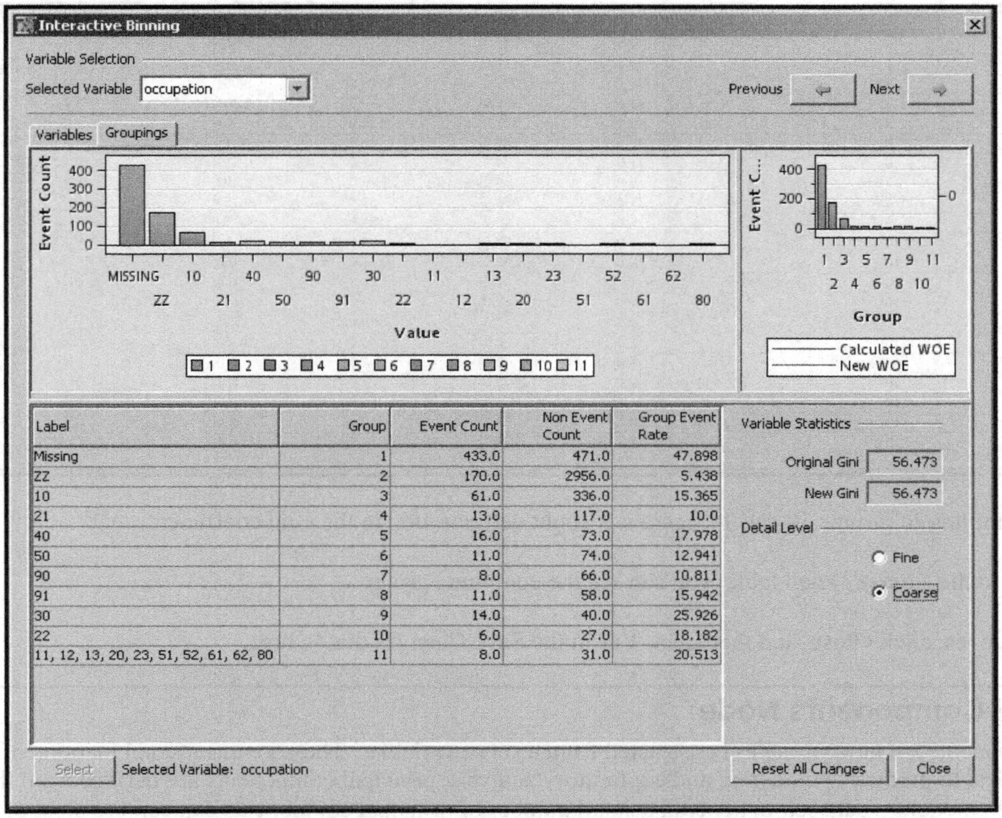

Now, to demonstrate the interactive binning process, let's combine Groups 10 and 11 by opening the Fine Detail table, as shown in Display 2.133.

Display 2.133

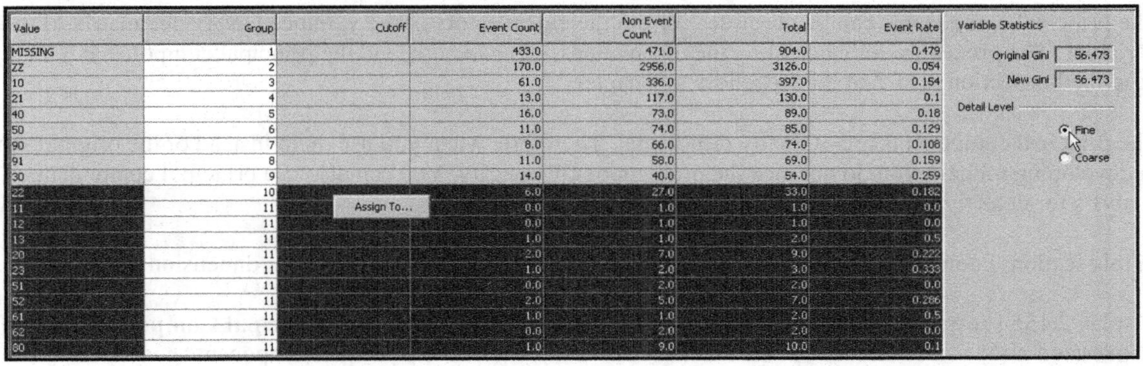

Select all occupation levels in groups 10 and 11 in the Fine Detail table, right-click, and select **Assign To**. In the Group Selection dialog box, select 10 and click **OK**.

We now have combined group 10 with group 11 and named the new group number 10, as seen in Display 2.134.

Display 2.134

Value	Group	Cutoff	Event Count	Non Event Count	Total	Event Rate	Variable Statistics
MISSING	1		433.0	471.0	904.0	0.479	Original Gini 56.473
ZZ	2		170.0	2956.0	3126.0	0.054	New Gini 56.472
10	3		61.0	336.0	397.0	0.154	
21	4		13.0	117.0	130.0	0.1	Detail Level
40	5		16.0	73.0	89.0	0.18	
50	6		11.0	74.0	85.0	0.129	⦿ Fine
90	7		8.0	66.0	74.0	0.108	
91	8		11.0	58.0	69.0	0.159	○ Coarse
30	9		14.0	40.0	54.0	0.259	
22	10		6.0	27.0	33.0	0.182	
11	10		0.0	1.0	1.0	0.0	
12	10		0.0	1.0	1.0	0.0	
13	10		1.0	1.0	2.0	0.5	
20	10		2.0	7.0	9.0	0.222	
23	10		1.0	2.0	3.0	0.333	
51	10		0.0	2.0	2.0	0.0	
52	10		2.0	5.0	7.0	0.286	
61	10		1.0	1.0	2.0	0.5	
62	10		0.0	2.0	2.0	0.0	
80	10		1.0	9.0	10.0	0.1	

Due to the combining of groups 10 and 11, there is a slight deterioration in the Gini coefficient.

If you go back to the Course Detail table, you can see the combined groups.

To save the changes, click **Close**, and then click **Yes** in the Save Changes dialog box.

2.9.5 Principal Components Node

Principal components are new variables constructed from a set of variables. They are linear combinations of the original variables. In predictive modeling and exploratory analysis, principal components are derived from the inputs in a given modeling data set. In general, a small number of principal components can capture most of the information contained in all the original inputs, where the information is measured in terms of the total variance. For example, if you have 100 inputs in your data set, you can construct 100 principal components, provided there are no linear dependencies among the inputs. However, most of the information contained in the 100 inputs can generally be captured by far fewer than 100 principal components.

The principal components can be constructed from the eigenvectors of the variance covariance matrix of the variables. The procedure for calculating the eigenvalues, eigenvectors, and the principal components is described in section 2.11.2 of the Appendix to Chapter 2.

The principal components are mutually orthogonal. Therefore, when you use them instead of the original inputs in developing models, you do not face the problem of collinearity. You can also use principal components analysis for detecting multidimensional outliers in a data set.

In this section, I demonstrate how you can use the **Principal Components** node for dimension reduction.

Display 2.135 shows the process flow diagram with an **Input Data** node and **Principal Components** node. The **Regression** node is also attached at the end in order to show how the principal components generated by the **Principal Components** node can be used as inputs into a logistic regression

Display 2.135

The list of variables included in the **Input Data** node is shown in Display 2.136.

Display 2.136

Name ／	Role	Level	Report	Order	Drop	Lowel
APR	Input	Interval	No		No	
BAL	Input	Interval	No		No	
CANCEL	Target	Binary	No		No	
creditline	Input	Interval	No		No	
CreditUtil	Input	Interval	No		No	
gendern	Input	Nominal	No		No	
good_bal	Input	Interval	No		No	
High_Credit	Input	Interval	No		No	
Margin	Input	Interval	No		No	
MaritalStatus	Input	Nominal	No		No	
MultLang	Input	Nominal	No		No	
num_tran_bal_GT_0	Input	Interval	No		No	
NumPD90in3M	Input	Interval	No		No	
Pct_unpaid_bal_4MPD	Input	Interval	No		No	
purch_amount	Input	Interval	No		No	
Purch_count_12M	Input	Interval	No		No	
TENURE	Input	Interval	No		No	

Since the role of the variable CANCEL is Target, it is not included in constructing the principal components.

Display 2.137 shows the property settings I use in this demonstration.

Display 2.137

Property	Value
General	
Node ID	PRINCOMP
Imported Data	
Exported Data	
Notes	
Train	
Variables	
Eigenvalue Source	Correlation
Interactive Selection	
Print Eigenvalue Source	No
Score	
Princomp Prefix	PC
☐ Eigenvalue Cutoff	
├ Cumulative	0.95
└ Increment	0.001
☐ Max Number Cutoff	
├ Apply Maximum Number	Yes
└ Maximum Number	20
Reject Original Input Variables	Yes
Hide Rejected Variables	Yes
Status	
Create Time	11/2/12 3:42 PM

In the Properties panel, the **Eigenvalue Source** property is set to Correlation. Therefore, the **Principal Components** node will use the correlation matrix of the inputs shown in display 2.136.

Run the **Principal Components** node. Display 2.138 shows that the weights used for calculating the first principal component.

Display 2.138

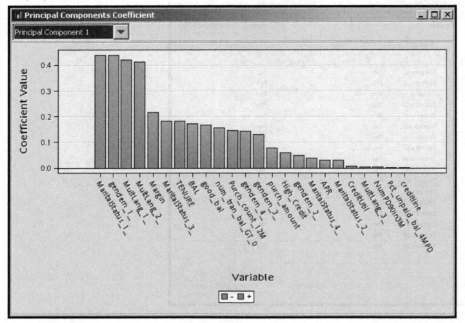

The principal components are calculated for each observation as weighted sums of the original variables (inputs), where the weights are the principal component coefficients that are shown in Display 2.138. The nominal variable MultLang has three levels: M (missing), N (no), and Y (yes). The **Principal Components** node has converted this categorical variable into three dummy variables: MultLang_1_, MultLang_2_, and MultLang_3_. Similarly, dummy variables are created for the other nominal variables, gender and MaritalStatus. Since these dummy variables are numeric, they could be used in the calculation of the correlation matrix.

As previously mentioned, the principal components are weighted sums (linear combinations), where the weights are the principal components coefficients. These coefficients are the eigenvalues of the correlation matrix of the inputs. (For a more detailed explanation, see Section 2.11.2 of the Appendix to Chapter 2)

The **Principal Components** node has created 20 principal components. Output 2.9shows the eigenvalues, the proportion of variance explained by each component, and the cumulative proportion of variation at each component. Since the Cumulative proportion eigenvalue cutoff of 0.95 (set by the EigenValue **Cutoff Cumulative** property) reached at the 17th principal component, the number of selected principal components is equal to 17. Display 2.139 shows the eigenvalue plot.

Output 2.9

```
The DMNEURL Procedure

            Eigenvalues of Correlation Matrix

        Eigenvalue    Difference    Proportion    Cumulative

    1   3.79264136    1.54675049      0.1580        0.1580
    2   2.24589088    0.38595927      0.0936        0.2516
    3   1.85993161    0.14273274      0.0775        0.3291
    4   1.71719887    0.10452492      0.0715        0.4007
    5   1.61267394    0.35416301      0.0672        0.4678
    6   1.25851093    0.06561368      0.0524        0.5203
    7   1.19289726    0.06504635      0.0497        0.5700
    8   1.12785091    0.06670172      0.0470        0.6170
    9   1.06114919    0.03543879      0.0442        0.6612
   10   1.02571041    0.02937314      0.0427        0.7039
   11   0.99633726    0.00071087      0.0415        0.7454
   12   0.99562639    0.03740089      0.0415        0.7869
   13   0.95822551    0.02592808      0.0399        0.8269
   14   0.93229743    0.18691015      0.0388        0.8657
   15   0.74538728    0.05383341      0.0311        0.8968
   16   0.69155386    0.09760714      0.0288        0.9256
   17   0.59394672    0.07583200      0.0247        0.9503
   18   0.51811472    0.03194528      0.0216        0.9719
   19   0.48616945    0.29828343      0.0203        0.9922
   20   0.18788601    0.18788601      0.0078        1.0000
   21   0.00000000    0.00000000      0.0000        1.0000
   22   0.00000000    0.00000000      0.0000        1.0000
   23   0.00000000    0.00000000      0.0000        1.0000
   24   0.00000000                    0.0000        1.0000
```

Display 2.139

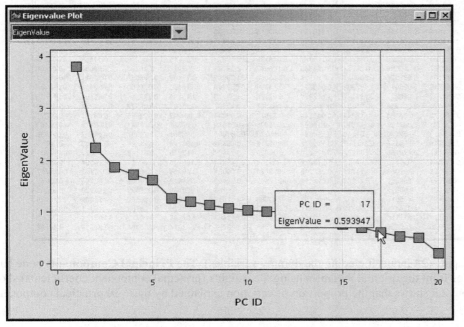

Select the first 10 principal components interactively by opening the Interactive Principal Components window.

Click next to the **Interactive Selection** property in the Properties panel (see Display 2.137). To select the first 10 principal components, change the number in the **Number of Principal Components to be exported** box to 10 as shown in Display 2.140, and click **OK**. Now, only 10 principal components will be exported to the successor node.

Display 2.140

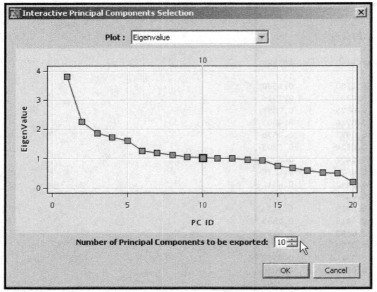

Display 2.141 shows the weights or principal components coefficients that will be used for calculating the principal components at each observation.

Display 2.141

Variable	Label	Principal Component 1	Principal Component 2	Principal Component 3	Principal Component 4	Principal Component 5	Principal Component 6	Principal Component 7	Principal Component 8	Principal Component 9	Principal Component 10
MaritalStatus_1_	.	-0.43985	-0.12381	0.132098	0.129939	0.069697	0.047377	0.190278	-0.13379	0.119116	0.049785
gendern_1_		-0.43985	-0.12381	0.132098	0.129939	0.069697	0.047377	0.190278	-0.13379	0.119116	0.049785
MultLang_1_	M	-0.41909	0.008196	0.075196	0.156185	-0.03655	-0.03606	-0.16418	0.214292	-0.04424	-0.05088
MultLang_2_	N	0.412513	0.006746	-0.08791	-0.14956	0.043538	0.064525	0.228663	-0.35751	0.080106	0.047121
Margin		0.216808	0.034806	-0.14488	0.460489	0.079663	0.010422	-0.06186	0.185666	-0.02144	-0.03236
MaritalStatus_3_	1	0.18104	-0.47379	0.011668	-0.02177	0.048739	-0.40873	-0.21719	-0.02229	0.11456	0.088884
TENURE		0.180707	-0.0089	-0.08024	0.317365	0.106943	-0.23003	0.399215	0.26297	-0.2036	-0.07008
BAL		0.17194	0.037527	0.600907	0.02527	-0.15966	0.020612	0.11402	0.090556	0.033311	0.03101
good_bal		0.167449	0.038019	0.597938	0.068517	-0.14396	0.063449	0.116465	0.035909	0.090081	0.066369
num_tran_bal_GT_0		0.154693	0.068824	0.132397	-0.17871	-0.07278	0.095175	-0.13195	0.232964	0.369104	0.077723
Purch_count_12M		0.146574	0.002818	-0.01487	0.504952	0.091561	0.193207	-0.26001	-0.09528	0.034374	-0.05928
gendern_4_	2	0.141978	-0.34586	-0.17944	0.043144	-0.55672	0.09044	0.060242	0.009545	-0.03132	-0.03037
gendern_3_	1	0.128381	0.092194	0.117612	-0.16668	0.704606	-0.01619	-0.07671	0.070941	0.006415	-0.00453
purch_amount		0.078032	-0.02195	0.07249	0.396624	0.055373	0.263001	-0.36861	-0.28307	0.164687	0.052493
High_Credit		0.058845	-0.15419	0.317897	0.142037	0.061714	-0.08452	0.086124	-0.14798	-0.44993	-0.20719
gendern_2_	0	-0.04717	0.470756	-0.01127	0.075479	-0.25826	-0.14637	-0.12716	-0.01118	-0.05722	0.011817
MaritalStatus_4_	2	0.038594	-0.0595	-0.07717	-0.10876	0.056516	0.749764	0.168706	0.30327	-0.14594	-0.10062
APR		0.030608	0.07068	-0.14737	0.297346	0.059179	-0.10051	0.49055	0.18201	0.335122	0.269777
MaritalStatus_2_	0	0.02997	0.59057	-0.034	0.021851	-0.12452	-0.09408	0.008818	-0.1008	-0.08753	-0.05282
CreditUtil		0.006024	-0.00322	0.039583	-0.00811	0.002387	-0.1288	0.060936	0.026029	0.245923	-0.54269
MultLang_3_	Y	-0.00495	-0.05913	0.056437	-0.01482	-0.03068	-0.11649	-0.27003	0.588566	-0.14669	0.011245
NumPD90in3M		0.003246	0.028941	0.030619	-0.00871	0.020852	-0.005	-0.06626	0.120768	0.212211	0.450635
Pct_unpaid_bal_4MPD		0.001127	0.011755	-0.02819	-0.00355	-0.04322	0.012462	-0.04787	0.08772	0.465111	-0.32273
creditline		-.000376	-0.00936	-0.00126	-0.00748	-0.03958	0.038857	-0.07014	-0.03994	-0.20759	0.47658

There are 16 original inputs (24 inputs including the dummy variables). The **Principal Components** node has created 10 new variables from the original variables using the weights (principal component coefficients) shown in Display 2.141. Output 2.9 shows that the proportion of variation explained by these 10 principal components is 0.7039.

Although we reduced the number of inputs from 24 to 10, we did sacrifice some information. Output 2.9 shows that the first 10 principal components explain only 70.39% of the total variance of all the inputs. But, in cases where the original inputs are highly intercorrelated, you may be able to capture a larger percentage of information from a few principal components.

I have attached a **Regression** node to the **Principal Components** node, as shown in the process flow diagram in Display 2.135. To see which variables are exported to the **Regression** node, select the **Regression** node in the process flow diagram. Click to the right of the **Variables** property. The Variables window appears, as shown in Display 2.142.

Display 2.142

Name	Use	Report	Role	Level
CANCEL	Yes	No	Target	Binary
PC_1	Default	No	Input	Interval
PC_10	Default	No	Input	Interval
PC_2	Default	No	Input	Interval
PC_3	Default	No	Input	Interval
PC_4	Default	No	Input	Interval
PC_5	Default	No	Input	Interval
PC_6	Default	No	Input	Interval
PC_7	Default	No	Input	Interval
PC_8	Default	No	Input	Interval
PC_9	Default	No	Input	Interval

The Variable table contains the target variable and the selected principal components. I ran the **Regression** node. The estimated equation is shown in output 2.10.

Output 2.10

```
                    Analysis of Maximum Likelihood Estimates

                            Standard        Wald                  Standardized
Parameter      DF   Estimate    Error    Chi-Square   Pr > ChiSq    Estimate    Exp(Est)

Intercept       1   -2.4459    0.0696     1235.15      <.0001                     0.087
PC_1            1   -0.4829    0.0261      341.15      <.0001        -0.5185      0.617
PC_10           1   -0.4028    0.0449       80.60      <.0001        -0.2249      0.668
PC_2            1   -0.0919    0.0311        8.75      0.0031        -0.0759      0.912
PC_3            1   -0.7340    0.0676      118.01      <.0001        -0.5519      0.480
PC_4            1    0.1120    0.0414        7.32      0.0068         0.0809      1.118
PC_5            1    0.0985    0.0413        5.69      0.0170         0.0689      1.103
PC_6            1   -0.1406    0.0429       10.73      0.0011        -0.0870      0.869
PC_7            1   -0.6398    0.0490      170.44      <.0001        -0.3853      0.527
PC_8            1    0.6077    0.0462      173.07      <.0001         0.3558      1.836
PC_9            1   -0.4090    0.0413       97.84      <.0001        -0.2323      0.664
```

2.9.6 Transform Variables Node

Sections 2.9.6.1 and 2.9.6.2 describe the different transformation methods available in the **Transform Variables** node. Sections 2.9.6.5 and 2.9.6.6 show an example of how to make the transformations, save the code, and pass the transformed variables to the next node.

2.9.6.1 Transformations for Interval Inputs

Best

If you set the **Interval Inputs** property to Best, the **Transform Variables** node selects the transformation that yields the best Chi-squared value for the target.

Multiple

If you set the **Interval Inputs** property to Multiple, the **Transform Variables** node makes several transformations for each input and passes them to the successor node. By connecting the **Regression** node to the **Transform Variables** node, you can use the stepwise selection method to select the best transformations and the best variables.

Simple Transformations

The simple transformations are Log, Log10, Square Root, Inverse, Square, Exponential, Range, Centering, and Standardize. They can be applied to any interval-scaled input. When a log or inverse transformation is done, the node automatically adds an Offset value (a constant) to the variable if the minimum value of the input is less than or equal to zero. A logarithm of zero or less does not exist, so SAS produces a missing value when we take the logarithm of zero or a negative number. By adding an offset value, we avoid generating missing values. You can set the **Offset Value** property to a number to avoid taking a logarithm of a negative number. The simple transformations can be used irrespective of whether the target is categorical or continuous.

For input, the simple transformations are: $\log(X)$, $\log_{10}(X)$, $X^{1/2}$, $1/X$, X^2, e^X, $\mathrm{Center}(X) = X - \mu_X$,

$\mathrm{Standardize}(X) = (X - \mu_X)/\sigma_X$, $\mathrm{Range}(X) = (X - \min(X))/(\max(X) - \min(X))$.

Binning Transformations

In SAS Enterprise Miner, there are three ways of binning an interval-scaled variable. To use these as default transformations, select the **Transform Variables** node, and set the value of **Interval Inputs** property to **Bucket**, **Quantile**, or **Optimal under Default Methods**.

Bucket

The Bucket option creates buckets by dividing the input into n equal-sized intervals, grouping the observations into the n buckets. The resulting number of observations in each bucket may differ from bucket to bucket. For example, if AGE is divided into the four intervals 0–25, 25–50, 50–75, and 75–100, then the number of observations in the interval 0–25 (bin 1) may be 100, the number of observations in the interval 25–50 (bin 2) may be 2000, the number of observations in the interval 50–75 (bin 3) may be 1000, and the number of observations in the interval 75–100 (bin 4) may be 200. The bins created by this option can be called equal-width bins.

Quantile

This option groups the observations into quintiles (bins) with an equal number of observations in each. If there are 20 quintiles, then each quintile consists of 5% of the observations.

Optimal Binning to Maximize the Relationship to the Target

This transformation is available for binary targets only. The input is split into a number of bins, and the splits are placed so as to make the distribution of the target levels (for example, response and non-response) in each bin significantly different from the distribution in the other bins. To help you understand optimal binning on the basis of relationship to target, consider two possible ways of binning. The first method involves binning the input variable recursively. Suppose you have an interval input X that takes on values ranging from 0 to 320. A new variable with 64 levels can be created very simply by dividing this range into 64 sub-intervals, as follows:

$$0 \leq X < 5, \ 5 \leq X < 10, \dots, 310 \leq X < 315, \ 315 \leq X \leq 320$$

The values the new variable can take are 5, 10, 15... 315, which can be called splitting *values*. I can split the data into two parts at each splitting value and determine the Chi-square value from a contingency table, with columns representing the splitting value of the input X and rows representing the levels of the target variable. Suppose the splitting value tested is 105. At that point, the contingency table looks like the following:

	$X < 105$	$X \geq 105$
Target level		
1	n_{11}	n_{12}
0	n_{01}	n_{02}

In the contingency table, n_{11}, n_{12}, n_{01}, and n_{02} denote the number of records in each cell.

The computation of the Chi-Square statistic is detailed in Chapter 4.

The splitting value that gives the maximum Chi-Square determines the first split of the data into two parts. Then each part is split further into two parts using the same procedure, resulting in two more splitting values and four partitions. The process continues until no more partitioning is possible. That is, there is no splitting value that gives a Chi-Square above the specified threshold. At the end of the process, the splitting values chosen define the optimal bins. This is illustrated in Display 2.143.

Display 2.143

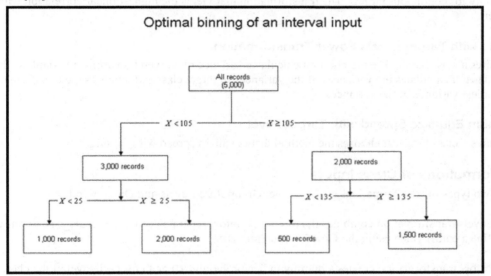

In Display 2.143, the optimal bins created for the input X are:

$$X < 25, \ 25 \leq X < 105, \ 105 \leq X < 135 \text{ and } 135 \leq X.$$

Four optimal bins are created from the initial 64 bins of equal size. The optimal bins are defined by selecting the best split value at each partition.

The second method involves starting with the 64 or so initial bins and collapsing them to a smaller number of bins by combining adjacent bins that have similar distribution of the target levels. SAS Enterprise Miner does not follow the steps outlined above exactly, but its method of finding optimal bins is essentially what is described in the second method above. There are other ways of finding the optimal bins. Note that in **Transform Variables** node, the optimal bins are created for a *one interval-scaled* variable at a time, in contrast to the **Decision Tree** node where the bins are created with *many inputs of different measurement scales*, interval as well as categorical.

Maximum Normal

The node selects the transformation that maximizes normality from X, $\log(X)$, $X^{1/2}$, e^X, $X^{1/4}$, X^2, and X^4, where X is the input.

To find the transformation that maximizes normality, sample quantiles from each of the transformations listed above are compared with the theoretical quantiles of a normal distribution. The transformation that yields quantiles that are closest to the normal distribution is chosen.

Suppose Y is obtained by applying one of the above transformations to X. For example, the 0.75-sample quantile of the transformed variable Y is that value of Y at or below which 75% of the observations in the data set fall. The 0.75-quantile for a standard normal distribution is 0.6745, given by $P(Z \leq 0.6745) = 0.75$, where Z is a normal random variable with mean 0 and standard deviation 1. The 0.75-sample quantile for Y is compared with 0.6745, and similarly the other quantiles are compared with the corresponding quantiles of the standard normal distribution.

Maximum Correlation

This is available only for continuous targets. The transformation that yields the highest linear correlation with the target is chosen.

Equalize Spread with Target Levels Power Transformation

This method requires a class target. The method first calculates variance of a given transformed variable within each target class. Then it calculates the variances of the variances of target classes. It chooses the transformation that yields the smallest variance of the variances.

Optimal Maximum Equalize Spread with Target Level

This method requires a class target. It chooses the method that equalizes spread with the target.

2.9.6.2 Transformations of Class Inputs

For class inputs, two types of transformations are available: Group Rare Levels and Dummy Indicators.

The Group Rare Levels transformation combines the rare levels into a separate group, _OTHER_. To define a rare level, you define a cutoff value using the **Cutoff Value** property.

The Dummy Indicators transformation creates a dummy indicator variable (0 or 1) for each level of the class variable.

To choose one of these available transformations, select the **Transform Variables** node and set the value of the **Class Inputs** property to the desired transformation.

By setting the **Class Inputs** property to Dummy Indicators Transformation, you get a dummy variable for each category of the class input. You can test the significance of these dummy variables using the **Regression** node. The groups that correspond to the dummy variables, which are not statistically significant, can be combined or omitted from the regression. If they are omitted from the regression, it implies their effects are captured in the intercept term.

2.9.6.3 Transformations of Targets

In SAS Enterprise Miner 12.1, the target variables can also be transformed by setting the **Interval Targets** and **Class Targets** properties. Click the Value column for either of these properties to select the desired property.

2.9.6.4 Selecting Default Methods

To select a default method for the interval inputs, select the **Transform Variables** node in the Diagram Workspace, and set the **Interval Inputs** property (under Default Methods section in the Properties panel) to one of the transformations described in Section 2.9.6.1. The selected method will be applied to all interval variables.

Similarly, set the **Class Inputs** property to one of the methods described in Section 2.9.6.2. SAS Enterprise Miner applies the chosen default method to all inputs of the same type.

2.9.6.5 Overriding the Default Methods

Display 2.144 shows the flow diagram.

Display 2.144

If you do not want to use the default method of transformation for an input, you can override it by right-clicking ▣ located on the right side of **Variables** property of the **Transform Variables** node, as shown in Display 2.145. This opens the Variables window.

Display 2.145

Property	Value
General	
Node ID	Trans
Imported Data	
Exported Data	
Notes	
Train	
Variables	
Formulas	
Interactions	
SAS Code	
⊟ Default Methods	
⊢ Interval Inputs	None
⊢ Interval Targets	None
⊢ Class Inputs	None
⊢ Class Targets	None
⊢ Treat Missing as Level	No
⊟ Sample Properties	
⊢ Method	Random
⊢ Size	Max
⊢ Random Seed	12345
⊟ Optimal Binning	
⊢ Number of Bins	4
⊢ Missing Values	Use in Search
⊟ Grouping Method	
⊢ Cutoff Value	0.1
⊢ Group Missing	No
⊢ Number of Bins	Variables
Add Minimum Value to Offset Value	Yes
Offset Value	1
Score	
Use Meta Transformation	Yes
Hide	No
Reject	No
Report	
Summary Statistics	Yes
Status	
Create Time	11/7/12 2:23 PM

In the Variables window, click the Method column in the variable VAR1N, and select Optimal Binning. Similarly, select the log transformation for the variable credscore (credit score). Transformations are not done for other variables since the default methods are set to NONE in the Properties panel (see Display 2.146).

Display 2.146

After you finish selecting the transformations, click **OK** and run the **TransformVariables** node.

2.9.6.6 Saving the SAS Code Generated by the Transform Variables Node

Right-click on the **Transform Variables** node and run it. Then click **Results**. In the Results window, click **View→Scoring →SAS Code** to see the SAS code, as shown in Display 2.147.

Display 2.147

```
2    * Computed Code;
3    *-----------------------------------------------------*;
4    *-----------------------------------------------------*;
5    * TRANSFORM: credscore , log(credscore + 1);
6    *-----------------------------------------------------*;
7    label LOG_credscore = 'Transformed: Credit score';
8    if credscore eq . then LOG_credscore = .;
9    else do;
10   if credscore + 1 > 0 then LOG_credscore = log(credscore + 1);
11   else LOG_credscore = .;
12   end;
13   *-----------------------------------------------------*;
14   * TRANSFORM: VAR1N , Optimal Binning(4);
15   *-----------------------------------------------------*;
16   label OPT_VAR1N = 'Transformed VAR1N';
17   length OPT_VAR1N $20;
18   if (VAR1N eq .) then OPT_VAR1N="02:4.5-17.5, MISSING";
19   else
20   if (VAR1N < 4.5) then
21   OPT_VAR1N = "01:low-4.5";
22   else
23   if (VAR1N >= 4.5 and VAR1N < 17.5) then
24   OPT_VAR1N = "02:4.5-17.5, MISSING";
25   else
26   if (VAR1N >= 17.5) then
27   OPT_VAR1N = "03:17.5-high";
28
```

You can save this code by clicking **File→Save As** and typing in the path of the file and the filename.

If you want to pass both the original and transformed variables to the next node, you must change the **Hide** and **Reject** properties to No in the Score area in the Properties panel, and then run the **Transform Variables** node. If you open the **Variables** property in the next node, which is the **Regression** node in this case, you see both the original and transformed variables passed to it from the **Transform Variables** node, as shown in Display 2.148.

Display 2.148

2.10 Utility Nodes

The following nodes are under the **Utility** tab: **Control Point**, **End Groups**, **Ext Demo**, **Metadata**, **Reporter**, **SAS Code**, **Score Code Export**, and **Start Groups**.

Since the **SAS Code** node is the most frequently used, I will focus on this node.

2.10.1 SAS Code Node

The **SAS Code** node is used to incorporate SAS procedures and external SAS code into the process flow of a project. You can also perform DATA step programming in this node. The **SAS Code** node is the starting point for creating custom nodes and extending the functionality of SAS Enterprise Miner 12.1.

The **SAS Code** node can be used at any position in the sequence of nodes in the process flow. For example, in Chapters 6 and 7, custom programming is done in the **SAS Code** node to create special lift charts.

We will explore the **SAS Code** node so you can become familiar with the SAS Code Editor window and with the macro variables and macros created by the **SAS Code** node. The macro variables refer to imported and exported data sets, libraries, etc. The macros are routines that can be used in the SAS code you write.

We will examine some of these macro variables and macros and write simple code and run it. For the purpose of this demonstration, I created a simple process flow that includes the **Input Data** node, the **Transform Variables** node, the **SAS Code** node, and the **Regression** node. This process flow diagram is shown in Display 2.149.

Display 2.149

The **Input Data** node imports the data from the directory that you entered in your project start-up code.

The SAS data set we use is called BinaryTarget_B and is located in the directory path specified by the libref, which we included in the project start-up code. On my computer, the data set is located in the directory C:\TheBook\EM12.1\Data\Chapter2. The **Input Data** node exports a data view of this data to the next node, which is the **Transform Variables** node. The exported data view is referred to as EMWS11.IDS_Data, where EMWS11 is a libref assigned by SAS Enterprise Miner and IDS stands for Input Data Source. Click ⊡ located to the right of the **Exported Data** property of the **Input Data** node, and then click Properties. You can see the properties of the data view that are exported by the **Input Data** node, shown in Display 2.150.

Display 2.150

Property	Value
Table	EMWS11.Ids_DATA
Member Type	VIEW
Description	
Data Set Type	DATA
Engine	SASDSV
Created	November 7, 2012 2:25:05 PM EST
Modified	November 7, 2012 2:25:05 PM EST
Number of Observations	Unknown
Number of Columns	10
Role	RAW

The libref EMWS11 can be different for different diagrams. As you will see later in the SAS Code node's Code Editor, a macro name &EM_LIB is given to this libref. In Display 2.149, the **Transform Variables** Node follows the **Input Data** node. The data imported into the **Transform Variables** node is EMWS11.Ids_DATA, and the file exported by the **Transform Variables** node to the following node (**SAS Code** node) is EMWS11.Trans_TRAIN, where Trans denotes that the data set or view is created by the **Transform Variables** node.

In this example, I transformed two variables using the **Transform Variables** node: VAR1N and credscore. Both variables are interval scaled. VAR1N is transformed to a categorical (nominal) variable and named OPT_VAR1N. OPT_VAR1N is obtained by binning the original variable VAR1N using the Optimal Binning method discussed earlier in Section 2.9.6.1 (Transformation for Interval Inputs). The second transformed variable is the natural logarithm of the variable credscore, and it is named LOG_credscore. Both the transformed and original variables are included in the data set exported by the **Transform Variables** node.

We can verify that the data set that is exported by the **Transform Variables** node to the **SAS code** node is EMWS11.Trans_TRAIN by first updating the path (to update the path, right click the SAS Code node and click "Update") and then clicking ⊡ located on the right of the the **Imported Data** property of the **SAS Code** node Properties panel.

Display 2.151 shows data set imported by the **SAS Code** node.

Display 2.151

Property	Value
Table	EMWS11.Trans_TRAIN
Member Type	VIEW
Description	
Data Set Type	DATA
Engine	SASDSV
Created	November 8, 2012 3:07:21 PM EST
Modified	November 8, 2012 3:07:21 PM EST
Number of Observations	Unknown
Number of Columns	12
Role	Train

Properties - EMWS11.Trans_TRAIN — Table | Variables

Display 2.152 shows the variables in the data set that are imported into to the SAS Code node.

Display 2.152

Variables - EMCODE

(none) | not | Equal to | | Apply | Reset

Columns: ☑ Label   ☑ Mining   ☑ Basic   ☑ Statistics

Name	Use	Role	Level	Type	Creator	Comment
LOG_credscore	Default	Input	Interval	Numeric	Trans	log(credscore + 1)
OPT_VAR1N	Default	Input	Nominal	Character	Trans	Optimal Binning(4)
VAR1N	Default	Input	Interval	Numeric		
VAR2O	Default	Input	Interval	Numeric		
VAR3N	Default	Input	Interval	Numeric		
VAR4C	Default	Input	Nominal	Character		
VAR5O	Default	Input	Interval	Numeric		
VAR6C	Default	Input	Nominal	Character		
VAR7C	Default	Input	Nominal	Character		
VAR8O	Default	Input	Interval	Numeric		
credscore	Default	Input	Interval	Numeric		
resp	Default	Target	Binary	Numeric		

Explore... | Update Path | OK | Cancel

Now we will open the **SAS Code** node's Code Editor window and become familiar with its components. Click ▣ located to the right of the **Code Editor** property in the Properties panel of the **SAS Code** node, as shown in Diplay 2.153.

Display 2.153

Property	Value
General	
Node ID	EMCODE
Imported Data	
Exported Data	
Notes	
Train	
Variables	
Code Editor	
Tool Type	Utility
Data Needed	No
Rerun	No
Use Priors	Yes
Score	
Advisor Type	Basic
Publish Code	Publish
Code Format	DATA step
Status	
Create Time	11/8/12 1:05 PM

The Code Editor window opens, as shown in Display 2.154.

Display 2.154

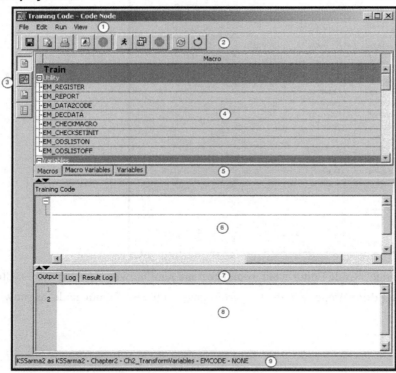

A detailed description of each component of the Code Editor user interface can be found in the Reference Help. The following descriptions of the components refer to the numbers in Display 2.154.

① Menu bar

② Tool Bar: For a complete description of all the buttons, refer to the SAS Enterprise 12.1 Reference Help.

There are two buttons on toolbar that I use to write and execute some simple programs in the SAS Code Editor. Press the **Run Code** button shown in Display 2.155 to execute the code you enter in the code window (indicated by ⑥ in Display 2.154).

Display 2.155

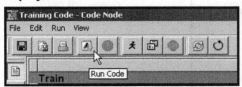

When you press the **Run Node** button shown in Display 2.156, the predecessor nodes also run if they were not executed before, along with the **SAS Code** node. If you create data sets in the **SAS Code** node and export them to the successor nodes, you should use the **Run Node** button.

Display 2.156

Refer to the Reference help for the use of all other buttons in the tool bar.

③ Use these four buttons to display the Training Code, the Score Code, the Report Code, or the Property Settings of the **SAS Code** node.
④ In this pane, you can display the list of macros, macro variables, or variables by clicking one of the tabs shown in ⑤.
⑤ Use these tabs to display the list of macros, macro variables, and variables.

Macros Tab: Under this tab there are two types of macros: (1) Utilities macros, which can be used in performing tasks such as preparing reports, registering data sets, etc. (2) Variables macros, which can be used to insert various strings into your SAS code. To illustrate this, I clicked on EM_INTERVAL_INPUT and dragged it into the Training Code pane indicated by ⑥. As a result, the string % EM_INTERVAL_INPUT is written into the SAS Code pane ⑥. I added "%put" in front of this string and a semi-colon at the end, and then clicked the **Run Code** button. I clicked the **Log** tab in ⑦. The log of the executed program appeared in the bottom window, as shown in Display 2.157. The interval inputs in the imported data set are Log_creditscore, VAR1N, and creditscore.

Display 2.157

The example shown here is trivial, but it does show how you can use the macros. For more complex examples, see the Reference Help.

Macro Variables Tab: Use this tab, shown in ⑤, to display and access the names of the data sets, libraries, etc. To illustrate this, I entered the code Proc means data= in the Training Code pane. Then I clicked the **Macro Variables** tab, selected the macro variable name EM_IMPORT_DATA, dragged it next to data=, and then added a semicolon in the data. Next, I typed the statement var, opened the **Macros** tab, and selected EM_INTERVAL_INPUT. I dragged that into the Training Code pane, placed it next to var in my SAS code, and clicked the **Run Code** button. Then I clicked the **Output** tab in ⑦ in order to see the output that was generated by the SAS code that I entered in the Training Code pane. Display 2.158 shows the SAS code and the output.

Display 2.158

2.11 Appendix to Chapter 2

2.11.1 The Type, the Measurement Scale, and the Number of Levels of a Variable

In Section 2.5 we discussed how to create a SAS data source. In Step 4 of that process, we selected a Metadata Advisor. The options were **Basic** or **Advanced**. If you selected Advanced, there were additional customization steps. In this section, I illustrate how the metadata options you select determine the measurement level (scale) of a variable.

The type of a variable indicates whether it is a character or numeric variable. The measurement scale, which is also referred to as the measurement level, indicates how SAS Enterprise Miner treats the variables. If the measurement scale is interval, then the variable is treated as a continuous variable. If the measurement scale is nominal, then the variable is treated as a categorical variable. The "Number of Levels" of a variable refers to the number of unique values that a nominal variable can take on. Display 2.159 shows the "Level" (sometimes called measurement scale) of a variable, "Number of Levels" of a nominal variable and the "Type" of each variable,

2.11.1.1 Basic Option

All numeric variables are assigned the measurement scale of interval and all character variables are assigned the measurement scale of nominal. The measurement scales are shown in the Level column in Display 2.159, the type of the variable is shown in the Type column, and the number of levels of a variable is shown in the Number of Levels column. The number of levels is shown only if the measurement scale is nominal. The number of levels is the number of unique values that a nominal variable can take on.

In Display 2.159, the variable DEPC is a nominal variable that takes the values N and Y, where N indicates that the customer does not own a department store credit card, and Y indicates that he does. Hence we say that the variable DEPC has two levels.

For another example, the variable DELINQ is numeric (see the Type column), and its measurement scale (see the Level column) is designated to be Interval. Since DELINQ has been assigned an Interval measurement scale, the Number of Levels column shows missing values for this variable. In general, the number of levels (unique values) of an interval scaled variable is not shown in this table.

Display 2.159

Name	Role	Level	Type	Length	Number of Levels	Rep
AGE	Input	Interval	Numeric	8	.	No
CRED	Input	Interval	Numeric	8	.	No
DELINQ	Input	Interval	Numeric	8	.	No
DEPC	Input	Nominal	Character	1	2	No
EMP_STA	Input	Nominal	Character	3	3	No
GENDER	Input	Nominal	Character	1	2	No
HEQ	Input	Interval	Numeric	8	.	No
INCOME	Input	Interval	Numeric	8	.	No
MFDU	Input	Interval	Numeric	8	.	No
MILEAGE	Input	Interval	Numeric	8	.	No
MOB	Input	Nominal	Character	1	2	No
MRTGI	Input	Nominal	Character	3	3	No
MS	Input	Nominal	Character	1	3	No
NUMTR	Input	Interval	Numeric	8	.	No
RESTYPE	Input	Nominal	Character	6	4	No
RES_STA	Input	Nominal	Character	3	2	No
cuscode	Input	Nominal	Character	5	513	No
resp	Target	Interval	Numeric	8	.	No

2.11.1.2 Advanced Option with default values

Here we select the **Advanced** advisor options and set the **Class Levels Count Threshold** property of the **Advanced** advisor option (see Display 2.11) to the default value of 20. Display 2.160 shows what measurement scale each variable is assigned. As mentioned before, with this setting, interval variables with less than 20 unique values (levels) are assigned the measurement scale of nominal.

For example, the variable DELINQ is a numberic variable indicating the number of delinquencies (late payments) by a customer. Since this variable has eight unique values (<20), it is assigned the measurement scale of Nominal, as shown in Display 2.160.

Display 2.160

Data Source Wizard -- Step 5 of 8 Column Metadata

Name	Role	Level	Type	Length	Number of Levels	Comment
AGE	Input	Interval	Numeric	8	.	
CRED	Input	Interval	Numeric	8	.	
DELINQ	Input	Nominal	Numeric	8	8	
DEPC	Input	Binary	Character	1	2	
EMP_STA	Input	Nominal	Character	3	3	
GENDER	Input	Binary	Character	1	2	
HEQ	Input	Nominal	Numeric	8	7	
INCOME	Input	Nominal	Numeric	8	7	
MFDU	Input	Binary	Numeric	8	2	
MILEAGE	Input	Interval	Numeric	8	.	
MOB	Input	Binary	Character	1	2	
MRTGI	Input	Nominal	Character	3	3	
MS	Input	Nominal	Character	1	3	
NUMTR	Input	Nominal	Numeric	8	8	
RESTYPE	Input	Nominal	Character	6	4	
RES_STA	Input	Binary	Character	3	2	
cuscode	Rejected	Nominal	Character	5	21	Exceeds maximum number of levels cutoff
resp	Target	Binary	Numeric	8	2	

2.11.1.3 Advanced Option with Customize Option

In this case, I selected the **Advanced** option, clicked **Customize**, and set the **Class Levels Count Threshold** property to 8. Display 2.161 shows how SAS Enterprise Miner assigns the measurement scales with this setting. For example, since the numeric variable DELINQ has eight unique values, it is now given the measurement scale of Inteval rather than Nominal because only those numeric variables that have fewer than eight unique variables are given the measurement scale of Nominal.

Display 2.161

Data Source Wizard -- Step 5 of 8 Column Metadata

Name	Role	Level	Type	Length	Number of Levels	
AGE	Input	Interval	Numeric	8	.	
CRED	Input	Interval	Numeric	8	.	
DELINQ	Input	Interval	Numeric	8	.	
DEPC	Input	Binary	Character	1	2	
EMP_STA	Input	Nominal	Character	3	3	
GENDER	Input	Binary	Character	1	2	
HEQ	Input	Nominal	Numeric	8	7	
INCOME	Input	Nominal	Numeric	8	7	
MFDU	Input	Binary	Numeric	8	2	
MILEAGE	Input	Interval	Numeric	8	.	
MOB	Input	Binary	Character	1	2	
MRTGI	Input	Nominal	Character	3	3	
MS	Input	Nominal	Character	1	3	
NUMTR	Input	Interval	Numeric	8	.	
RESTYPE	Input	Nominal	Character	6	4	
RES_STA	Input	Binary	Character	3	2	
cuscode	Rejected	Nominal	Character	5	21	Exceeds maximum number of levels cutoff
resp	Target	Binary	Numeric	8	2	

Two models built using the same variables are in general be different when different measurement scales are assigned to their variables. Hence it is important to be careful in assigning measurement scales. However, you can easily override the initial measurement scales assigned by SAS Enterprise Miner and manually assign different measurement scales to specific variables at a later stage in an Enterprise Miner Project, if necessary.

2.11.2 Eigenvalues, Eigenvectors, and Principal Components

In this section, I show how principal components are estimated from the eigenvectors of the variance-covariance matrix of a set of inputs. To clarify the concepts, I have created a small data set with 100 observations on three inputs. The inputs are X1, X2, and X3. The first 10 records of the input file are shown in Display 2.162.

Display 2.162

Obs	X1	X2	X3
1	250.760	40.6118	50.5603
2	251.219	40.5779	50.7204
3	249.387	39.7867	54.1581
4	253.917	41.4915	51.4365
5	251.026	40.4353	51.8280
6	249.710	39.8552	52.9012
7	251.993	41.0823	51.3885
8	248.292	39.3695	53.7557
9	250.691	40.4590	52.0699
10	249.571	39.8590	51.8386

Raw Data

The mean vector of the inputs is (249.7654 39.9048 52.1560).

By subtracting the means from the raw inputs, we get the adjusted inputs, which are shown in Display 2.163.

Display 2.163

Obs	X1-Mean(X1)	X2-Mean(X2)	X3-Mean(X3)
1	0.99498	0.70699	-1.59569
2	1.45390	0.67312	-1.43559
3	-0.37859	-0.11814	2.00212
4	4.15118	1.58663	-0.71949
5	1.26048	0.53051	-0.32792
6	-0.05526	-0.04965	0.74527
7	2.22728	1.17749	-0.76745
8	-1.47304	-0.53531	1.59973
9	0.92600	0.55419	-0.08608
10	-0.19453	-0.04579	-0.31733

Adjusted Data

Variance-Covariance Matrix

$$\Sigma = \begin{pmatrix} 2.3204 & 1.0002 & -1.1465 \\ 1.0002 & 0.4553 & -0.4885 \\ -1.1465 & -0.4885 & 0.8795 \end{pmatrix}$$

Extraction of Eigenvalues

We extract the eigenvalues by solving this equation:

$$\det(\Sigma - \lambda I) = 0 \qquad (2.1),$$

where det stands for the determinant,

λ is a scalar, and I is a 3×3 identity matrix.

When you expand equation (2.1) you will get a polynomial of the third degree in λ When you solve this polynomial, you get three roots that are the eigenvalues of Σ. In this example, the eigenvalues are:

$\lambda_1 = 3.3772$, $\lambda_2 = 0.2580$ and $\lambda_3 = 0.02$. The sum of the eigenvalues gives the total variance of the inputs.

Total Variance of All Inputs

The total variance of the inputs X1, X2, and X3 is therefore =
$$\lambda_1 + \lambda_2 + \lambda_3 = 3.3772 + 0.2580 + 0.0200 = 3.6552.$$

This can be easily verified by directly calculating the total variance of the inputs as the sum of the diagonal elements of Σ (the trace of Σ). The sum of the diagonal elements of the variance-covariance matrix $\Sigma =$ 2.3204 + 0.4553 + 0.8795= 3.6552, which is exactly identical to the sum of the eigenvalues.

Computation of Eigenvectors

The eigenvectors[6] are obtained from the eigenvalues by solving this equation:

$$(\Sigma - \lambda_i I) U_i = 0 \qquad (2.2),$$

where i=1, 2, 3 and U_i is the three-element eigenvector

corresponding to the eigenvalue λ_i.

The solution of (2.2) is not unique. Therefore, one of the elements of U_i is arbitrarily set to 1, and then the other two elements are adjusted so as to preserve the direction of the vector. Also, each eigenvector is normalized by dividing each element of it by the length of the eigenvector. So the eigenvector corresponding to the eigenvalue λ_i is $U_i = (u_{i1}, u_{i2}, u_{i3})$, and the normalized eigenvector is obtained by dividing each element of U_i by the length of the vector U_i, which is $= \sqrt{u_{11}^2 + u_{12}^2 + u_{13}^2}$. The normalized eigenvector is $V_i = (v_{i1}, v_{i2}, v_{i3})$.

In our example, the eigenvalue-eigenvector pairs are:

$\lambda_1 = 3.3772$ and

$V_1 = (0.8211, 0.3557, -0.4465);$

$\lambda_2 = 0.2580$ and

$V_2 = (0.3953, 0.2099, 0.8942);$

$\lambda_3 = 0.02$ and

$V_3 = (-0.4118, 0.9107, -0.0317);$

Computation of Principal Components

The principal components of the set of adjusted inputs are calculated as the weighted sums of the inputs using the elements of the eigenvectors as weights.

The first principal component for the jth observation is calculated as:

$$PC1_j = 0.8211\tilde{X}_{1j} + 0.3557\tilde{X}_{2j} - 0.4465\tilde{X}_{3j}$$

where

$$\tilde{X}_{1j} = X_{1j} - \bar{X}_1,$$

$$\tilde{X}_{2j} = X_{2j} - \bar{X}_2,$$

$$\tilde{X}_{3j} = X_{3j} - \bar{X}_3 \text{ and}$$

$\bar{X}_1, \bar{X}_2$ and $\bar{X}_3$ are the means of the variables X_1, X_2 and X_3

The second principal component is calculated as:

$$PC2_j = 0.3953\tilde{X}_{1j} + 0.2099\tilde{X}_{2j} + 0.8942\tilde{X}_{3j}$$

The third principal component is calculated as:

$$PC3_j = -0.4118\tilde{X}_{1j} + 0.9107\tilde{X}_{2j} - 0.0317\tilde{X}_{3j}$$

Display 2.164[7] shows the principal components for the first ten observations.

Display 2.164

Principal Components			
Obs	PC1	PC2	PC3
1	1.78084	-0.88520	0.28473
2	2.07411	-0.56773	0.05983
3	-1.24674	1.61591	-0.01515
4	4.29398	1.33068	-0.24161
5	1.37004	0.31641	-0.02551
6	-0.39577	0.63419	-0.04608
7	2.59021	0.44135	0.17953
8	-2.11409	0.73586	0.06836
9	0.99587	0.40541	0.12613
10	-0.03433	-0.37028	0.04846

Proportion of Variance Explained by the Principal Components

Earlier I showed that the total variance in the inputs = $\lambda_1 + \lambda_2 + \lambda_3$ =3.3772+0.2580+0.002=3.6552. It can also be shown that the variance of the first principal component is λ_1 (3.3772), the variance of the second principal component is λ_2 (0.2580), and the variance of the third principal component is λ_3 (0.002). Hence in our example, the proportion of variance explained by the first principal component is = 3.3772/3.6552 = 0.924 or 92.4%. The first principal component explains 92.4% of the total variance. Similarly we can verify that the proportion of variance explained by the second principal component is 0.2580/3.6552 or 7.1% and the proportion of the variance explained by the third principal component is 0.002/3.6552 or 0.5%. Since most of the variance is explained by the first principal component, we can safely use just one component in this case.

Thus one of the main uses of the **Principal Component** node is to reduce the number of inputs needed to explain and predict our target variable.

2.11.3 Cramer's V

Cramer's V measures the strength of relationship between two categorical variables. Suppose you have a categorical target Y with K distinct categories $Y_1, Y_2, ... Y_K$, and a categorical input X with M distinct categories $X_1, X_2, ... X_M$. If there are n_{ij} observations in the ith category in Y, and j^{th} category in X, then you can make a contingency table of the following type:

$X \setminus Y$	Y_1	Y_2	....	Y_K	Row Total
X_1	n_{11}	n_{12}	....	n_{1K}	$n_1.$
X_2	n_{11}	n_{11}	....	n_{2K}	$n_2.$
..					
X_M	n_{M1}	n_{M2}	....	n_{MK}	$n_{M.}$
Column Total	$n_{.1}$	$n_{.2}$	....	$n_{.K}$	N

To calculate Cramer's V, you must first calculate the Chi-Square Statistics under the null hypothesis that there is no association between X and Y.

$$\chi^2 = \sum_{i=1}^{M} \sum_{j=1}^{K} \frac{(n_{ij} - E_{ij})^2}{E_{ij}}, \text{ where } E_{ij} = n_{.j}\left(\frac{n_{i.}}{N}\right), \text{ for } j = 1, .. K.$$

Cramer's V $= \sqrt{\dfrac{\chi^2}{N \min(M-1, K-1)}}$, which takes a value between 0 to 1 for all tables larger than 2x2. For a 2x2 table, Cramer's V takes a value between -1 to +1.

Suppose that Y is a binary variable with two levels, and X is a 16-level categorical variable. Then Cramer's V $= \sqrt{\dfrac{\chi^2}{N}}$.

2.11.4 Calculation of Chi-Square Statistic and Cramer's V for a Continuous Input

The AGE variable is divided into five groups (bins). Table 2.1 shows the number of responders (RESP=1) and non-responders (RESP=0).

Table 2.1

Age group	RESP 0	1	Total
LOW-32.4	3830	1959	5789
32.4 -46.8	4952	2251	7203
46.8 -61.2	6915	3181	10096
61.2 -75.6	3882	1597	5479
75.6 - HIGH	946	391	1337
Total	20525	9379	29904
Overall Response Rate	0.686363	0.313637	

Table 2.2 shows the first step in calculating the Chi-Square value. It shows the expected number of observations in each cell under the null hypothesis of independence between rows and columns.

Table 2.2

Age group	RESP		
	0	1	Total
LOW-32.4	3973	1816	5789
32.4 -46.8	4944	2259	7203
46.8 -61.2	6930	3166	10096
61.2 -75.6	3761	1718	5479
75.6 - HIGH	918	419	1337

The expected number of observations is obtained by applying the overall response rate to each row total given in Table 2.1.

The Chi-Square for each cell is calculated as $\dfrac{(O-E)^2}{E}$, where O is the observed frequency as shown in Table 2.1, and E is the expected frequency (under the null hypothesis) as given in Table 2.2. The Chi-Square values for each cell are shown in Table 2.3.

Table 2.3

Age group	RESP	
	0	1
LOW-32.4	5.172157	11.318747
32.4 -46.8	0.013360	0.029237
46.8 -61.2	0.030430	0.066592
61.2 -75.6	3.920158	8.578872
75.6 - HIGH	0.874759	1.914323

By summing the cell Chi-Squares shown in Table 2.3, you get the overall Chi-Square (χ^2) of 31.9186 for the

AGE variable. Since there are 29904 observations in the sample, Cramer's V can be calculated $\sqrt{\dfrac{\chi^2}{29904}}$

which is equal to 0.0326706.

You can calculate these cell frequencies using the **StatExplore** node, as shown in Display 2.165.

Display 2.165

In order to calculate Chi-Square statistic and Cramer's V for an interval variable such as Age, you have to bin it first. If you set the **Chi-Square** and the **Interval Variables** properties (in the Chi-Square Statistics properties group of the **StatExplore** node) to Yes, interval variables are binned and the Chi-Square statistic and Cramer's V are calculated.

After running the **StatExplore** node, the cell-specific Chi-Square values can be retrieved from the Results window by clicking **View→Summary Statistics→Cell Chi_Square** in the menu bar. These cell-specific Chi-Square values are shown in Display 2.166.

Display 2.166

Target	Input	Target: Formatted Value	Input: Formatted Value	Frequency Count	Target: Numeric Value	Input: Numeric Value	Chi-Square
resp	AGE	0	32.4 -46.8	4952	0	33	0.01336
resp	AGE	0	46.8 -61.2	6915	0	47	0.03043
resp	AGE	0	61.2 -75.6	3882	0	62	3.920158
resp	AGE	0	75.6 - HIGH	946	0	76	0.874759
resp	AGE	0	LOW-32.4	3830	0	18	5.172157
resp	AGE	1	32.4 -46.8	2251	1	33	0.029237
resp	AGE	1	46.8 -61.2	3181	1	47	0.066592
resp	AGE	1	61.2 -75.6	1597	1	62	8.578872
resp	AGE	1	75.6 - HIGH	391	1	76	1.914323
resp	AGE	1	LOW-32.4	1959	1	18	11.31875

By selecting the Chi-Square Plot (or clicking in it), and then selecting the Table in the Results window, you can see the Chi-Square and Cramer's V for all variables. A partial view of this table is shown in Display 2.167.

Display 2.167

Target	Input	Cramer's V	Prob	Chi-Square	Df
resp	RESTYPE	0.165349	<.0001	817.5808	3
resp	MFDU	0.162414	<.0001	788.8147	1
resp	NUMTR	0.13795	<.0001	569.0763	7
resp	MRTGI	0.112498	<.0001	378.4571	2
resp	INCOME	0.06085	<.0001	110.7257	6
resp	AGE	0.032671	<.0001	31.9186	4
resp	DELINQ	0.03132	0.0001	29.3346	7
resp	CRED	0.027549	0.0001	22.6953	4
resp	EMP_STA	0.022864	0.0004	15.6328	2
resp	MILEAGE	0.020033	0.0173	12.0011	4
resp	HEQ	0.019458	0.0789	11.3215	6
resp	RES_STA	0.016075	0.0054	7.7274	1
resp	DEPC	0.011413	0.0484	3.8952	1
resp	GENDER	0.008698	0.1325	2.2624	1
resp	MS	0.007168	0.4638	1.5366	2
resp	MOB	0.002769	0.6320	0.2294	1

Note that the Cramer's V and the Chi-Square values for the variable AGE shown in Displays 2.166 and 2.167 are virtually identical to those we manually calculated and presented in Tables 2.1, 2.2, and 2.3.

2.12 Exercises

1. Create a Project Called CreditCards.
2. Create a data source using the SAS data set Ch2_Clus_Data2.
   a. Select **Basic** for the Metadata Advisor Options.
   b. Change the role of the variable Cancel to Target.
   c. How many inputs have the Measurement Level of Interval?
   d. How many inputs have the Measurement Level of Nominal?
   e. Rename the data source as Option1.
3. Create a data source using the SAS data set Ch2_Clus_Data2.
   a. Select **Advanced** for the Metadata Advisor Options.
   b. Change the role of the variable Cancel to Target.
   c. Select No to the question **Do You want to build models based on the values of the decisions?**
   d. How many inputs have the Measurement Level of Interval?
   e. How many inputs have the Measurement Level of Nominal?
   f. Rename the data scource as Option 2.

4. Create a data source using the SAS data set Ch2_Clus_Data2.
   a. Select **Advanced** for the Metadata Advisor Options and click **Customize**.
   b. Change the **Class Levels Count Threshold** property to 3 and click **OK**.
   c. Change the role of the variable Cancel to Target
   d. Select No to the question **Do You want to build models based on the values of the decisions?**
   e. How many inputs have the Measurement Level of Interval?
   f. How many inputs have the Measurement Level of Nominal?
   g. Rename the data source as Option 3.
5. Create a diagram and name it Explore.
   a. The first node in the diagram is **Input Data** (use the data source Option 2), and the second node is **StatExplore**.
   b. Select the **StatExplore** node. In the Interval Variable group, set the **Interval Variable** property to Yes
   c. Run the **StatExplore** node and open the results window.
   d. What are the top five variables that are most closely related to the target variable?
6. Create a diagram and name it VarSel1.
   a. The first node in the diagram is **Input Data** (use the data source Option 2), and the second node is **Variable Selection**.
   b. Select the **Variable Selection** node and set the **Target Model** property to R-Square.
   c. Set the **Use Aov16 Variables** property to Yes.
   d. Run the **Variable Selection** node.
   e. Based on this analysis, what are the best five inputs?
7. Create a diagram and name it VarSel2.
   a. The first node in the diagram is **Input Data** (use the data source Option 2), and the second node is **Variable Selection**.
   b. Select the **Variable Selection** node and set the **Target Model** property to Chi-Square.
   c. Set the **Use Aov16 Variables** property to Yes.
   d. Run the **Variable Selection** node.
   e. Based on this analysis, what are the best five inputs?
8. Create a diagram and name it VarSel3.
   a. The first node in the diagram is **Input Data** (use the data source Option 2), and the second node is **Variable Selection**.
   b. Select the **Variable Selection** node and set the **Target Model** property to R and Chi-Square.
   c. Set the **Use Aov16 Variables** property to Yes.
   d. Run the **Variable Selection** node.
   e. What are the top five variables that are most closely related to the target variable?

Notes

1. All the illustrations presented here are from sessions in which SAS Enterprise Miner 12.1 is installed on a personal computer.
2. I use the terms "measurement scale" and "measurement level" interchangeably.
3. Alternatively I could have first appended the data set Data2 to the data set Data1 using the **Append** node. Then I could have passed the combined data set to the **Transform Variables** node for making the transformations in the combined data set. Since my intention is to highlight the fact that the **Append** node can be used at any point in the process flow, I made the transformations by connecting the **Transform Variables** node separately to each data set first. Then I combined the resulting output data sets using the **Append** node as shown in Display 2.69. These two sequences produce identical results.
4. The steps outlined here are not exact representation of the algorithm used by SAS Enterprise Miner to do variable clustering.
5. The terms "class interval," "bucket," "group," and "bin" are used synonymously.
6. For examples of how to calculate eigenvalues and eigenvectors, see Anton, Howard, "Elementary Linear Algebra," Wiley 2010, pages 295-301.
7. Due to rounding, the Principal Components presented here may differ slightly from those you get by using the formulas presented.

Chapter 3: Variable Selection and Transformation of Variables

3.1 Introduction

This chapter familiarizes you with SAS Enterprise Miner's **Variable Selection** and **Transform Variables** nodes, which are essential for building predictive models. Since variable selection can also be done by the **Variable Clustering** and **Decision Tree** nodes, I have included some illustrations in this section of how to use these two nodes as well. Let us start with the **Variable Selection** node.

The **Variable Selection** node is useful when you want to make an initial selection of inputs or eliminate irrelevant inputs. It can also help identify non-linear relationships between the inputs and the target.

Unlike the **Variable Selection** node, the **Variable Clustering** node selects inputs without reference to the target variable. It is used primarily to identify groups (called clusters) of input variables that are similar looking and then to either select a representative variable from each cluster or create a new variable that is a linear combination of the inputs in a cluster. Replacing the variables in a cluster by the single variable that was selected or created to represent that cluster reduces the number of inputs and thereby reduces the severity of collinearity of the inputs in the estimation of the models.

The **Decision Tree** node can also be used for selecting important inputs. The inputs selected by the **Decision Tree** node are the inputs that contribute to the segmentation of the data set into homogeneous groups. The Decision Tree node is sometimes used after making an initial selection of inputs using the **Variable Selection** node.

The Transform Variables node provides a wide variety of transformations that can be applied to the inputs for improving the precision of the predictive models.

3.2 Variable Selection

In predictive modeling and data mining, we are often confronted with a large number of inputs (explanatory variables). The number of potential inputs to choose from may be 2000 or higher. Some of these inputs may not have any relation to the target. An initial screening can eliminate irrelevant variables and keep the number of inputs to a manageable size. A final selection can then be made from the remaining variables using either a stepwise regression or a decision tree.

The **Variable Selection** node performs the initial selection as well as a final variable selection. This section describes how the **Variable Selection** node handles interval and nominal inputs, as well as the selection criteria for categorical and continuous targets. The discussion focuses on four common situations (Cases 1–6):

- Case 1: The target is continuous (interval-scaled) and the inputs are numeric, consisting of interval-scaled variables.
- Case 2: The target is continuous and the inputs are categorical and nominal-scaled (These are sometimes referred to as class variables).
- Case 3: The target is binary and the inputs are numeric interval-scaled.
- Case 4: The target is binary and the inputs are categorical and nominal-scaled.
- Case 5: If the target is continuous and inputs are mixed, you have to specify how you want the Variables Selection node to handle categorical variables and continuous variables by appropriate property settings. The options considered in Case 1 and Case 2 are available even if you have mixed inputs. I did not include explicit examples for this case in this section because most of the examples included in subsequent chapters are with mixed inputs.
- Case 6: If the target is binary and inputs are mixed, you have to specify how you want the Variables Selection node to handle categorical variables and continuous variables by appropriate property settings. The options considered in Case 3 and Case 4 are available even if you have mixed inputs.

To focus on how each case is handled, use separate data sets for illustration purposes. In addition, to keep the discussion general, I indicate all numeric inputs as NVAR1, NVAR2, etc., and all nominal (character) inputs as CVAR1, CVAR2, etc.

Here are some examples of a continuous target:

- change in savings deposits of a hypothetical bank after a rate change
- customer's asset value obtained from a survey of a brokerage company
- loss frequency of an auto insurance company

All the data I use here is simulated, so it does not refer to any real bank or company.

The binary target used in Case 3 and Case 4 represents a response to a direct mail campaign by a hypothetical auto insurance company. The binary target, as its name suggests, takes only two values—response (represented by 1), and no response (represented by 0).

In SAS Enterprise Miner, both ordinal and nominal inputs are treated as class variables.

3.2.1 Continuous Target with Numeric Interval-scaled Inputs (Case 1)

In this example, a hypothetical bank wants to measure the effect of a promotion that involves an increase in the interest rate offered on savings deposits in the bank. The target variable, called DEPV in the modeling data set, is the change in a customer's deposits in response to the increase.

When the target is continuous (interval-scaled), the **Variable Selection** node uses R-Square as the default criterion of selection. For each interval-scaled input, the **Variable Selection** node calculates two measures of correlation between each input and the target. One is the R-Square between the target and the original input. The other is the R-Square between the target and the binned version of the input variable. The binned variable is a categorical variable created by the **Variable Selection** node from each continuous (interval-scaled) input. The levels (categories) of this categorical variable are the bins. In SAS Enterprise Miner, this binned variable is referred to as an AOV16 variable. The number of levels or categories of the binned variable (AOV16) is at most 16, corresponding to 16 intervals that are equal in width. This does not mean that the number of records in each bin is equal.

In general I recommend using AOV16 variables along with original variables and let the Regression node select the best one. If both are selected, you can use them both.

To get a better picture of these binned variables, let us take an example. Suppose we are binning the AGE variable, and it takes on values from 18–81. The first bin includes all individuals with ages 18–21, the second with ages 22–25, and so on. If there are no cases in a given bin, that bin is automatically eliminated, and the number of bins will be fewer than 16. The binned variable from AGE is called AOV16_AGE. The binned variable for INCOME is called AOV16_INCOME, and so on.

For this demonstration, I created a data source by selecting **Advanced** for the Metadata Advisor Option. Then I customized the properties, as shown in Display 3.1.

Display 3.1

The SAS data set used in the data source in this example is NUMERI_NTARG_B. The settings shown in Display 3.1 resulted in the following metadata.

Display 3.2

You can see that there are 227 interval scaled inputs and one binary input.

The **Variable Selection** node created 227 binned variables (AOV16 variables) from the interval-scaled variables, and then included all 455 variables in the selection process. At the end of the variable selection process, the **Variable Selection** node selected two original variables, NVAR133 and NVAR239, and three binned variables, AOV16_NVAR133, AOV16_NVAR049, and AOV16_NVAR239.

The following is a description of the variables that were selected.

NVAR133   average balance in the customer's checking account for the 12-month period prior to the promotional increase in the interest rate at the hypothetical bank.

NVAR049   number of transactions during the past 12 months by the customer.

NVAR239   other deposit balance 3 months prior to the promotional rate change.

Note that variables NVAR133 and NVAR239 are selected in both their original form and in their binned form. As discussed in Chapter 2, only AOV16 variables are passed to the next node if you set the **Use AOV16 Variables** property to Yes. Because the **Use AOV16 Variables** property to **Yes** in this example, NVAR133 and NVAR239 are not passed to the next node as inputs. If you set the **Hide Rejected Variables** property (in the Score section of the Properties panel) to No, you can see the selected AOV16 variables and all the original variables in the list of variables passed to the successor node.

As we discussed in Chapter 2, Section 2.8.6.1, the R-Square method uses a two-step selection procedure. In the first step, a preliminary selection is made, based on **Minimum R-Square**. For the original variables, the R-Square is calculated from a regression of the target on each input; for the binned variables, it is calculated from a one-way Analysis of Variance (ANOVA).

The default value of the **Minimum R-Square** property is 0.005, but you can set this property to a different value. The inputs that have R-Square values above the lower bound set by the **Minimum R-Square** property are included in the second step of the variable selection process.

In the second step, a sequential forward selection process is used. This process starts by selecting the input variable that has the highest correlation coefficient with the target. A regression equation (model) is estimated with the selected input. At each successive step of the sequence, an additional input variable that provides the largest incremental contribution to the Model R-Square is added to the regression. If the lower bound for the incremental contribution to the Model R-Square is reached, the selection process stops. You can specify the lower bound for the incremental contribution to the Model R-Square by setting the **Stop R-Square** property to the desired value.

In this example, the **Minimum R-Square** property is set to 0.05 and the **Stop R-Square** property is set to 0.005. Ninety-two variables, some original and some binned, were selected in the first step and, of these, only five variables were selected in the second step.

The smaller the values to which **Minimum R-Square** and **Stop R-Square** properties are set, the larger the number of variables passed to the next node with the role of **Input**. Letting more variables into the next node makes more sense if there is a **Decision Tree** or **Regression** node downstream, as both these nodes can themselves perform rigorous methods of selection. Variable selection is automatic in the **Decision Tree** node but needs to be explicitly specified for the **Regression** node.

To use the **Variable Selection** node for preliminary variable selection only, set the minimum for the **Stop R-Square** property to a very low value such as 0.00001. This, in effect, bypasses Step 2 of the variable selection process, which is okay since Step 2 can always be done in the **Decision Tree** or **Regression** node.

The R-Square for the original raw inputs captures the linear relationship between the target and the original input, while the R-Square between a binned (**AOV16**) variable and the target captures the non-linear relationship. In our example, NVAR049 was not selected in its original form, but the binned version of it (AOV16_NVAR049) was selected. By plotting the target mean for each bin against the binned variable, you can see whether a non-linear relationship exists between the input and the target.

In the following pages, I show how to set up the properties of the **Variable Selection** node in SAS Enterprise Miner 12.1. Display 3.3 shows the process flow for the variable selection.

Display 3.3

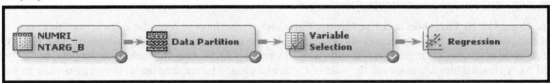

The first tool in the process flow is the **Input Data** node, which reads the SAS modeling data set. The data set is then partitioned into Training, Validation, and Test data sets. The partitioning is done in such a way that 50% of the observations are allocated to Training, 30% to Validation, and 20% to Test. The **Variable Selection** node selects variables using only the Training data set. The selected variables are then passed to a **Regression** node, with the role of Input.

The selection criterion is specified by setting the **Target Model** property. Display 3.4 shows the properties panel of the **Variable Selection** node. In the case of a continuous target variable, only the R-Square method is available. When the **Variable Selection** node detects that the target is continuous, it selects the R-Square method.

Display 3.4

Property	Value
General	
Node ID	Varsel
Imported Data	
Exported Data	
Notes	
Train	
Variables	
Max Class Level	100
Max Missing Percentage	50
Target Model	R-Square
Manual Selector	
Rejects Unused Input	Yes
⊟ Bypass Options	
Variable	None
Role	Input
⊟ Chi-Square Options	
Number of Bins	50
Maximum Pass Number	6
Minimum Chi-Square	3.84
⊟ R-Square Options	
Maximum Variable Number	3000
Minimum R-Square	0.05
Stop R-Square	0.0050
Use AOV16 Variables	Yes
Use Group Variables	Yes
Use Interactions	No
Use SPD Engine Library	Yes
Print Option	Default
Score	
Hides Rejected Variables	No
Hides Unused Variables	Yes
Status	
Create Time	11/9/12 1:25 PM

Because **Use AOV16 Variables** and **Use Group Variables** are both set to Yes, the AOV16 variables are created and passed to the next node, and all class variables are grouped (their levels collapsed). The **Variable Selection** node considers both the original and AOV16 variables in the two-step selection process, but only the selected AOV16 variables are assigned the role of Input and passed to the next node. If you also want to see the all the original inputs along with the selected AOV16 variables in the successor node, you should set the **Hides Rejected Variables** property in the Score section to No.

Output 3.1 shows a section of the output window in the **Results** window of the **Variable Selection** node. The output shows the variables selected in Step 1.

Output 3.1

```
The DMINE Procedure

        R-Squares for Target Variable: DEPV

Effect          DF      R-Square

Var:    NVAR133  1       0.214804
Class:  NVAR125  1       0.170772
AOV16:  NVAR239  15      0.168009
AOV16:  NVAR133  9       0.152164
AOV16:  NVAR009  10      0.141193
AOV16:  NVAR010  9       0.140946
AOV16:  NVAR008  9       0.139130
AOV16:  NVAR007  10      0.137466
AOV16:  NVAR135  12      0.137062
AOV16:  NVAR011  9       0.135750
AOV16:  NVAR134  11      0.135647
AOV16:  NVAR136  10      0.131550
AOV16:  NVAR138  11      0.131436
AOV16:  NVAR235  11      0.131436
AOV16:  NVAR139  11      0.129268
AOV16:  NVAR241  11      0.128856
AOV16:  NVAR137  11      0.127621
Var:    NVAR011  1       0.125407
Var:    NVAR009  1       0.125106
Var:    NVAR007  1       0.124150
```

Output 3.1 (*continued*)

108	Var:	NVAR198	1	0.082589	
109	Var:	NVAR199	1	0.082346	
110	Var:	NVAR196	1	0.082067	
111	Var:	NVAR200	1	0.080211	
112	Var:	NVAR084	1	0.076044	
113	AOV16:	NVAR033	12	0.073751	
114	AOV16:	NVAR012	11	0.067226	
115	AOV16:	NVAR014	10	0.066869	
116	AOV16:	NVAR013	10	0.064891	
117	AOV16:	NVAR049	13	0.064249	
118	AOV16:	NVAR166	8	0.061181	
119	AOV16:	NVAR169	12	0.057662	
120	Var:	NVAR033	1	0.057612	
121	AOV16:	NVAR178	12	0.057351	
122	AOV16:	NVAR179	12	0.056720	
123	AOV16:	NVAR177	8	0.051881	
124	AOV16:	NVAR180	11	0.051642	
125	AOV16:	NVAR167	12	0.050377	
126	AOV16:	NVAR173	12	0.049710	R2 < MINR2
127	AOV16:	NVAR207	12	0.048313	R2 < MINR2
128	AOV16:	NVAR086	12	0.048313	R2 < MINR2
129	AOV16:	NVAR168	9	0.048211	R2 < MINR2

A number of variables have met the **Minimum R-Square** criterion in the preliminary variable selection (Step 1). Those that have an R-Square below the minimum threshold are not included in Step 2 of the variable selection process.

In Step 2, all the variables that are selected in Step 1 are subjected to a sequential forward regression selection process, as discussed before. In the second step, only a handful of variables met the **Stop R-Square** criterion, as can be seen from Output 3.2. This output contains a table taken from the output window in the **Results** window of the **Variable Selection** node.

Output 3.2

The DMINE Procedure

Effects Chosen for Target: DEPV

Effect		DF	R-Square	F Value	p-Value	Sum of Squares	Error Mean Square
Var:	NVAR133	1	0.214804	398.860949	<.0001	854.145803	2.141463
AOV16:	NVAR133	9	0.382606	153.008124	<.0001	1521.393734	1.104802
AOV16:	NVAR239	15	0.024565	6.212188	<.0001	97.678241	1.048243
AOV16:	NVAR049	13	0.012379	3.700517	<.0001	49.222481	1.023194
Var:	NVAR239	1	0.008199	32.572863	<.0001	32.603923	1.000954

The variables NVAR133 and NVAR239 are selected both in their original form and in their binned forms. All other continuous inputs were selected in only their binned forms. Although many variables were selected in the preliminary variables assessment (Step 1), most of them did not meet the **Stop R-Square** criterion used in Step 2.

The next node in the process flow is the **Regression** node. I right-clicked on it and selected **Update** so that the path is updated. This means that the variables selected by the **Variable Selection** node are now passed to the **Regression** node with the role of Input. To verify this, I click ⬛, located to the right of the **Variables** property of the **Regression** node. The Variables window opens, as shown in Display 3.5.

Display 3.5

Name	Use	Report	Role	Level
AOV16_NVAR049	Default	No	Input	Ordinal
AOV16_NVAR133	Default	No	Input	Ordinal
AOV16_NVAR239	Default	No	Input	Ordinal
DEPV	Yes	No	Target	Interval
MATCHKEY	Default	No	Rejected	Nominal
NVAR001	Default	No	Rejected	Interval
NVAR002	Default	No	Rejected	Interval
NVAR003	Default	No	Rejected	Interval
NVAR004	Default	No	Rejected	Interval
NVAR005	Default	No	Rejected	Interval
NVAR006	Default	No	Rejected	Interval
NVAR007	Default	No	Rejected	Interval
NVAR008	Default	No	Rejected	Interval
NVAR009	Default	No	Rejected	Interval

At the top of this window, the selected AOV16 variables are listed. In the Role column, the variables are given the role of Input. To see the role assigned to the variables NVAR133 and NVAR239, scroll down the Variables window. As you scroll, NVAR133 becomes visible. Scroll further to show NVAR239. Both these variables have the role Rejected, despite the fact that they were selected by the **Variable Selection** node. This is because I set the **Use AOV16 Variables** property to Yes in the **Variable Selection** node. Note also that, as Output 3.1 and Output 3.2 show, AOV16_NVAR133 and AOV16_NVAR239 have met both the **Minimum R-Square** and **Stop R-Square** criteria and hence are given the role of Input in the **Regression** node.

3.2.2 Continuous Target with Nominal-Categorical Inputs (Case 2)

The target variable here is the same as in Case 1, and all the inputs are nominal-scaled categorical.[1]

Display 3.6 shows the process flow diagram created for demonstrating Case 2.

Display 3.6

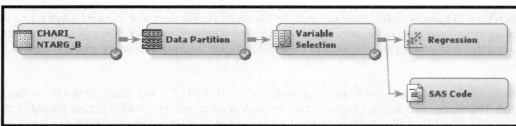

In the case of nominal-scaled categorical inputs with a continuous target, R-Square is calculated using one-way ANOVA. Here you have the option of using either the original or the grouped variables. I decided to use only grouped class variables. If a variable cannot be grouped, it is still included in the selection process. Hence, I have set the **Use Group Variables** property to Yes in the Properties panel (Display 3.7). I made this choice because some of the class variables have too many categories. The term *grouped variables* refers to those variables whose categories (levels) are collapsed or combined. For example, suppose there is a categorical (nominal) variable called LIFESTYLE, which indicates the lifestyle of the customer. It can take values such as Urban Dweller, Foreign Traveler, etc. If the variable LIFESTYLE has 100 levels or categories, it can be collapsed to fewer levels or categories by setting the **Group Variables** property to Yes.

Following are descriptions of some of the nominal-scaled class inputs in my example data set:

CVR14	Customer segment, based on the value of the customer, where value is based on factors such as interest revenue generated by the customer. This takes only two values: High or Low.
CVAR03	A segment, based on the type of accounts held by the customer.
CVR13	A variable indicating whether the customer uses the Internet.
CVAR07	A variable indicating whether the customer contacts by phone.
CVAR06	A variable indicating whether the customer visits the branch frequently.
CVAR11	A variable indicating whether the customer holds multiple accounts.

Display 3.7

Property	Value
General	
Node ID	Varsel
Imported Data	...
Exported Data	
Notes	...
Train	
Variables	...
Max Class Level	100
Max Missing Percentage	50
Target Model	Default
Manual Selector	...
Rejects Unused Input	Yes
⊟Bypass Options	
├Variable	None
└Role	Input
⊟Chi-Square Options	
├Number of Bins	50
├Maximum Pass Number	6
└Minimum Chi-Square	3.84
⊟R-Square Options	
├Maximum Variable Number	3000
├Minimum R-Square	0.0050
├Stop R-Square	5.0E-4
├Use AOV16 Variables	No
├Use Group Variables	Yes
└Use Interactions	No
Use SPD Engine Library	Yes
Print Option	Default
Score	
Hides Rejected Variables	Yes
Hides Unused Variables	Yes
Status	
Create Time	11/10/12 2:45 PM

Output 3.3 shows the variables selected in Step 1 of the **Variable Selection** node.

Output 3.3

```
The DMINE Procedure

          R-Squares for Target Variable: DEPV

Effect              DF        R-Square

Class: CVR14         1        0.173336
Class: CVAR03       10        0.048790
Group: CVAR03        4        0.048027
Class: CVR13         2        0.020582
Group: CVR13         1        0.020577
Class: CVAR07        2        0.014605
Group: CVAR07        1        0.014560
Class: CVAR06        2        0.014177
Group: CVAR06        1        0.013946
Class: CVAR11        2        0.007594
Class: CVAR16        4        0.004338     R2 < MINR2
Class: CVAR05        5        0.004122     R2 < MINR2
Group: CVAR05        4        0.004120     R2 < MINR2
Class: CVR08         4        0.003097     R2 < MINR2
Group: CVR08         3        0.003067     R2 < MINR2
Class: CVAR10        2        0.000041720   R2 < MINR2
```

To see how SAS Enterprise Miner groups or collapses categorical variables, let us consider the nominal scale variable CVAR03. By using PROC FREQ in the **SAS Code** node, we can see that the nominal-scaled variable CVAR03 has 11 categories, including the "missing" or unlabeled category, as shown in Output 3.4.

Output 3.4

```
The FREQ Procedure

                                 Cumulative    Cumulative
CVAR03   Frequency    Percent    Frequency      Percent
-------------------------------------------------------------
            101         8.65        101          8.65
01           52         4.45        153         13.10
02           60         5.14        213         18.24
03          126        10.79        339         29.02
04          228        19.52        567         48.54
05          156        13.36        723         61.90
06          148        12.67        871         74.57
07          130        11.13       1001         85.70
08          104         8.90       1105         94.61
09           54         4.62       1159         99.23
10            9         0.77       1168        100.00
```

The **Variable Selection** node created a new variable called G_CVAR03 by collapsing the eleven categories of CVAR03 into five categories. Using PROC FREQ again, we can see the frequency distribution of the categories of G_CVAR03, which is shown in Output 3.5. Output 3.5A shows a cross tabulation of CVAR03 by G_CVAR03.

Output 3.5

```
                    Grouped Levels for CVAR03

                                    Cumulative      Cumulative
G_CVAR03      Frequency      Percent   Frequency      Percent
-------------------------------------------------------------
      0           436        37.33         436         37.33
      1           257        22.00         693         59.33
      2           237        20.29         930         79.62
      3           186        15.92        1116         95.55
      4            52         4.45        1168        100.00
```

Output 3.5A

```
Table of CVAR03 by G_CVAR03

CVAR03      G_CVAR03(Grouped Levels for CVAR03)

Frequency|      0|      1|      2|      3|      4|  Total
---------+--------+--------+--------+--------+--------+
         |      0 |    101 |      0 |      0 |      0 |    101
---------+--------+--------+--------+--------+--------+
01       |      0 |      0 |      0 |      0 |     52 |     52
---------+--------+--------+--------+--------+--------+
02       |      0 |      0 |      0 |     60 |      0 |     60
---------+--------+--------+--------+--------+--------+
03       |      0 |      0 |      0 |    126 |      0 |    126
---------+--------+--------+--------+--------+--------+
04       |      0 |      0 |    228 |      0 |      0 |    228
---------+--------+--------+--------+--------+--------+
05       |      0 |    156 |      0 |      0 |      0 |    156
---------+--------+--------+--------+--------+--------+
06       |    148 |      0 |      0 |      0 |      0 |    148
---------+--------+--------+--------+--------+--------+
07       |    130 |      0 |      0 |      0 |      0 |    130
---------+--------+--------+--------+--------+--------+
08       |    104 |      0 |      0 |      0 |      0 |    104
---------+--------+--------+--------+--------+--------+
09       |     54 |      0 |      0 |      0 |      0 |     54
---------+--------+--------+--------+--------+--------+
10       |      0 |      0 |      9 |      0 |      0 |      9
---------+--------+--------+--------+--------+--------+
Total         436      257      237      186       52     1168
```

The SAS code that generated Outputs 3.4, 3.5, and 3.5A is shown below:

```
proc freq data=&EM_IMPORT_DATA;
  table CVAR03 G_CVAR03 / missing;
  table CVAR03*G_CVAR03/ norow  nocol nopercent  missing;
run;
```

Display 3.8 shows which categories of CVAR03 are grouped (collapsed) to create the variable G_CVAR03.

Display 3.8

```
*************************************************;
*** Begin Scoring Code from PROC DMINE ***;
*************************************************;


length _WARN_ $ 4;
label _WARN_ = "Warnings";



/*----G_CVAR03 begin----*/
length _NORM2 $ 2;
%DMNORMCP( CVAR03 , _NORM2 )
drop _NORM2;
select(_NORM2);
  when("  " ) G_CVAR03 = 1;
  when("01" ) G_CVAR03 = 4;
  when("02" ) G_CVAR03 = 3;
  when("03" ) G_CVAR03 = 3;
  when("04" ) G_CVAR03 = 2;
  when("05" ) G_CVAR03 = 1;
  when("06" ) G_CVAR03 = 0;
  when("07" ) G_CVAR03 = 0;
  when("08" ) G_CVAR03 = 0;
  when("09" ) G_CVAR03 = 0;
  when("10" ) G_CVAR03 = 2;
  otherwise substr(_WARN_, 2, 1) = 'U';
end;
label G_CVAR03="Grouped Levels for CVAR03";
/*----G_CVAR03 end----*/
```

The grouping procedure that SAS Enterprise Miner uses in collapsing the categories of a class variable can be illustrated by a simple example. Suppose a class variable has five categories: A, B, C, D, and E. For each level, the mean of the target for the records in that level is calculated. In the case of a binary target, if I represent a response by 1 and non-response by 0, then the target mean for a given level of an input variable is the proportion of responders within that level. The levels (categories) are then sorted in descending order by their target means. That is, the category that has the highest mean for the target becomes the first category, the one with the next highest target mean is second, and so on. Suppose this ordering resulted in this ranking: C, E, D, A, and B. Then the procedure combines the categories starting from the top. First it calculates the R-Square between the input and the target prior to grouping. Suppose the R-square is 0.20. Next, the procedure combines the categories C and E and recalculates the R-Square. Now the input has four categories instead of five, and the R-Square between the input and the target is now less than 0.20. If the reduction in R-Square is less than 5%, then C and E remain combined. A reduction of less than 5% in an R-Square of 0.20 means that the newly grouped input must have an R-square of at least 0.19 with the target. If not, C and E are not combined. Then an attempt is made to combine E and D using this same method, and so on. Output 3.3 shows the R-Square between the target and the original inputs before grouping (labeled Class) as well as the R-square between the target and the grouped version of the inputs (labeled Group).

The nominal-scaled variable CVAR03 has an R-Square of 0.048790 with the target before its categories are combined or grouped, and an R-Square of 0.048027 after collapsing the categories (grouping) from 11 to 5. The reduction in R-Square due to grouping is 0.000763 or 1.56%, which is less than 5%.

Both G_CVAR03 and CVAR03 met the R-Square criterion and were selected in Step 1. However, only the grouped variable is kept in the final Step 2 selection, as can be seen in Output 3.6. The R-Square shown in

Output 3.6 is an incremental or sequential R-Square, as distinct from the bivariate R-Square (coefficient of determination) shown in Output 3.3.

Output 3.6

```
The DMINE Procedure


                       Effects Chosen for Target: DEPV


                                                           Sum of      Error Mean
Effect            DF      R-Square      F Value    p-Value  Squares        Square

Class: CVR14       1      0.173336    244.487591   <.0001   544.655917    2.227745
Group: CVAR03      4      0.009016      3.203197   0.0125    28.329505    2.211033
Group: CVR13       1      0.008175     11.725295   0.0006    25.687918    2.190812
Group: CVAR07      1      0.006723      9.715610   0.0019    21.126479    2.174488
Group: CVAR06      1      0.003592      5.210083   0.0226    11.288295    2.166625
```

The first variable, CVR14, is a two-level class variable, and hence its levels are not collapsed (see Output 3.3, and note that no Group version of CVR14 appears). As before, the final selection is based on the stepwise forward selection procedure. The procedure starts with the best variable, i.e., the variable with the highest correlation, followed by the other variables according to their contribution to the improvement of the stepwise R-Square.

Next the **Regression** node is connected to the **Variable Selection** node. Display 3.9 shows the variables that are passed on to the **Regression** node with the role of Input.

Display 3.9

3.2.3 Binary Target with Numeric Interval-scaled Inputs (Case 3)

In this example, a hypothetical bank wants to cross-sell a mortgage refinance product. The bank sends mail to its existing customers. The target variable is response to the mail, taking the value of 1 for response and 0 for non-response.

The following is a description of some of the inputs in the data set used in this example.

NVAR004   credit card balance

NVAR014   auto loan balance

NVAR016   auto lease balance

NVAR030   mortgage loan balance

NVAR044	other loan balances
NVAR075	value of monthly transactions in checking and savings accounts
NVAR137	number of direct payments by the bank per month
NVAR148	available credit
NVAR187	number of investment-related transactions per month
NVAR193	percentage of households with children under age 10 in the block group
NVAR195	percentage of households living in rented homes in the block group
NVAR210	length of residence
NVAR286	age of the customer
NVAR315	number of products owned by the customer
NVAR318	number of teller-transactions per month

In the case of binary targets, inputs can be selected by using either the R-Square criterion, as in Case 1, or the Chi-Square criterion.

3.2.3.1 R-Square Criterion

When the target is binary then Chi-Square criterion is more appropriate than the R-square criterion especially if your goal is to estimate a logistic regression with the selected inputs. However, the continuous (interval-scaled inputs) have to be "binned" in order to calculate the Chi-Square statistics. This may be time consuming if there are many continuous inputs. Binning may also result in some loss of information. You can avoid binning the continuous variables if you use the R-Square criterion. In practice there is usually overlap between the variables selected under the Chi-Square and R-square criteria. So it is best to pass the union of the two sets of variables selected by these two methods to the next node (such as the Regression node or Decision Tree node) where further variable selection takes place.

When the R-Square criterion is used, the variable selection is similar to the case in which the target is continuous (Case 1). The target variable is treated like a continuous variable, and the R-Square with the target is computed for each original input and for each binned input (**AOV16** variable.) Then the usual two-step procedure is followed: In the first step, an initial variable set is selected based on the **Minimum R-Square** criterion; in the second step, a sequential forward selection procedure is used to select variables from those selected in Step 1 on the basis of a **Stop R-Square** criterion. Inputs that meet both these criteria receive the role of Input in the subsequent modeling tool, such as the **Regression** node.

3.2.3.2 Chi-Square Criterion

The Chi-Square criterion can be used only when the target is binary. When the Chi-Square criterion is selected, the selection process does not have two distinct steps, as in the case of the **R-square** criterion. Instead, a tree is constructed. The inputs selected in the construction of the tree are passed to the next node with the assigned role of **Input**. In this example, there are 302 inputs, and the **Variable Selection** node selected 19 inputs. This selection is done by developing a tree based on the Chi-Square criterion, as explained below. A tree constructed using the Chi-Square criterion is sometimes called CHAID (Chi-Squared Automatic Interaction Detection).

The tree development proceeds as follows. First, the records of the data set are divided into two groups. Each group is further divided into two more groups, and so on. This process of dividing is called *recursive partitioning*. Each group created during the recursive partitioning is called a *node*. The nodes that are not divided further are called *terminal nodes,* or *leaf nodes,* or *leaves* of the tree. The nodes that are divided further are called *intermediate nodes*. The node with all the records of the data set is called the *root node*. All these nodes form the tree. Displays 3.10 and 3.11 show such a tree.

In order to split the records in a given node into two new nodes, the algorithm evaluates all inputs (302 in this example) one at a time, and finds the best split for each input. To find the best split for an interval input the algorithm first divides the input into intervals, or bins, of equal length. The number of bins into which the interval input is divided is determined by the value to which the **Number of Bins** property is set. The default value is 50. However, suppose you set it to 10 and the interval input to be binned is **income**, with values that range from $10,000 to $110,000 in the data set. In that case, the 10 equal intervals will be $10,000– $20,000, $20,000–$30,000, $30,000–$40,000...and $100,000–$110,000. Any node split based on this input will be partitioned at one of the split values of $20,000, $30,000, $40,000, etc. A 2x2 contingency table is created for each of these nine split values, and a Chi-Square statistic is computed for each split. In this example, there are nine Chi-Square values. The split value with the highest Chi-Square value is the best split based on the input **income**. Then the algorithm does a similar calculation to find the best split of the records on the basis of each of the other available inputs. After finding the best split for each input, the algorithm compares the Chi-Square values of all these best splits and picks the one that gives the highest Chi-Square value. It then splits the records into two nodes on the basis of the input that gave the best split. This process is repeated at each node. All those inputs that are selected in this process, i.e. those that are chosen to be the basis of the split at any node of the tree, are passed to the next node with the assigned role of **Input**.

The tree generated by the **Variable Selection** node for our example data is shown in Displays 3.10 and 3.11. Display 3.10 shows the root node and the half of the tree. Display 3.11 shows the root node and the right half of the tree.

Display 3.10

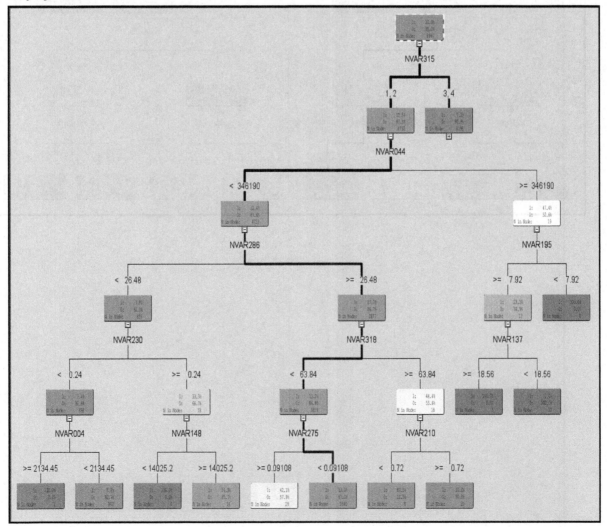

Display 3.11

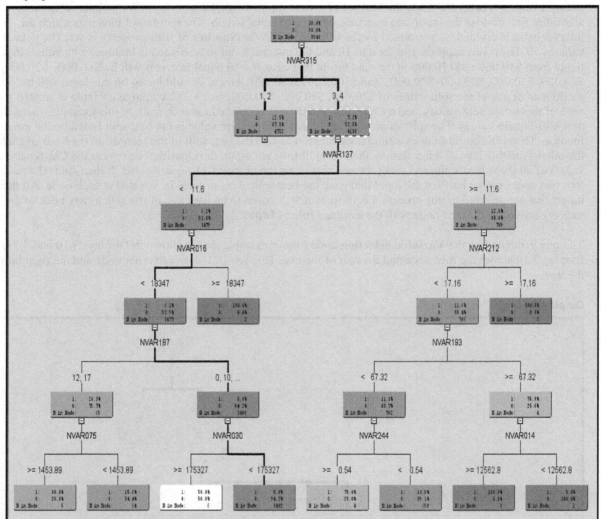

Display 3.12 shows the relative importance of the selected variables.

Display 3.12

Variable	Label	Number of Rules in Tree	Relative Importance
NVAR315	NVAR315	1	1
NVAR137	NVAR137	2	0.823072
NVAR210	NVAR210	1	0.648884
NVAR195	NVAR195	1	0.61899
NVAR212	NVAR212	1	0.606703
NVAR044	NVAR044	1	0.604833
NVAR148	NVAR148	1	0.600437
NVAR286	NVAR286	1	0.56696
NVAR075	NVAR075	1	0.550585
NVAR016	NVAR016	1	0.527099
NVAR318	NVAR318	1	0.525406
NVAR244	NVAR244	1	0.507789
NVAR193	NVAR193	1	0.504904
NVAR275	NVAR275	1	0.501803
NVAR187	NVAR187	1	0.471972
NVAR230	NVAR230	1	0.432374
NVAR030	NVAR030	1	0.429983
NVAR004	NVAR004	1	0.367988
NVAR014	NVAR014	1	0.343943

The variables that appear in Displays 3.10 and 3.11 are passed to the next node with the role of **Input**. You can verify this by first connecting a **Regression** node to the **Variable Selection** node, as shown in Display 3.13.

Display 3.13

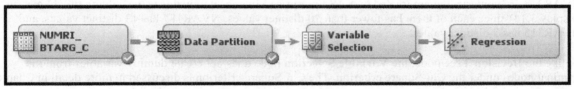

Then, if you select the **Regression** node, update the path and click ⬚ to the right of the **Variables** property in the Properties panel, a window appears displaying the inputs passed to the **Regression** node. This is shown in Display 3.14.

Display 3.14

Name	Use	Report	Role	Level	Type	Format
NVAR004	Default	No	Input	Interval	Numeric	
NVAR014	Default	No	Input	Interval	Numeric	
NVAR016	Default	No	Input	Interval	Numeric	
NVAR030	Default	No	Input	Interval	Numeric	6.0
NVAR044	Default	No	Input	Interval	Numeric	
NVAR075	Default	No	Input	Interval	Numeric	
NVAR137	Default	No	Input	Interval	Numeric	
NVAR148	Default	No	Input	Interval	Numeric	
NVAR187	Default	No	Input	Nominal	Numeric	
NVAR193	Default	No	Input	Interval	Numeric	
NVAR195	Default	No	Input	Interval	Numeric	
NVAR210	Default	No	Input	Interval	Numeric	
NVAR212	Default	No	Input	Interval	Numeric	
NVAR230	Default	No	Input	Interval	Numeric	
NVAR244	Default	No	Input	Interval	Numeric	
NVAR275	Default	No	Input	Interval	Numeric	
NVAR286	Default	No	Input	Interval	Numeric	
NVAR315	Default	No	Input	Nominal	Numeric	
NVAR318	Default	No	Input	Interval	Numeric	
resp	Yes	No	Target	Binary	Numeric	

Display 3.14 shows that the role assigned to each of these selected variables is **Input**. Note that NVAR187 and NVAR315 are numeric, but are treated by SAS Enterprise Miner as nominal (see the column titled Level in Display 3.14) since each of them has fewer than 20 distinct values. NVAR187 has 13 distinct values, and NVAR315 has 7 distinct values.

Unlike the **Decision Tree** node, the **Variable Selection** node does not create dummy variables from the terminal nodes under the Chi-Square criterion. The Chi-Square criterion is discussed in more detail in Chapter 4 in the context of decision trees.

Display 3.15 shows the properties panel where the properties for the **Chi-Square** criterion are set.

Display 3.15

Property	Value
General	
Node ID	Varsel
Imported Data	...
Exported Data	...
Notes	...
Train	
Variables	...
Max Class Level	100
Max Missing Percentage	50
Target Model	Chi-Square
Manual Selector	...
Rejects Unused Input	Yes
⊟Bypass Options	
├Variable	None
└Role	Input
⊟Chi-Square Options	
├Number of Bins	50
├Maximum Pass Number	6
└Minimum Chi-Square	3.84
⊟R-Square Options	
├Maximum Variable Number	3000
├Minimum R-Square	0.0050
├Stop R-Square	5.0E-4
├Use AOV16 Variables	No
├Use Group Variables	Yes
└Use Interactions	No
Use SPD Engine Library	Yes
Print Option	Default
Score	
Hides Rejected Variables	Yes
Hides Unused Variables	Yes
Status	
Create Time	11/11/12 1:27 PM

3.2.4 Binary Target with Nominal-scaled Categorical Inputs (Case 4)

In this case, the target variable is the same as in Case 3, and all the inputs are nominal-scaled categorical. You can use either the R-Square or Chi-Square criterion to select the variables. Here are descriptions of some of the nominal-scaled class inputs in the data set:

CVAR08	dwelling size code
CVAR11	income code
CVAR20	indicator of home ownership
CV30	demographic cluster
CVAR34	indicator of Internet banking
CVAR35	customer type
CVAR40	life cycle code
CVAR47	indicator of Internet usage

Display 3.16 shows the process flow used to demonstrate the examples in this section.

Display 3.16

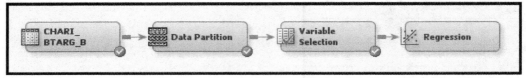

3.2.4.1 R-Square Criterion

With the R-Square criterion, the selection procedure is similar to the procedure in Case 2 in which two steps are needed. I set the **Target Model** property to **R-Square** and ran the **Variable Selection** node. The variables selected in Step 1 are shown in Output 3.7.

Output 3.7

```
The DMINE Procedure

            R-Squares for Target Variable: resp

Effect                  DF        R-Square

Class: CVAR47            1        0.014025
Class: CVAR40           14        0.002226
Group: CVAR40            6        0.002188
Class: CVAR35            6        0.002118
Group: CVAR35            3        0.002084
Class: CV30             10        0.001715     R2 < MINR2
Group: CV30              6        0.001694     R2 < MINR2
Class: CVAR14            2        0.001028     R2 < MINR2
```

Output 3.8 shows the results of the forward selection process (Step 2).

Output 3.8

```
The DMINE Procedure

                        Effects Chosen for Target: resp

                                                              Sum of      Error Mean
Effect            DF      R-Square      F Value    p-Value     Squares        Square

Class: CVAR47      1      0.014025    127.141908    <.0001    11.284807      0.088758
Group: CVAR40      6      0.002005      3.033106    0.0058     1.613065      0.088637
Group: CVAR35      3      0.001777      5.385909    0.0011     1.430059      0.088506
```

When you compare Outputs 3.7 and 3.8, you can see that the class variables CVAR40 and CVAR35 met the **Minimum R-Square** criterion in the first step in their original and grouped form. But in the second step, only the grouped variables survived. Display 3.17 shows the variables window of the next node, which is the **Regression** node.

Display 3.17

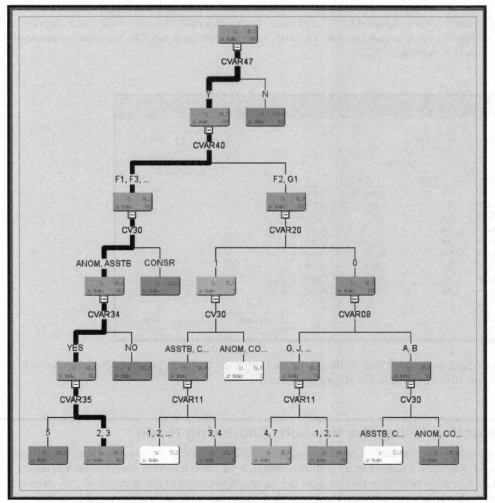

3.2.4.2 Chi-Square Criterion

Using the same data set, I selected the Chi-Square criterion by setting the value of the **Target Model** property to **Chi-Square**.

Display 3.18 shows the tree constructed by this method.

Display 3.18

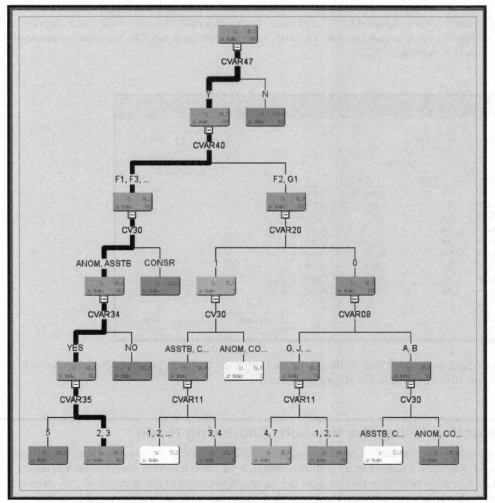

The names of the variables selected appear below the intermediate (non-terminal) nodes of the tree. Display 3.19 shows the variable importance.

Display 3.19

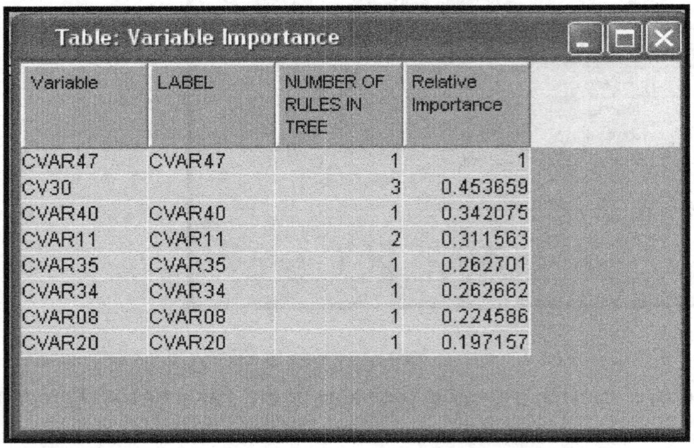

Variable	LABEL	NUMBER OF RULES IN TREE	Relative Importance
CVAR47	CVAR47	1	1
CV30		3	0.453659
CVAR40	CVAR40	1	0.342075
CVAR11	CVAR11	2	0.311563
CVAR35	CVAR35	1	0.262701
CVAR34	CVAR34	1	0.262662
CVAR08	CVAR08	1	0.224586
CVAR20	CVAR20	1	0.197157

When you compare the variables in Display 3.19 with those in Outputs 3.7 and 3.8, you can see that CVAR47 emerges as the most important variable by both the R-Square and Chi-Square criteria. As you may recall, this variable is an indicator of Internet usage by the customer in the data set.

The variables in Display 3.19 are exported to the next node. In this example, the next node is the **Regression** node. Open its Variables window to see the same variables with the role of Input. The variables window of the **Regression** node is shown in Display 3.20.

Display 3.20

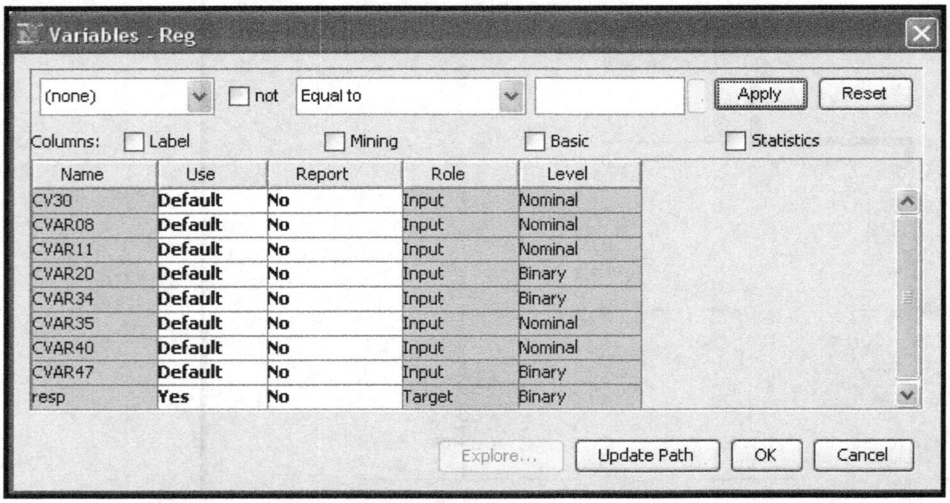

Name	Use	Report	Role	Level
CV30	Default	No	Input	Nominal
CVAR08	Default	No	Input	Nominal
CVAR11	Default	No	Input	Nominal
CVAR20	Default	No	Input	Binary
CVAR34	Default	No	Input	Binary
CVAR35	Default	No	Input	Nominal
CVAR40	Default	No	Input	Nominal
CVAR47	Default	No	Input	Binary
resp	Yes	No	Target	Binary

Eight variables were selected—those that produced the best splits in the development of the tree. The selected variables are given the role of **Input** in the **Regression** node.

3.3 Variable Selection Using the Variable Clustering Node

In this section, I show how the **Variable Clustering** node is used for variable (input) selection. The variable selection done by the **Variable Clustering** node differs from the variable selection done by the **Variable Selection** node. The **Variable Selection** node selects the inputs on the basis of the strength of their relationship with the target (dependent) variable, whereas the **Variable Clustering** node selects the variables without reference to the target variable. The **Variable Clustering** node first creates clusters of variables, by assigning variables that look similar to the same cluster. Then it selects one representative variable from each cluster. You can also ask for a linear combination of all variables in a cluster, instead of one representative variable from each cluster. The linear combination of the variables that the **Variable Clustering** node creates from each cluster is called the component of the cluster.

In Chapter 2 we discussed in detail how these clusters are created and how the components are calculated from the clusters. Whether you are using representative variables from the clusters or using the cluster components, you are in effect reducing the number of inputs substantially. In this section, I demonstrate both these methods of input reduction and develop a predictive model with the reduced set of inputs, and compare the resulting models. The model development and model comparison are discussed in detail in Chapters 4, 5, 6 and 7. I hope that the exercise in this section provides a preview of these tasks without causing confusion.

To do this demonstration, I constructed a modeling data set for a sample of 15,000 customers. The data set has 15,000 observations, one for each customer, and 85 columns, 84 of which refer to the inputs or explanatory variables and one of which refers to the target variable named Event. The Event variable shows whether the customer canceled (closed) his/her account during the 3-month observation window. The 84 inputs in the data set include variables that describe the customer's behavior prior to the observation window, as well as other characteristics such as age and marital status of the customer. If the variable Event takes the value 1 for an account, it means that the account closed during the observation window (0 indicates that the account remained open). From past experience we know that the event (cancellation) rate is 2.4% per 3 months, but since the sample we are using is biased, it shows an event rate of around 16%.

We use the **Variable Clustering** node to create a number of clusters of inputs. In our example, we use the default settings for all of the properties of the **Variable Clustering** node—except the **Variable Selection** property, which we set first to Best Variables and then to Cluster Component.

The Best Variables setting selects one representative input from each cluster, while the Cluster Component setting selects the cluster component (a linear combination of the inputs in a given cluster) of each cluster.

The selected variables or cluster components are then passed to the **Regression** node, which estimates a logistic regression using the inputs passed to it and the target variable Event. Then we compare the logistic regressions estimated from these two methods of selection.

The **Regression** Node is discussed in detail in Chapter 6. Here I use it only to illustrate how the variables selected by the **Variable Clustering** node are passed to the next node and how they can be used in the next node.

Display 3.21 shows the process flow for demonstrating variable selection by the **Variable Clustering** node.

Display 3.21

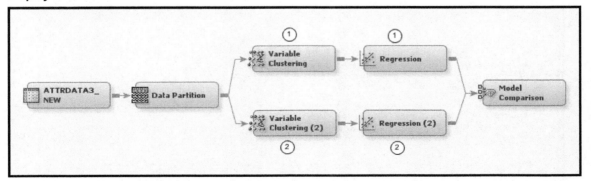

Display 3.21 shows both of the approaches to variable clustering that I want to demonstrate. In the first approach, shown in the upper segment of Display 3.21, I set the **Variable Selection** property to Best Variables. All other properties are set to their default values. The default value for the **Include Class Variables** property is

No. With these settings, the **Variable Clustering** node creates clusters of variables, identifies the best variable in each cluster, and passes the best variable from each cluster on to the next node. (Recall from Chapter 2 that the best variable in any given cluster is the one that has the minimum 1-RSquare ratio.) Since the **Include Class Variables** property is set to No, all of the nominal (class) variables among the inputs are passed to the next node along with the best variables. If the **Include Class Variables** property is set to Yes instead, the variable clustering algorithm creates dummy variables for each category of each class variable and includes these dummy variables in the process of creating the clusters.

3.3.1 Selection of the Best Variable from Each Cluster

Display 3.22 shows the properties of the **Variable Clustering** node in the upper section of Display 3.21.

Display 3.22

Property	Value
General	
Node ID	VarClus
Imported Data	
Exported Data	
Notes	
Train	
Variables	
Clustering Source	Correlation
Keeps Hierarchies	Yes
Includes Class Variables	No
Two Stage Clustering	Auto
Stopping Criteria	
Maximum Clusters	Default
Maximum Eigenvalue	Default
Variation Proportion	Default
Print Option	Short
Suppress Sampling Warning	No
Score	
Variable Selection	Best Variables
Interactive Selection	
Hides Rejected Variables	Yes
Status	
Create Time	11/24/12 9:12 PM

The clusters and the variables in each cluster can be seen in the output window of the **Variable Clustering** node, as shown in Output 3.9.

Output 3.9

```
Total variation explained = 55.25598 Proportion = 0.6425
```

| 26 Clusters | | R-squared with | | |
| | | Own | Next | 1-R**2 |
Cluster	Variable	Cluster	Closest	Ratio
Cluster 1	NUMRPD6M	0.9329	0.2682	0.0917
	NUMSPCL	0.7337	0.2642	0.3619
	NUMTOTOPN	0.8319	0.1818	0.2055
	OpenCards	0.8556	0.1976	0.1800
	PastDue4PF	0.1306	0.0541	0.9191
	TOTBAL3M2	0.5764	0.2557	0.5690
Cluster 2	NEGRTPD3	0.7565	0.3259	0.3613
	NEGRTPD6	0.7401	0.2952	0.3687
	NUMMAJ	0.6127	0.1149	0.4376
	PDR3	0.8047	0.4164	0.3346
	PDR3M	0.8895	0.4935	0.2182
	PDR4	0.8236	0.4130	0.3006
	PDR4M	0.8914	0.4706	0.2052
Cluster 3	Bal_Actl	0.7940	0.2544	0.2763
	GOODBAL	0.8936	0.2074	0.1342
	PPSCORE6	0.2924	0.1644	0.8467
	TOTBAL	0.9223	0.3374	0.1173
Cluster 4	APR	0.0283	0.0424	1.0148
	NUM30PDIN24M	0.7112	0.1600	0.3438
	NUMNEG30PD	0.7315	0.0690	0.2883
	NUMNEG60PD	0.5140	0.0465	0.5097
	NUMNEGPD	0.8904	0.1806	0.1338
	NUMOVLMFEE	0.0058	0.0038	0.9980
	PDAMTPCT	0.7522	0.2822	0.3452
Cluster 5	BALAVG3M	0.8372	0.1876	0.2003
	MtgBal	0.9271	0.0716	0.0785
	MtgInd	0.6655	0.0769	0.3623
	TOTAMNT3M	0.9487	0.0921	0.0565
Cluster 6	CREDLINE	0.8030	0.0484	0.2070
	HCREDIT	0.9488	0.0477	0.0537
	MAXCRED	0.9488	0.0477	0.0537
	TENURE	0.0633	0.1051	1.0468

Output 3.9 (cont'd)

Cluster 7	POINTEARND	0.8193	0.0286	0.1860
	POINTSRDMD	0.9615	0.0090	0.0389
	Redeemed	0.9615	0.0089	0.0388
Cluster 8	EARNEDTD	0.0843	0.1589	1.0887
	PURCHAMNT	0.9436	0.1374	0.0654
	SALESAMT	0.9436	0.1374	0.0654
	SALESFRQ	0.7432	0.1788	0.3127
Cluster 9	NUMOPNCC6M	0.4745	0.1110	0.5911
	PCTRVOPN6M	0.6291	0.0335	0.3837
	PCTTROPN12M	0.6919	0.1888	0.3798
	newest	0.4569	0.1208	0.6177
Cluster 10	Bal_Act2	0.6817	0.2149	0.4055
	OBSID	0.0177	0.0848	1.0733
	PCTHBAL3M	0.5464	0.2451	0.6009
	PPSCORE7	0.8173	0.1287	0.2097
	REVPERF	0.8065	0.1219	0.2203
	TOTMPAY	0.2130	0.0587	0.8361
Cluster 11	INCOME	0.4535	0.0091	0.5515
	MBBYR	0.4754	0.0230	0.5370
	NUMADULT	0.6100	0.1005	0.4336
	NUMCAR	0.1712	0.0617	0.8832
Cluster 12	FRQFIN12M	0.4459	0.1916	0.6854
	REV	0.9472	0.2512	0.0705
	TOTFC	0.7649	0.2797	0.3265
	TOTFEE	0.3832	0.3086	0.8921
Cluster 13	AGE	0.0512	0.3335	1.4236
	L2BNKENQ	0.8893	0.1740	0.1340
	L2PRMENQ	0.8841	0.0658	0.1241
Cluster 14	OpenCardsPct	0.5141	0.0599	0.5168
	PCTNUMTR	0.3918	0.0583	0.6458
	PCTOPNTOTTR	0.8185	0.3051	0.2611
Cluster 15	INQ12M	0.8606	0.1793	0.1698
	INQ24M	0.8512	0.2302	0.1933
	NUMOPN7M12M	0.2768	0.1026	0.8058
	RETCHECK	0.0025	0.0015	0.9990
Cluster 16	DEQVAR	0.4135	0.0118	0.5935
	PDSEVERE	0.3591	0.0165	0.6517
	PPSCORE8	0.5077	0.1364	0.5701

Output 3.9 (cont'd)

```
Cluster 17    NUM6MPDIN24M    0.6403    0.3041    0.5168
              PCTPDBAL3M      0.3133    0.0911    0.7555
              RNEG120PD       0.4501    0.1986    0.6862
              TOTAMTPD3M2     0.6739    0.3481    0.5002
----------------------------------------------------------------
Cluster 18    FAMCOMP         1.0000    0.0296    0.0000
----------------------------------------------------------------
Cluster 19    INSFEE          0.5092    0.0069    0.4942
              PCTTOTTR        0.5092    0.0500    0.5167
----------------------------------------------------------------
Cluster 20    PMNTAMNT        1.0000    0.1412    0.0000
----------------------------------------------------------------
Cluster 21    MKTVAL          0.5096    0.0911    0.5395
              PURCHADJ        0.5096    0.0008    0.4908
----------------------------------------------------------------
Cluster 22    BalTranAmt      1.0000    0.0108    0.0000
----------------------------------------------------------------
Cluster 23    FRQPURCH12M     1.0000    0.1654    0.0000
----------------------------------------------------------------
Cluster 24    REVPERFM        1.0000    0.0492    0.0000
----------------------------------------------------------------
Cluster 25    BalTranFrq      0.5095    0.0175    0.4993
              INTRCHNGR       0.5095    0.0007    0.4909
----------------------------------------------------------------
Cluster 26    TOTAMTCOL       1.0000    0.0535    0.0000
```

Display 3.23 shows the selected variables.

Display 3.23

Cluster	Variable	R-Square With Own Cluster Component	R-Square with Next Cluster Component	1-R2 Ratio
CLUS1	NUMRPD6M	0.932905	0.268224	0.091687
CLUS10	PPSCORE7	0.817335	0.128735	0.209655
CLUS11	NUMADULT	0.609987	0.10048	0.433579
CLUS12	REV	0.947197	0.251216	0.070518
CLUS13	L2PRMENQ	0.884063	0.065836	0.124108
CLUS14	PCTOPNTOTTR	0.818543	0.305109	0.26113
CLUS15	INQ12M	0.860621	0.179277	0.169824
CLUS16	PPSCORE8	0.507661	0.136402	0.570102
CLUS17	TOTAMTPD3M2	0.673936	0.348136	0.500202
CLUS18	FAMCOMP	1	0.029616	0
CLUS19	INSFEE	0.509168	0.006872	0.494228
CLUS2	PDR4M	0.891361	0.470634	0.205225
CLUS20	PMNTAMNT	1	0.14119	0
CLUS21	PURCHADJ	0.509584	.0007625	0.49079
CLUS22	BALTRANAMT	1	0.010775	0
CLUS23	FRQPURCH12M	1	0.165403	0
CLUS24	REVPERFM	1	0.049172	0
CLUS25	INTRCHNGR	0.509469	.0006982	0.490874
CLUS26	TOTAMTCOL	1	0.053479	0
CLUS3	TOTBAL	0.922259	0.337439	0.117334
CLUS4	NUMNEGPD	0.890364	0.180646	0.133807
CLUS5	TOTAMNT3M	0.94867	0.092133	0.05654
CLUS6	MAXCRED	0.948838	0.047688	0.053724
CLUS7	REDEEMED	0.961525	0.008905	0.03882
CLUS8	SALESAMT	0.943627	0.137387	0.065351
CLUS9	PCTTROPN12M	0.691928	0.188837	0.37979

Next we connect a **Regression** node to the **Variable Clustering** node in the upper section of Display 3.21. The variables shown in Display 3.23 and the nominal variables that were excluded from clustering are passed to the **Regression** node.

In the **Regression** node, we make a further selection of these variables by setting the **Selection Model Property** to Stepwise and the **Selection Criterion** property to Validation Error. With these settings, the **Regression** node tests all the variables passed to it and selects the variables to keep based on the Stepwise selection criterion. During this process of variable selection, the **Regression** node creates a number of models and selects the model that has the smallest error when applied to the Validation sample.

Display 3.24 shows the property settings of the **Regression** node shown in the upper segment of Display 21.

Display 3.24

Property	Value
General	
Node ID	Reg
Imported Data	
Exported Data	
Notes	
Train	
Variables	
⊟Equation	
├Main Effects	Yes
├Two-Factor Interactions	No
├Polynomial Terms	No
├Polynomial Degree	2
├User Terms	No
└Term Editor	
⊟Class Targets	
├Regression Type	Logistic Regression
└Link Function	Logit
⊟Model Options	
├Suppress Intercept	No
└Input Coding	Deviation
⊟Model Selection	
├Selection Model	Stepwise
├Selection Criterion	Validation Error
├Use Selection Defaults	Yes

Output 3.10 shows the estimated logistic regression.

Output 3.10

```
                     Analysis of Maximum Likelihood Estimates

                              Standard       Wald                   Standardized
Parameter     DF   Estimate      Error   Chi-Square   Pr > ChiSq       Estimate    Exp(Est)

Intercept      1     5.3067     0.8030        43.67       <.0001                    201.681
FRQPURCH12M    1     0.0591     0.0117        25.53       <.0001         0.1316       1.061
INQ12M         1     0.0392     0.0190         4.27       0.0387         0.0517       1.040
INTRCHNGR      1    -0.1818     0.0758         5.76       0.0164        -0.0501       0.834
L2PRMENQ       1     0.0433    0.00909        22.70       <.0001         0.1010       1.044
NUMNEGPD       1     0.0817     0.0263         9.64       0.0019         0.0754       1.085
NUMRPD6M       1    -0.0113    0.00311        13.28       0.0003        -0.1194       0.989
PCTOPNTOTTR    1    -0.0105    0.00243        18.77       <.0001        -0.1295       0.990
PCTTROPN12M    1     0.0102    0.00491         4.27       0.0388         0.0480       1.010
PPSCORE7       1   0.000114   0.000045         6.29       0.0121         0.0624       1.000
PPSCORE8       1    -0.0773    0.00988        61.17       <.0001        -0.2117       0.926
REV            1     0.0175    0.00208        70.82       <.0001         0.3192       1.018
REVPERFM       1    -0.1715     0.0700         6.00       0.0143        -0.0553       0.842
SALESAMT       1   -0.00537   0.000571        88.45       <.0001        -1.5838       0.995
TOTBAL         1   -0.00049   0.000039       161.68       <.0001        -0.7967       1.000
```

The logistic regression presented in Output 3.10 is part of the output sub-window of the Results window. The inputs included in the logistic regression shown in Output 3.10 can be considered the best of the best because variables entered into the logistic regression were passed through two selection processes. The **Variable Clustering** node first selected the best interval-scaled input from each cluster and passed them to the

Regression node along with all the nominal variables present in the data set. The **Regression** node then selected the best of these variables through the stepwise selection. Hence, the inputs included in the regression equation presented in Output 3.10 can be considered the best of the best in a statistical sense. In general, you should pay careful attention to the sign of the parameter estimates and make sure that they make business sense.

In order to help assess the regression equation, the Results window also produces a number of charts. You can use two charts described below to compare the logistic regressions estimated from the alternative variable selections that we have made. The first chart is called the Lift chart. In order to create the Lift chart, the **Regression** node first applies the model to a data set that contains all the inputs used in the estimated regression. Then, for each observation (customer), it uses the estimated regression coefficients and specification to calculate the probability that each customer will close his/her account or "attrit" (cancellation is sometimes referred to as attrition). Then, once this estimated probability of attrition is calculated for each customer, the **Regression** node sorts the data set in descending order of the calculated probabilities. That is, it ranks all the customers in the database by their probability to cancel their accounts, with those most likely to cancel at the top of the list. Then it divides the data set into 20 equal segments called Percentiles. Since the percentiles are created from the sorted data set based on the computed probabilities, the first percentile (called the top percentile) has the customers with the highest mean probability of cancellation. If the model is accurate, then the first percentile in this ranked list should also have the highest *actual* observed cancellation rate, where the cancellation rate is defined as the number of cancellations in a given percentile divided by the number of observations in the percentile. The **Regression** node calculates the observed cancellation rates for each of the 20 percentiles created as described above.

The lift in a given percentile is the actual observed cancellation rate in that percentile divided by the overall actual cancellation rate for the data set. Suppose the top percentile has an actual observed cancellation rate of 55.1% and the overall cancellation rate is 16%; the lift in the top percentile is 55.1/16, or 3.44. Similarly, if the actual observed cancellation rate is 30.67% in the second percentile, then the lift in the second percentile is 30.67/16.0, or 1.92. These lift numbers provide a good measure of the ability of the estimated model to predict which customers will cancel and which will not.

The **Regression** node calculates the lift for the Training and Validation data sets. The partition of the data into Training, Validation, and Test is done by the **Data Partition** node, which is the second node in the process flow diagram shown in Display 3.21.

In addition to calculating the lift for each percentile separately, the **Regression** node calculates the cumulative lift. The cumulative lift calculated above for the first percentile is 3.44. The cumulative lift at the second percentile is the combined lift of the first and second percentiles. The first and second percentiles together contain 10% of the observations in the data set. If the cancellation rate in the first and second percentiles together is 42.89%, then the cumulative lift at the second percentile is 42.89/16.0, or 2.68. If the cancellation rate for the first, second and third percentiles together is 37.63%, then the cumulative lift at the third percentile is 37.63/16.0, or 2.35. The cumulative lift is calculated for each of the 20 percentiles.

Output 3.11 shows the response rate (which is the cancellation rate or attrition rate in this example), cumulative response rate (cumulative cancellation rate), lift, cumulative lift, and other statistics calculated from the Training data set. Output 3.12 shows these statistics calculated from the Validation data set.

Output 3.11

```
Assessment Score Rankings

Data Role=TRAIN Target Variable=EVENT Target Label=' '
```

Depth	Gain	Lift	Cumulative Lift	% Response	Cumulative % Response	Number of Observations	Mean Posterior Probability
5	258.333	3.58333	3.58333	57.3333	57.3333	300	0.52144
10	190.625	2.22917	2.90625	35.6667	46.5000	300	0.32111
15	152.778	1.77083	2.52778	28.3333	40.4444	300	0.28427
20	118.750	1.16667	2.18750	18.6667	35.0000	300	0.25998
25	103.333	1.41667	2.03333	22.6667	32.5333	300	0.23920
30	98.264	1.72917	1.98264	27.6667	31.7222	300	0.22186
35	90.476	1.43750	1.90476	23.0000	30.4762	300	0.20605
40	83.333	1.33333	1.83333	21.3333	29.3333	300	0.18982
45	73.843	0.97917	1.73843	15.6667	27.8148	300	0.17521
50	67.083	1.06250	1.67083	17.0000	26.7333	300	0.16131
55	58.144	0.68750	1.58144	11.0000	25.3030	300	0.14682
60	52.083	0.85417	1.52083	13.6667	24.3333	300	0.13036
65	45.032	0.60417	1.45032	9.6667	23.2051	300	0.11174
70	37.500	0.39583	1.37500	6.3333	22.0000	300	0.08864
75	29.444	0.16667	1.29444	2.6667	20.7111	300	0.06334
80	22.656	0.20833	1.22656	3.3333	19.6250	300	0.04127
85	16.299	0.14583	1.16299	2.3333	18.6078	300	0.02327
90	10.417	0.10417	1.10417	1.6667	17.6667	300	0.01070
95	5.044	0.08333	1.05044	1.3333	16.8070	300	0.00342
100	0.000	0.04167	1.00000	0.6667	16.0000	300	0.00019

Output 3.12

```
Data Role=VALIDATE Target Variable=EVENT Target Label=' '
```

Depth	Gain	Lift	Cumulative Lift	% Response	Cumulative % Response	Number of Observations	Mean Posterior Probability
5	244.444	3.44444	3.44444	55.1111	55.1111	225	0.50468
10	168.056	1.91667	2.68056	30.6667	42.8889	225	0.31610
15	135.185	1.69444	2.35185	27.1111	37.6296	225	0.28063
20	113.889	1.50000	2.13889	24.0000	34.2222	225	0.25705
25	98.889	1.38889	1.98889	22.2222	31.8222	225	0.23769
30	88.889	1.38889	1.88889	22.2222	30.2222	225	0.21957
35	79.762	1.25000	1.79762	20.0000	28.7619	225	0.20328
40	70.833	1.08333	1.70833	17.3333	27.3333	225	0.18952
45	66.975	1.36111	1.66975	21.7778	26.7160	225	0.17557
50	61.667	1.13889	1.61667	18.2222	25.8667	225	0.16110
55	57.576	1.16667	1.57576	18.6667	25.2121	225	0.14731
60	50.926	0.77778	1.50926	12.4444	24.1481	225	0.13244
65	44.231	0.63889	1.44231	10.2222	23.0769	225	0.11320
70	37.500	0.50000	1.37500	8.0000	22.0000	225	0.08977
75	29.444	0.16667	1.29444	2.6667	20.7111	225	0.06357
80	22.569	0.19444	1.22569	3.1111	19.6111	225	0.04081
85	16.176	0.13889	1.16176	2.2222	18.5882	225	0.02286
90	10.031	0.05556	1.10031	0.8889	17.6049	225	0.00998
95	4.678	0.08333	1.04678	1.3333	16.7485	225	0.00286
100	0.000	0.11111	1.00000	1.7778	16.0000	225	0.00018

You can find Output 3.11 and 3.12 if you scroll down the output sub-window of the Results window of the **Regression** node shown in the upper part of Display 3.21.

Display 3.25 shows the graphs of the cumulative lift provided by the model when it is applied to the Training data set (see Output 3.11) and to the Validation data set (see Output 3.12.) These charts show the predictive performance of the logistic regression that was based on the best variables selected from each cluster.

Display 3.25

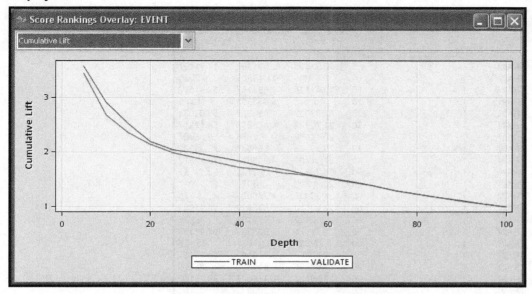

Another measure of the model's power to discriminate between the bad (cancellations) and the good (no cancellation) is the Cumulative % Captured Response. The Cumulative % Captured Response at a given percentile is the percentage of observed cancellations (events) in that percentile and in all the preceding percentiles. In other words, it is the number of cancellations observed in that percentile and in all the preceding percentiles combined, divided by the total number of cancellations in the data set.

Display 3.26 shows the Cumulative Capture Rate by percentile from the training and validation data sets.

Display 3.26

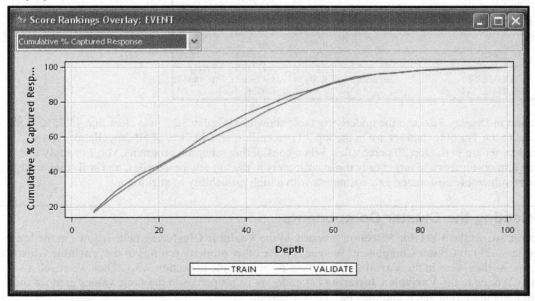

By selecting Cumulative % Captured Response (see Display 3.26) chart in the Results window and clicking the table icon, you can see the table that shows the numbers behind the chart in Display 3.26. A partial view of this table is shown in Display 3.27.

Display 3.27

Target Variable	Data Role	Event	Depth	Cumulative % Response	% Captured Response	Cumulative % Captured Response
EVENT	TRAIN	1	5	57.33333	17.91667	17.91667
EVENT	TRAIN	1	10	46.5	11.14583	29.0625
EVENT	TRAIN	1	15	40.44444	8.854167	37.91667
EVENT	TRAIN	1	20	35	5.833333	43.75
EVENT	TRAIN	1	25	32.53333	7.083333	50.83333
EVENT	TRAIN	1	30	31.72222	8.645833	59.47917
EVENT	TRAIN	1	35	30.47619	7.1875	66.66667
EVENT	TRAIN	1	40	29.33333	6.666667	73.33333
EVENT	TRAIN	1	45	27.81481	4.895833	78.22917
EVENT	TRAIN	1	50	26.73333	5.3125	83.54167
EVENT	TRAIN	1	55	25.30303	3.4375	86.97917
EVENT	TRAIN	1	60	24.33333	4.270833	91.25
EVENT	TRAIN	1	65	23.20513	3.020833	94.27083
EVENT	TRAIN	1	70	22	1.979167	96.25
EVENT	TRAIN	1	75	20.71111	0.833333	97.08333
EVENT	TRAIN	1	80	19.625	1.041667	98.125
EVENT	TRAIN	1	85	18.60784	0.729167	98.85417
EVENT	TRAIN	1	90	17.66667	0.520833	99.375
EVENT	TRAIN	1	95	16.80702	0.416667	99.79167
EVENT	TRAIN	1	100	16	0.208333	100
EVENT	VALIDATE	1	5	55.11111	17.22222	17.22222
EVENT	VALIDATE	1	10	42.88889	9.583333	26.80556
EVENT	VALIDATE	1	15	37.62963	8.472222	35.27778
EVENT	VALIDATE	1	20	34.22222	7.5	42.77778
EVENT	VALIDATE	1	25	31.82222	6.944444	49.72222
EVENT	VALIDATE	1	30	30.22222	6.944444	56.66667
EVENT	VALIDATE	1	35	28.7619	6.25	62.91667
EVENT	VALIDATE	1	40	27.33333	5.416667	68.33333
EVENT	VALIDATE	1	45	26.71605	6.805556	75.13889
EVENT	VALIDATE	1	50	25.86667	5.694444	80.83333
EVENT	VALIDATE	1	55	25.21212	5.833333	86.66667
EVENT	VALIDATE	1	60	24.14815	3.888889	90.55556
EVENT	VALIDATE	1	65	23.07692	3.194444	93.75
EVENT	VALIDATE	1	70	22	2.5	96.25
EVENT	VALIDATE	1	75	20.71111	0.833333	97.08333
EVENT	VALIDATE	1	80	19.61111	0.972222	98.05556
EVENT	VALIDATE	1	85	18.58824	0.694444	98.75
EVENT	VALIDATE	1	90	17.60494	0.277778	99.02778
EVENT	VALIDATE	1	95	16.74854	0.416667	99.44444
EVENT	VALIDATE	1	100	16	0.555556	100

From the chart in Display 3.26 and the underlying table shown in Display 3.27, it is clear that 43.75% of all cancellations in the Training data set are in the top 20 percentiles, and 42.78% of all cancellations in the Validation data set are in the top 20 percentiles. When considering retention programs, you can apply this model to current customers and target only those customers in the top few percentiles, and still reach a large proportion of vulnerable customers, i.e., customers with a high probability of attrition.

3.3.2 Selecting the Cluster Components

In this section, we set the **Variable Selection** property of the **Variable Clustering** node (shown in the lower part of Display 3.21) to Cluster Component. We keep all the other property settings of the **Variable Clustering** node the same as they were in the **Variable Clustering** node example in section 3.3.1. Then we attach a **Regression** (2) node to the **Variable Clustering** (2) node, set its properties to the same values used for the **Regression** node in section 3.3.1, and then run the whole path.

After running the **Variable Clustering** node, all the components (denoted by Clus1, Clus2…, Clus26) from the 26 clusters plus any nominal variables that were excluded from the clustering process were passed to the **Regression** node. The stepwise selection procedure of the **Regression** node resulted in the selection of a logistic regression equation with some of the cluster components, as can be seen in Output 3.13.

Output 3.13

```
                    Analysis of Maximum Likelihood Estimates

                           Standard       Wald                Standardized
Parameter    DF   Estimate   Error    Chi-Square   Pr > ChiSq   Estimate   Exp(Est)

Intercept    1    -2.7337    0.1027     708.40      <.0001                  0.065
Clus1        1    -0.2049    0.0571      12.86       0.0003     -0.1129     0.815
Clus10       1     0.1873    0.0492      14.48       0.0001      0.1033     1.206
Clus12       1     0.4134    0.0667      38.42      <.0001       0.2279     1.512
Clus13       1     0.3230    0.0417      59.96      <.0001       0.1781     1.381
Clus14       1    -0.3216    0.0470      46.83      <.0001      -0.1773     0.725
Clus15       1     0.1504    0.0499       9.08       0.0026      0.0829     1.162
Clus16       1     0.2361    0.0416      32.14      <.0001       0.1302     1.266
Clus18       1    -0.1314    0.0404      10.60       0.0011     -0.0725     0.877
Clus20       1     0.1881    0.0371      25.74      <.0001       0.1037     1.207
Clus21       1    -0.4272    0.0440      94.17      <.0001      -0.2356     0.652
Clus23       1     0.2035    0.0509      16.00      <.0001       0.1122     1.226
Clus25       1    -0.1118    0.0387       8.33       0.0039     -0.0617     0.894
Clus3        1    -1.1596    0.0925     157.17      <.0001      -0.6393     0.314
Clus4        1     0.1718    0.0517      11.03       0.0009      0.0947     1.187
Clus5        1    -0.1626    0.0498      10.67       0.0011     -0.0896     0.850
Clus8        1    -3.3794    0.3315     103.92      <.0001      -1.8632     0.034
Clus9        1     0.0893    0.0433       4.26       0.0391      0.0492     1.093
```

In order to compare the two models, i.e. the model based on the Best Variables option and the model based on the Cluster Component option, a **Model Comparison** node is attached at the end of the process flow (see Display 3.21). When we run the **Model Comparison** node, it calculates the cumulative lifts and cumulative capture rates by percentile for both the models.

Display 3.28 shows the Cumulative Lift charts and Display 3.29 shows the Cumulative Capture Rates.

Display 3.28

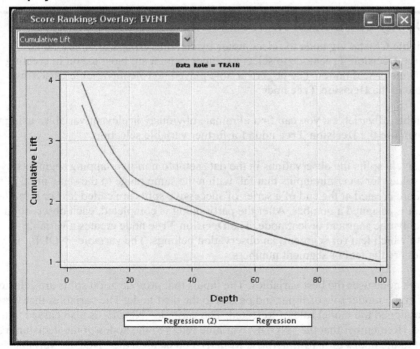

In Display 3.28, the line labeled Regression refers to the logistic regression based on the Best Variables selection, and Regression (2) refers to the logistic regression based on the cluster components. To make it easier, let us call the logistic regression based on the Best Variables option Model 1, and the logistic regression based on the Cluster Components option Model 2.

Display 3.29

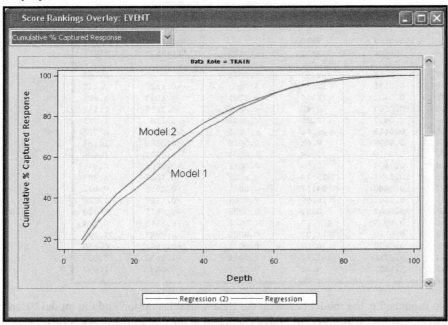

Display 3.29 shows that the cumulative capture rate of Model 2 is above that of Model 1 at each percentile. But the difference between these two models is very small. Since Model 1 uses actual inputs rather than the cluster components that were constructed from the actual inputs, it is easier to interpret the coefficients of Model 1 than those of Model 2.

3.4 Variable Selection Using the Decision Tree Node

In Section 3.3 we used the **Variable Clustering** node to select inputs by creating *groups of inputs* called clusters. In this section, we use the **Decision Tree** node to select the inputs which are most useful in creating *groups of customers* called segments or leaf nodes. The target variable played no role in variable clustering, whereas it plays an important role in the **Decision Tree** node.

When you have a very large number of variables, you can first eliminate obviously irrelevant variables using the **Variable Selection** node, and then use the **Decision Tree** node for further variable selection.

The **Decision Tree** node successively splits the observations in the data set into non-overlapping segments by assigning all observations with values for a certain inputs that fall within the same range to the same node of the tree. The ultimate segments that are created at the end of a series of successive splits are called leaf or terminal nodes. Each leaf node or segment is assigned a number. After the partitioning is completed, each observation in the data set belongs to one and only one segment or leaf node. The **Decision Tree** node creates a variable named _NODE_ that indicates to which leaf (or segment) an observation belongs. The variable _NODE_ is a categorical variable whose values are the leaf or segment numbers.

For creating the segments, the tree node uses the best variables. The inputs that provide good splits are selected. These selected variables are given the model role of Input and passed to the next node. The variables that were not used in the partitioning are assigned the model role of Rejected. The variable _NODE_ is also passed to the next SAS Enterprise Miner node. (Remember that the _NODE_ variable refers to the nodes of the decision tree that you have just estimated, not to the nodes of SAS Enterprise Miner.) You can use the _NODE_ variable as an additional input or use it only for identifying the segment.

In this section, I show how you can use the variables selected by the **Decision Tree** node in the next node, which is again the **Regression** node in this example. We develop two Logistic Regression models. In the first model, we use the variables selected by the **Decision Tree** node, but not the special variable _NODE_, as an input. This is Model A. In the second model, we use the special variable _NODE_ as in input, in addition to all the variables selected by the **Decision Tree** node. This is Model B. Then, as in Section 3.3, we compare these two models using the **Model Comparison** node.

Display 3.30 shows the process flow for this example. Model A is created by the sequence of nodes in the upper segment of Display 3.30, and Model B uses the nodes on the bottom.

Display 3.30

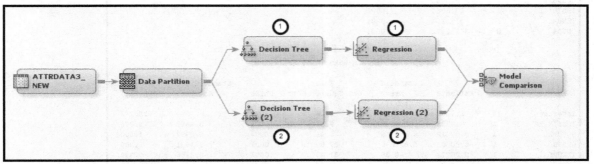

For Model A, the **Leaf Role** property of the **Decision Tree** node is set to Segment. This setting results in treating the variable _NODE_ as a segment variable (an identifier of the node), and not as a variable that can be used as an input in the subsequent nodes. The **Regression** node that follows the **Decision Tree** node will not use the variable _NODE_ as an input.

Output 3.14 shows the equation estimated without using the variable _NODE_ (Model A).

Output 3.14

Analysis of Maximum Likelihood Estimates							
Parameter	DF	Estimate	Standard Error	Wald Chi-Square	Pr > ChiSq	Standardized Estimate	Exp(Est)
Intercept	1	6.5389	0.9661	45.81	<.0001		691.541
FRQFIN12M	1	0.0819	0.0107	58.38	<.0001	0.2048	1.085
GOODBAL	1	-0.00079	0.000063	157.94	<.0001	-1.2249	0.999
MKTVAL	1	-0.00007	5.111E-6	174.15	<.0001	-0.4593	1.000
OBSID	1	-0.00005	4.677E-6	133.64	<.0001	-0.2874	1.000
PPSCORE8	1	-0.0741	0.0118	39.63	<.0001	-0.2032	0.929
TENURE	1	-0.00601	0.000998	36.28	<.0001	-0.1681	0.994

For Model B, the **Leaf Role** property of the **Decision Tree** node is set to Input. This setting results in treating the variable _NODE_ as an input. The **Regression** node that follows the **Decision Tree** node will include the variable _NODE_ as a nominal variable in selecting variables and estimating the logistic regression equation.

Output 3.15 shows the equation estimated using the variable _NODE_ (Model B).

Output 3.15

```
        Type 3 Analysis of Effects

                        Wald
Effect          DF    Chi-Square    Pr > ChiSq

FRQFIN12M        1      9.4532        0.0021
GOODBAL          1      8.4080        0.0037
OBSID            1     27.6885        <.0001
TENURE           1     31.4264        <.0001
_NODE_           8   4561.2972        <.0001


              Analysis of Maximum Likelihood Estimates

                          Standard      Wald                    Standardized
Parameter       DF   Estimate   Error  Chi-Square  Pr > ChiSq    Estimate   Exp(Est)

Intercept        1    2.7243   5.3384     0.26       0.6098                  15.245
FRQFIN12M        1    0.0736   0.0240     9.45       0.0021       0.1842      1.076
GOODBAL          1   -0.00019  0.000066   8.41       0.0037      -0.2997      1.000
OBSID            1   -0.00003  5.831E-6  27.69       <.0001      -0.1631      1.000
TENURE           1   -0.00774  0.00138   31.43       <.0001      -0.2166      0.992
_NODE_    5      1   11.4633  42.6591     0.07       0.7881                 999.000
_NODE_    7      1   -6.0923   5.3376     1.30       0.2537                   0.002
_NODE_    9      1   -4.6324   5.3470     0.75       0.3863                   0.010
_NODE_   10      1   -0.2015   5.3627     0.00       0.9700                   0.818
_NODE_   14      1   -0.3497   5.3493     0.00       0.9479                   0.705
_NODE_   19      1   -3.2252   5.3372     0.37       0.5457                   0.040
_NODE_   24      1   -0.9099   5.3721     0.03       0.8655                   0.403
_NODE_   25      1   -4.0584   5.3763     0.57       0.4503                   0.017
_NODE_   30      1   11.6492      .          .          .            .       999.000
```

In both models, the properties of the **Regression** nodes are set such that they make a further selection of variables from the sets of variables that are passed to them from the **Decision Tree** nodes that precede them. Each **Regression** node selects, from the sequence of models it creates in the Stepwise selection process, the model that has the smallest validation error. These settings for the properties are shown in Display 3.31.

Display 3.31

Model Selection	
Selection Model	Stepwise
Selection Criterion	Validation Error
Use Selection Defaults	Yes
Selection Options	[...]

Display 3.32 shows the Captured % Response for both models.

Display 3.32

It appears that Model B, which makes use of the special variable _NODE_, is slightly better than Model A in the sense that the Cumulative %Captured Cancellations curve of Model B is higher than that for Model A.

3.5 Transformation of Variables

3.5.1 Transform Variables Node

The **Transform Variables** node can make a variety of transformations of interval-scaled variables. These were described in Section 2.9.6.1. In addition, there are two types of transformations for the categorical (class) variables.

- Group Rare Levels
- Dummy Indicators Transformation for the categories (see Section 2.9.6.2)

If you want to transform variables prior to variable selection, you can set up the path as **Input Data** node → **Data Partition** node → **Transform Variables** node → **Variable Selection** node, as shown in the upper path in Display 3.33. If you do not want to reject inputs on the basis of linear relationship alone, you can use this process to capture non-linear relationships between the inputs and the target. However, with large data sets this process may become quite tedious.

When there are a large number of inputs (2000 or more), it may be more practical to eliminate irrelevant variables first, and then define transformations on important variables only. The process flow for this is **Input Data** node → **Data Partition** node → **Variable Selection** node→ **Transform Variables** node (shown in the lower path of Display 3.33). This section examines the results of both of these sequences with alternative transformations using a small data set. The process flow shown in Display 3.33 does not include the **Impute** node, but in any given project you might need to use this node before applying transformations or selecting variables.

Display 3.33

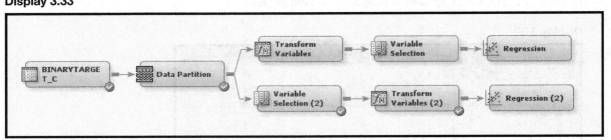

To demonstrate the transformations, I have selected a small number of inputs and the binary target RESP, which represents the response to a solicitation by a hypothetical bank. Here are these inputs, shown in Display 3.34:

VAR1N	number of on-line inquiries by the customer per month
VAR2O	number of ATM deposits during the past three months
VAR3N	number of ATM withdrawals during the past three months
VAR4C	lifestyle indicator
VAR5O	number of asset types
VAR6C	an indicator of wealth
VAR7C	life stage segment
VAR8O	total number of bank products owned by the customer
CREDSCORE	credit score
RESP	target variable; indicates response to an offer

Display 3.34

Name	Role	Level	Type	Number of Levels	Drop	Report
VAR1N	Input	Interval	Numeric	,	No	No
VAR2O	Input	Nominal	Numeric	,	No	No
VAR3N	Input	Interval	Numeric	,	No	No
VAR4C	Rejected	Nominal	Character	68	No	No
VAR5O	Input	Nominal	Numeric	,	No	No
VAR6C	Rejected	Nominal	Character	121	No	No
VAR7C	Rejected	Nominal	Character	43	No	No
VAR8O	Input	Nominal	Numeric	,	No	No
credscore	Input	Interval	Numeric	,	No	No
resp	Target	Binary	Numeric	2	No	No

3.5.2 Transformation before Variable Selection

In the upper path of Display 3.33, the variables are transformed using the **Transform Variables** node and then passed to the **Variable Selection** node.

In the **Transform Variables** node, I selected the Maximum Normal transformation as the default method for all continuous variables by setting the **Interval Inputs** property to Maximum Normal. The node generates different power transformations of the variable and picks the transformation that transforms the input into a new variable whose distribution is closest to normal, bell-shaped, distribution. (Section 2.9.6.1 gives the details of this process). For class inputs, I selected the Dummy Indicators transformation by setting the **Class Inputs** property to Dummy Indicators.

This data set has three interval-scaled variables: VAR1N, VAR3N, and credscore. If a numeric variable has only a few values, then SAS Enterprise Miner sets its measurement level to Nominal if you select **Advanced** for the Metadata Advisor Option when you created the data source (see Displays 2.10 and 2.11). VAR2O, VAR5O, and VAR8O are numeric variables and are set to Nominal by SAS Enterprise Miner. For the variables set to interval-scaled, the node searches and finds the Maximum Normal, as I specified earlier. For the rest of the variables, which are now nominal scaled, it creates a dummy variable for each level. The transformed variables are then passed to the next node in the process flow. In this example, the **Variable Selection** node follows the **Transform Variables** node.

By running the **Transform Variables** node and opening the Results window, you can see the Transformation Statistics. Display 3.35 shows a partial view of the Transformation Statistics table.

Display 3.35

Source	Method	Variable Name	Formula	Number of Levels
Input	Original	VAR2O		8
Input	Original	VAR3N		
Input	Original	VAR5O		8
Input	Original	VAR8O		7
Input	Original	credscore		
Output	Computed	LOG_credscore	log(max(credscore-417, 0.0)/9582 + 1)	
Output	Computed	SQRT_VAR3N	sqrt(max(VAR3N-1, 0.0)/18)	
Output	Computed	TI_VAR2O1	Dummy	2
Output	Computed	TI_VAR2O2	Dummy	2
Output	Computed	TI_VAR2O3	Dummy	2
Output	Computed	TI_VAR2O4	Dummy	2
Output	Computed	TI_VAR2O5	Dummy	2
Output	Computed	TI_VAR2O6	Dummy	2
Output	Computed	TI_VAR2O7	Dummy	2
Output	Computed	TI_VAR2O8	Dummy	2

When you open the Variables window of the **Variable Selection** node, you can see the variables passed to it by the **Transform Variables** node. A partial listing of these variables is shown in Display 3.36.

Display 3.36

Name	Use	Role	Level	Type	Format	Informat
LOG_credscore	Default	Input	Interval	Numeric		
SQRT_VAR3N	Default	Input	Interval	Numeric		
TI_VAR2O1	Default	Input	Binary	Numeric		
TI_VAR2O2	Default	Input	Binary	Numeric		
TI_VAR2O3	Default	Input	Binary	Numeric		
TI_VAR2O4	Default	Input	Binary	Numeric		
TI_VAR2O5	Default	Input	Binary	Numeric		
TI_VAR2O6	Default	Input	Binary	Numeric		
TI_VAR2O7	Default	Input	Binary	Numeric		
TI_VAR2O8	Default	Input	Binary	Numeric		
TI_VAR5O1	Default	Input	Binary	Numeric		
TI_VAR5O2	Default	Input	Binary	Numeric		
TI_VAR5O3	Default	Input	Binary	Numeric		
TI_VAR5O4	Default	Input	Binary	Numeric		
TI_VAR5O5	Default	Input	Binary	Numeric		
TI_VAR5O6	Default	Input	Binary	Numeric		
TI_VAR5O7	Default	Input	Binary	Numeric		
TI_VAR5O8	Default	Input	Binary	Numeric		

You can see that a log transformation of the credscore variable and a square root transformation of VAR3N made them closest to the normal distribution. No transformation could be found for VAR1N; hence, it is passed to the next node as is. If a transformation is made for a variable, then by default only the transformed variable is passed to the next node, not the original variable. If you set the **Hide** property to No and the **Reject** property to No, then both the original and transformed variables are passed to the next node that has the role set to Input.

Some of the dummy variables created for the class variable levels are also shown in Display 3.36. For example, VAR2O has eight categories. They are 0, 1, 2, 3, 4, 5, 6, and 8. Eight dummy variables (binary variables) are created (TI_VAR2O1 through TI_VAR2O8). The variable VAR2O has some missing values but no separate category is created for it. Therefore, no dummy variable is created for the missing values. The dummy variable TI_VAR2O1 takes the value of 1 if VAR2O is 0, and 0 otherwise. TI_VAR2O2 takes the value of 1 if VAR2O is 1 and 0 otherwise. TI_VAR2O3 takes the value of 1 if VAR2O is 2 and 0 otherwise, etc.

The variables selected by the **Variable Selection** node are shown in Output 3.16. Note that only three variables were selected, including two group variables created by the transformation. The **Variable Selection** node created the variable AOV16_SQRT_VAR3N because I set the **Use AOV16 Variables** property of the **Variable Selection** node to Yes.

Output 3.16

```
The DMINE Procedure


                          Effects Chosen for Target: resp


                                                                Sum of      Error Mean
Effect                   DF    R-Square    F Value    p-Value    Squares        Square

AOV16: SQRT_VAR3N        13    0.010966    6.088148    <.0001    7.057081      0.089165
Group: VAR80              2    0.001515    5.472450    0.0042    0.974685      0.089054
Group: VAR50             3    0.000809    1.948886    0.1194    0.520460      0.089018
```

3.5.3 Transformation after Variable Selection

If you have a very large number of variables in your data set, you may want to first eliminate those that have extremely low linear correlation with the target. To achieve this, I set the values of the **Minimum R-Square** and **Stop R-Square** properties of the **Variable Selection** node to 0.005 and 0.0005, respectively.

The lower segment of Display 3.33 shows the process flow for transforming variables after variable selection. The **Transform Variables** (2) node follows the **Variable Selection** (2) node and precedes the **Regression** node.

After running the **Variable Selection** (2) node, open the Results window. In the Output sub-window, you can see the variables selected in Step 1 and Step 2. The variables selected in Step 1 of the variable selection are shown in Output 3.17

Output 3.17

```
The DMINE Procedure

            R-Squares for Target Variable: resp

Effect                DF         R-Square

AOV16: VAR3N          14         0.009215
Class: VAR80           6         0.008606
Group: VAR80           2         0.008591
Class: VAR20           8         0.008559
Group: VAR20           4         0.008554
Class: VAR50           8         0.008393
Group: VAR50           3         0.008282
Var:   VAR3N           1         0.005742
AOV16: VAR1N          15         0.004799    R2 < MINR2
Var:   VAR1N           1         0.002278    R2 < MINR2
AOV16: credscore       3         0.000198    R2 < MINR2
Var:   credscore       1    0.000078951       R2 < MINR2
```

Output 3.18 shows the variables selected in Step 2.

Output 3.18

```
The DMINE Procedure


                         Effects Chosen for Target: resp


                                                               Sum of      Error Mean
Effect               DF      R-Square      F Value    p-Value   Squares        Square

AOV16: VAR3N         14      0.009215     4.741433    <.0001    5.930119      0.089336
Group: VAR80          2      0.002902    10.479059    <.0001    1.867351      0.089099
Group: VAR50          3      0.001040     2.505555    0.0572    0.669305      0.089043
Group: VAR20          4      0.000554     1.000345    0.4059    0.356294      0.089043
```

Display 3.37 shows the variables passed by the **Variable Selection** (2) node and the transformations done by the **Transform Variables** (2) node.

Display 3.37

Source	Method	Variable Name	Formula	Label
Input	Original	AOV16_VAR3N		
Input	Original	G_VAR2O		Grouped Levels for VAR2O
Input	Original	G_VAR5O		Grouped Levels for VAR5O
Input	Original	G_VAR8O		Grouped Levels for VAR8O
Output	Computed	TI_AOV16_VAR3N1	Dummy	AOV16_VAR3N:1
Output	Computed	TI_AOV16_VAR3N10	Dummy	AOV16_VAR3N:10
Output	Computed	TI_AOV16_VAR3N11	Dummy	AOV16_VAR3N:11
Output	Computed	TI_AOV16_VAR3N12	Dummy	AOV16_VAR3N:12
Output	Computed	TI_AOV16_VAR3N13	Dummy	AOV16_VAR3N:13
Output	Computed	TI_AOV16_VAR3N14	Dummy	AOV16_VAR3N:15
Output	Computed	TI_AOV16_VAR3N15	Dummy	AOV16_VAR3N:16
Output	Computed	TI_AOV16_VAR3N2	Dummy	AOV16_VAR3N:2
Output	Computed	TI_AOV16_VAR3N3	Dummy	AOV16_VAR3N:3
Output	Computed	TI_AOV16_VAR3N4	Dummy	AOV16_VAR3N:4
Output	Computed	TI_AOV16_VAR3N5	Dummy	AOV16_VAR3N:5
Output	Computed	TI_AOV16_VAR3N6	Dummy	AOV16_VAR3N:6
Output	Computed	TI_AOV16_VAR3N7	Dummy	AOV16_VAR3N:7
Output	Computed	TI_AOV16_VAR3N8	Dummy	AOV16_VAR3N:8
Output	Computed	TI_AOV16_VAR3N9	Dummy	AOV16_VAR3N:9
Output	Computed	TI_G_VAR2O1	Dummy	G_VAR2O:0
Output	Computed	TI_G_VAR2O2	Dummy	G_VAR2O:1
Output	Computed	TI_G_VAR2O3	Dummy	G_VAR2O:2
Output	Computed	TI_G_VAR2O4	Dummy	G_VAR2O:3
Output	Computed	TI_G_VAR2O5	Dummy	G_VAR2O:4
Output	Computed	TI_G_VAR5O1	Dummy	G_VAR5O:0
Output	Computed	TI_G_VAR5O2	Dummy	G_VAR5O:1
Output	Computed	TI_G_VAR5O3	Dummy	G_VAR5O:2
Output	Computed	TI_G_VAR5O4	Dummy	G_VAR5O:3
Output	Computed	TI_G_VAR8O1	Dummy	G_VAR8O:0
Output	Computed	TI_G_VAR8O2	Dummy	G_VAR8O:1
Output	Computed	TI_G_VAR8O3	Dummy	G_VAR8O:2

Display 3.37 shows that the **Variable Selection** node created both the AOV16 variables and grouped class variables (indicated by the prefix G_), while the **Transform Variables** node created the dummy variables (indicated by the prefix TI) for each level of these AOV16 and grouped class variables.

Display 3.38 shows the variables passed to the **Regression** node.

Display 3.38

Because I set the **Hide** and **Reject** properties of the **Transform Variables** node to Yes, the original values of the transformed variables are not passed to the next node.

3.5.4 Passing More Than One Type of Transformation for Each Interval Input to the Next Node

3.5.4.1 Passing Two Types of Transformations Using the Merge Node

You can pass two types of transformations for each interval input to the **Variable Selection** or **Regression** node by applying two instances of the **Transform Variables** node. In this example, I pass Maximum Normal and Optimal Binning transformations for each interval input. In the first instance (represented by **Transform Variables** in Display 3.39) of the **Transform Variables** node, I transform all the interval inputs using the Maximum Normal transformation by setting the **Interval Inputs** property to Maximum Normal. In the second instance (**Transform Variables (2)** in Display 3.39), I transform all interval inputs using the Optimal Binning transformation by setting the **Interval Inputs** property to Optimal Binning.

I also set the **Node ID** property for each instance of the **Transform Variable** node. The **Node ID** for the first instance is Trans, and the **Node ID** for the second instance is Trans2.

I use the **Merge** node to merge the data sets created by the **Transform Variables** nodes so that I can pass both types of transformation to the **Regression** node for final selection of the variables considering both types of transformation.

Display 3.39

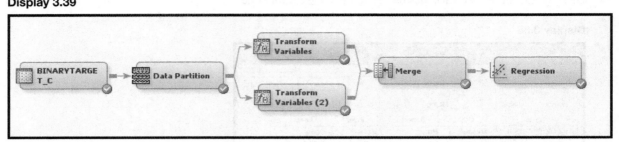

Display 3.40 shows the transformations done by the **Transform Variables** node in the upper segment of Display 3.39.

Display 3.40

Source	Method	Variable Name	Formula
Input	Original	VAR3N	
Input	Original	credscore	
Output	Computed	LOG_credscore	log(max(credscore-417, 0.0)/9582 + 1)
Output	Computed	SQRT_VAR3N	sqrt(max(VAR3N-1, 0.0)/18)

Display 3.41 shows the transformations done by the **Transform Variables** node in the lower segment of Display 3.39.

Display 3.41

Source	Method	Variable Name	Formula	Number of Levels
Input	Original	VAR1N		.
Input	Original	VAR3N		.
Input	Original	credscore		.
Output	Computed	OPT_VAR1N	Optimal Binning(4)	2
Output	Computed	OPT_VAR3N	Optimal Binning(4)	4
Output	Computed	OPT_credscore	Optimal Binning(4)	4

Display 3.42 shows the property settings of the **Merge** node used to combine the data sets created by the **Transform Variables** nodes **Trans** and **Trans2**.

Display 3.42

Property	Value
General	
Node ID	Merge
Imported Data	...
Exported Data	...
Notes	...
Train	
Variables	...
Merging	One-to-One
By Ordering	...
Overwrite Variables	Yes
Variables Group	
Segment	No
Assess	No
Classification	No
Predicted or Posterior	No
Residual	No
Status	
Create Time	12/1/12 5:20 PM

The **Merging** property is set to One-to-One because there is a one-to-one correspondence between the records of the data sets merged. Note that the **Node ID** property of the first instance of the **Transform Variables** node is **Trans**, and that of the second instance is **Trans2**.

The Variables window of the **Regression** node is shown in Display 3.43.

Display 3.43

Name	Use	Report	Role	Level	Type	Format	Informat
LOG_credscore	Default	No	Input	Interval	Numeric		
OPT_VAR1N	Default	No	Input	Nominal	Character		
OPT_VAR3N	Default	No	Input	Nominal	Character		
OPT_credscore	Default	No	Input	Nominal	Character		
SQRT_VAR3N	Default	No	Input	Interval	Numeric		
VAR1N	Default	No	Input	Interval	Numeric	2.0	2.0
VAR2O	Default	No	Input	Nominal	Numeric		
VAR3N	Default	No	Input	Interval	Numeric		
VAR4C	Default	No	Rejected	Nominal	Character		
VAR5O	Default	No	Input	Nominal	Numeric		
VAR6C	Default	No	Rejected	Nominal	Character		
VAR7C	Default	No	Rejected	Nominal	Character		
VAR8O	Default	No	Input	Nominal	Numeric		
credscore	Default	No	Input	Interval	Numeric	4.0	4.0
resp	Yes	No	Target	Binary	Numeric		

The transformations done by both the **Transform Variables** nodes are now available to the **Regression** node. In Display 3.43 you can see that the variable credscore has two types of transformation. The first one is LOG_credscore, which is created by transforming the variable credscore using the Maximum Normal method by the **Transform Variables** node **Trans**. The second transformation is OPT_credscore, created by the second instance of the **Transform Variables** node **Trans2**. In addition, the original interval variable credscore is also passed to the **Regression** node with its role assigned to Input since I set the **Hide** and **Reject** properties to No in both **Trans** and **Trans2**.

We can also see that the variable VAR1N has only one kind of transformation, and that it is an Optimal Binning transformation. The Maximum Normal method did not produce any transformation of VAR1N because the original variable is closer to normal than any power transformation of the original.

Output 3.19 shows the variables selected by the **Regression** node. Variable selection by the **Regression** node is covered in Chapter 6.

Output 3.19

```
            Type 3 Analysis of Effects

                              Wald
Effect              DF    Chi-Square    Pr > ChiSq

OPT_VAR1N            1        4.8127       0.0283
OPT_VAR3N           3       11.5980       0.0089
OPT_credscore       3       22.4899       <.0001
```

Note that VAR80 is a nominal variable. The **Class Inputs** property for this case was set to None, so the variable was not transformed.

3.5.4.2 Multiple Transformations using the Multiple Method Property

You can send more than one type of transformation for each interval input to the **Variable Selection** node or the **Regression** node by setting the **Interval Inputs** property to Multiple. I set the **Interval Inputs** property to Multiple, the **Class Inputs** property to Dummy Indicators, the **Hide** property to No, and the **Reject** property to No in the **Transform Variables** node shown in Display 3.44.

Display 3.44

If you click [...] located on the right of Variables property of the **Regression** node, you can see a list of the transformed and original variables passed to the **Regression** node. The inputs selected by the Stepwise Selection method of the **Regression** node are shown in Output 3.20.

Output 3.20

```
            Type 3 Analysis of Effects

                              Wald
Effect              DF    Chi-Square    Pr > ChiSq

OPT_VAR3N           3       10.0145       0.0184
OPT_credscore       3       23.7712       <.0001
TI_VAR802           1        6.5693       0.0104
```

Note that TI_VAR802 shown in Output 3.20 is a dummy variable created from the variable VAR80.

3.5.5 Saving and Exporting the Code Generated by the Transform Variables Node

You can save and export the SAS code generated by the **Transform Variables** node. Simply open the **Results** window and select **View** → **Scoring** → **SAS Code** from the menu bar.

To save the SAS code, select **File** → **Save As,** and then select a name for the saved file. A partial listing of the SAS code saved in my example is shown in Display 3.45.

Display 3.45

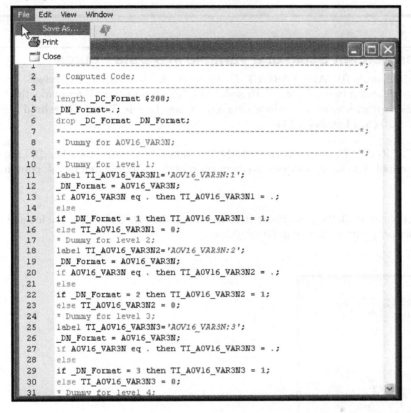

3.6 Summary

- This chapter showed how the **Variable Selection** node selects categorical and continuous variables to be used as inputs in the modeling process when the target is binary or continuous.

- Two criteria are available for variable selection when the target is binary: R-Square and Chi-Square.

- The R-Square criterion can be used with binary as well as continuous targets, but the Chi-Square criterion can be used only with binary targets.

- The **Variable Selection** node transforms inputs into binned variables, called AOV16 variables, which are useful in capturing non-linear relationships between inputs and targets.

- The **Variable Selection** node also simplifies categorical variables by collapsing the categories.

- The **Variable Clustering** node can be used for variable selection by grouping the variables that look similar in the data set, and selecting a representative variable from each group. It can also be used to combine variables that look similar.

- The **Decision Tree** node can also be used for variable selection. However, if you have a large number of inputs, you should make a preliminary selection by using the **Variable Selection** node, and then use the **Decision Tree node** for further selection.

- The **Transform Variables** node offers a wide range of choices for transformations, including multiple transformations of an input, and the option of creating dummy variables for the categories of categorical variables.

- If there are a large number of inputs, you can use the **Variable Selection** node to first make a preliminary variable selection, and then do transformations of the selected variables. If the number of inputs is relatively small, then you can transform all of the inputs prior to variable selection.

- The code generated by the **Variable Selection** node and the **Transform Variables** node can be saved and exported.

3.7 Appendix to Chapter 3

3.7.1 Changing the Measurement Scale of a Variable in a Data Source

A data source is created with the SAS table BINARYTARGET_D. To create the data source, I selected **Advanced** for the Metadata Advisor Option (see Displays 2.10 and 2.11). SAS Enterprise Miner marks the interval variables with less than 20 distinct levels as Nominal, unless the **Class Levels Count Threshold** property is set to a value other than 20 (See Display 2.11).

In the example data set BINARYTARGET_D, the variable VAR3N is a numeric variable, and it has 18 unique values. Hence, it is marked as Nominal. To change the measurement scale of this variable to interval, complete the following steps.

Under Data Sources in the Project panel, click the table name BINARYTARGET_D. Then click ⊡ located in the Value column of the **Variables** property, as shown in Display 3.46.

Display 3.46

Property	Value
ID	binarytargetd
Name	BINARYTARGET_D
Variables	⊡
Decisions	⊡
Role	Raw
Notes	⊡
Library	THEBOOK
Table	BINARYTARGET_D
Sample Data Set	
Size Type	
Sample Size	
Type	DATA
No. Obs	17880
No. Cols	10
No. Bytes	1303552
Segment	
Created By	sasdemo
Create Date	12/7/12 8:44 PM
Modified By	sasdemo
Modify Date	12/7/12 8:44 PM

The Variables window appears, as shown in Display 3.47.

Display 3.47

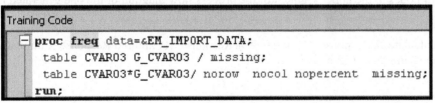

Name	Role	Level	Type	Number of Levels	Percent Missing	Drop
VAR1N	Input	Interval	Numeric	.	40.94519	No
VAR2O	Input	Nominal	Numeric	8	0.956376	No
VAR3N	Input	Nominal	Numeric	18	0.956376	No
VAR4C	Rejected	Nominal	Character	21	0	No
VAR5O	Input	Nominal	Numeric	8	0.083893	No
VAR6C	Rejected	Nominal	Character	20	21.35948	No
VAR7C	Rejected	Nominal	Character	20	23.34288	No
VAR8O	Input	Nominal	Numeric	7	0	No
credscore	Input	Interval	Numeric	.	36.63311	No
resp	Input	Binary	Numeric	2	0	No

For variable VAR3N, click in the Level cell and select **Interval**, and then click **OK**. The measurement scale of the input VAR3N is changed from Nominal to Interval.

For a full discussion of the **Metadata Advisor Options** when creating a data source, see Section 2.5.

3.7.2 SAS Code for Comparing Grouped Categorical Variables with the Ungrouped Variables

CVAR03 has 11 categories, including the "missing" or unlabeled category, as shown in Output 3.4.

The categories of CVAR03 are grouped to create the variable G_CVAR03 with only five categories. You can compare the frequency distribution of the original variable with the grouped variables and see the mapping of the categories of CVAR03 into G_CVAR03 using the SAS code shown in Display 3.48.

Display 3.48

```
Training Code

 proc freq data=&EM_IMPORT_DATA;
   table CVAR03 G_CVAR03 / missing;
   table CVAR03*G_CVAR03/ norow  nocol nopercent  missing;
 run;
```

Exercises

1. Create a data source using the SAS data set Attrdata3_New.
2. Select **Basic** for the Metadata Advisor Option.
3. Set the model role of the variable Event to Target, and set its measurement level to Binary.
4. Open the **Prior Probabilities** tab of the Decision Configuration step and enter the following prior probabilities in the Adjusted Prior column: 0.024 for level 1 and 0.976 for level 0.
5. Create the following process flow diagram.

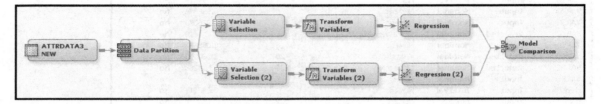

In the **Data Partition** node, allocate 50% of the observations for Training and 50% for Validation.

In the **Variable Selection** node located in the upper segment of the process flow diagram, set the **Target Model** property to R-Square and the **Use AOV16 Variables** property to No.

In the **Transform Variables** node located in the upper segment of the process flow diagram, set the **Interval Inputs** property to Optimal Binning, and set the **Class Inputs** property to Dummy Indicators.

In the **Variable Selection** node located in the lower segment of the process flow diagram, set the **Target Model** property to R-Square and the **Use AOV16 Variables** property to Yes.

In the **Transform Variables** node located in the lower segment of the process flow diagram, set the **Interval Inputs** property to Best, and set the **Class Inputs** property to Dummy Indicators.

In both the **Regression** nodes, set the **Selection Model** property to Stepwise and the **Selection Criterion** to Validation Error.

Run all the nodes in sequence. Then open the Results window of the **Model Comparison** node, examine the Cumulative % Captured Response plot, and get a preliminary assessment of which model is better—the one generated by the sequence of nodes in the upper segment of the process flow diagram or the one generated by the lower segment?

Using the following table, calculate % Response, Cumulative % Response, %Captured Response, Cumulative Captured Response, Lift and Cumulative Lift by Bin/Percentile.

Bin	Percentile	Number Of Responses	Number Of Non-responses	Number Of Observations
1	5	95	281	376
2	10	14	361	375
3	15	12	363	375
4	20	9	366	375
5	25	9	366	375
6	30	8	366	374
7	35	8	367	375
8	40	6	369	375
9	45	5	370	375
10	50	5	370	375
11	55	3	371	374
12	60	1	374	375
13	65	0	374	374
14	70	1	373	374
15	75	1	374	375
16	80	0	375	375
17	85	1	374	375
18	90	0	375	375
19	95	0	375	375
20	100	0	374	374

Note

1. The SAS manuals often use the term "Class" instead of "Categorical." I use these two terms synonymously.

Chapter 4: Building Decision Tree Models to Predict Response and Risk

4.1 Introduction

This chapter shows you how to build decision tree models to predict a categorical target and how to build *regression tree models* to predict a continuous target. Two examples are presented. The first example shows how to build a *decision tree model* to predict *response* to direct mail. In this example, the target variable is binary, taking on the values *response* and *no response*. The second example shows how to build a regression tree model to forecast a continuous (but interval-scaled) target often used in the auto insurance industry, namely *loss frequency* (described in Chapter 1). *Loss frequency* can also be modeled as a categorical target variable if it takes on only a few values but in this example it is treated as a continuous target.

4.2 An Overview of the Tree Methodology in SAS Enterprise Miner

4.2.1 Decision Trees

A decision tree represents a hierarchical segmentation of the data. The *original segment* is the entire data set, and it is called the *root node* of the tree. The original segment is first partitioned into two or more *segments* by applying a series of simple rules. Each rule assigns an observation to a segment based on the value of an *input* (*explanatory variable*) for that observation. In a similar fashion, each resulting segment is further partitioned into *sub-segments* (segments within a segment); each sub-segment is further partitioned into more sub-segments, and so on. This process continues until no more partitioning is possible. This process of segmenting is called *recursive partitioning,* and it results in a hierarchy of segments within segments. The hierarchy is called a *tree*, and each segment or sub-segment is called a *node*.

Sometimes the term *parent node* is used to refer to a segment that is partitioned into two or more sub-segments. The sub-segments are called *child nodes*. When a child node itself is partitioned further, it becomes a parent node.

Any segment or sub-segment that is further partitioned into more sub-segments can also be called an *intermediate node*. A node with all its successors forms a *branch* of the tree. The final segments that are not partitioned further are called *terminal nodes* or *leaf nodes* or *leaves* of the tree. Each leaf node is defined by a unique combination of ranges of the values of the input variables. The leaf nodes are disjoint subsets of the original data. There is no overlap between them, and each record in the data set belongs to one and only one leaf node. In this book, I sometimes refer to a leaf node as a *group* or *segment*.

4.2.2 Decision Tree Models

A decision tree model is composed of several parts:

- node definitions, or rules, to assign each record of a data set to a leaf node
- posterior probabilities of each leaf node
- the assignment of a target level to each leaf node

Node definitions are developed using the Training data set and are stated in terms of input ranges. Posterior probabilities are calculated for each node using the Training data set. The assignment of the target level to each node is also done during the training phase using the Training data set.

Posterior probabilities are *observed proportions* of target levels within each node in the training data set. Take the example of a binary target. A binary target has two levels, which can be represented by response and no response, or 1 and 0. The *posterior probability* of response in a node is the proportion of records with the target level equal to response, or 1, within that node. Similarly, the *posterior probability* of no response of a node is the proportion of records with the target level equal to no response, or 0, within that node. These posterior probabilities are determined during the training of the tree and they become part of the decision tree model. As mentioned above, they are calculated using the training data.

The assignment of a target level to an individual record, or to a node as a whole, is called a *decision*. The decisions are made in order to maximize profit, minimize cost, or minimize misclassification error. To illustrate a decision based on profit maximization, consider the example of a binary target. In order to maximize profits, you need a profit matrix. For illustrative purposes, I devised the profit matrix given in Table 4.1.

Table 4.1

Actual target level/class	Decision1	Decision2
1	$10	0
0	-$1	0

In Table 4.1, Decision1 means assigning a record to the target level 1 (response), and Decision2 means assigning a record to target level 0 (non-response). This profit matrix indicates that if a true responder is correctly classified as a responder, then the profit earned is $10. If a true non-responder is classified as responder, a loss of $1 is incurred. If a true responder is classified as non-responder, then the profit is zero.[1] If a true non-responder is classified as a non-responder, then also the profit is zero.

Suppose E_1 represents the expected profit under Decision1, and E_2 represents the expected profit under Decision2. If $E_1 > E_2$ then Decision1 is taken, and the record is assigned to the target level 1. If $E_2 > E_1$ then Decision2 is taken, and the record is assigned to the target level 0. Given a profit matrix, E_1 and E_2 are based on the *posterior probabilities* of the node to which the record belongs. In general, each record is assigned the posterior probabilities of the node in which it resides. The expected profits (E_1 and E_2) are calculated as

$$E_1 = 10 * \hat{P}_1 - 1 * \hat{P}_0 \text{ and } E_2 = 0 * \hat{P}_1 + 0 * \hat{P}_0$$ for each record, where $\hat{P}_1$ and $\hat{P}_0$ represent the posterior probabilities of response and no response for the record.

It follows from these profit equations that all the records in a node have the same posterior probabilities, and are therefore assigned to the same target level.

An example of a tree is shown in Figure 4.1. The tree shown in Figure 4.1 is handcrafted[2] in order to highlight the process of assigning target levels to the nodes.

Figure 4.1

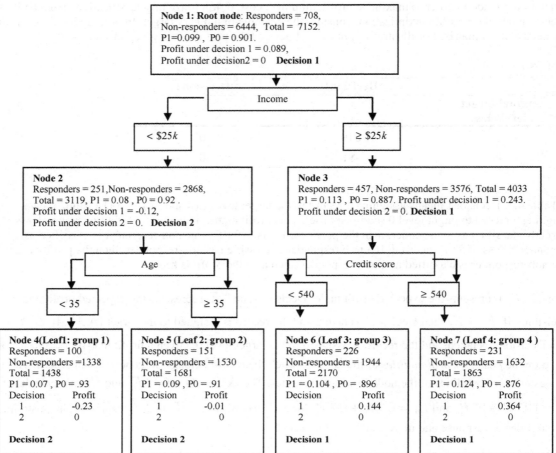

Figure 4.1 is an example of a tree for the target variable response. This variable has two levels: 1 for response and 0 for no response. Each node gives the information on the number of responders and non-responders, the proportion of responders (P1) and non-responders (P0), the decision regarding the assignment of a target level to the node, and the profit under each decision for that node.

Section 4.3.5 gives a step-by-step illustration of how SAS Enterprise Miner derives the information given in each node of the tree.

4.2.3 Decision Tree Models vs. Logistic Regression Models

A decision tree model is composed of a set of rules that can be applied to partition the data into disjoint groups. Unlike the logistic regression model, the tree model does not have any equations or coefficients. Instead, for each disjoint group or leaf node, it contains posterior probabilities which are themselves used as predicted probabilities. These posterior probabilities are developed during the *training* of the tree.

4.2.4 Applying the Decision Tree Model to Prospect Data

The SAS code generated by the **Tree** node can be applied to the prospect data set for purposes of prediction and classification. The code places each record in one of the predefined leaf nodes. The predicted probability of a target level (such as response) for each record is the posterior probability associated with the leaf node into which the record is placed. Since a target level is assigned to each leaf node, all records that fall into a leaf node are assigned the same target level. For example, the target level might be responder or non-responder.

4.2.5 Calculation of the Worth of a Tree

There are situations in which you must calculate the *worth* of a tree. The worth of a tree can be calculated using the Validation data set, Test data set, or any other data set where the target levels are known and the inputs necessary for defining the leaf nodes are available. Only leaf nodes are used for calculating the worth.

In SAS Enterprise Miner, the worth of a tree is calculated by using the validation data set to compare trees of different sizes in order to pick the tree with the optimum number of leaves. The worth of a tree can also be calculated using the test data in order to compare the performance of the decision tree model with other models. In both the cases, the method of calculating the worth of a tree is the same.

The following example calculates the worth of a tree using the Validation data set. The example uses a response model with a binary target. The binary target has two levels or classes: response (indicated by 1) and no response (indicated by 0). I use profit as a measure of worth, and show how the profits of the leaf nodes of a tree are calculated using the profit matrix introduced in Table 4.1. The calculation is a two-step procedure.

In Step 1, each record from the Validation data set is assigned to a leaf node based on rules or node definitions that are developed during the training phase. As noted above, all records that are placed in a particular leaf node are assigned the same posterior probabilities that were computed for that leaf node during the training phase using the training data set. Similarly, all records that fall into a particular leaf node are assigned the same target class/level (responder or non-responder) that was determined for that leaf node during training.

In Step 2, the profit is calculated for each leaf node of the tree based on the *actual* value of the target in each record of the validation data set.

If a leaf node is classified as a responder node (assigned the target level of 1), having n_1 records with the actual target level of 1 and n_0 records with the actual target level of 0, then the profit of that leaf node is $n_1 * \$10 + n_0 * (-\$1)$. If, on the other hand, the leaf node is classified as a non-responder node, then the profit of that node is $n_1 * \$0 + n_0 * \$0$. Following this procedure, the profit of each leaf node of the tree is calculated and then summed to calculate the total profit of the tree. The *average profit* of the tree is the total profit divided by the total number of records in the tree.

You can also calculate the total profit and average profit of a decision tree model using the Test data set. This is normally done when you want to compare the performance of a decision tree model to other models.

In SAS Enterprise Miner, you can enter the profit matrix when you create a data source, as shown in Chapter 2, Displays 2.14, 2.15, 2.16, and 2.17. You can also enter it by selecting the **Input Data** node and clicking ⊡ located to the right of the **Decisions** property. Click the **Decision Weights** tab and enter the profit matrix, shown in Display 4.1.

Display 4.1

Display 4.1 also shows that I selected the decision function **Maximize**. The next step is to tell the **Decision Tree** node to use the profit matrix and the decision function I selected. I select the **Decision Tree** node. In the Subtree section in the Properties panel of the **Decision Tree** node (shown in Display 4.2), I set the **Method** property to Assessment and the **Assessment Measure** property to Decision.

Display 4.2

Subtree	
Method	Assessment
Number of Leaves	1
Assessment Measure	Decision
Assessment Fraction	0.25

In SAS Enterprise Miner, profit is not the only criterion that can be used for calculating the worth of a tree. You can see the available criteria or measures by clicking in the Value column of the **Assessment Measure** property. The assessment measures are **Decision**, **Average Square Error**, **Misclassification**, and **Lift**.

Following is a brief description of each measure of model performance. For further details, see the SAS Enterprise Miner Help.

Decision

If you enter a profit matrix and select the decision function **Maximize** (as in Display 4.1), then SAS Enterprise Miner calculates average profit as a measure of the worth of a tree and selects a tree so as to maximize that value. If you select the decision function **Minimize**, then SAS Enterprise Miner minimizes cost in selecting a tree.

Average Squared Error

This is the average of the square of the difference between the predicted outcome and the actual outcome. This measure is more appropriate when the target is continuous.

Misclassification

Using the example of a response model, if the model classifies a responder as a non-responder or non-responder as responder, then the result is a misclassification. The misclassification rate is inversely related to the worth of a tree. Selecting this option instructs SAS Enterprise Miner to select a tree so as to minimize the number of misclassifications.

Lift

In the case of a response model, which has a binary target, the lift is calculated as the ratio of the actual response rate of the top *n*% of the ranked observations in the validation data set to the overall response rate of the validation data set. The ranking is done by the predicted probability of response for each record in the validation data set.

The percentage of ranked data (*n*%) to be used as the basis for the lift calculation is specified by setting the **Assessment Fraction** property of the **Decision Tree** node. The default value is 0.25 (25%).

4.2.6 Roles of the Training and Validation Data in the Development of a Decision Tree

To develop a tree model, you need two data sets. The first is for training the model, and the second is for pruning or fine-tuning the model. A third data set is optionally needed for an independent assessment of the model. These three data sets are referred to in SAS Enterprise Miner as Training, Validation, and Test, respectively. You can decide what percentage of the model data set is to be allocated to each of these three purposes. While there are no uniform rules for allocation, you can follow certain rules of thumb. If the data available for modeling is large, you can allocate it equally between the three data sets. If that is not the case, use something like 40%, 30%, and 30%, or 50%, 25%, and 25%. This partitioning is done by specifying the values for the **Training, Validation,** and **Test** properties of the **Data Partition** node. Reserving more data for the training generally results in more stable parameter estimates.

4.2.6.1 Training Data

The *Training* data set is used to perform three main tasks:

- developing rules to assign records to nodes (*node definitions*)
- calculating the posterior probabilities (the proportion of cases or records in each level of the target) for each node
- assigning each node to a target level (decision)

4.2.6.2 Validation Data

The *Validation* data set is used to *prune* the tree, i.e., select the right-sized tree or optimal tree. The initial tree is usually large. This is called the *maximal tree*. Some call it the *whole tree*. In this book, I use these two terms interchangeably.

By removing certain branches of the maximal tree, I can create smaller trees. By removing a different branch, I get a different tree. The smallest of these trees has only one leaf, called the *root node*, and the largest is the initial *maximal tree*, which can have many *leaves*. Thus removing different branches produces a number of possible sub-trees. You must select one of them. Validation data is used for selecting a sub-tree.

As explained in Section 4.2.5, the worth of each tree of a different size is calculated using the Validation data set, and one of the measures of worth is profit. Since profit is calculated using records of the Validation data set, profit is referred to as the *validation profit* in SAS Enterprise Miner.

Validation profit is used to select the optimal (right-sized) tree. The *size* of a tree is defined as the number of leaves in the tree. An *optimal tree* is defined as that tree that yields higher profit than any smaller tree (tree with fewer leaves) and equal or higher profit than any larger tree.

Since it is easy to travel from node to node and construct trees that end up with, say, ten leaves, there may be more than one tree of a given size within the maximal (or whole) tree. SAS Enterprise Miner searches through all the possible trees to find the one with the highest validation profit. Details of this search process are discussed in section 4.3.

4.2.6.3 Test Data

The *Test* data set is used for an independent assessment of the final model, particularly when you want to compare the performance of a decision tree model to the performance of other models.

4.2.7 Regression Tree

When the target is continuous, the tree is called a *regression tree*. The regression tree model consists of the rules (node definitions) for partitioning the data set into leaf nodes and also the mean of the target for each leaf node.

4.3 Development of the Tree in SAS Enterprise Miner

Construction of a decision tree or regression tree involves two major steps. The first step is growing a large tree. As mentioned before, you can then create many smaller trees, called sub-trees, of different sizes by removing one or more branches of the large tree. The second step is selecting the optimal sub-tree. SAS Enterprise Miner offers a number of choices for performing these steps. The following sections show how to use these methods by setting various properties.

4.3.1 Growing an Initial Tree

As described in Section 4.2.1, growing a tree involves successively partitioning the data set into segments using the method of *recursive partitioning*. Here I discuss how the inputs are selected at each step of the sequence of recursive partitioning. I focus on two-way, or binary, splits of the inputs, although SAS Enterprise Miner can also perform three-way and multi-way splits. In addition, the discussion assumes that the target is binary.

4.3.1.1 How a Node is Split Using a Binary Split Search

In a binary split, two branches emerge from each node. When an input such as household income is used to partition the records into two groups, a *splitting value* such as the household income value of $10,000 may be chosen. The records with income less than the splitting value are sent to the left branch, and records with income greater than or equal to the splitting value are sent to the right branch. In a multi-way split, more than two branches may emerge from any node. For example, the income variable could be divided into ranges such as $0–$10,000, $10,001–$20,000, $20,001–$30,000, etc. As mentioned earlier, for the purposes of simplicity and clarity, I confine my discussion to two-way splits across an input.

In order to split any segment or sub-segment of the data set at a node, SAS Enterprise Miner calculates the worth of all candidate splitting values for each of the inputs and selects the split with the highest worth, which involves finding the best splitting value for each of the available inputs and then comparing all of the inputs to see which one's split is best. You can select the method of calculating the worth of the split. These methods are discussed in the following sections.

The process of selecting the best split consists of two steps. In the first step, the best splitting value for each input is determined. In the second step, the best input among all inputs is selected by comparing the worth of the best split of each variable with the best splits of the others, and selecting the input whose best split yields the highest worth. The node is split at the best splitting value on the best input. The following example illustrates this process.

Suppose there are 100 inputs, represented by $X_1, X_2, ..., X_{100}$. The tree algorithm starts with input X_1 and examines all candidate splits of the form $X_1 < C$, where C is a splitting value somewhere between the minimum and maximum value of X_1. All records that have $X_1 < C$ go to the left child node and all records that have $X_1 \geq C$ go to the right child node. The algorithm goes through all candidate splitting values on the same input and selects the best splitting value. Let the best splitting value on input X_1 be C_1. The algorithm repeats the same process on the next input X_2. This process is repeated on inputs $X_2, X_3, X_4, ..., X_{100}$ and the corresponding best splitting values $C_2, C_3, C_4, ..., C_{100}$ are found. Having found the best splitting value for each input, the algorithm then compares these splits to find the input whose best splitting value gives a better split of the data than the best splitting value of any other input. Suppose C_{10} is the best *splitting value* for input

X_{10} and suppose X_{10} is chosen as the best input upon which to base a split of the node. Consequently, the node is partitioned using the input X_{10}. All records with X_{10} less than C_{10} are sent to the left child node and all records with $X_{10} \geq C_{10}$ are sent to the right child node. This process is repeated for each node. A different input may be chosen at a different node. How are the splits compared to determine the best splits? To answer this, it is necessary to examine how we measure the worth of each split.

4.3.1.2 Measuring the Worth of a Split (Splitting Criterion Property of the Decision Tree Node in SAS Enterprise Miner)

The splitting method used for partitioning the data is determined by the value of the **Criterion** property in the Splitting Rule section of the Properties panel. If your target variable is nominal, then click in the column to the right of the **Nominal Criterion** property to see the various methods of splitting, as shown in Display 4.3.

Display 4.3

Splitting Rule	
Interval Criterion	ProbF
Nominal Criterion	ProbChisq ▾
	ProbChisq
	Entropy
	Gini

A nominal variable is a variable that takes on values such as Response and No Response. The values that a nominal variable takes on are sometimes referred to as levels or categories. In this example, the target variable, Resp, takes on only two values—response and no response. A nominal variable can take on more than two values. For example, a nominal variable Color can take the values such as Red, Blue, Green, etc. The important feature of a nominal variable is that the values that it takes on cannot be ranked according to any measurement scale. That is, Response is not higher or lower than No response; Blue is not higher or lower than Green. In general, we can say that a nominal variable is a categorical variable having unordered levels or categories. When a nominal variable has only two levels, it can be referred to as a binary variable. In our example, since we are considering a target variable that takes the values Response and No response, which do not represent any order, our target variable is a nominal target. Since it takes only two values, we can call it a binary variable. We can also call it a categorical variable having unordered scales since the levels or categories or the values that it takes on are not ordered.

If your target is a categorical variable with ordered scales, then it is called an ordinal variable. An example of an ordinal variable is Risk, which can take on values such as High, Medium, and Low. Since High represents higher risk than Medium and Medium represents higher risk than Low, Risk is an ordinal variable.

If your target variable is ordinal, then click in the column to the right of the **Ordinal Criterion** property to see the methods of splitting that apply to the ordinal variable situation. The methods are Entropy and Gini.

A variable such as our Savings Account Balances variable is called an interval scaled variable.

If your target variable is an interval scaled variable, then click in the column to the right of the **Interval Criterion** property to see the methods of splitting that are appropriate for this situation, which are Variance and ProbF.

When the target variable is binary or categorical with more than two levels, there are two approaches for measuring the worth of a split:

- the degree of separation achieved by the split. In SAS Enterprise Miner, this is measured by the *p*-value of the Pearson Chi-Square test.
- the impurity reduction achieved by the split. In SAS Enterprise Miner, this is measured either by Entropy reduction or by Gini reduction.

When the target is continuous, SAS Enterprise Miner measures the worth of a split by an F-test, which tests for the degree of separation of the child nodes. If you set the **Interval Criterion** to Variance, Variance Reduction is used to measure the worth of a split. This chapter looks at these measures case by case.

4.3.1.2A Measuring the Worth of a Split When the Target is Binary

4.3.1.2A.1 Degree of Separation

Any two-way split partitions a parent node into two child nodes. *Logworth* is a measure of how different these child nodes are from each other and from the parent node, in terms of the composition of the classes. The greater the difference between the two child nodes, the greater the degree of separation achieved by the split, and the better the split is considered to be.

To see how logworth is calculated at each splitting value, consider this example. Let the target be *response*, and let household income be an input (explanatory variable). Let each row of the data set represent a record (or observation) from the data set. Table 4.2 shows a partial view of the data set, showing only selected records from the data set and only two columns. The records are in increasing order of income, to help discuss the notion of splitting the records on the basis of income.

Table 4.2

Observation (record)	Response	Household income
1	1	$8,000
2	0	$10,000
...	...	...
100	1	$12,000
...	...	...
1200	0	$14,000
...	...	...
...	...	...
10,000	1	$150,000

The data shown in Table 4.2 can be split at different values of household income. At each splitting value, a 2x2 contingency table can be constructed, as shown in Table 4.3. Table 4.3 shows an example of one split. The columns represent the two child nodes that result from the split, and rows represent the levels of the target.

Table 4.3

	Income $< \$10,000$	Income $\geq \$10,000$	Total
Responders	n_{11}	n_{12}	$n_{1\bullet}$
Non-responders	n_{01}	n_{02}	$n_{0\bullet}$
Total	$n_{\bullet 1}$	$n_{\bullet 2}$	n

To assess the degree of separation achieved by a split, you must calculate the value of the Chi-Square (χ^2) statistic and test the null hypothesis that the proportion of responders among those with income less than $10,000 is not different from those with income greater than or equal to $10,000. This can be written as:

$$H_0 : P_1 = P_2 = P$$

$$\text{Let } \hat{P}_1 = \frac{n_{11}}{n_{\bullet 1}} \ , \ \hat{P}_2 = \frac{n_{12}}{n_{\bullet 2}} \text{ and } \hat{P} = \frac{n_{1\bullet}}{n}$$

Under the null hypothesis, the expected values are:

	Income $< \$10,000$	**Income** $\geq \$10,000$
Responders	$E_{11} = \hat{P} \times n_{\bullet 1}$	$E_{12} = \hat{P} \times n_{\bullet 2}$
Non-responders	$E_{01} = (1 - \hat{P}) \times n_{\bullet 1}$	$E_{02} = (1 - \hat{P}) \times n_{\bullet 2}$

The Chi-Square statistic is calculated as follows:

$$\chi^2 = \sum_{i=0}^{1} \sum_{j=1}^{2} \frac{(n_{ij} - E_{ij})^2}{E_{ij}}$$

The *p*-value of χ^2 is found by solving the equation:

$P(\chi^2 > \text{calculated } \chi^2 \mid \text{null hypothesis is true}) = p\text{-value}$. And the *logworth* is simply calculated as $Logworth = -log_{10}(p - value)$.

The larger the *logworth* (and therefore the smaller the *p*-value), the better the split.

Let the logworth calculated from the first split be LW_1 . Another split is made at the next level of income (for example \$12,000), a contingency table is made, and the logworth is calculated in the same way. I call this LW_2 . If there are 100 distinct values for income in the data set, 99 contingency tables can be created, and the logworth computed for each. The calculated values of the logworth are LW_1 , LW_2 ,..., LW_{99} . The split that generates the highest logworth is selected. Suppose the best splitting value for income is \$15,000, with a logworth of 20.5. Now consider the next variable, Age. If there are 67 distinct values of Age in the data set, 66 splits are considered; the split with the maximum logworth is selected for Age. Suppose the best splitting value for Age is 35, with a logworth of 10.2. If age and income are the only inputs in the data set, then income is selected to split the node because it has the best logworth value. In other words, the best split on income (\$15,000) has a higher logworth (20.5) than the best split on Age (35) with a logworth of 10.2. Hence the data set would be split at \$15,000 of income. This split can be called the best of the best.

If there were 200 inputs in the data set, the process of finding the best split is performed 200 times (once for each input), and repeated again at each node that is further split. Each input is examined, the best split found, and the one with the highest logworth chosen as the best of the best.

To use the Chi-Square method of choosing a split, set the **Nominal Criterion** property in the Splitting Rule section of the **Decision Tree** node to ProbChisq.

4.3.1.2A.2 Impurity Reduction as a Measure of Goodness of a Split

In SAS Enterprise Miner, you can choose maximizing *impurity reduction* as an alternative to maximizing the degree of separation for selecting the splits on a given input, and for selecting the best input. The *impurity* of a node is the degree of heterogeneity with respect to the composition of the levels of the target variable. If node v is split into child nodes a and b, and if ω_a and ω_b are the proportions of records in nodes a and b, then the decrease in impurity is $i(v) - \omega_a i(a) - \omega_b i(b)$, where $i(v)$ is the impurity index of node v, and $i(a)$ and $i(b)$ are the impurity indexes of the child nodes a and b, respectively. In SAS Enterprise Miner there are two measures of impurity, one measure called the *Gini Impurity Index*, and the other called *Entropy*.

To split the node v into child nodes a and b based on the splitting value of an input X_1, the Tree algorithm examines all candidate splits of the form $X_1 < C_j$ and $X_1 \geq C_j$, where C_j is a real number between the minimum and maximum values of X_1. Those records that have $X_1 < C_j$ go to the left child node and those with $X_1 \geq C_j$ go to the right. Suppose there are 100 candidate splitting values on the input X_1. The candidate splitting values are $C_j, j = 1,2,....100$. The algorithm compares impurity reduction on these 100 splits and selects the one that achieves the greatest impurity reduction as the best split.

4.3.1.2A.2.1 Gini Impurity Index

If p_1 is the proportion of responders in a node, and p_0 is the proportion of non-responders, the *Gini Impurity Index* of that node is defined as $i(p) = 1 - p_1^2 - p_0^2$. If two records are chosen at random (with replacement) from a node, the probability that both are responders is p_1^2, while the probability that both are non-responders is p_0^2, and the probability that they are either both responders or both non-responders is $p_1^2 + p_0^2$. Hence $1 - p_1^2 - p_0^2$ can be interpreted as the probability that any two elements chosen at random (with replacement) are different. For binary targets, the Gini Index simplifies to $2p_1(1 - p_1)$. A pure node has a Gini Index of zero. It has a maximum value of $1/2$ when both classes are equally represented. To use this criterion when your target variable is an unordered categorical variable, set the **Nominal Criterion** property in the Splitting Rule section to Gini. If your target variable is ordinal (ordered categorical), set the **Ordinal Criterion** property to Gini.

4.3.1.2A.2.2 Entropy

Entropy is another measure of the impurity of a node. It is defined as $i(p) = -\sum_{i=0}^{1} p_i \log_2(p_i)$ for binary targets. A node that has larger entropy than another is more heterogeneous and therefore less pure.

The rarity of an event is measured as $-\log_2(p_i)$. If an event is rare, it means the probability of its occurrence is low. Consider the event Response. Suppose the probability of response in a node is 0.005. Then the rarity of response is $-\log_2(0.005) = 7.644$. This is a rare event. The probability of non-response is conversely 0.995, hence the rarity of non-response is $-\log_2(0.995) = 0.0072$. A node that has a response rate of 0.005 is less impure than the node that has equal proportions of responders and non-responders. Thus $-\log_2(p_i)$ is high when the rarity is high and low when the rarity of the event is low. The entropy of this node is as follows:

$$i(p) = -[0.005 * \log_2(0.005) + 0.995 * \log_2(0.995)] = 0.0454.$$

Consider another node in which the probability of response = probability of non-response = 0.5. The entropy of this node is $i(p) = -[0.5 * \log_2(0.5) + 0.5 * \log_2(0.5)] = 1$. The node that is predominantly non-responders (with a proportion of 0.995) has an entropy value of 0.0454. A node with an equal distribution of responders and non-responders has entropy equal to 1. A node that has all responders or all non-responders has an entropy equal to zero. Thus, entropy ranges between 0 to 1, with 0 indicating maximum purity and 1 indicating maximum impurity.

To use the Entropy measure when your target variable is an unordered categorical variable, go to the Splitting Rule section, and set the **Nominal Criterion** property to Entropy. If your target variable is ordinal (ordered categorical), rather than nominal, set the **Ordinal Criterion** property to Entropy.

4.3.1.3 Measuring the Worth of a Split When the Target is Categorical with More than Two Categories

If the target is categorical with more than two categories (levels), the procedures are the same. The Chi-Square statistics are calculated from $r \times b$ contingency tables, where b is the number of child nodes being created on the basis of a certain input, and r is the number of target levels (categories). The p-values are calculated from the Chi-Square distribution with degrees of freedom equal to $(r-1)(b-1)$.

The Gini Index and Entropy measures of impurity can also be applied in this case. They are simply extended for more than two levels (categories) of the target variable.

4.3.1.4 Measuring the Worth of a Split When the Target is Continuous

If the target is continuous, an F-test is used to measure the degree of separation achieved by a split. Suppose the target is loss frequency and the input is income having 100 distinct values. There are 99 possible two-way splits on income. At each split value, two groups are formed. One group has income level below the split value, and the other group has income greater than or equal to the split value. Consider a parent node with n records and a split value of $10,000. All records with income less than $10,000 are sent to the left node and those with income greater than or equal to $10,000 are sent to the right node, as shown in Figure 4.2. The mean of the target is calculated for each node.

Figure 4.2

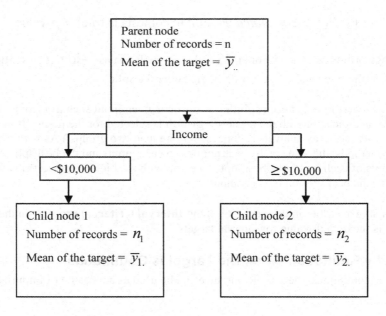

In order to calculate the worth of this split, a one-way ANOVA is first performed, and the logworth is then calculated from the F-test.

The Sum of Squares between the groups (child nodes) is given by:

$$SS_{between} = \sum_{i=1}^{2} n_i (\overline{y}_{i.} - \overline{y}_{..})^2,$$ where $\overline{y}_{i.}$ is the mean of the target in the ith child node, and

$\overline{y}_{..}$ is the mean of the target in the parent node.

The Sum of Squares within the groups (child nodes) is given by:

$$SS_{within} = \sum_{i=1}^{2} \sum_{j=1}^{n_i} (y_{ij} - \overline{y}_{i.})^2,$$ where y_{ij} is the value of the target variable for

the jth record in the ith child node.

The Total Sum of Squares is given by

$$SS_{total} = \sum_{i=1}^{2} \sum_{j=1}^{n_i} (y_{ij} - \overline{y}_{..})^2,$$ and the F-statistic is computed as

$$F = \left(\frac{SS_{between} / (2-1)}{SS_{within} / (n-2)} \right).$$ This statistic has an F-distribution with 1 and n − 2 degress

of freedom under the null hypohesis.

The null hypothesis says that there is no difference in the target mean between the child nodes.

The p-value is given by $P(F > calculated\ F \mid$ null hypothesis is true$) = p$-value.

As before, the logworth is calculated from the p-value of the F-test as $-\log_{10}(p\text{-value})$. The larger the logworth (and therefore the smaller the p-value), the better the split.

One-way analysis is performed at each splitting value on a given input such as income, the logworth of each split is calculated, and the split with the highest logworth is selected for the input. (If income is split into more than two groups, then one-way analysis is done with more than two groups.) As with the Chi-Square method and all the other methods, the same analysis is applied to each input, and the best split for each input is found. From the best splits of each input, the best input (the one with the highest logworth) is found, and that is the best of the best, that is, the best split on the best input.

To use the F-test shown earlier in this section, set the **Interval Criterion** property in the Splitting Rule section to ProbF, which is the default value for interval targets.

4.3.1.5 Impurity Reduction When the Target is Continuous

When the target is continuous, the sample variance can be used as a measure of impurity.

The impurity of the tth node is $i(t) = \dfrac{1}{n_t} \sum_{j=1}^{n_t} \left(y_{jt} - \overline{y}_t \right)^2$,where

$i(t)$ = impurity of the tth node.

n_t = number observations in the tth node.

y_{jt} = value of the target variable in the jth observation in the tth node.

$\overline{y}_t$ = mean of the target variable in the tth node.

The impurity of the root node (0) is $i(0) = \dfrac{1}{n_0} \sum_{j=1}^{n_0} \left(y_{j0} - \overline{y}_0 \right)^2$. If the root node is split into two child nodes 1 and 2, then the impurity reduction due the split is calculated as

$$\Delta i = i(0) - \left\{ \frac{n_1}{n_0} i(1) + \frac{n_2}{n_0} i(2) \right\}$$ where n_0 is the number of observations in the root node, n_1 is the number of

observations in the first child node and n_2 is the number of observations in the second child node. Impurity reduction can be used to measure the worth of a split. To use impurity reduction when the target is continuous, set the **Interval Criterion** in the Splitting Rule section to Variance.

4.3.2 *P*-value Adjustment Options

Using the default settings, SAS Enterprise Miner uses adjusted *p*-values in comparing splits on the same input and splits on different inputs. Two types of adjustments are made to *p*-values : Bonferroni adjustment and Depth adjustment.

4.3.2.1 Bonferroni Adjustment Property

When you are comparing the splits from different inputs, *p*-values are adjusted to take into account the fact that not all inputs have the same number of levels. In general, some inputs are binary, some are ordinal, some are nominal, and others are interval-scaled. For example, a variable such as Mail Order Buyer takes on the value of 1 or 0, 1 being mail order buyer and 0 being not a mail order buyer. I will call this the Mail Order Buyer variable. For this variable, only one split is evaluated, only one contingency table is considered, and only one test is performed. An input like "number of delinquencies in the past 12 months" can take any value from 0 to 10 or higher. I will call this the Delinquency variable. This is an ordinal variable. Suppose it takes values from 0 to 10. SAS Enterprise Miner compares the split values 0.5, 1.5, 2.5,..., 9.5. Ten contingency tables are constructed, and ten Chi-Square values are calculated. In other words, ten tests are performed on this input for selecting the best split.

Suppose the ith split on the Delinquency variable has a *p*-value $= \alpha_i$, which means that

$P(\chi^2 \geq calculated\ \chi_i^2 \mid$ null hypothesis is true)$= \alpha_i$. In other words, the probability of finding a Chi-Square

greater than or equal to the calculated χ_i^2 at random is α_i under the null hypothesis. The probability that, out of the ten Chi-Square tests on the delinquency variable, at least one of the tests yields a false positive decision

(where we reject the null hypothesis when it is actually true) is $1 - \prod_{i=1}^{10} (1 - \alpha_i)$. This family-wise error rate is

much larger than the individual error rate of α_i. For example, if the individual error rate (α_i) on each test

=0.05, then the family-wise error rate is $1 - 0.95^{10} = 0.401$. This means that when you have multiple χ_i^2 comparisons (one for each possible split), the *p*-values underestimate the risk of rejecting the null hypothesis when it is true. Clearly the more possible splits a variable has, the less accurate the *p*-values will be. Therefore, when comparing the best split on the Delinquency variable with the best split on the Mail Order Buyer variable, the logworths need to be adjusted for the number of splits, or tests, on each variable. In the case of the Mail Order Buyer variable, there is only one test, and hence no adjustment is necessary. But in the case of the

Delinquency Variable, the best split is chosen from a set of ten splits. Therefore, $\log_{10}(10)$ is subtracted from the logworth of the delinquency variable's best split. In general, if an input has *m* splits, then $\log_{10}(m)$ is subtracted from the logworth of each split on that input. This adjustment is called the *Bonferroni Adjustment*.

To apply the Bonferroni Adjustment, set the **Bonferroni Adjustment** property to Yes, which is the default.

4.3.2.2 Time of Kass Adjustment Property

If you set the value of the **Time of Kass Adjustment** property to After, then the *p*-values of the splits are compared without Bonferroni Adjustment. If you set the property to Before, then the *p*-values of the splits are compared with Bonferroni Adjustment.

4.3.2.3 Depth Adjustment

In SAS Enterprise Miner, if you set the **Split Adjustment** property of the **Decision Tree** node to Yes, then *p*-values are adjusted for the number of ancestor splits. You can call the adjustment for the number of ancestor splits the *depth adjustment*, because the adjustment depends on the depth of the tree at which the split is done. Depth is measured as the number of branches in the path from the current node, where the splitting is taking place, to the root node. The calculated *p*-value is multiplied by a depth multiplier, based on the depth in the tree of the current node, to arrive at the depth-adjusted *p*-value of the split. For example, let us assume that prior to the current node, there were four splits (four splits were required to get from the root node to the current node) and that each split involved two branches (using binary splits). In this case, the depth multiplier is 2x2x2x2. In general, the depth multiplier for binary splits is 2^d, where *d* is the depth of the current node. The calculated *p*-value is adjusted by multiplying by the depth multiplier. This means that at a depth of 4, if the calculated *p*-value is 0.04, the depth-adjusted *p*-value would be 0.04x16=0.64. Without the depth adjustment, the split would have been considered statistically significant. But after the adjustment, the split is not statistically significant.

The depth adjustment can also be interpreted as dividing the threshold *p*-value by the depth multiplier. If the threshold *p*-value specified by the **Significance Level** property is 0.05, then it is adjusted to be 0.05/16 =0.003125. Any split with a *p*-value above 0.003125 is rejected. In general, if α is the significance level specified by the **Significance Level** property, then any split which has a *p*-value above $\alpha / depth\ multiplier$ is rejected. The effect of the depth adjustment is to increase the threshold value of the logworth by $\log_{10}(2^d) = d\log_{10}(2)$. Hence, the deeper you go in a tree, the stricter the standard becomes for accepting a split as significant. This leads to the rejection of more splits than would have been rejected without the depth adjustment. Hence, the depth adjustment may also result in fewer partitions, thereby limiting the size of the tree.

4.3.2.4 Controlling Tree Growth through the Threshold Significance Level Property

You can control the initial size of the tree by setting the threshold *p*-value. In SAS Enterprise Miner the threshold *p*-value (*significance level*) is specified by setting the **Significance Level** property of the **Decision Tree** node to the desired *p*-value. From this *p*-value, SAS Enterprise Miner calculates the threshold logworth. For example, if you set the **Significance Level** property to 0.05, then the threshold logworth is given by $-\log_{10}(0.05)$, or 1.30. If, at any node, none of the inputs has a split with logworth higher than or equal to the threshold, then the node is not partitioned further. By decreasing the threshold *p*-value, you increase the degree to which the two child nodes must differ in order to consider the given split to be significant. Thus, tree growth can be controlled by setting the threshold *p*-value.

4.3.2.5 Controlling Tree Growth through the Split Adjustment Property

As previously mentioned, if you set the **Split Adjustment** property of the **Decision Tree** node to Yes, then *p*-values are adjusted for the number of ancestor splits. In particular, if α is the significance level specified by the **Significance Level** property, then any split that has a *p*-value above $\alpha / depth\ multiplier$ is rejected. Hence, the deeper you go in a tree, the stricter the standard becomes for accepting a split as significant. This leads to the rejection of more splits than without the adjustment, resulting in fewer partitions. So switching on the **Split Adjustment** property is another way of controlling tree growth.

4.3.2.6 Controlling Tree Growth through the Leaf Size Property

You can control the growth of the tree by setting the **Leaf Size** property. If you set the value of the property to, say, 100, then if a split results in the creation of a leaf with fewer than 100 records, that split should not be performed. Hence, the growth stops at the current node.

4.3.3 Controlling Tree Growth: Stopping Rules

4.3.3.1 Controlling Tree Growth through the Split Size Property

If the value of the **Split Size** property is set to a number such as 300, then if a node has fewer than 300 records, it should not be considered for splitting. The default value of this property is twice the leaf size, which is specified by the **Leaf Size** property.

4.3.3.2 Controlling Tree Growth through the Maximum Depth Property

You can also choose a value for the **Maximum Depth** property. This determines the maximum number of generations of nodes. The root node is generation 0, the children of the root node are the first generation, etc. The default setting of this property is 6. You can set the **Maximum Depth** property between 1 and 50.

4.3.4 Pruning: Selecting the Right-Sized Tree Using Validation Data

Having grown the largest possible tree (*maximal tree*) under the constraints of stopping rules described in the previous section, you now need to prune it to the right size. The pruning process proceeds as follows. Start with the *maximal tree*, and eliminate one split at each step. If the maximal tree has M leaves and if I remove one split at a given point, I will get a sub-tree of size M-1. If one split is removed at a different point, I will get another sub-tree of size M-1. Thus, there can be a number of sub-trees of size M-1. From these, SAS Enterprise Miner selects the best sub-tree of size M-1. Similarly, by removing two splits from the maximal tree, I get several sub-trees of size M-2. From these, the best sub-tree of size M-2 is selected. This process continues until there is a tree with only one leaf. At the end of this process, there is a sequence of trees of sizes M, M-1, M-2, M-3,...1. (Display 4.5 shows a plot of the Average Profit associated with the sequence of best of the sub-trees of the same size.) Thus, the sequence is an optimum sequence. From the optimum sequence, the best tree is selected on the basis of average profit, or accuracy or, alternatively, the lift (I describe some of these methods later in this chapter).

Figure 4.3 shows how sub-trees are created by pruning a large tree at different points.

Figure 4.3

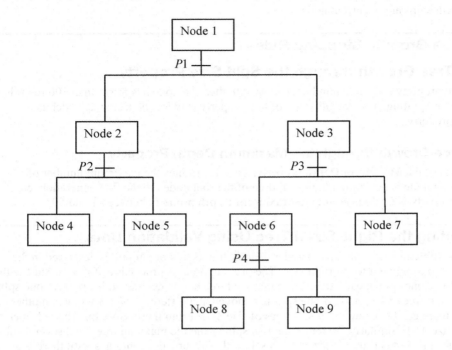

Figure 4.3 shows a hypothetical tree. It has nine nodes. In the maximal tree there are five leaf nodes (Nodes 4, 5, 7, 8, and 9). By removing a split at the point *P4*, I get one tree with four leaves (Nodes 4, 5, 6, and 7). By removing the split at *P2*, I get a tree with four leaves with Nodes 2, 8, 9, and 7 as its leaves. Thus, there are two sub-trees of size 4.

By removing the splits at *P3* and *P4*, I get a tree with three leaves (consisting of Nodes, 4, 5, and 3 as its leaves). By removing the splits at *P2* and *P4*, I will get another tree with three leaves consisting of Nodes 2, 6, and 7 as its leaves. Thus, there are two sub-trees of size 3.

By removing the splits at *P2*, *P3*, and *P4*, I get one tree with two leaves consisting of Nodes 2 and 3 as its leaves. By removing the splits at *P1*, *P2*, *P3*, and *P4*, I get one tree with only one leaf, with Node 1 as its leaf.

Thus the sequence consists of one tree with five leaves, two trees with four leaves, two trees with three leaves, one tree with two leaves, and one with a single leaf.

Since there are two sub-trees of size 4, only one is selected. Similarly, one sub-tree of size 3 is selected. This selection is based on the user-supplied criterion discussed below.

The final sequence consists of one sub-tree of each size. This sequence, along with the profit or other assessment measure used to choose the best sub-tree of each size, is displayed in a chart in the Results window of the decision tree. (As mentioned previously, Display 4.5 shows an example of such a chart.) To choose the final model, SAS Enterprise Miner selects the best tree from the sequence.

What is the criterion used for selecting the best tree from a number of trees of the same size at each step of the sequence described above? One possible criterion is to compare the profit of the sub-trees in each step of building the sequence. The profit associated with each tree is calculated using the Validation data set.

Alternative criteria for selecting the best tree include cost minimization, minimization of miscalculation rate, minimization of average squared error, or maximization of lift. In the case of a continuous target, minimization of average squared error is used for selecting the best tree.

In SAS Enterprise Miner, set the **Method** property in the Subtree section in the Properties panel of the **Decision Tree** node to Assessment, and the **Assessment Measure** property to Decision, Average Square Error, Misclassification, or Lift. These options are described Section 4.2.5.

If you set the value of the **Assessment Measure** property to Decision, you must enter information on the **Decision Weights** tab when you create the **Data Source** (or later by selecting the **Input Data** node and click ⊡ located to the right of the **Decisions** property in the Columns section of the Properties panel). Click the **Decision Weights** tab. Enter the profit matrix, and select **Maximize**, as shown in Figure 4.4.

Display 4.4

If you enter costs rather than profits as decision weights, you should select the **Minimize** option.

If a profit or loss matrix is not defined, then assessment is done by Average Square Error if the target is interval, and Misclassification if the target is categorical.

As noted in Section 4.2.5, you can set the **Assessment Measure** property to Decision, Average Square Error, Misclassification, or Lift. The value Decision calls for the assessment to be done by maximizing profit on the basis of the profit matrix entered by the user. This is illustrated in Section 4.3.5. The Misclassification criterion is illustrated in Section 4.3.7, and the Average Square Error criterion of assessment is demonstrated in an example in Section 4.3.8.

As the number of leaves increases, the profit may increase initially. However, beyond some point there is no additional profit. The profit may even decline. The point at which marginal, or incremental, profit becomes insignificant is the optimum size of the tree.

Following is a step-by-step illustration of pruning a tree.

4.3.5 Step-by-Step Illustration of Growing and Pruning a Tree

The tree is grown using the Training data set with 7152 records. The rules of partition are developed and the nodes are classified (so as to maximize profit or other criterion chosen) using the Training data set.

The Validation data set that is used for pruning consists of 5364 records. The rules of partition and the node classifications that were developed on the Training data set are then applied to all 5364 records, in effect reproducing the tree with the Validation data set. The node definitions and classification of the nodes are the same, but the records in each node are from the validation data set.

The process begins by displaying the maximal tree constructed from the training data set:

Step IA: A tree is developed using the training data set

The initial tree is developed using the Training data set. This tree is shown in Figure 4.4.

Figure 4.4

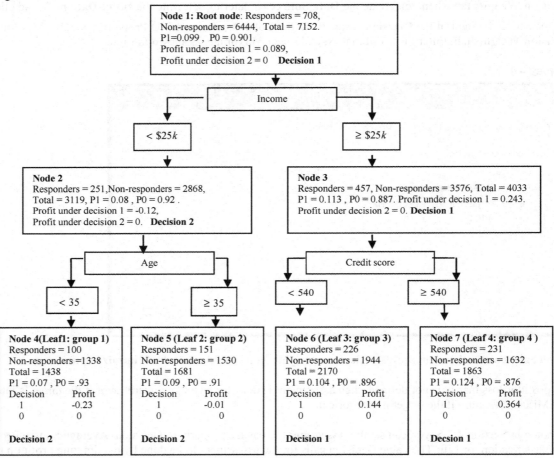

Figure 4.4 shows the tree developed from the training data. The above tree diagram gives information on: the node identification, the leaf identification (if the node is a terminal node), the number of responders in the node,

the number of non-responders, the total number of records in each node, proportion of responders (posterior probability of response), proportion of non-responders (posterior probability of non-response), profit under **Decision1**, profit under **Decision2**, and the label of the decision under which the profits are maximized for the node.

The tree consists of the rules or definitions of the nodes. Starting from the root node and going down to a terminal node, you can read the definition of the leaf node of the tree. These definitions are stated in terms of the input ranges. The inputs selected by the tree algorithm in this example are age, income, and credit score. The rules or definitions of the leaf nodes are

Leaf 1:   If income < \$25,000 and age < 35 years, then the individual record goes to *group 1 (leaf 1)*.

Leaf 2:   If the individual's income < \$25,000 and age ≥ 35 years, then the record goes to *group 2 (leaf 2)*.

Leaf 3:   If an individual's income ≥ \$25,000 and credit score < 540, then the record goes to *group 3 (leaf 3)*.

Leaf 4:   If the individual's income is ≥ \$25,000 and credit score ≥ 540, then the record goes to *group 4 (leaf 4)*.

Step 1B: Posterior probabilities for the leaf nodes from the training data

When the target is response (which is a binary target), the *posterior probabilities* are the proportion of responders and the proportion of non-responders in each node. In applying the model to prospect data, these posterior probabilities are used as predictions of the probabilities. All customers (records) in a leaf are attributed the same predicted probability of response.

Step 1C: Classification of records and nodes by maximization of profits

Classification of records or nodes is the name given to the task of assigning a target level to a record. All records within a node are assigned the same target level. The decision to assign a class or target level to a record or node is based on both the posterior probabilities (Step 1B) and the profit matrix. As described later, you can use a criterion other than profit maximization to classify a node or the records within a node. I first demonstrate how classification is done using profit when the target is response. In Figure 4.4, each node contains a label: **Decision1** or **Decision2**. A label of **Decision1** means the node is classified as a responder node. A label of **Decision2** means the node is classified as a non-responder node. In arriving at these decisions, the following profit matrix is used:

Table 4.4

Profit Matrix		
Target	Decision	
	1	0
1	\$10	0
0	-\$1	0

According to this profit matrix, if I correctly assign the target level 1 (**Decision1**) to a true responder (target = 1), then I make \$10 of profit. If I incorrectly assign the target level 1 to a non-responder, I lose \$1 (a profit of −\$1). For example, in Node 4 the posterior probabilities are 0.07 for response and 0.93 for no response. Under **Decision1** the expected profit = 0.07*\$10 + 0.93*(−\$1) = −\$0.23. Under **Decision2** (that is, if I decide to assign target level 0, or classify Node 4 as a non-responder node) the expected profit = 0.07*0+0.93*0 = \$0. Since \$0 > −\$0.23, I take **Decision2,** that is, I classify Node 4 as a non-responder node. All records in a node are assigned the same target level since they all have the same posterior probabilities and the same profit matrix. Figure 4.4 shows the posterior probabilities for each node as P1 and P0. The following calculations show how these posterior probabilities, together with the profit matrix, are used to assign each node to a target level.

Node 1 (root node)

Under **Decision1** the expected profit is = $10*0.099 + (–$1)*0.901 = $0.089. Under **Decision2** the expected profit is = 0. Since the profit under **Decision1** is greater than the profit under **Decision2**, **Decision1** is taken. That is, this node is assigned to the target level 1. All members of this node are classified as responders.

Node 2

Under **Decision1** the expected profit is $10*0.08+ (–$1)*0.92 = – $0.12. Under **Decision2** the expected profit is 0*0.08 + 0*0.92 = $0. Since the profit under **Decision2** is greater than the profit under **Decision1**, this node is assigned to the target level 0. All members of this node are classified as non-responders.

Node 3

Under **Decision1** the expected profit =$10*0.113+ (–$1)*0.887 = $0.243. Under **Decision2** the expected profit is 0*0.113 + 0*0.887= $0. Since profit under **Decision1** is greater than the profit under **Decision2**, this node is assigned the target level 1. All members of this node are classified as responders.

Node 4

Under **Decision1** the expected profit =$10*0.07+ (–$1)*0.93 = –$0.23. Under **Decision2** the expected profit is 0*0.07 + 0*0.93= $0. Since profit under **Decision2** is greater than the profit under **Decision1**, this node is assigned the target level 0. All members of this node are classified as non-responders.

Node 5

Under **Decision1** the expected profit =$10*0.09+ (–$1)*0.91 = – $0.01. Under **Decision2** the expected profit is 0*0.09 + 0*0.91= $0. Since profit under **Decision2** is greater than the profit under **Decision1**, this node is assigned the target level 0. All members of this node are classified as non-responders.

Node 6

Under **Decision1** the expected profit =$10*0.104+ (–$1)*0.896 = $0.144. Under **Decision2** the expected profit is 0*0.104 + 0*0.896= $0. Since profit under **Decision1** is greater than the profit under **Decision2**, this node is assigned the target level 1. All members of this node are classified as responders.

Node 7

Under **Decision1** the expected profit =$10*0.124+ (–$1)*0.876 = $0.364. Under **Decision2** the expected profit is 0*0.124 + 0*0.876= $0. Since profit under **Decision1** is greater than the profit under **Decision2**, this node is assigned the target level 1. All members of this node are classified as responders.

Step 2: Calculation of validation profits and selection of the right-sized tree (pruning)

First, the node definitions are used to partition the validation data into different nodes. Since each node is already assigned a target level based on the posterior probabilities and the profit matrix, I can calculate the total profit of each node of the tree using the validation data set. Since the profits are calculated from the Validation data set, they are called *validation profits*. Figure 4.5 shows the application of the tree to the Validation data set.

Figure 4.5

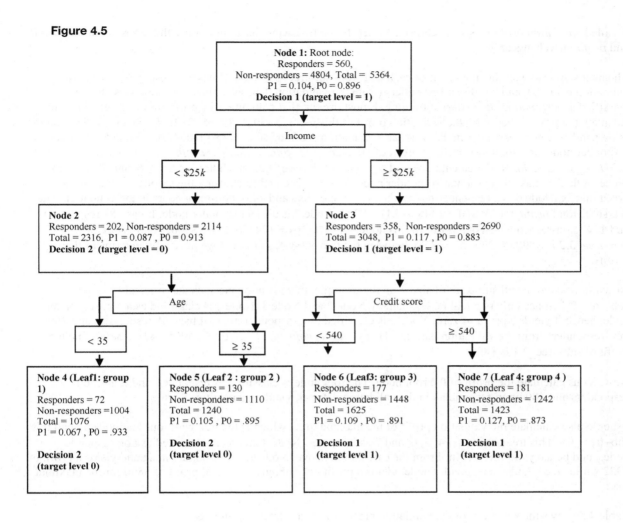

After applying the node definitions to the validation data set, I get the tree shown in Figure 4.5. Comparing the tree from the validation data (Figure 4.5) with the tree from the training data (Figure 4.4), notice that the decision labels (**Decision1** or **Decision2**) of each node are exactly the same in both the diagrams. This occurs because these decisions are based on the node definitions and the posterior probabilities generated during the training of the tree (using the training data set). They become part of the model and do not change when they are applied to a new data set.

Using these node definitions and this classification of the nodes, I will now demonstrate how an optimal tree is selected.

The tree in Figure 4.5 has four leaf nodes, and I refer to this tree as the *maximal tree*. However, within this tree there are several sub-trees of different sizes. There are two trees of size 3. That is, if I prune Nodes 6 and 7, I get a sub-tree with Nodes 3, 4, and 5 as its leaves. I call this sub-tree_3.1, indicating that it has three leaves and that it is the first one of the three-leaf trees. If I prune Nodes 4 and 5, I get another sub-tree of size 3, this time with Nodes 2, 6, and 7 as its leaves. I call this sub-tree_3.2. Altogether, there are two sub-trees of size 3. From these two, I must select the one that yields the higher profit. Since the profit is calculated using validation data set, it is called *validation profit* in SAS Enterprise Miner. In the following discussion, I use the terms *validation profit* and *profit* interchangeably.

The next step is to examine the profit of each leaf from sub-tree_3.1. Node 3 has 358 responders (from the validation data set), and 2690 non-responders (from the validation data set). Since this node was already classified as a *responder node* during training, I must use the first column of the profit matrix given above to calculate the profit. Therefore, the validation profit of this leaf is $890 (= 358*$10+2690*($–1)). Node 4, on the other hand, is a non-responder node, so it has a validation profit of $0 (=72*$0+1004*$0). Node 5 is also a non-responder node, and so, it also yields a profit of $0. *Hence, the total validation profit of all the leaves of sub_tree_3.1 is $890.* Now I calculate the total *profit* of sub–tree_3.2, which has Nodes 2, 6, and 7 as its leaves. Since Node 2 is classified as a non-responder node, its profit is equal to $0 using the second column of the profit matrix. Node 6 is a responder node; it has177 responders and 1448 non-responders in the validation data set, so its total *profit* = 177*$10 + 1448*($–1) = $322. Node 7 is also a responder node. It has 181 responders and 1242 non-responders. Total profit of this node = 181*($10) + 1242*($–1) = $568. *Total profit of sub–tree_3.2 is $890 ($322 + $568.)* So, between the two sub-trees of size 3 we have a tie in terms of their profits.

Now I must examine all sub-trees of size 2. In the example, there is only one such sub-tree, and I call it sub–tree_2.1. It has only two leaf nodes, namely Node 2 and Node 3. Node 2 is classified as a non-responder node; hence it yields a profit of zero. Node 3 is classified as a responder node. It has 358 responders and 2690 non-responders from the validation data set. This yields a *profit = 358*($10) + 2690*(–$1) = $890.* The total profit of sub–tree_2.1 is $890.

Now, let us calculate the profit of the root node. The root node is classified as responder, and it has 560 responders and 4804 non-responders in the validation data set, yielding a profit of $796.

I should also calculate the validation profit of the maximal tree, which is the tree with four leaf nodes. I call this sub–tree_4.1. This tree has Nodes 4, 5, 6, and 7 as its leaves. Node 4 and 5 are classified as non-responder nodes, and hence yield zero profit. From the calculation above, Node 6 is a responder node and yields a profit of $322 while Node 7, also a responder node, yields a profit of $568, giving a total profit of $890 for the maximal tree.

Table 4.5 shows the summary of the validation profits of all sub-trees and the tree:

Table 4.5

Summary of Validation Profits		
Tree	Number of leaves	Validation Profit
Root node	1	$796
sub_tree_2.1 (nodes 2 and 3)	2	$890
sub_tree_3.1 (nodes 3,4 and 5)	3	$890
Sub_tree_3.2 (nodes 2,6 and 7)	3	$890
Sub_tree_4.1 (nodes 4,5,6, and 7)	4	$890

The optimal tree is sub-tree_2.1 with two leaves, because all the trees with equally high profits are trees of larger size. As you go down the columns "Number of leaves" and "Validation profit", notice that, as the number of leaves increases, there is no increase in profit. Another way to think of this comparison with larger trees is to say that beyond sub–tree_2.1 the marginal profit = 0. That is, the incremental profit per additional leaf is zero. Hence sub-tree_2.1 is the optimum, or right-sized, tree. Note that sub–tree_4.1 is same as the maximal tree.

4.3.6 Average Profit vs. Total Profit for Comparing Trees of Different Sizes

In the above example I calculated the profit of a tree (or sub-tree) as the sum of the profits of its leaf nodes. The validation data has 5364 records, of which 560 are responders. Therefore *average profit* equals *total profit* divided by 5364. Since the denominator (number of records) is the same for all the trees compared, regardless of their number of leaves, I arrive at the same optimal tree using either average or total profits for comparison. Likewise, the sum of responders in all leaf nodes is equal to the total number of responders in the validation data set, which is 560 in this example. Hence calculating average profit as *Total profit*/560 for each (sub-)tree produces the same optimal tree.

4.3.7 Accuracy /Misclassification Criterion in Selecting the Right-sized Tree (Classification of Records and Nodes by Maximizing Accuracy)

Thus far, I have described how to classify each node and how to prune a tree using profit maximization as the criterion. Another often used criterion for both node classification and tree pruning is *validation accuracy*. In order to use this criterion for sub-tree selection or pruning in SAS Enterprise Miner, you must set the **Assessment Measure** property to Misclassification. Since misclassification rate = 1– validation accuracy, I use the terms *validation accuracy* and *misclassification* interchangeably.

When the Misclassification criterion is selected for assessment of the trees in pruning, the nodes are classified (assigned a target level) according to maximum posterior probability. This assignment is done during the training of the tree using the training data set.

4.3.7.1 Classification of the Nodes by Maximum Posterior Probability / Accuracy

This can be thought of as using a profit matrix of the following type:

Table 4.6

Accuracy		
Target	*Decision*	
	1	0
1	1	0
0	0	1

In this matrix, if a responder is correctly classified as a true responder, then one unit of accuracy is attained. If a non-responder is correctly classified as non-responder, then one unit of accuracy is gained. Otherwise, there is no gain.

As before, the nodes are classified as responder or non-responder nodes based on the posterior probabilities calculated from the training data set and the above profit matrix. In the root node of the training data set the proportion of responders is 9.9%, and the proportion of non-responders is 90.1%. Hence, if the root node is classified as a responder node, the accuracy would be 9.9%; if it is classified as non-responder node, the accuracy would be 90.1%. Hence, the optimal decision is to classify the root node as a non-responder node.

Similarly nodes 2, 3, 4, 5, 6, and 7 are all classified as non-responder nodes, since they all have more non-responders than responders.

4.3.7.2 Calculation of the Misclassification Rate / Accuracy Rate and Selection of the Right-sized Tree

Since the misclassification rate is 1 – validation accuracy, maximizing the validation accuracy is the same as minimizing the misclassification rate.

Table 4.7 shows the accuracy of the different sub-trees when applied to the validation data set. Using rules and partitioning definitions derived from the Training data set, all the nodes are classified as non-responder nodes. The tree has three sub-trees (not counting the root node as a tree): sub–tree_3.1 (with nodes 3, 4, and 5), sub-tree_3.2 (with nodes 2, 6, and 7), and sub-tree_2.1 (with nodes 2 and 3). The validation accuracy of sub-tree_3.1 = (2690+1004+1110)/5364 = 89.6%, and the misclassification rate is 10.4%. The validation accuracy of sub–tree_3.2 = (2114+1448+1242)/5364 = 89.6% and the corresponding misclassification rate is 10.4%. The validation accuracy of sub–tree_2.1 = (2114+2690)/5364 = 89.6%. The validation accuracy of the root node is (4804/5364) = 89.6%, and the misclassification rate is 10.4%. The validation accuracy of the maximal tree (with nodes 4, 5, 6, and 7) = (1004+1110+1448+1242)/5364= 89.6% with a misclassification rate of 10.4%. There is no gain in accuracy by going beyond the root node. Hence, the optimal or right-sized tree is a tree with the root node as its only leaf. As the number of leaves increased, there was no gain in accuracy.

Table 4.7

Summary of Validation Accuracy		
Tree	Number of leaves	Validation Accuracy
Root node	1	89.6%
Sub_tree_2.1 (nodes 2 and 3)	2	89.6%
Sub_tree_3.1 (nodes 3,4 and 5)	3	89.6%
Sub_tree_3.2 (nodes 2,6 and 7)	3	89.6%
Sub_tree_4.1 (nodes 4,5,6 and 7)	4	89.6%

4.3.8 Assessment of a Tree or Sub-tree Using Average Square Error

When the target is continuous, the assessment of a tree is done using sums of squared errors. As in the case of a binary target, I use the validation data set to perform the assessment. Therefore, I begin by applying the node definitions or rules of partitioning developed using the training data set to the validation data set. The worth of a node is measured (inversely) by the Average Square Error for that node. For the tth node, the Average Square

Error $= \dfrac{1}{n_t} \sum_{i=1}^{n_t} (y_{it} - \overline{y}_t)^2$, where y_{it} is the actual value of the target variable for the ith record in the tth node,

and $\overline{y}_t$ is the expected value of the target for the tth node. The expected value $\overline{y}_t$ is calculated from the

Training data set, but the observations y_{it} are from the Validation data set. The node definitions are constructed from the Training data set.

The overall Average Square Error for a tree or sub-tree is the weighted average of the average squared errors of the individual leaves, which can be given as follows:

$$ \sum_{t=1}^{T} \frac{n_t}{n} \left[\frac{1}{n_t} \sum_{i=1}^{n_t} (y_{it} - \overline{y}_t)^2 \right] . $$

Classification of nodes is not necessary for calculating the assessment value using Average Square Error (because the target in this case is continuous).

4.3.9 Selection of the Right-sized Tree

Display 4.5 gives another illustration of the selection of the right-sized tree from a sequence of optimal trees. In the graph shown in Display 4.5, the horizontal axis shows the number of leaves in each of the trees in the optimal sequence of trees. The vertical axis shows the average profit of each tree. The display clearly shows that the optimal tree has seven leaves. That is, a tree with seven leaf nodes has the same average validation profit as any tree with more leaves. Hence the tree with seven leaves is selected by the algorithm.

Display 4.5 shows the initially constructed maximal tree of size 14, before pruning it to a tree of size 7 using the technique described earlier.

Display 4.5

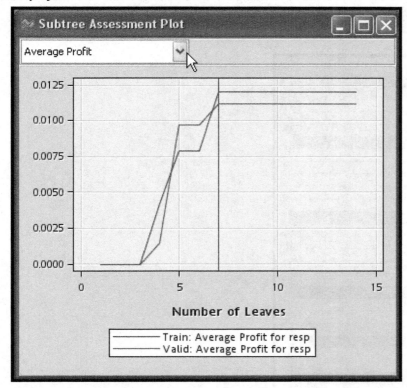

To get the Subtree Assessment plot as shown in Display 4.5, run the **Decision Tree** node as shown in Display 4.8 and open the Results window. Then click **View** →**Model** → **Subtree Assessment Plot**. A small chart window opens. Click on the down-arrow in the box located above the chart and select **Average Profit**. You will see a chart such as the one shown in Display 4.5. I used the profit matrix shown in Display 4.8 to calculate the average profit.

4.4 A Decision Tree Model to Predict Response to Direct Marketing

In this section, I develop a response model using the **Decision Tree** node. The model is based on simulated data representing the direct mail campaign of a hypothetical insurance company. The main purpose of this section is to show how to develop a response model using the **Decision Tree** node, and to make you familiar with the various components of the Results window.

Since this model predicts an individual's response to direct mailing, it is called a *response model*. In this model, the target variable (the variable that I want to predict) is binary: it takes on values of response and no response.

The process flow diagram for this model is shown in Display 4.6.

Display 4.6

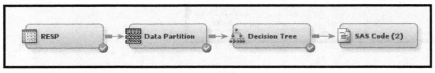

In the process flow diagram, the first node is the **Input Data** node. The properties of this node are shown in Display 4.7. You set the initial measurement levels of the variables by using the advanced settings of the

Metadata Advisor Options. In this example, we are not overriding the initial settings of the variables. The role of the variable Resp, however, is changed to Target.

Display 4.7

Property	Value	
General		
Node ID	Ids3	
Imported Data		[...]
Exported Data		[...]
Notes		[...]
Train		
Output Type	View	
Role	Raw	
Rerun	No	
Summarize	No	
Drop Map Variables	Yes	
Columns		
Variables		[...]
Decisions		[...]
Refresh Metadata		[...]
Advisor	Advanced	
Advanced Options		[...]
Data		
Data Selection	Data Source	
Sample	Default	
Sample Options		[...]
Data Source		
Data Source	RESP	[...]
Data Source Properties		[...]
New Table		
Table Name		[...]
Variable Validation	Strict	
New Variable Role	Reject	
Metadata		
Table	RESP	
Library	THEBOOK	

I enter a profit matrix for the **Decisions** property of the **Input Data** node. This matrix is shown in Display 4.8. This matrix is used for making decisions about node classification and for making assessments of sub-trees.

Display 4.8

Decision Processing - RESP

Targets | Prior Probabilities | Decisions | Decision Weights

Select a decision function:
- ⦿ Maximize
- ◯ Minimize

Enter weight values for the decisions.

Level	DECISION1	DECISION2
1 ...	5.0	0.0
0 ...	-1.0	0.0

OK | Cancel

The next node is the **Data Partition** node. Display 4.9 shows the settings of the properties of this node.

Display 4.9

Property	Value
General	
Node ID	Part2
Imported Data	
Exported Data	
Notes	
Train	
Variables	
Output Type	Data
Partitioning Method	Simple Random
Random Seed	12345
Data Set Allocations	
Training	60.0
Validation	30.0
Test	10.0
Report	
Interval Targets	Yes
Class Targets	Yes
Status	
Create Time	12/29/12 6:10 AM
Run ID	ea624b7f-03f5-4b4c-b799-31d32c7ca363

In this example, I allocated 60% of the data for training, 30% for validation, and 10% for testing. This allocation is quite arbitrary. You can use different proportions and then see how those proportions affect the results.

The third node in the process flow diagram is the **Decision Tree** node. The properties of the **Decision Tree** node are shown in Display 4.10.

Display 4.10

Property	Value
Node ID	Tree2
Imported Data	
Exported Data	
Notes	
Train	
Variables	
Interactive	
Use Frozen Tree	No
Use Multiple Targets	No
Precision	4
⊟ Splitting Rule	
Interval Criterion	ProbF
Nominal Criterion	ProbChisq
Ordinal Criterion	Entropy
Significance Level	0.2
Missing Values	Use in search
Use Input Once	No
Maximum Branch	2
Maximum Depth	6
Minimum Categorical Size	5
Split Precision	4
⊟ Node	
Leaf Size	5
Number of Rules	5
Number of Surrogate Rules	0
Split Size	.
⊟ Split Search	
Use Decisions	No
Use Priors	No
Exhaustive	5000
Node Sample	5000
⊟ Subtree	
Method	Assessment
Number of Leaves	1
Assessment Measure	Decision
Assessment Fraction	0.25

I kept the default values for all the properties. As a result, the nodes are split using the Chi-Square criterion. The threshold *p*-value for a split to be considered significant is set at 0.2. The sub-tree selection is done by comparing average profit since the **Sub Tree Method** property is set to Assessment, the **Assessment Measure** property is set to Decision, and a profit matrix is entered.

After running the **Decision Tree** node, you can open the Results window shown in Display 4.11.

Display 4.11

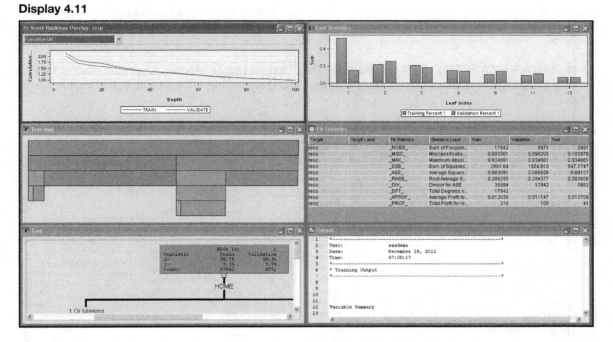

If you maximize the Tree pane in the Results window, you can see the tree as shown in Display 4.12.

Display 4.12

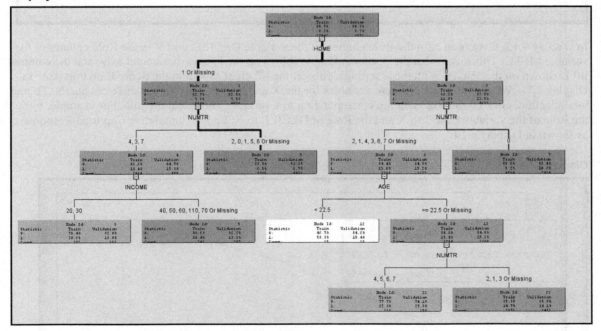

When you view the tree, the intensity of the color of the node indicates the response rate of the customers included in the node. In the tree above, Node 12 has a white background (least intensity of color), which indicates the highest response rate (in the Training data set). The customers in this node do not own a home (HOME= 0), they own one or more credit cards (NUMTR=1, 2, 3, 4, 6, 7) and they are young (AGE < 22.5). This node also includes those customers with missing values for the variable NUMTR.

The Cumulative Lift chart is shown in the Score Rankings Overlay pane in the top left-hand section of the Results window. To see a list of the available charts, click the arrow on the box where the chart type is displayed.

An alternative way of making charts is through the Data Options Dialog window. Open this window by right-clicking in the chart area, and selecting **Data Options**. The Data Options Dialog window is shown in Display 4.13.

Display 4.13

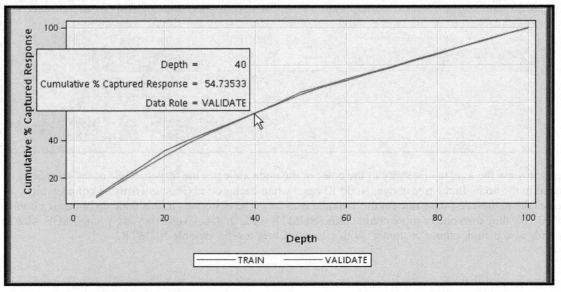

In Display 4.13, there is an X in the Role column of the variable DECILE, and Y in the Role column of the variable LIFTC. This means that the decile ranking is shown on the X-axis (horizontal axis) and the cumulative lift is shown on the Y axis. With these settings, you get the lift chart shown in the Score Rankings pane in Display 4.11. When you select different variables for the X and Y axes, you can get different charts. To make this selection, click on the Role column corresponding to a variable, and select the role. For example, by setting the Role of the variable CAPC to Y and the Role of DECILE to X, I get a Cumulative Captured Response chart, as shown in Display 4.14.

Display 4.14

According to this chart, if I target the top 40% of the customers ranked by the predicted probability of response, then I will capture 54.7% percent of all responders. These capture rates are not impressive, but they are sufficient for demonstrating the tool.

To get a view of the pruning process displayed by the Subtree Assessment plot, select **View→Model→ Subtree Assessment Plot** in the Results window, as shown in Display 4.15.

Display 4.15

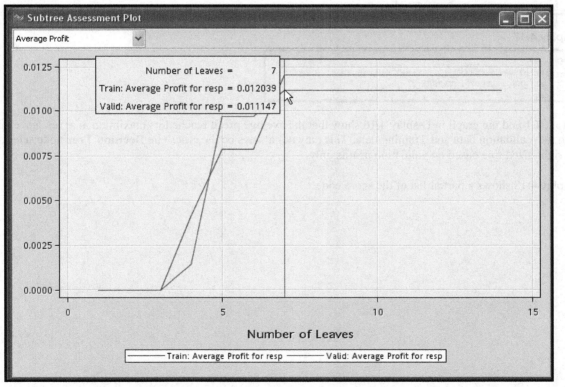

The plot generated by choosing the settings in Display 4.15 is shown in Display 4.16.

Display 4.16

Display 4.16 shows the average profit for trees of different sizes. On the horizontal axis, the number of leaves is shown indicating the size of the tree. On the vertical axis, the average profit of the tree is shown. The calculation of the average profit is illustrated in Section 4.2.5 and in Step 2 of Section 4.3.5.

The graph shown in Display 4.16 depicts the pruning process. An initial tree with 14 leaves is grown with the training data set, and a sub-tree of only 7 leaves is selected using the validation data set. The selected tree is shown in Display 4.12.

The data behind these charts can be retrieved by clicking the **Table** icon on the toolbar in the Results window. The data is stored in a SAS data set in the project directory. Output 4.1 is printed from the saved data set, and it shows the data underlying the graph shown in Display 4.16. The SAS code shown in Display 4.17 is run from the **SAS code** node for generating Output 4.1.

Output 4.1

Number of Leaves	Train: Average Profit for resp	Valid: Average Profit for resp
1	0.000000	0.000000
2	0.000000	0.000000
3	0.000000	0.000000
4	0.004180	0.001449
5	0.007859	0.009698
6	0.007859	0.009698
7	0.012039	0.011147
8	0.012039	0.011147
9	0.012039	0.011147
10	0.012039	0.011147
11	0.012039	0.011147
12	0.012039	0.011147
13	0.012039	0.011147
14	0.012039	0.011147

Display 4.17

```
Training Code

  proc print data=&EM_LIB..tree2_outseq label noobs;
    var _NW_ _APROF_ _VAPROF_;
  run;
```

Output 4.1 and the graph in Display 4.16 show that the average profit reached its maximum at seven leaves for both the Validation data and Training data. This may not always be the case. The **Decision Tree** node selects the right-sized tree based on validation profits only.

Display 4.18 shows a partial list of the score code.

Display 4.18

```
 28   ******              ASSIGN OBSERVATION TO NODE              ******;
 29   _ARBFMT_12 = PUT( HOME , BEST12.);
 30   %DMNORMIP( _ARBFMT_12);
 31   IF _ARBFMT_12 IN ('0' ) THEN DO;
 32     _ARBFMT_12 = PUT( NUMTR , BEST12.);
 33     %DMNORMIP( _ARBFMT_12);
 34     IF _ARBFMT_12 IN ('0' ) THEN DO;
 35       _NODE_   =                   7;
 36       _LEAF_   =                   7;
 37       P_resp1  =       0.09239130434782;
 38       P_resp0  =       0.90760869565217;
 39       Q_resp1  =       0.09239130434782;
 40       Q_resp0  =       0.90760869565217;
 41       V_resp1  =       0.10622406639004;
 42       V_resp0  =       0.89377593360995;
 43       I_resp   = '0' ;
 44       U_resp   =                   0;
 45     END;
 46     ELSE DO;
 47       IF  NOT MISSING(AGE ) AND
 48         AGE  <                   22.5 THEN DO;
 49         _NODE_   =                 12;
 50         _LEAF_   =                  4;
 51         P_resp1  =     0.53333333333333;
 52         P_resp0  =     0.46666666666666;
 53         Q_resp1  =     0.53333333333333;
 54         Q_resp0  =     0.46666666666666;
 55         V_resp1  =     0.15384615384615;
 56         V_resp0  =     0.84615384615384;
 57         I_resp   = '1' ;
 58         U_resp   =                  1;
 59       END;
 60       ELSE DO;
 61         _ARBFMT_12 = PUT( NUMTR , BEST12.);
```

The segment of the code included in Display 4.18 shows how an observation is assigned to Leaf Node 7. Note that P_resp1 = 0.09239130434782 is computed from the Training data set and it is called the *posterior probability of response*. P_resp0 = 0.90760869565217 is also computed from the Training data set. It is called the *posterior probability of non-response*. These posterior probabilities are the same as $\hat{P}_1$ and $\hat{P}_0$ discussed in Section 4.2.2. Any observation that meets the definition of Leaf Node 1 is assigned these posterior probabilities. This, in essence, is how a decision tree model predicts the probability of response. This code can be applied to score the target population, as long as the data records in the prospect data set are consistent with the sample used for developing the model. Display 4.19 shows another segment of the score code, which shows the decision-making process in classifying customers as responders or non-responders.

204

Display 4.19

```
135   *****   DECISION VARIABLES *******;
136
137   *** Decision Processing;
138   label D_RESP = 'Decision: resp' ;
139   label EP_RESP = 'Expected Profit: resp' ;
140
141   length D_RESP $ 9;
142
143   D_RESP = ' ';
144   EP_RESP = .;
145
146   *** Compute Expected Consequences and Choose Decision;
147   _decnum = 1; drop _decnum;
148
149   D_RESP = '1' ;
150   EP_RESP = P_resp1 * 5 + P_resp0 * -1;
151   drop _sum;
152   _sum = P_resp1 * 0 + P_resp0 * 0;
153   if _sum > EP_RESP + 2.273737E-12 then do;
154      EP_RESP = _sum; _decnum = 2;
155      D_RESP = '0' ;
156   end;
157
158
159   *** End Decision Processing ;
160
161   ***************************************************************;
```

This code shows how decisions are made with regard to assigning an observation to a target level (such as response and non-response). The decision-making process is also discussed in Section 4.2.2 and Step 1C of Section 4.3.5.

The decision process is clear from the code segments in Displays 4.18 and 4.19. In any data set, an observation is assigned to a node first, as shown in Display 4.18. Then, based on the posterior probabilities (P_resp1 and P_resp0) of the assigned node and the profit matrix, expected profits are computed under **Decision1** and **Decision2**. The expected profit under **Decision1** is given by the SAS statement EP_RESP = P_resp1*5 + p_resp0*(-1). Note that the numbers 5 and -1 are from the first column of the profit matrix. If you classify a true responder as a responder, then you make a profit of $5, and if you classify a non-responder as a responder, the profit is −$1.0. Hence, EP_RESP represents the consequences of classifying an observation as a responder. The alternative decision is to classify an observation as a non-responder, and this is given by the SAS statement _sum = P_resp1*0 + P_resp0* 0. The two numbers 0 and 0 are the second column of the profit matrix. The code shows that if _sum > EP_RESP, then the decision is to assign the target level 0 to the observation. The variable that indicates the decision is D_RESP, which takes the value 0 if the observation is assigned to the target level 0, or 1 if the observation is assigned to the target level 1. Since observations that are assigned to the same node will all have the same posterior probabilities, they are all assigned the same target level.

4.4.1 Testing Model Performance with a Test Data Set

I used the Training data set for developing the node definitions and posterior probabilities, and I used the Validation data set for pruning the tree to find the optimal tree. Now I need an independent data set on which to evaluate the model. The test data set is set aside for this assessment. I connect the **Model Comparison** node to the **Decision Tree** node and run it. The Results window of the **Model Comparison** node shows the lift charts for all three data sets. These charts are shown in Display 4.20.

Display 4.20

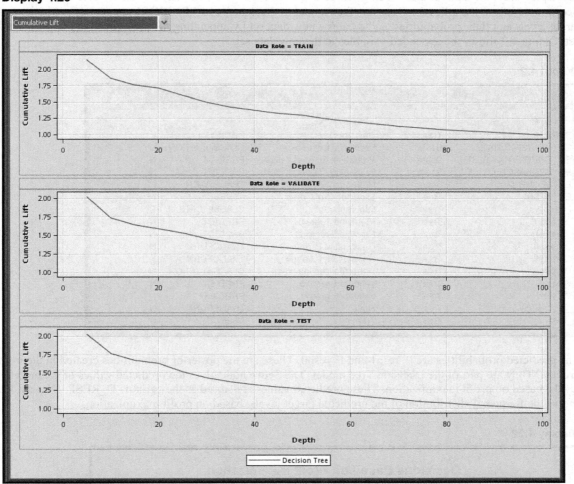

The performance of the model on the test data is very similar to its performance on the Training and Validation data sets. This comparison shows that the model is robust.

4.4.2 Applying the Decision Tree Model to Score a Data Set

This section shows the application of the decision tree model developed earlier to score a prospect data set. The records in the prospect data set do not have the value of the target variable. The target variable has two levels, response and no response. I use the model to predict the probability of response for each record in the prospect data set.

I use the **Score** node to apply the model and make the predictions. I add an **Input Data** node for the prospect data and a **Score** node in the process flow, as shown in Display 4.21.

Display 4.21

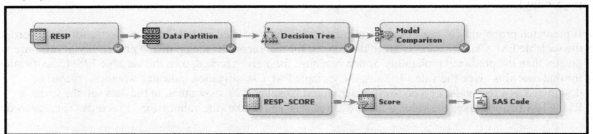

In Display 4.21, the data set that is being scored is named RESP_SCORE, and its role is set to Score. When the **Score** node is run, it applies the Decision Tree model to the data set RESP_SCORE, generates predictions of the target levels and their probabilities, and creates a new data set by appending the predictions to the prospect data set. Output 4.2 shows the appended variables in the new data set.

Output 4.2

Variable Name	Creator	Variable Label	Function	Type
D_RESP	Tree	Decision: resp	DECISION	C
EM_CLASSIFICATION	Score	Prediction for resp	CLASSIFICATION	C
EM_EVENTPROBABILITY	Score	Probability for level 1 of resp	PREDICT	N
EM_PROBABILITY	Score	Probability of Classification	PREDICT	N
EM_SEGMENT	Score	Node	TRANSFORM	N
EP_RESP	Tree	Expected Profit: resp	ASSESS	N
I_resp	Tree	Into: resp	CLASSIFICATION	C
P_resp0	Tree	Predicted: resp=0	PREDICT	N
P_resp1	Tree	Predicted: resp=1	PREDICT	N
Q_resp0	Tree	Unadjusted P: resp=0	TRANSFORM	N
Q_resp1	Tree	Unadjusted P: resp=1	TRANSFORM	N
U_resp	Tree	Unnormalized Into: resp	CLASSIFICATION	N
V_resp0	Tree	Validated: resp=0	PREDICT	N
V_resp1	Tree	Validated: resp=1	PREDICT	N
NODE	Tree	Node	TRANSFORM	N
WARN	Tree	Warnings	ASSESS	C

The predicted probabilities are P_resp1 and P_resp0. These are the posterior probabilities created in the training process. They are part of the Decision Tree model. The **Score** node gives the predicted values of the target levels, based on profit maximization. These predicted values are stored as the variable D_RESP. Display 4.22 shows the frequency distribution of the predicted target levels, based on profit maximization.

Display 4.22

Decisions Based on Profit Maximization

The FREQ Procedure

Decision: resp		
D_RESP	Frequency	Cumulative Frequency
Decision 0: Profit under decision 1 < Profit under decision 0	28791	28791
Decision 1: Profit under decision 1 >= Profit under decision 0	1113	29904

Display 4.22 shows that, under profit maximization, 1113 out of the total 29904 customers are designated as responders. The remaining 28791 customers are designated as non-responders. The score code gives an alternative prediction of the target levels based on predicted probability (which is the same as the posterior probability).

If predicted probability of non-response is greater than the predicted probability of response for any record, then the variable EM_Classification is given the value 0 for that record. If the predicted probability of response is greater than the predicted probability of non-response for a given record, then the variable EM_Classification for that record is given the value 1. Thus, the variable EM_Classification indicates whether a record is designated as a responder or a non-responder record. Display 4.23 shows that, in the data set, the variable EM_Classification takes the value 0 for 29873 records, and it takes the value 1 for 31 records. Thus, according to the decisions based on error minimization, only 31 customers (records) are designated as responders, while 29867 customers are designated as non-responders.

The frequency distribution of EM_Classification is shown in Display 4.23.

Display 4.23

Decisions Based on Maximum Posterior Probability			
The FREQ Procedure			
Frequency	Table of CL by EM_CLASSIFICATION		
		EM_CLASSIFICATION(Prediction for resp)	
CL	0	1	Total
Decision 0: P_resp1 < P_resp0	29873	0	29873
Decision 1: P_resp1 >= P_resp0	0	31	31
Total	29873	31	29904

The program that generated Displays 4.22 and 4.23 is shown in Display 4.24. It was run in the **SAS code** node.

Display 4.24

```
proc format ;
 value clfmt
 1 = 'Decision 1: P_resp1 >= P_resp0'
 0 = 'Decision 0: P_resp1 <  P_resp0';
 value $pfmt
 '1' = 'Decision 1: Profit under decision 1 >= Profit under decision 0'
 '0' = 'Decision 0: Profit under decision 1 <  Profit under decision 0' ;
run;
/*
data TheBook.Ch4_score_tables;
 set &EM_LIB..score_SCORE;
 CL = P_resp1 ge p_resp0;
run;
*/
 ods html file='C:\Thebook\EM12.1\Reports\Chapter4_Dec1.html' ;
 proc freq data=TheBook.Ch4_score_tables;
  table D_resp /nopercent norow nocol ;
  title 'Decisions Based on Profit Maximization' ;
 format d_resp $pfmt.;
 run;
 ods html close;

 ods html file='C:\Thebook\EM12.1\Reports\Chapter4_Dec2.html' ;
 proc freq data=TheBook.Ch4_score_tables;
  table cl*EM_CLASSIFICATION /nopercent norow nocol ;
  title 'Decisions Based on Maximum Posterior Probability' ;
 format cl clfmt.;
 run;
 ods html close;
```

Displays 4.25A and 4.25B show segments of the Flow code generated by the **Score** node, which shows how the E_RESP and EM_Classification are calculated. To see the Flow code, click **Score node → Results → SAS Results → Flow Code**. By scrolling down in this window, you can see the code shown in Displays 4.25A and 4.25B.

Display 4.25A

```
 Flow Code                                                    _ □ ⊠
201    *** Compute Expected Consequences and Choose Decision;
202    _decnum = 1; drop _decnum;
203
204    D_RESP = '1' ;
205    EP_RESP = P_resp1 * 5 + P_resp0 * -1;
206    drop _sum;
207    _sum = P_resp1 * 0 + P_resp0 * 0;
208    if _sum > EP_RESP + 2.273737E-12 then do;
209      EP_RESP = _sum; _decnum = 2;
210      D_RESP = '0' ;
211    end;
212
213    *** Decision Matrix;
214    array TREEdema [2,2] _temporary_ (
215    /* row 1 */  5 0
216    /* row 2 */  -1 0
217    );
```

Display 4.25B

```
 Flow Code                                                    _ □ ⊠
248    *-------------------------------------------------------*;
249    * Score: Creating Fixed Names;
250    *-------------------------------------------------------*;
251    LABEL EM_SEGMENT = 'Node';
252    EM_SEGMENT = _NODE_;
253    LABEL EM_EVENTPROBABILITY = 'Probability for level 1 of resp';
254    EM_EVENTPROBABILITY = P_resp1;
255    LABEL EM_PROBABILITY = 'Probability of Classification';
256    EM_PROBABILITY =
257    max(
258    P_resp1
259    ,
260    P_resp0
261    );
262    LENGTH EM_CLASSIFICATION $%dmnorlen;
263    LABEL EM_CLASSIFICATION = "Prediction for resp";
264    EM_CLASSIFICATION = I_resp;|
265    LABEL EM_CLASSTARGET = 'Target Variable: resp';
266    EM_CLASSTARGET = F_resp;
```

4.5 Developing a Regression Tree Model to Predict Risk

In this section, I develop a decision tree model to predict risk. In general, *risk* is defined as the expected amount of loss to the company. An auto insurance company incurs losses resulting from such events as accidents or thefts. An auto insurance company is often interested in predicting the expected loss frequency of its customers or prospects. The insurance company can set the premium for a customer according to the predicted loss frequency. *Loss frequency* is measured as the number of losses per car year, where car year is defined as the duration of the insurance policy multiplied by the number of cars insured under the policy. If the insurance policy has been in effect for four months and one car is covered, then the number of car years is 0.3333. If there is one loss during this four months then the loss frequency is $1/(4/12) = 3$. If, during these four months, two cars are covered under the same policy, then the number of car years is 0.6666; and if there is one loss, the loss frequency is 1.5.

The target variable in this model is the LOSSFRQ (loss frequency), which is continuous. When the target variable is continuous, the tree constructed is called a regression tree. A *regression tree*, like a decision tree, partitions the records (customers) into segments or *nodes*. The *regression tree model* consists of the leaf nodes.

While posterior probabilities are calculated for each leaf node in a decision tree model, the expected value or mean of the target variable is calculated for each leaf node in a regression tree model. In the current example, the **Tree** calculates the expected frequency of losses for each leaf node or segment. These target means are calculated from the training data.

The process flow for modeling risk is shown in Display 4.26.

Display 4.26

In this example, the data source is created from the SAS table called LossFrequency, and I used Advanced Metadata Options. I set the **Class Levels Count Threshold** to 5.

In this example, the modeling data set is partitioned such that 60% of the records are allocated for training, 30% for validation, and 10% for testing. The selection method used for this partitioning is simple random sampling.

The property settings of the **Decision Tree** node are shown in Display 4.27.

Display 4.27

Splitting Rule	
Interval Criterion	ProbF
Nominal Criterion	ProbChisq
Ordinal Criterion	Entropy
Significance Level	0.2
Missing Values	Use in search
Use Input Once	No
Maximum Branch	2
Maximum Depth	6
Minimum Categorical Size	5
Split Precision	4
Node	
Leaf Size	5
Number of Rules	5
Number of Surrogate Rules	0
Split Size	.
Split Search	
Use Decisions	No
Use Priors	No
Exhaustive	5000
Node Sample	20000
Subtree	
Method	Assessment
Number of Leaves	1
Assessment Measure	Average Square Error
Assessment Fraction	0.25

The main difference between trees with binary targets and trees with continuous targets is in the test statistics used for evaluating the splits and the calculation of the worth of the trees. For the continuous target, the test statistic has an F-distribution. The model assessment is done by using the Error Sums of Squares. The **Splitting Rule Interval Criterion** property is ProbF, which is default for interval targets. The **Assessment Measure** property for assessment is set to Average Square Error. Section 4.3.1.4 discusses the ProbF criterion for calculating the worth of a split. Section 4.3.8 discusses how Average Square Error is calculated.

Displays 4.28A through 4.28C show the Tree model developed by the **Decision Tree** node.

Display 4.28A

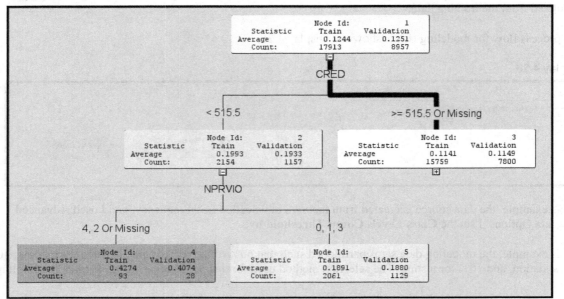

Display 4.28B

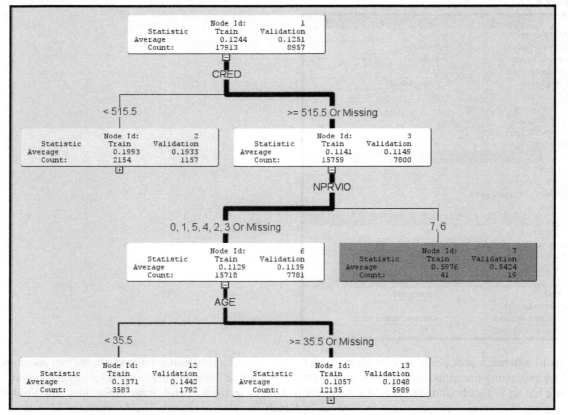

Display 4.28C

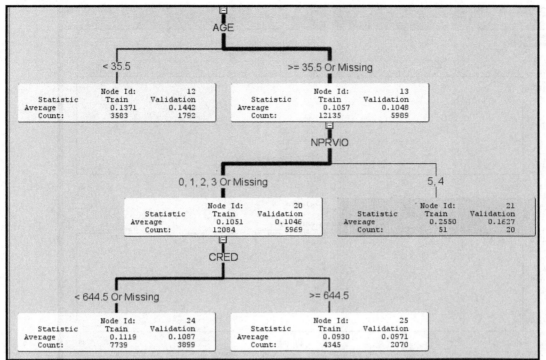

The variables defining the nodes of the tree are NPRVIO (number of prior violations), CRED (Credit score), and AGE (Age).

Display 4.29, which is generated by the **Model Comparison** node, shows the mean of predicted loss frequency in the Training, Validation and Test data sets. You need to compare the predicted means with the actual means to assess how well the model succeeds in ranking the various risk groups. Displays 4.29A, 4.29B and 4.29C show the actual and predicted mean of the target variable (Loss Frequency) by percentile for the Training, Validation, and Test data sets.

Display 4.29

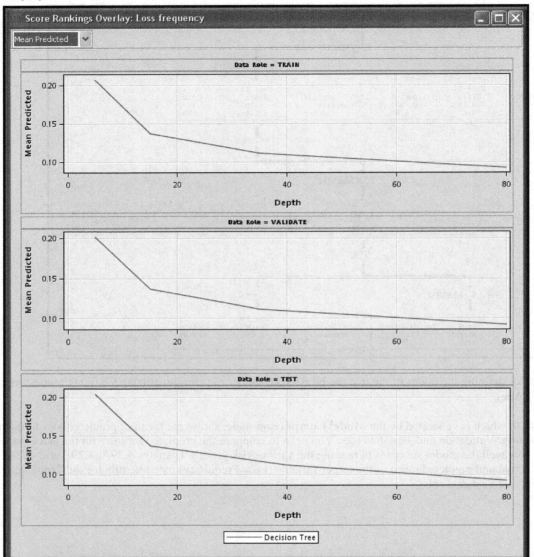

The percentiles computed from ordered values of the predicted loss frequency are shown on the horizontal axis, with those customers with the highest loss frequency in the first percentile and those with the lowest losses in the tenth percentile. The vertical axis shows the mean of predicted loss frequency for each percentile.

Display 4.29A

Comparison of Predicted and Actual Mean Loss Frequency by Decile Training Dataset

Depth	Mean Predicted	Mean Target
5	0.20787	0.20787
15	0.13708	0.13708
35	0.11187	0.11187
80	0.09303	0.09303

Display 4.29B

Comparison of Predicted and Actual Mean Loss Frequency by Decile
Validation Dataset

Depth	Mean Predicted	Mean Target
5	0.20222	0.19835
15	0.13708	0.14419
35	0.11187	0.10866
80	0.09303	0.09706

Display 4.29C

Comparison of Predicted and Actual Mean Loss Frequency by Decile
Test Dataset

Depth	Mean Predicted	Mean Target
5	0.20447	0.18704
15	0.13708	0.11251
40	0.11187	0.09934
80	0.09303	0.09570

Displays 4.29A to 4.29C show that, for the Training, Validation and Test data sets, the mean of the actual loss frequency (Mean Target) is highest for the 5th percentile, the mean of the actual loss frequency for the 15th percentile is lower than that of the 5th percentile, the mean of the actual loss frequency for the 40th percentile is lower than that of the 15th percentile, etc. The pattern of Mean Target exhibited in Displays 4.29A to 4.29C shows that the model is rating the risk groups correctly.

To generate the tables presented in Displays 4.29A to 4.29C, you can go to the Code Editor window of the **SAS Code** node, type in SAS code similar to the code in Display 4.29D, and run it.

Display 4.29D

```
Training Code
      ods html file='C:\TheBook\EM12.1\Reports\Chapter4\Ch4_Risk_Train.html';
   proc print data=&EM_IMPORT_RANK label noobs    ;
     var decile  _MEANP_  _TARGETMEAN_ ;
     where DATAROLE ="TRAIN" ;
     Title "Comparison of Predicted and Actual Mean Loss Frequency by Decile";
     Title2 "Training Dataset";
   run;
     ods html close;
     ods html file='C:\TheBook\EM12.1\Reports\Chapter4\Ch4_Risk_Valid.html';
   proc print data=&EM_IMPORT_RANK label noobs  ;
     var decile  _MEANP_  _TARGETMEAN_ ;
     where DATAROLE ="VALIDATE" ;
     Title "Comparison of Predicted and Actual Mean Loss Frequency by Decile";
     Title2 "Validation Dataset";
   run;
     ods html close;

     ods html file='C:\TheBook\EM12.1\Reports\Chapter4\Ch4_Risk_Test.html';
   proc print data=&EM_IMPORT_RANK label noobs  ;
     var decile  _MEANP_  _TARGETMEAN_ ;
     where DATAROLE ="TEST" ;
     Title "Comparison of Predicted and Actual Mean Loss Frequency by Decile";
     Title2 "Test Dataset";
   run;
     ods html close;
```

You can see a plot of the actual and predicted means for the training and validation data sets in the Results window of the Decision Tree Node .

4.5.1 Summary of the Regression Tree Model to Predict Risk

Although the regression tree shown in Displays 4.28A-4.28C is developed from simulated data, it provides a realistic illustration of how this method can be used to identify the profiles of high-risk groups.

We used the regression tree model developed in this section to group the records (in this case, the customers) in the data set into seven disjoint groups—one for each of the seven leaf nodes or terminal nodes on the tree—as shown in Display 4.30. The tree model also reports the mean of the target variable for each group, calculated from the Training data set. In this example, these means are used as predictions of the target variable loss frequency for each of the ten groups. Since the mean of the target variable for each group can be used for prediction, I refer to it as Predicted Loss Frequency in Display 4.30. The predicted loss frequency of a group reflects the level of risk associated with the group. The groups are shown in decreasing order of their risk levels. Display 4.30 also shows the profiles of the different risk groups, that is, their characteristics with regard to previous violations, credit score, and age.

Display 4.30

Profiles of customers with different risk ranks from the Regression Tree Model				
Input Ranges	Node ID	Number of Observations	Predicted Loss Frequency	Risk Rank
NPRVIO IS ONE OF: 7, 6 AND CRED >= 515.5 or MISSING	7	41	0.5976	1
NPRVIO IS ONE OF: 4, 2 or MISSING AND CRED < 515.5	4	93	0.4274	2
NPRVIO IS ONE OF: 5, 4 AND CRED >= 515.5 or MISSING AND AGE >= 35.5 or MISSING	21	51	0.2550	3
NPRVIO IS ONE OF: 0, 1, 3 AND CRED < 515.5	5	2061	0.1891	4
NPRVIO IS ONE OF: 0, 1, 5, 4, 2, 3 or MISSING AND CRED >= 515.5 or MISSING AND AGE < 35.5	12	3583	0.1371	5
NPRVIO IS ONE OF: 0, 1, 2, 3 or MISSING AND CRED < 644.5 AND CRED >= 515.5 or MISSING AND AGE >= 35.5 or MISSING	24	7739	0.1119	6
NPRVIO IS ONE OF: 0, 1, 2, 3 or MISSING AND CRED >= 644.5 AND AGE >= 35.5 or MISSING	25	4345	0.0930	7
NPRVIO=Number of prior violations CRED = Credit Score AGE = Age of the insured				

In an effort to prevent unfair or discriminatory practices by insurance companies, many state agencies that oversee the insurance industry have decreed that certain variables cannot be used for risk rating, rate setting, or both. The variables that appear in Display 4.30 are not necessarily legally permitted variables. The table is given only for illustrating the Regression Tree model.

4.6 Developing Decision Trees Interactively

You can develop decision and regression trees either autonomously or interactively. We developed autonomous trees in Sections 4.4 and 4.5. In this section, I show how to develop decision trees interactively. You can use the **Interactive** property of the **Decision Tree** node either to modify a tree that was developed previously or to develop an entire tree starting from the root node. To demonstrate the Interactive facility of the **Decision Tree** node, I created the process flow diagram shown in Display 4.31. This process flow diagram includes three instances of the **Decision Tree** node. I use the first instance of the **Decision Tree** node, denoted by ①, to demonstrate how to modify a previously created tree in interactive mode. I use the second instance of the **Decision Tree** node, denoted by ②, to develop an entire tree interactively starting from the root node. Finally, I use the **Decision Tree** node denoted by ③ to develop a maximal tree in an interactive mode. The size of the maximal tree is controlled by the values of the **Significance Level**, **Maximum Depth**, **Leaf Size**, and **Split Adjustment** properties.

Display 4.31

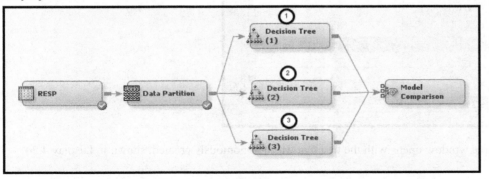

4.6.1 Interactively Modifying an Existing Decision Tree

I first created a tree autonomously by running the **Decision Tree** node ① located at the top of the process flow diagram. The resulting tree is shown in Display 4.32.

Display 4.32

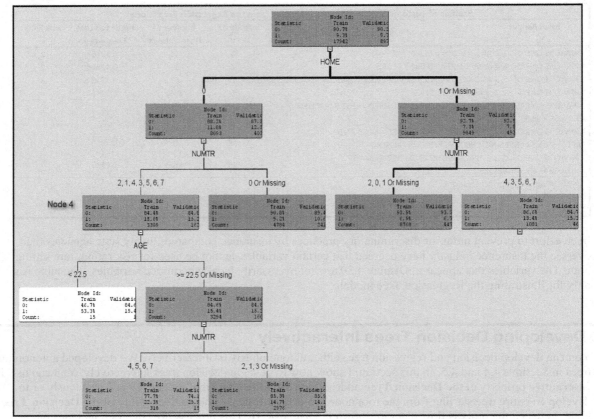

I switched to the interactive mode by clicking ⎡⋯⎤ located at the right of the **Interactive** property of the **Decision Tree** node, as shown in Display 4.33.

Display 4.33

General	
Node ID	Tree3
Imported Data	⋯
Exported Data	⋯
Notes	⋯
Train	
Variables	⋯
Interactive	⋯
Use Frozen Tree	No
Use Multiple Targets	No
Precision	4

The Interactive Tree window opens with the tree that was autonomously created, shown in Display 4.34.

Display 4.34

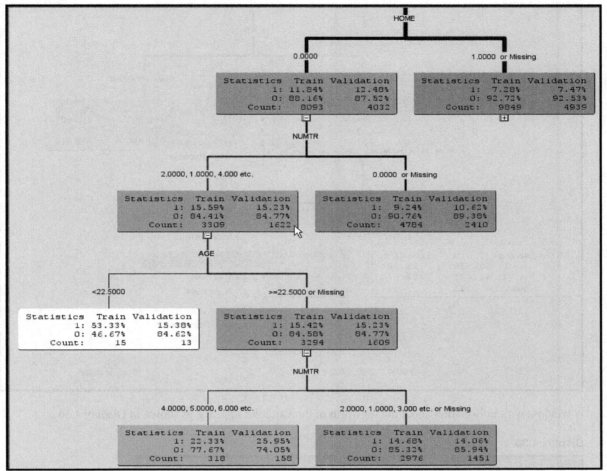

4.6.1.1. Changing the Split Point of an Interval-Scaled Variable

To modify this tree, select the node whose definitions you want to change, as shown by the mouse-pointer in Display 4.34. In this example, I select Node 4 (see Display 4.32) for modification. You can see in Display 4.32 that this node is split by using the AGE variable, with the split point at 22.5 years. To change the split point for this node (the terms *split point* and *split value* are used interchangeably) to some other value, right-click on Node 4 and select **Split Node**, as shown in Display 4.35.

Display 4.35

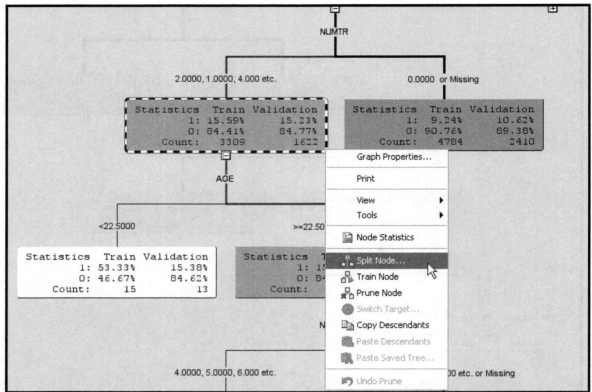

A window with information about the logworth of the variables appears, as shown in Display 4.36.

Display 4.36

Split Node 4				✕
Target Variable: resp				
Variable	Variable Description	-Log(p)	Branches	
AGE	AGE	1.8274	2	
NUMTR	NUMTR	1.4651	2	

Edit Rule...

OK   Cancel   Apply   Refresh

The AGE variable has a logworth of 1.8274. Since it has the highest logworth of any input variable listed in Display 4.36, it was chosen for splitting the node, and the node was split at 22.5 years of age. To change the split points, click **Edit Rule**. The Split Rule window opens, as shown in Display 4.37.

Display 4.37

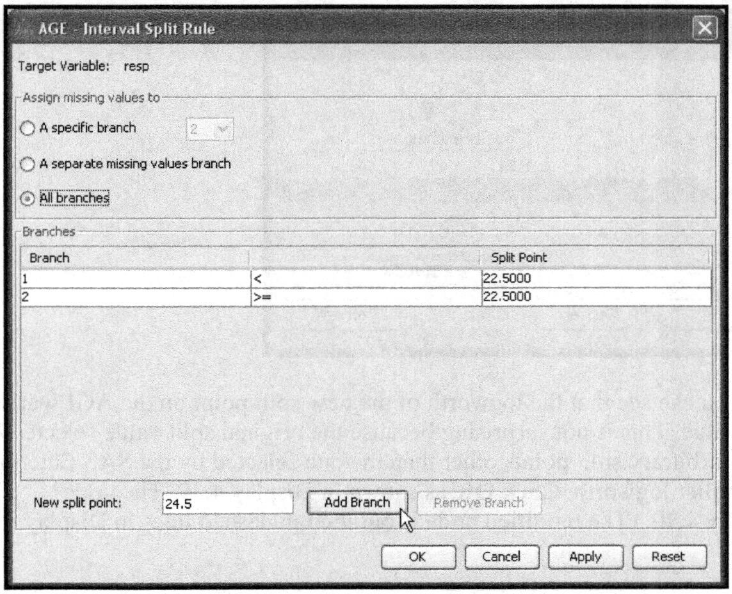

Type the new split value (24.5) in the **New split point** box and click **Add Branch**. A new branch 3 is added, as shown in Display 4.38. To remove the old split point, highlight branch 1and click **Remove Branch**. The branches are renumbered so that 2 and 3 are now 1 and 2, respectively. The split rule is changed.

Display 4.38

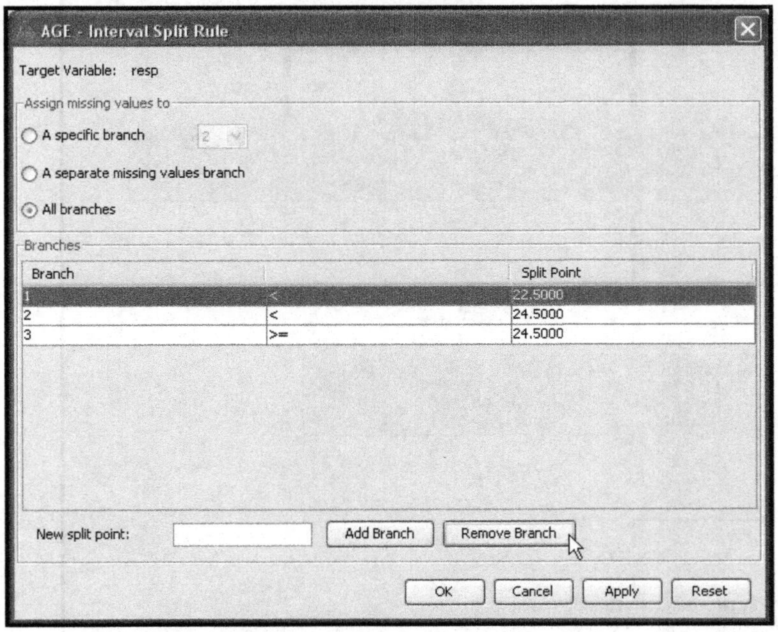

Click **Apply**, and then click **OK**. Close the Split Node window by clicking OK. The logworth of the new split value is calculated, which is shown in Display 4.39.

Display 4.39

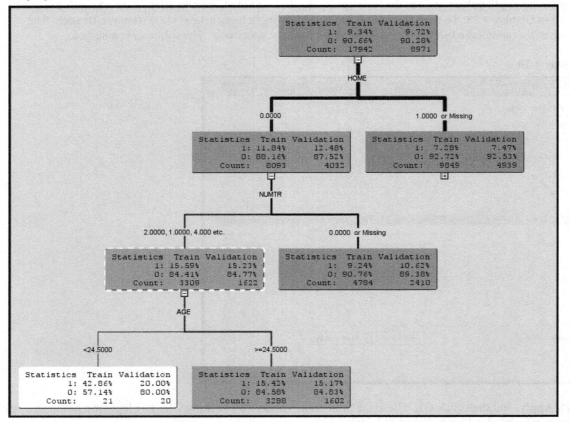

By comparing Displays 4.36 and 4.39, you can see that the logworth of the new split point on the AGE variable is smaller than that of the original split value. This is not surprising because the original split value selected by the tree algorithm was the best one. Any arbitrary split points other than the one selected by the SAS Enterprise Miner Tree algorithm tends to have a smaller logworth. Click **OK** as shown in Display 4.39. The node definition is modified as shown in Display 4.40. (The modified node is outlined in dashed lines in Display 4.40.)

Display 4.40

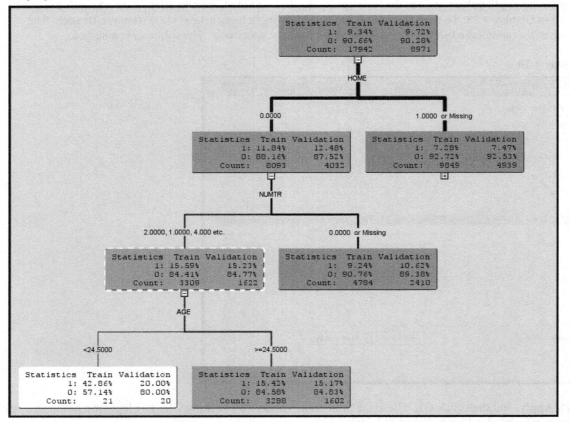

To further split the node with age >=24.5, right-click on it and select **Train Node** from the menu. The node is split, as shown in Display 4.41.

Display 4.41

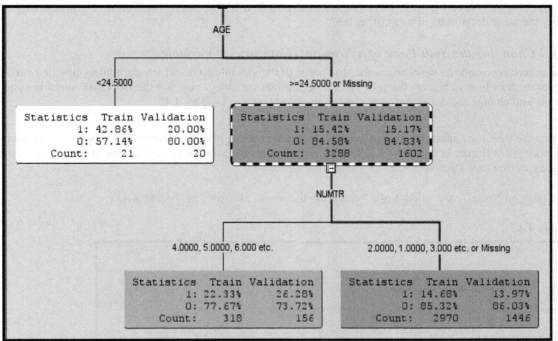

Close the Interactive Tree window by clicking the X in the top right-hand corner and open the Results window where the interactively created tree appears as shown in Display 4.42.

Display 4.42

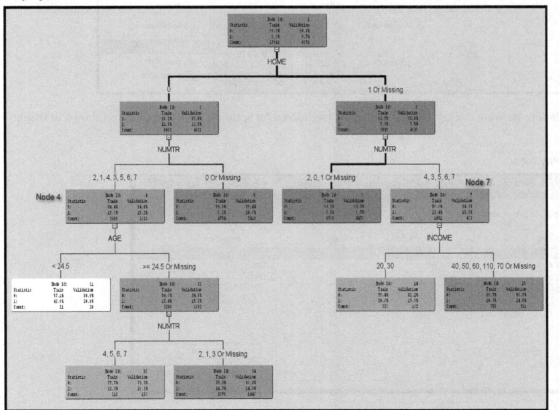

By comparing the tree in Display 4.42 with the tree in Display 4.32, you see that we have successfully changed the point at which Node 4 is split. The change we have made is trivial, but it is sufficient to demonstrate how to modify the node definitions of an existing tree.

4.6.1.2 Changing the Split Point of a Nominal (Categorical) Variable

In the preceding example, we changed the split value of the variable AGE, which is an interval scaled variable. In order to show how to change the split value of a nominal variable, I now use the INCOME variable as an example and change the definition of Node 7 of the tree shown in Display 4.42

To open the tree in an interactive mode, click located at the right of the **Interactive** property of the **Decision Tree** node ① as shown in Display 4.33. The Interactive Tree window opens with the tree we modified in the previous section (see Display 4.42).

Right click on Node 7, and select **Split Node** from the menu, as shown in Display 4.43.

Display 4.43

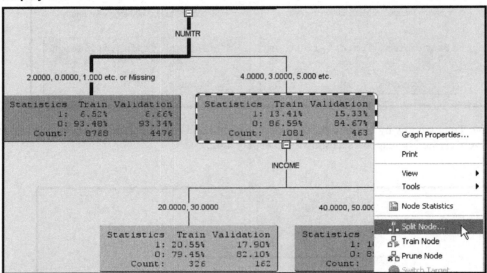

A window showing the logworth of the inputs considered for splitting Node 7 appears, as shown in Display 4.44.

Display 4.44

Split Node 7				
Target Variable: resp				
Variable	Variable Description	-Log(p)		Branches
INCOME	INCOME		2.8186	2
GENDER	GENDER		2.1062	2
female	female		2.1062	2

Edit Rule...

OK   Cancel   Apply   Refresh

Since the INCOME variable is used for splitting Node 7, it is highlighted. Click **Edit Rule**. The Split Rule window opens.

Just for illustration, we assign the value 40 to Branch 1. To do this, click on the row corresponding to value 40, enter 1 in the **Assign to branch** box, and click **Assign**, as shown in Display 4.45.

Display 4.45

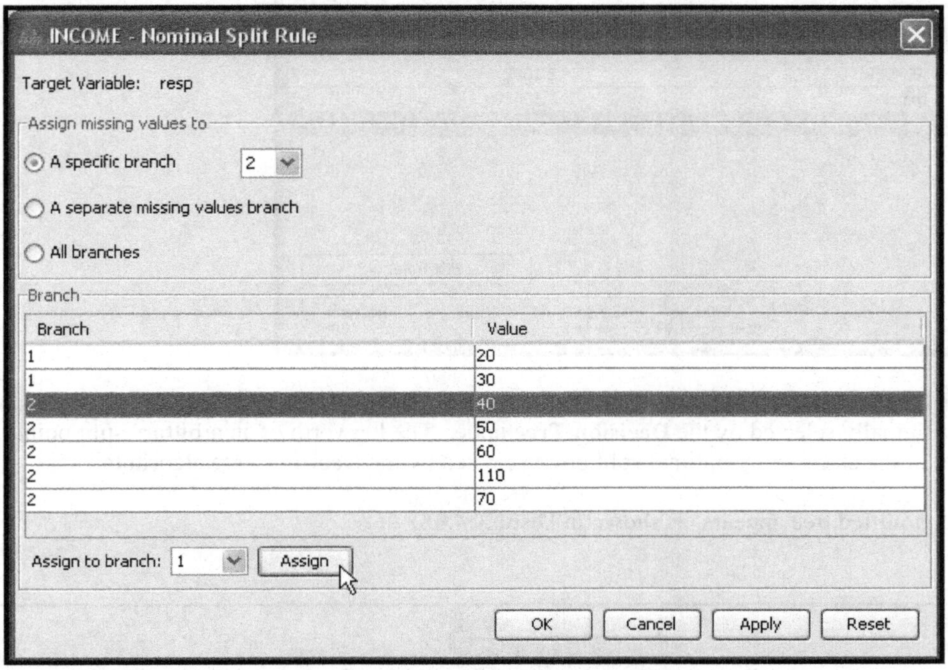

The split rule is changed as shown in Display 4.46.

Display 4.46

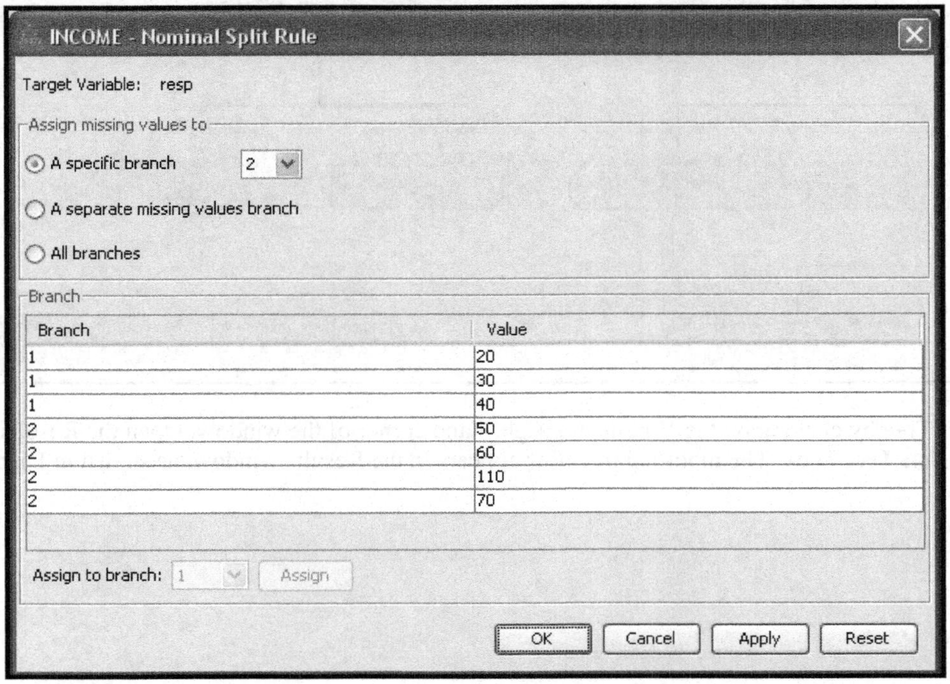

Click **Apply**, and then click **OK**. Close the Split Node window by clicking OK. The logworth at the new split point appears in the logworth table, as shown in Display 4.47.

Display 4.47

	Split Node 7			✕
Target Variable:	resp			

Variable	Variable Description	-Log(p)	Branches
GENDER	GENDER	2.1062	2
female	female	2.1062	2
INCOME	INCOME	0.3502	2

Edit Rule...

OK   Cancel   Apply   Refresh

From Display 4.47 it is clear that the logworth of INCOME variable with the new split value is smaller than the split value that was originally selected by the **Decision Tree** node. The logworth of an arbitrary split point is in general less than that of the statistically optimal split point selected by the Decision Tree algorithm.

If you click **OK**, the modified tree appears, as shown in Display 4.48.

Display 4.48

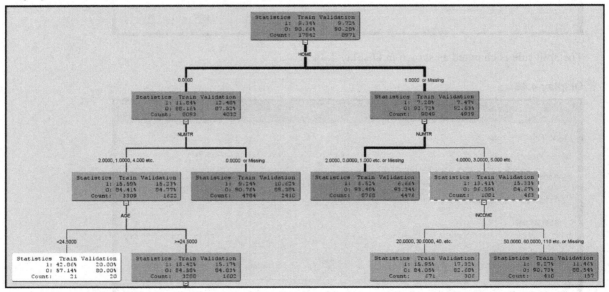

Close the Interactive Tree by clicking on the X in the top right-hand corner of the window. Open the Results window of the **Decision Tree** node. The modified tree now appears in the Results window, as shown in Display 4.49.

Display 4.49

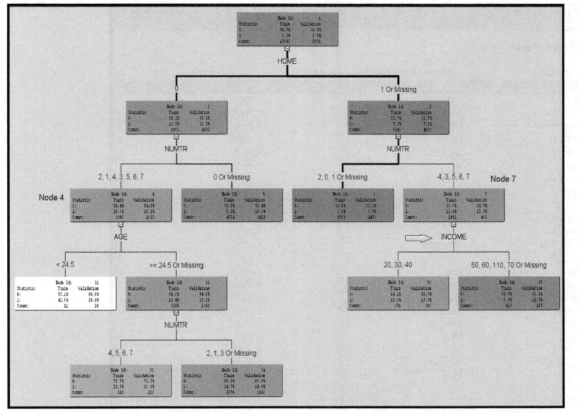

Compare Display 4.49 with Display 4.42 to confirm that Node 7 is split at a new value of income.

In general, arbitrary constraints diminish the predictive performance of any model, and a Decision Tree model is no exception. In real life, either for marketing convenience or other business reasons, you must impose constraints on a model. I leave it to you to compare the original autonomous tree and the modified tree with respect to such measures as lift and cumulative lift, etc. from the Results window.

4.6.2 Growing a Tree Interactively Starting from the Root Node

In this section we grow a tree starting from the root node. Click the **Decision Tree** node ② shown in Display 4.31. In the Properties panel, click located to the right of the **Interactive** property. The Interactive window opens, as shown in Display 4.50.

Display 4.50

At this point the tree has only the root node. To split the root node into two child nodes, right-click on it and select **Split Node**. The Split Node window opens, as shown in Display 4.51.

Display 4.51

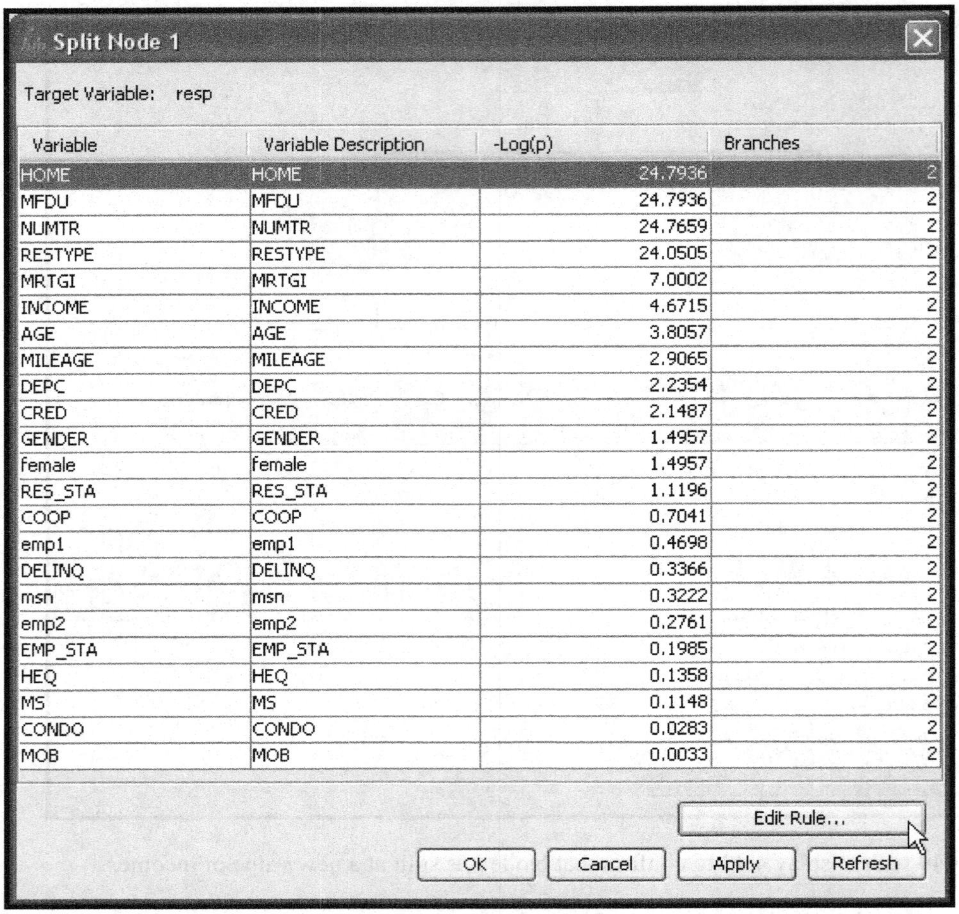

Variable	Variable Description	-Log(p)	Branches
HOME	HOME	24.7936	2
MFDU	MFDU	24.7936	2
NUMTR	NUMTR	24.7659	2
RESTYPE	RESTYPE	24.0505	2
MRTGI	MRTGI	7.0002	2
INCOME	INCOME	4.6715	2
AGE	AGE	3.8057	2
MILEAGE	MILEAGE	2.9065	2
DEPC	DEPC	2.2354	2
CRED	CRED	2.1487	2
GENDER	GENDER	1.4957	2
female	female	1.4957	2
RES_STA	RES_STA	1.1196	2
COOP	COOP	0.7041	2
emp1	emp1	0.4698	2
DELINQ	DELINQ	0.3366	2
msn	msn	0.3222	2
emp2	emp2	0.2761	2
EMP_STA	EMP_STA	0.1985	2
HEQ	HEQ	0.1358	2
MS	MS	0.1148	2
CONDO	CONDO	0.0283	2
MOB	MOB	0.0033	2

Target Variable: resp

Split Node 1

Edit Rule... | OK | Cancel | Apply | Refresh

Each variable has a number of candidate split points. The algorithm first selects the best split point of each variable. The logworth of the selected best split point for each variable is listed in the table in Display 4.51. Next, the algorithm selects the best split point on the best variable. The best variable is the one that has the highest logworth at its best split point. In this example, both the variables HOME and MFDU are the best variables, because they have the same logworth at their respective best split points, but the algorithm selected the variable HOME for splitting the node. To see exactly what the best of the best split points is, click **Edit Rule**. The Split Rule window opens, as shown in Display 4.52.

Display 4.52

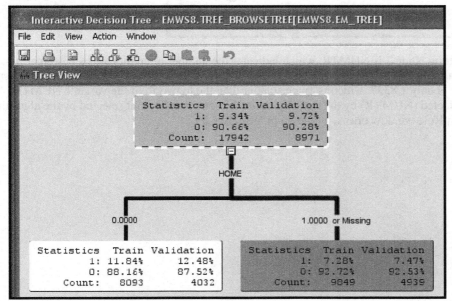

Display 4.52 shows that the variable HOME takes on the value 0 or 1, 0 indicating that the customer is a renter and 1 indicating the customer is a home owner. The node is split in such a way that all customers with HOME = 0 are placed into the left branch (branch 1) and those with HOME =1 are placed in the right branch (branch 2). By default, customers with missing values of the variable HOME are placed in branch 2. You can change this rule by changing the value next to **A specific branch** to 1. You may also choose to assign the records with missing values to a separate branch or distribute them to all branches. In this example, I am leaving the default assignment of the observations with missing values to branch 2. Then I click **Apply**, and then **OK**. Close the Split Node window by clicking OK. A tree is grown with two branches, as shown in Display 4.53.

Display 4.53

The variable HOME stands for home ownership status. It indicates the type of home a customer owns. MFDU indicates whether a customer lives in a multi-family dwelling unit. These two variables are very similar, hence they have the same logworth. In parametric models such as the Logistic Regression model, the inclusion of both these variables would have resulted in unstable estimates of the coefficients or other undesirable results due to collinearity. But in the case of decision trees, collinearity is not a problem.

To grow a larger tree by adding more splits, select the node that is at the bottom left of the tree in Display 4.53, right-click on it and select **Split Node** as we did before.

The Split node window opens, as shown in Display 4.54.

Display 4.54

Split Node 3				
Target Variable: resp				
Variable	Variable Description	-Log(p)	Branches	
NUMTR	NUMTR	15.0738		2
AGE	AGE	1.8337		2
RES_STA	RES_STA	1.411		2
DEPC	DEPC	1.2807		2
MILEAGE	MILEAGE	0.93		2
CONDO	CONDO	0.4389		2
DELINQ	DELINQ	0.3877		2
CRED	CRED	0.3364		2
RESTYPE	RESTYPE	0.1277		2
HEQ	HEQ	0.1163		2
COOP	COOP	0.0131		2
EMP_STA	EMP_STA	0.0		2
GENDER	GENDER	0.0		2
INCOME	INCOME	0.0		2
MOB	MOB	0.0		2
MRTGI	MRTGI	0.0		2
MS	MS	0.0		2
emp1	emp1	0.0		2
emp2	emp2	0.0		2
female	female	0.0		2
msn	msn	0.0		2

Edit Rule...

OK   Cancel   Apply   Refresh

The best variable for splitting Node 3 is NUMTR, which indicates the number of credit cards a customer has. You can select another variable for splitting. Although the next best variable is AGE, the logworth of the best split on the AGE variable is only 1.8337, which is much smaller than the logworth of the variable NUMTR. So we accept the variable selected (NUMTR) by the tree algorithm. To see the split point selected by the algorithm, click **Edit Rule**. The Split Rule window opens, as shown in Display 4.55.

Display 4.55

Since I do not have any particular reason for changing the split values selected by the algorithm (see Display 4.55), I click **Apply** and **OK**. Display 4.56 shows the expanded tree.

Display 4.56

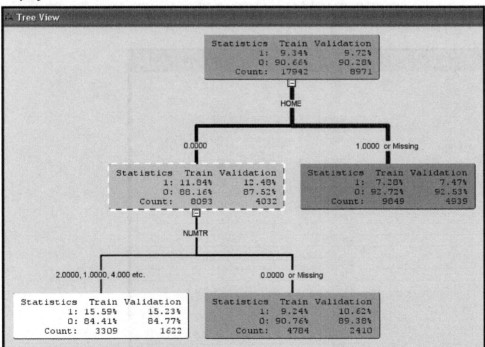

Now let us split the second branch (right-hand side) from the root node. Right-click on this node and select **Split Node** from the menu.

Display 4.57 shows all the inputs available for splitting this node and the logworth of the best split point for each of them.

Display 4.57

Variable	Variable Description	-Log(p)	Branches
NUMTR	NUMTR	13.3106	2
INCOME	INCOME	9.9117	2
AGE	AGE	2.2217	2
CRED	CRED	2.0996	2
GENDER	GENDER	1.9447	2
female	female	1.9447	2
DEPC	DEPC	0.7475	2
MRTGI	MRTGI	0.5976	2
msn	msn	0.254	2
emp2	emp2	0.1678	2
emp1	emp1	0.1413	2
MILEAGE	MILEAGE	0.1213	2
DELINQ	DELINQ	0.0	2
EMP_STA	EMP_STA	0.0	2
HEQ	HEQ	0.0	2
MOB	MOB	0.0	2
MS	MS	0.0	2
RES_STA	RES_STA	0.0	2

Split Node 4 — Target Variable: resp — Edit Rule... | OK | Cancel | Apply | Refresh

As Display 4.57 shows, the statistically best variable is NUMTR. The second best variable is INCOME. For purposes of illustration, we override the statistically best variable NUMTR with the second best variable INCOME by selecting it and clicking **Edit Rule**. Display 4.58 shows the new split rule.

Display 4.58

INCOME - Nominal Split Rule

Target Variable: resp

Assign missing values to
- ⦿ A specific branch [2 ⌄]
- ◯ A separate missing values branch
- ◯ All branches

Branch

Branch	Value
1	20
1	30
2	40
2	50
2	60
2	110
2	70

Assign to branch: [1 ⌄] [Assign]

OK | Cancel | Apply | Reset

According to this split rule, the records with INCOME less than or equal to $30,000 are sent to Branch 1, and those with income greater than $30,000 are sent to Branch 2. To accept this split rule, click **Apply** and **OK**. Close the Split Node window by clicking OK. The resulting tree is shown in Display 4.59.

Display 4.59

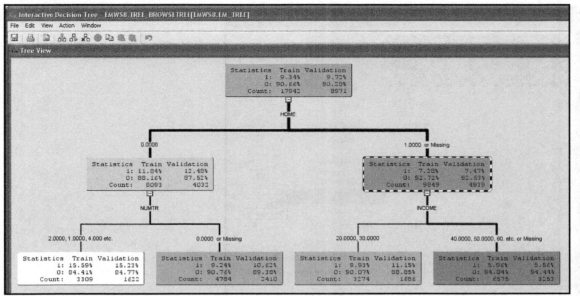

Close the Interactive Decision Tree window by clicking the X in the top right-hand corner of the window. Next, right-click on the **Decision Tree** node indicated by ② in the process flow diagram shown in Display 4.31, select **Results**, and expand the Tree sub-window. The Tree that we have grown interactively appears, as shown in Display 4.60.

Display 4.60

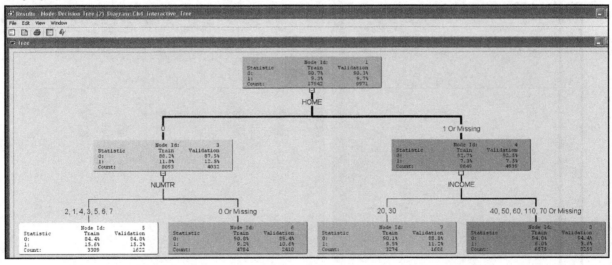

4.6.3 Developing the Maximal Tree in Interactive Mode

In this section, we grow a maximal tree starting from the root node. The maximal tree is the largest tree that can be grown, subject to the stopping rules defined by the Decision Tree properties, which were discussed earlier.

Select the **Decision Tree** node ③ shown in Display 4.31 by clicking on it. In the Properties panel, click located to the right of the **Interactive** property. The Interactive window opens showing a tree with only the root node. Right click on the root node, and select **Train Node**, as shown in Display 4.61.

Display 4.61

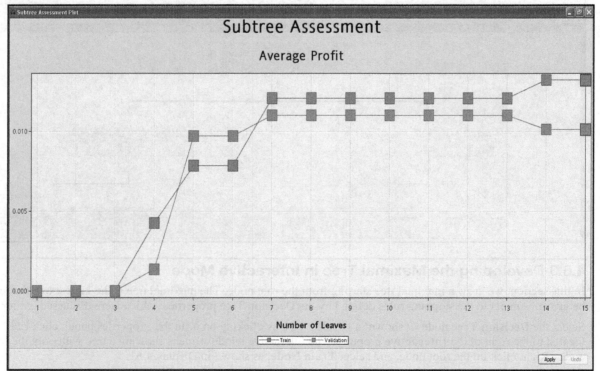

The maximal tree is created. Since the tree is too large, I am not including it here. Instead I include the Subtree Assessment plot. To view the Subtree Assessment plot, click **View** on the menu bar, and select **Subtree Assessment Plot**. The plot is shown in Display 4.62.

Display 4.62

As you can see from Display 4.62, the average profit does not increase with the number of leaves beyond seven leaves. Hence the optimal tree is the one with seven leaves, or terminal nodes. Close the Interactive Tree window.

Run the **Model Comparison** node to compare the three trees we created in Sections 4.6.1, 4.6.2, and 4.6.3. Display 4.63 shows the ROC charts for these three trees.

Display 4.63

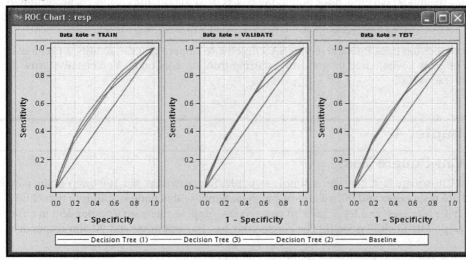

Display 4.63 shows that the maximal tree performs slightly better than Decision Trees 1 and 2, on the basis of the ROC curves constructed from the Training data. But when we compare the ROC curves constructed from the Validation and Test data sets, the maximal tree performs no better than the other trees.

4.7 Summary

- *Growing a decision tree* means arriving at certain rules for segmenting a given data set, stated in terms of ranges of values for inputs, and assigning a target class[3] (such as response or non-response) to each segment that is created.

- The *Training data set* is used for developing the rules for the segmentation, for calculating the *posterior probabilities* of each segment (or the target mean of each segment when the target variable is continuous), and for assigning each segment to a target class when the target is categorical.

- Node definitions, target means, posterior probabilities, and the assignment of the target levels to nodes do not change when the tree is applied to the Validation, or Prospect, data set.

- SAS Enterprise Miner offers several criteria for splitting a node; you can select the criterion suitable for the type of model being developed by setting the **Splitting Rule Criterion** property to an appropriate value.

- In general, there are a number of sub-trees within the initial tree that you need to examine during the tree development. The Training data set is used to create this initial tree, while the *Validation data set* is used to select the best sub-tree by applying an assessment measure that you choose. Several measures of assessment are available in SAS Enterprise Miner, and you can select one of them by setting the **Sub-tree Assessment Measure** property to a certain value. Selecting the best sub-tree is often referred to as *pruning*.

- SAS Enterprise Miner offers different methods for selecting a sub-tree: instead of applying an assessment measure to select a sub-tree, you can also select the tree with the maximum number of leaves, or the one with a specified number of leaves (regardless of whether the tree has maximum profit, or minimum cost, or meets any other criterion of assessment.)

- A *decision tree model* consists finally of only the *leaf* nodes (also known as the terminal nodes) of the selected tree, the attributes of which can be used for scoring prospect data, etc.

- The definition of the leaf nodes, the target classes assigned to the nodes, and their posterior probabilities can all be seen in the SAS code (score code) generated by SAS Enterprise Miner. You can export the SAS code if the scoring is to be applied to external data.

- The two models developed in this chapter, one for a binary target and one for a continuous target, serve as examples of how to use the **Decision Tree** node: how to interpret the results given in the Output window and how to interpret the various types of graphs and tables associated with each of them.

- The Regression Tree model developed for predicting risk is used to rank the customers according to their degree of risk. Each risk group is then profiled and represented as an exclusive segment of the tree model.

- You can develop decision trees interactively in SAS Enterprise Miner. You can modify a previously developed tree or grow a whole tree interactively, starting from the root node. You can also grow a maximal tree interactively.

4.8 Appendix to Chapter 4

4.8.1 Pearson's Chi-Square Test

As an illustration of the Chi-square test, consider a simple example. Suppose there are 100 cases in the parent node (τ), of which 10 are responders and 90 are non-responders. Suppose this node is split into two child nodes, A and B. Let A have 50 cases, and let B have 50 cases. Also, suppose there are 9 responders in child node A, and there is 1 responder in child node B. The parent node has 10 responders and 90 non-responders.

Here is a 2 x 2 contingency table representing the split of parent node τ:

	Child node A	Child node B
Responders	9	1
Non-responders	41	49

The null hypothesis is that the child nodes A and B are not different from their parent node in terms of class composition (i.e., their mix of *responders* and *non-responders*). Under the null hypothesis, each child node is similar to the parent node, and contains 10% responders and 90% non-responders.

Expected frequencies under the null hypothesis are as follows:

	Child node A	Child node B
Responders	5	5
Non-responders	45	45

$$\chi^2 = \sum \frac{(O-E)^2}{E} = \frac{(4)^2}{5} + \frac{(-4)^2}{45} + \frac{(-4)^2}{5} + \frac{(4)^2}{45} = 7.1$$

A Chi-square value of 7.1 implies the following values for *p*-value and for logworth:

p-value = 0.00766 and logworth = $\log_{10}(p\text{-value})$ = 2.11546.

The *p*value is the probability of the Chi_Square statistic taking a value of 7.1 or higher when there is no difference in the child nodes. In this case it is very small, 0.00766. This indicates that in this case, the child nodes do differ significantly with respect to their composition of responders and non-responders.

4.8.2 Adjusting the Predicted Probabilities for Over-sampling

When modeling rare events, it is a common practice to include all the records having the rare event, but only a fraction of the non-event records in the modeling sample. This practice is often referred to as over-sampling, case-control sampling, or biased sampling.

For example, in response modeling, if the response rate is very low, you can include in the modeling sample all the responders available and only a randomly selected fraction of non-responders. The bias introduced by such over-sampling is corrected by adjusting the predicted probabilities with prior probabilities. Display 2.15 in Chapter 2 shows how to enter the prior probabilities.

Adjustment of the predicted probabilities can be illustrated by a simple example. Suppose, based on a biased sample, the *unadjusted* probabilities of response and no-response calculated by the decision tree process for a particular node T are $\hat{P}_{1T}$ and $\hat{P}_{0T}$. If the prior probabilities, that is, the proportion of responders and non-responders in the entire population, are known to be π_1 and π_0, and if the proportion of responders and the non-responders in the biased sample are ρ_1 and ρ_0, respectively, then the *adjusted* predictions of the probabilities are given by

$$P_{1T} = \frac{\hat{P}_{1T}(\pi_1 / \rho_1)}{\hat{P}_{0T}(\pi_0 / \rho_0) + \hat{P}_{1T}(\pi_1 / \rho_1)} \text{, and}$$

$$P_{0T} = \frac{\hat{P}_{0T}(\pi_0 / \rho_0)}{\hat{P}_{0T}(\pi_0 / \rho_0) + \hat{P}_{1T}(\pi_1 / \rho_1)},$$

where P_{1T} is the adjusted probability of response, and P_{0T} is the adjusted probability of no-response. It can be easily verified that $P_{1T} + P_{0T} = 1$. If there are n_{1T} responders and n_{oT} non-responders in node T, prior to adjustment, then the adjusted number of responders and non-responders will be $\left(\pi_1 \middle/ \rho_1 \right) n_{1T}$ and $\left(\pi_0 \middle/ \rho_0 \right) n_{0T}$, respectively.

The effect of adjustment on the number of responders and non-responders, and on the expected profits for node T, can be seen from a numerical example.

Let the proportions of response and non-response for the entire population be $\pi_1 = 0.05$ and $\pi_0 = 0.95$, while the proportions for the biased sample are $\rho_1 = 0.5$ and $\rho_0 = 0.5$. If the predicted proportions for node T are $\hat{P}_{1T} = 0.4$ and $\hat{P}_{0T} = 0.6$, and if the predicted number of responders and non-responders are $n_{1T} = 400$ and $n_{oT} = 600$, then the adjusted probability of response (P_{1T}) is equal to 0.0339, and the adjusted probability of no-response (P_{0T}) equals 0.9661. In addition, the adjusted number of responders is 40, and the adjusted number of non-responders is 1140.

One more measure that may need adjustment in the case of a sample biased due to over-sampling is expected profits. To show expected profits under alternative decisions with adjusted and unadjusted probabilities, I use the profit matrix given in Table 4.8.

Table 4.8

	Decision1	Decision2
Actual target level/class		
1	$10	-$2
0	-$1	$1

4.8.3 Expected Profits Using Unadjusted Probabilities

Using the unadjusted predicted probabilities in my example, under Decision1 the expected profits are ($10*0.4) + (−$1*0.6) = $3.4. Under Decision2, the expected profits are (−$2*0.4) + ($1*0.6) = −$0.2.

4.8.4 Expected Profits Using Adjusted Probabilities

Using the adjusted probabilities, under Decision1 the expected profits are ($10*0.0339) + (−$1*0.9661) = −$0.6271. Under Decision 2, the expected profits are (−$2*0.0339) + ($1*0.9661) = $0.8983.

4.9 Exercises

Exercise 1

1. Create a data source with the SAS dataset BookData1. Use the Advanced Metadata Advisor Options to create the metada. Customize the metada by setting the **Class Levels Count Threshold** property to 5. Set the Role of the variable EVENT to Target. On the **Prior Probabilities** tab of the Decision Configuration step, set Adjusted Prior to 0.052 for level 1 and 0.948 for level 0.

2. Create a process flow diagram as shown below:

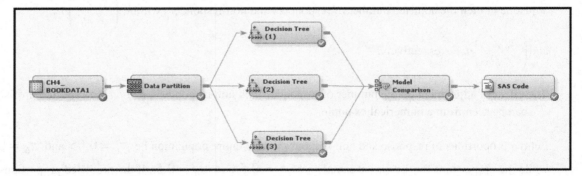

   Partition the data such that 60% of the records are used for training, 30% for validation, and 10% for test.

3. Set the **Nominal Criterion** property to ProbChisq in **Decision Tree (1)** node, set it to Entropy in **Decision Tree (2)** node , and set it to Gini in **Decision Tree (3)** node. Set the **Subtree Method** property to Assessment and the **Assessment Measure** property to Decision in all three **Decision Tree** nodes. Use the Decision Weights shown in Display 4.1.

4. Run the three **Decision Tree** nodes and then the **Model Comparison** node.

5. Open the Results window of the **Model Comparison** node and compare the Cumulative % Captured Response of the three models.

6. Run the **SAS Code** node. Open the Code Editor window and type the following code:

```
Training Code
Data Train;
   set &EM_LIB..MDLCOMP_EMRANK;
   keep model decile cap capc;
   if Datarole = "TRAIN";
run;
Data Validate;
   set &EM_LIB..MDLCOMP_EMRANK;
   keep model decile cap capc;
   if Datarole = "VALIDATE";
run;
Data TEST;
   set &EM_LIB..MDLCOMP_EMRANK;
   keep model decile cap capc;
   if Datarole = "TEST";
run;
```

7. Type in additional SAS lines needed to print the three data sets Train, Validate, and Test. Run the code and examine the results.

8. Which decision tree is better in terms of the Cumulative % Captured Response?

Exercise 2

An auto insurance company has developed a decision tree model in order to classify the prospective customers as likely responders and non-responders. The target variable is RESP.

1. The decision tree assigns all customers who have income less than \$25,000 and Age greater than or equal to 35 years to the Leaf Node 1. The decision tree also calculated the posterior probabilities as $\hat{P}_1$ and $\hat{P}_0$. These posterior probabilities indicate that the probability that a customer who is assigned to the Leaf Node 1 will respond to a solicitation to apply for auto insurance is $\hat{P}_1 = 0.02$ and the probability that the customer will not respond is $\hat{P}_0 = 0.98$.

2. Given the posterior probabilities given in Step 1 and using the profit matrix shown in Display 4.8, classify the customers who belong to Leaf Node 1 as Responders or Non-responders (Hint: Compare the expected profits under Decision 1 and Decision 2, where Decision 1 means labeling the customers in Leaf Node1 as responders and Decision 2 means labeling them as non-responders.)

3. After applying the decision tree to the test data set, it is found that 200 customers in the Test data set ended up in Leaf Node 1. Of these 200 customers, 6 were found to be actual responders and 194 customers were non-responders.

4. Suppose you apply the decision tree model and score a Prospect data set. If a customer from the scored data set is randomly selected and is assigned to Leaf Node 1 by the Decision Tree model, then what is the predicted probability that the customer will respond to an invitation to apply for auto insurance?

5. Given the decision you arrived at in Step 2, would you send an invitation to the customer selected in Step 4 to apply for auto insurance? Why or why not?

Notes

1. There may be a loss when a true responder is classified as a non-responder. However, in this example, I am assuming it is zero for simplicity. This assumption can be relaxed, and all the procedures shown here would apply.

2. The decision tree shown in Figures 4.1, 4.3, and 4.4 is developed using the **Decision Tree** node. However, the tree diagrams shown are handcrafted for illustrating various steps of the tree development process.

3. The terms *target class* and *target level* are used synonymously.

Chapter 5: Neural Network Models to Predict Response and Risk

5.1 Introduction

I now turn to using SAS Enterprise Miner to derive and interpret neural network models. This chapter develops two neural network models using simulated data. The first model is a response model with a binary target, and the second is a risk model with an ordinal target.

5.1.1 Target Variables for the Models

The target variable for the response model is RESP, which takes the value of 0 if there is no response and the value of 1 if there is a response. The neural network produces a model to predict the probability of response based on the values of a set of input variables for a given customer. The output of the **Neural Network** node gives probabilities $\Pr(RESP = 0 \mid X)$ and $\Pr(RESP = 1 \mid X)$, where X is a given set of inputs.

The target variable for the risk model is a discrete version of *loss frequency,* defined as the number of losses or accidents per car-year, where car-year is defined as the duration of the insurance policy multiplied by the number of cars insured under the policy. If the policy has been in effect for four months and one car is covered under the policy, then the number of car-years is 4/12 at the end of the fourth month. If one accident occurred during this four-month period, then the loss frequency is approximately $1/(4/12) = 3$. If the policy has been in effect for 12 months covering one car, and if one accident occurred during that 12-month period, then the loss frequency is $1/(12/12) = 1$. If the policy has been in effect for only four months, and if two cars are covered under the same policy, then the number of car-years is (4/12) x2. If one accident occurred during the four-month period, then the loss frequency is $(1/(8/12) = 1.5$. Defined this way, *loss frequency* is a continuous variable. However, the target variable LOSSFRQ, which I will use in the model developed here, is a discrete version of loss frequency defined as follows:

$$LOSSFRQ = 0 \text{ if loss frequency } = 0,$$
$$LOSSFRQ = 1 \text{ if } 0 < \text{loss frequency } < 1.5, \text{ and}$$
$$LOSSFRQ = 2 \text{ if loss frequency } \geq 1.5.$$

The three levels 0, 1, and 2 of the variable LOSSFRQ represent low, medium, and high risk, respectively.

The output of the **Neural Network** node gives:

$$\Pr(LOSSFRQ = 0 \mid X) = \Pr(\text{loss frequency} = 0 \mid X),$$
$$\Pr(LOSSFRQ = 1 \mid X) = \Pr(0 < \text{loss frequency} < 1.5 \mid X),$$
$$\Pr(LOSSFRQ = 2 \mid X) = \Pr(\text{loss frequency} \geq 1.5 \mid X).$$

5.1.2 Neural Network Node Details

Neural network models use mathematical functions to map inputs into outputs. When the target is categorical, the outputs of a neural network are the probabilities of the target levels. The neural network model for risk (in the current example) gives formulas to calculate the probability of the variable LOSSFRQ taking each of the values 0, 1, or 2, whereas the neural network model for response provides formulas to calculate the probability of response and non-response. Both of these models use the complex nonlinear transformations available in the **Neural Network** node of SAS Enterprise Miner.

In general, you can fit logistic models to your data directly by first transforming the variables, selecting the inputs, combining and further transforming the inputs, and finally estimating the model with the original combined and transformed inputs. On the other hand, as I show in this chapter, these combinations and transformations of the inputs can be done within the neural network framework. A neural network model can be thought of as a complex nonlinear model where the tasks of variable transformation, composite variable creation, and model estimation (estimation of weights) are done simultaneously in such a way that a specified error function is minimized. This does not rule out transforming some or all of the variables before running the neural network models, if you choose to do so.

In the following pages, I make it clear how the transformations of the inputs are performed and how they are combined through different "layers" of the neural network, as well as how the final model that uses these transformed inputs is specified.

A neural network model is represented by a number of layers, each layer containing computing elements known as *units* or *neurons*. Each unit in a layer takes inputs from the preceding layer and computes outputs. The outputs from the neurons in one layer become inputs to the next layer in the sequence of layers.

The first layer in the sequence is the *input layer*, and the last layer is the *output,* or *target,* layer. Between the input and output layers there can be a number of *hidden* layers. The units in a hidden layer are called *hidden units.* The hidden units perform *intermediate calculations* and pass the results to the next layer.

The calculations performed by the hidden units involve combining the inputs they receive from the previous layer and performing a mathematical transformation on the combined values. In SAS Enterprise Miner, the formulas used by the hidden units for combining the inputs are called *Hidden Layer Combination Functions.* The formulas used by the hidden units for transforming the combined values are called *Hidden Layer Activation Functions.* The values generated by the combination and activation operations are called the *outputs*. As noted, outputs of one layer become inputs to the next layer.

The units in the target layer or output layer also perform the two operations of combination and activation. The formulas used by the units in the target layer to combine the inputs are called *Target Layer Combination Functions*, and the formulas used for transforming the combined values are called *Target Layer Activation Functions*. Interpretation of the outputs produced by the target layer depends on the Target Layer Activation Function used. Hence, the specification of the Target Layer Activation Function cannot be arbitrary. It should reflect the business question and the theory behind the answer you are seeking. For example, if the target variable is response, which is binary, then you should select Logistic as the **Target Layer Activation Function** value. This selection ensures that the outputs of the target layer are the probabilities of response and non-response.

The combination and activation functions in the hidden layers and in the target layer are key elements of the *architecture* of a neural network. SAS Enterprise Miner offers a wide range of choices for these functions. Hence, the number of combinations of hidden layer combination function, hidden layer activation function, target layer combination function, and target layer activation function to choose from is quite large, and each produces a different neural network model.

5.2 A General Example of a Neural Network Model

To help you understand the models presented later, here is a general description of a neural network with two hidden layers[1] and an output layer.

Suppose a data set consists of n records, and each record contains, along with a person's response to a direct mail campaign, some measures of demographic, economic, and other characteristics of the person. These measures are often referred to as *explanatory variables* or *inputs*. Suppose there are p inputs $x_{i1}, x_{i2}, \ldots .. x_{ip}$

for the ith person. In a neural network model, the input layer passes these inputs to the next layer. In the input layer, there is one input unit for each input. The units in the input layer do not make any computations other than optionally standardizing the inputs. The inputs are passed from the input layer to the next layer. In this illustration, the next layer is the first hidden layer. The units in the first hidden layer produce *intermediate outputs*. Suppose there are three hidden units in this layer. Let H_{i11} denote the intermediate output produced by the first unit of the first hidden layer for the ith record. In general, H_{ikj} stands for the output of the jth unit in the kth hidden layer for the ith record in the data set.

Similarly, the intermediate outputs produced by the second and third units are H_{i12} and H_{i13}. These intermediate outputs are passed to the second hidden layer. Suppose the second hidden layer also contains three units. The outputs of these three hidden units are H_{i21}, H_{i22} and H_{i23}. These intermediate outputs become inputs to the *output* or *target layer*, the last layer in the network. The output produced by this layer is the predicted value of the target, which is the probability of response (and non-response) in the case of the response model (a binary target), and the probability of the number of losses occurring in the case of the risk model (an ordinal target with three levels). In the case of a continuous target, such as value of deposits in a bank or household savings, the output of the final layer is the expected value of the target.

In the following sections, I show how the units in each layer calculate their outputs.

The network just described is shown in Display 5.0A.

Display 5.0A

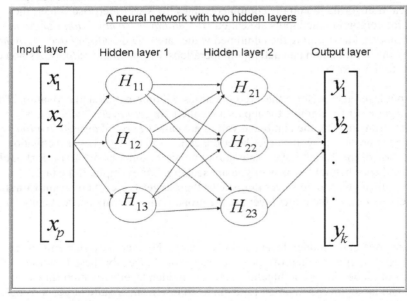

5.2.1 Input Layer

This layer passes the inputs to the next layer in the network either without transforming them or, in the case of interval-scaled inputs, after standardizing the inputs. Categorical inputs are converted to dummy variables. For each record, there are p inputs: $x_1, x_2, \ldots \ldots x_p$.

5.2.2 Hidden Layers

There are three hidden units in each of the two hidden layers in this example. Each hidden unit produces an intermediate output as described below.

5.2.2.1 Hidden Layer 1: Unit 1

A weighted sum of the inputs for the ith record for the first unit in the first hidden layer is calculated as

$$\eta_{i11} = w_{011} + w_{111}x_{i1} + w_{211}x_{i2} + \ldots + w_{p11}x_{ip}. \tag{5.1}$$

where $w_{111}, w_{211}, \ldots\ldots w_{p11}$ are the *weights* to be estimated by the iterative algorithm to be described later, one weight for each of the p inputs. w_{011} is called *bias*. η_{i11} is the weighted sum for the ith person[2] or record in the data set. Equation 5.1 is called the *combination function*. Since the combination function given in Equation 5.1 is a linear function of the inputs, it is called a *linear combination function*. Other types of combination functions are discussed in Section 5.7.

A nonlinear transformation of the weighted sum yields the output from unit 1 as:

$$H_{i11} = \tanh(\eta_{i11}) = \frac{\exp(\eta_{i11}) - \exp(-\eta_{i11})}{\exp(\eta_{i11}) + \exp(-\eta_{i11})} \tag{5.2}$$

The effect of this transformation is to map values of η_{i11}, which can range from $-\infty$ to $+\infty$, into the narrower range of -1 to $+1$. The transformation shown in Equation 5.2 is referred to as a *hyperbolic tangent activation function*.

To illustrate this, consider a simple example with only two inputs, Age and CREDIT, both interval-scaled. The input layer first standardizes these variables and passes them to the units in Hidden Layer 1. Suppose, at some point in the iteration, the weighted sum presented in Equation 5.1 is
$\eta_{i11} = 0.01 + 0.4 * Age + 0.7 * CRED,$ where Age is the age and CRED is the credit score of the person reported in the ith record. The value of η_{i11} (the combination function) can range from -5 to $+5$. Substituting the value of η_{i11} in Equation 5.2, I get the value of the activation function H_{i11} for the ith record. If I plot these values for all the records in the data set, I get the chart shown in Display 5.0B.

Display 5.0B

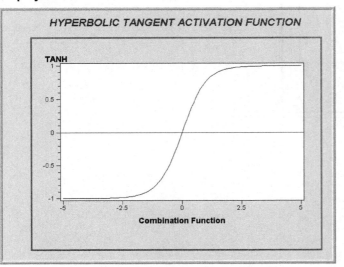

In SAS Enterprise Miner, a variety of activation functions are available for use in the hidden layers. These include Arc Tangent, Elliot, Hyperbolic Tangent, Logistic, Gauss, Sine, and Cosine, in addition to several other types of activation functions discussed in Section 5.7.

The four activation functions, Arc Tangent, Elliot, Hyperbolic Tangent and Logistic, are called Sigmoid functions; they are *S*-shaped and give values that are bounded within the range of 0 to 1 or –1 to 1.

The hyperbolic tangent function shown in Display 5.0B has values bounded between –1 and 1. Displays 5.0C, 5.0D, and 5.0E show Logistic, Arc Tangent, and Elliot functions, respectively.

Display 5.0C

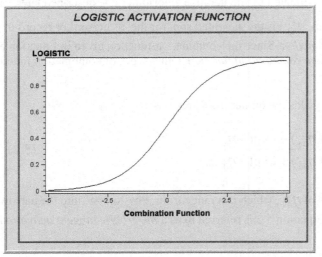

Display 5.0D

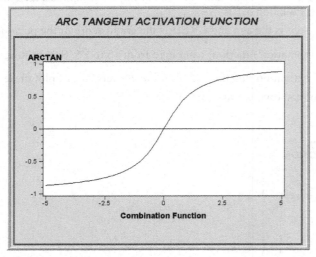

Display 5.0E

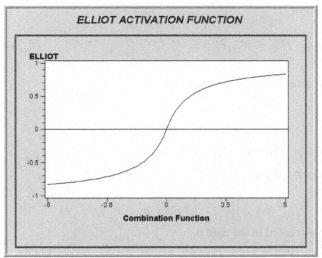

In general, w_{mkj} is the weight of the mth input in the jth neuron (unit) of the kth hidden layer, and w_{0kj} is the corresponding bias. The weighted sum and the output calculated by the jth neuron in the kth hidden layer are η_{ikj} and H_{ikj}, respectively, where i stands for the ith record in the data set.

Hidden Layer 1: Unit 2
The calculation of the output of unit 2 proceeds in the same way as in unit 1, but with a different set of weights and bias. The weighted sum of the inputs for the ith person or record in the data is

$$\eta_{i12} = w_{012} + w_{112}x_{i1} + w_{212}x_{i2} + ...w_{p12}x_{ip}. \tag{5.3}$$

The output of unit 2 is:

$$H_{i12} = \tanh(\eta_{i12}) = \frac{\exp(\eta_{i12}) - \exp(-\eta_{i12})}{\exp(\eta_{i12}) + \exp(-\eta_{i12})} \tag{5.4}$$

Hidden Layer 1: Unit 3
The calculations in unit 3 proceed exactly as in units 1 and 2, but with a different set of weights and bias. The weighted sum of the inputs for i^{th} person in the data set is

$$\eta_{i13} = w_{013} + w_{113}x_{i1} + w_{213}x_{i2} + ...w_{p13}x_{ip}, \tag{5.5}$$

and the output of this unit is:

$$H_{i13} = \tanh(\eta_{i13}) = \frac{\exp(\eta_{i13}) - \exp(-\eta_{i13})}{\exp(\eta_{i13}) + \exp(-\eta_{i13})} \tag{5.6}$$

5.2.2.2 Hidden Layer 2
There are three units in this layer, each producing an intermediate output based on the inputs it receives from the previous layer.

Hidden Layer 2: Unit 1

The weighted sum of the inputs calculated by this unit is:

$$\eta_{i21} = w_{021} + w_{121}H_{i11} + w_{221}H_{i12} + w_{321}H_{i13} \tag{5.7}$$

η_{i21} is the weighted sum for the ith record in the data set in unit 1 of layer 2.

The output of this unit is:

$$H_{i21} = \tanh(\eta_{i21}) = \frac{\exp(\eta_{i21}) - \exp(-\eta_{i21})}{\exp(\eta_{i21}) + \exp(-\eta_{i21})} \tag{5.8}$$

Hidden Layer 2: Unit 2

The weighted sum of the inputs for the ith person or record in the data is

$$\eta_{i22} = w_{022} + w_{122}H_{i11} + w_{222}H_{i12} + w_{322}H_{i13}, \tag{5.9}$$

and the output of this unit is:

$$H_{i22} = \tanh(\eta_{i22}) = \frac{\exp(\eta_{i22}) - \exp(-\eta_{i22})}{\exp(\eta_{i22}) + \exp(-\eta_{i22})} \tag{5.10}$$

Hidden Layer 2: Unit 3

The weighted sum of the inputs for i^{th} person in the data set is

$$\eta_{i23} = w_{023} + w_{123}H_{i11} + w_{223}H_{i12} + w_{323}H_{i13}, \tag{5.11}$$

and the output of this unit is:

$$H_{i23} = \tanh(\eta_{i23}) = \frac{\exp(\eta_{i23}) - \exp(-\eta_{i23})}{\exp(\eta_{i23}) + \exp(-\eta_{i23})} \tag{5.12}$$

5.2.3 Output Layer or Target Layer

This layer produces the final output of the neural network—the predicted values of the targets. In the case of a response model, the output of this layer is the predicted probability of response and non-response, and the output layer has two units. In the case of a categorical target with more than two levels, there are more than two output units. When the target is continuous, such as a bank account balance, the output is the expected value. Display 5.0A shows k outputs, and they are: $y_1, y_2, \ldots\ldots, y_k$. For the response model k=2, and for the loss frequency (risk) model k=3.

The inputs into this layer are H_{i21}, H_{i22}, and H_{i23}, which are the intermediate outputs produced by the preceding layer. These intermediate outputs can be thought of as the nonlinear transformations and combinations of the original inputs performed through successive layers of the network, as described in Equations 5.1 through 5.13. H_{i21}, H_{i22}, and H_{i23} can be considered *synthetic variables* constructed from the inputs.

In the output layer, these transformed inputs (H_{i21}, H_{i22}, and H_{i23}) are used to construct predictive equations of the target. First, for each output unit, a linear combination of H_{i21}, H_{i22}, and H_{i23}, called a linear predictor (or combination function), is defined and an activation function is specified to give the desired predictive equation.

In the case of response model, the first output unit gives the probability of response represented by y_1, while the second output unit gives the probability of no response represented by y_2. In the case of the loss frequency model, there are three outputs: y_1, y_2, and y_3. The first output unit gives the probability that LOSSFRQ is 0 (y_1), the second output unit gives the probability that LOSSFRQ is 1 (y_2), and the third output gives the probability that LOSSFRQ is (y_3).

The linear combination of the inputs from the previous layer is calculated as:

$$\eta_{i31} = w_{031} + w_{131}H_{i21} + w_{231}H_{i22} + w_{331}H_{i23} \tag{5.13}$$

Equation 5.13 is the linear predictor (or linear combination function), because it is linear in terms of the weights.

5.2.4 Activation Function of the Output Layer

The *activation function* is a formula for calculating the target values from the inputs coming from the final hidden layer. The outputs of the final hidden layer are first combined as shown in Equation 5.13. The resulting combination is transformed using the activation function to give the final target values. The activation function is chosen according to the type of target and what you are interested in predicting. In the case of a binary target I use a logistic activation function in the final output or target layer wherein the ith person's probability of response (π_i) is calculated as:

$$\pi_i = \frac{\exp(\eta_{i31})}{1 + \exp(\eta_{i31})} = \frac{1}{1 + \exp(-\eta_{i31})} \tag{5.14}$$

Equation 5.14 is called the *logistic activation* function, and is based on a logistic distribution. By substituting Equations 5.1 through 5.13 into Equation 5.14, I can easily verify that it is possible to write the output of this node as an explicit nonlinear function of the weights and the inputs. This can be denoted by:

$$\pi_i(W, X_i) \tag{5.15}$$

In Equation 5.15, W is the vector whose elements are the weights shown in Equations 5.1, 5.3, 5.5, 5.7, 5.9, 5.11, and 5.13, and X_i is the vector of inputs $x_{i1}, x_{i2}, \ldots x_{ip}$ for the ith person in the data set.

The neural network described above is called a multilayer perceptron or MLP. In general, a network that uses linear combination functions and sigmoid activation functions in the hidden layers is called a multilayer perceptron. In SAS Enterprise Miner, if you set the **Architecture** property to MLP, then you get an MLP with only one hidden layer.

5.3 Estimation of Weights in a Neural Network Model

The weights are estimated iteratively using the training data set in such a way that the error function specified by the user is minimized. In the case of a response model, I use the following Bernoulli error function:

$$E = -2\sum_{i=1}^{n}\left\{ y_i \ln \frac{\pi(W, X_i)}{y_i} + (1-y_i)\ln \frac{1-\pi(W, X_i)}{1-y_i}\right\} \qquad (5.16)$$

where π is the estimated probability of response. It is a function of the vector of weights, W, and the vector of explanatory variables for the ith person, X_i. The variable y_i is the observed response of the ith person, in which $y_i = 1$ if the ith person responded to direct mail and $y_i = 0$ if the ith person did not respond. If $y_i = 1$ for the ith observation, then its contribution to the error function (Equation 5.16) is $-2\log \pi(W, X_i)$; if $y_i = 0$, then the contribution is $-2\log(1 - \pi(W, X_i))$.

The total number of observations in the Training data set is n. The error function is summed over all of these observations. The inputs for the ith record are known, and the values for the weights W are chosen so as to minimize the *error function*. A number of iterative methods of finding the optimum weights are discussed in Bishop.[3]

In general, the calculation of the optimum weights is done in two steps:

Step 1: Finding the error-minimizing weights from the Training data set

In the first step, an iterative procedure is used with the training data set to find a set of weights that minimize the error function given in Equation 5.16. In the first iteration, a set of initial weights is used, and the error function Equation 5.16 is evaluated. In the second iteration, the weights are changed by a small amount in such a way that the error is reduced. This process continues until the error cannot be further reduced, or until the specified maximum number of iterations is reached. At each iteration, a set of weights is generated. If it takes 100 iterations to find the error-minimizing weights for the Training data set, then 100 sets of weights are generated and saved. Thus, the training process generates 100 models in this example. The next task is to select the best model out of these 100 models. This is done using the Validation data set.

Step 2: Finding the optimum weights from the Validation data set.

In this step, the user-supplied Model Selection Criterion is applied to select one of the 100 sets of weights generated in step 1. This selection is done using the Validation data set. Suppose I set the **Model Selection Criterion** property to Average Error. Then the average error is calculated for each of the 100 models using the Validation data set. If the Validation data set has m observations, then the error calculated from the weights generated at the k_{th} iteration is

$$E_{(k)} = -2\sum_{i=1}^{m}\left\{ y_i \ln \frac{\pi(W_{(k)}, X_i)}{y_i} + (1-y_i)\ln \frac{1-\pi(W_{(k)}, X_i)}{1-y_i}\right\} \qquad (5.17)$$

where $y_i = 1$ for a responder, and 0 for a non-responder. The average error at the k_{th} iteration is $\dfrac{E_{(k)}}{2m}$. The set of weights with the smallest average error is used in the final predictive equation. Alternative model selection criteria can be chosen by changing the **Model Selection Criterion** property.

As pointed out earlier, Equations 5.1, 5.3, 5.5, 5.7, 5.9, 5.11, and 5.13 are called combination functions, and Equations 5.2, 5.4, 5.6, 5.8, 5.10, 5.12, and 5.14 are called activation functions.

By substituting Equations 5.1 through 5.13 into Equation 5.14, I can write the neural network model as an explicit function of the *inputs*. This is an arbitrary nonlinear function, and its complexity (and hence the degree of nonlinearity) depends on network features such as the number of hidden layers included, the number of units in each hidden layer, and the combination and activation functions in each hidden unit.

In the neural network model described above, the final layer inputs (H_{i21}, H_{i22}, and H_{i23}) can be thought of as final transformations of the original inputs made by the calculations in different layers of the network.

In each hidden unit, I used a linear combination function to calculate the weighted sum of the inputs, and a hyperbolic tangent activation function to calculate the intermediate output for each hidden unit. In the output layer, I specified a linear combination function and a logistic activation function. The error function I specified is called the Bernoulli error function. A neural network with two hidden layers and one output layer is called a three-layered network. In characterizing the network, the input layer is not counted, since units in the input layer do not perform any computations except standardizing the inputs.

5.4 A Neural Network Model to Predict Response

This section discusses the neural network model developed to predict the response to a planned direct mail campaign. The campaign's purpose is to solicit customers for a hypothetical insurance company. A two-layered network with one hidden layer was chosen. Three units are included in the hidden layer. In the hidden layer, the combination function chosen is linear, and the activation function is hyperbolic tangent. In the output layer, a logistic activation function and Bernoulli error function are used. The logistic activation function results in a logistic regression type model with non-linear transformation of the inputs, as shown in Equation 5.14 in Section 5.2.4. Models of this type are in general estimated by minimizing the Bernoulli error functions shown in Equation 5.16. Minimization of the Bernoulli error function is equivalent to maximizing the likelihood function.

Display 5.1 shows the process flow for the response model. The first node in the process flow diagram is the **Input Data** node, which makes the SAS data set available for modeling. The next node is **Data Partition**, which creates the Training, Validation, and Test data sets. The Training data set is used for preliminary model fitting. The Validation data set is used for selecting the optimum weights. The **Model Selection Criterion** property is set to Average Error.

As pointed out earlier, the estimation of the weights is done by minimizing the error function. This minimization is done by an iterative procedure. Each iteration yields a set of weights. Each set of weights defines a model. If I set the **Model Selection Criterion** property to Average Error, the algorithm selects the set of weights that results in the smallest error, where the error is calculated from the Validation data set.

Since both the Training and Validation data sets are used for parameter estimation and parameter selection, respectively, an additional holdout data set is required for an independent assessment of the model. The Test data set is set aside for this purpose.

Display 5.1

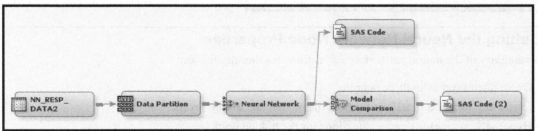

Input Data Node

I create the data source for the **Input Data** node from the data set NN_RESP_DATA2. I create the metadata using the Advanced Advisor Options, and I customize it by setting the **Class Levels Count Threshold** property to 8, as shown in Display 5.2

Display 5.2

Property	Value
Missing Percentage Threshold	50
Reject Vars with Excessive Missing Values	Yes
Class Levels Count Threshold	8
Detect Class Levels	Yes
Reject Levels Count Threshold	20
Reject Vars with Excessive Class Values	Yes
Database Pass-Through	Yes

I set adjusted prior probabilities to 0.03 for response and 0.97 for non-response, as shown in Display 5.3.

Display 5.3

Targets | Prior Probabilities | Decisions | Decision Weights

Do you want to enter new prior probabilities?
⦿ Yes ○ No   Set Equal Prior

Level	Count	Prior	Adjusted Prior
1	9379	0.3136	0.03
0	20525	0.6864	0.97

Data Partition Node

The input data is partitioned such that 60% of the observations are allocated for training, 30% for validation, and 10% for Test, as shown in Display 5.4.

Display 5.4

Data Set Allocations	
Training	60.0
Validation	30.0
Test	10.0

5.4.1 Setting the Neural Network Node Properties

Here is a summary of the neural network specifications for this application:

- One hidden layer with three neurons
- Linear combination functions for both the hidden and output layers
- Hyperbolic tangent activation functions for the hidden units
- Logistic activation functions for the output units
- The Bernoulli error function
- The **Model Selection Criterion** is Average Error

These settings are shown in Displays 5.5–5.7.

Display 5.5 shows the Properties panel for the **Neural Network** node.

Display 5.5

Property	Value	
General		
Node ID	Neural	
Imported Data		...
Exported Data		...
Notes		...
Train		
Variables		...
Continue Training	No	
Network		...
Optimization		...
Initialization Seed	12345	
Model Selection Criterion	Average Error	
Suppress Output	No	
Score		
Hidden Units	No	
Residuals	Yes	
Standardization	No	
Status		
Create Time	1/8/13 7:59 AM	

To define the network architecture, click ⬚ located to the right of the **Network** property. The Network Properties panel opens, as shown in Display 5.6.

Display 5.6

Property	Value
Architecture	User
Direct Connection	No
Number of Hidden Units	3
Randomization Distribution	Normal
Randomization Center	0.0
Randomization Scale	0.1
Input Standardization	Standard Deviation
Hidden Layer Combination Function	Linear
Hidden Layer Activation Function	Hyperbolic Tangent
Hidden Bias	Yes
Target Layer Combination Function	Linear
Target Layer Activation Function	Logistic
Target Layer Error Function	Bernoulli
Target Bias	Yes
Weight Decay	0.0

Architecture

[OK] [Cancel]

Set the properties as shown in Display 5.6 and click **OK**.

To set the iteration limit, click ⬚ located to the right of the **Optimization** property. The Optimization Properties panel opens, as shown in Display 5.7. Set **Maximum Iterations** to 100.

Display 5.7

After running the **Neural Network** node, you can open the Results window, shown in Display 5.8. The window contains four windows: Score Rankings Overlay, Iteration Plot, Fit Statistics, and Output.

Display 5.8

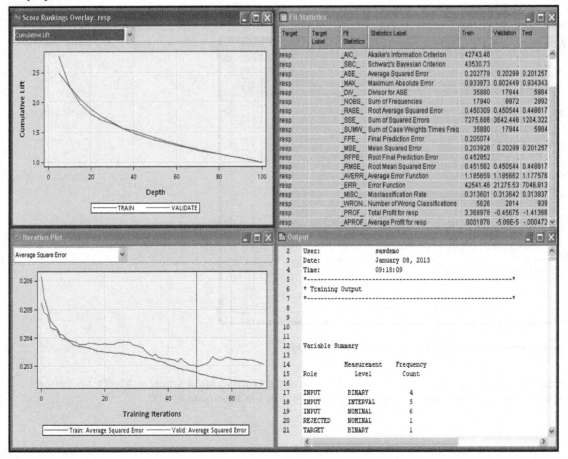

The Score Rankings Overlay window in Display 5.8 shows the cumulative lift for the Training and Validation data sets. Click the down arrow next to the text box to see a list of available charts that can be displayed in this window.

Display 5.9 shows the iteration plot with Average Squared Error at each iteration for the Training and Validation data sets. The estimation process required 70 iterations. The weights from the 49th iteration were selected. After the 49th iteration, the Average Squared Error started to increase in the Validation data set, although it continued to decline in the Training data set.

Display 5.9

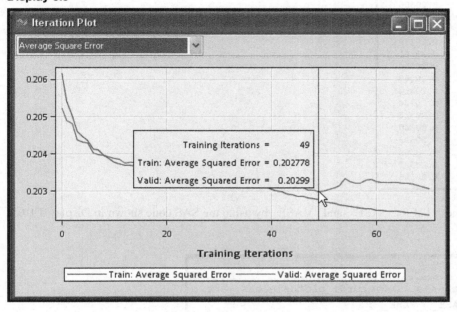

You can save the table corresponding to the plot shown in Display 5.9 by clicking the **Tables** icon and then selecting **File→ Save As**. Table 5.1 shows the three variables _ITER_ (iteration number), _ASE_ (Average Squared Error for the Training data), and _VASE_ (Average Squared Error from the Validation data) at iterations 41-60.

Table 5.1

Training Iterations	Train: Average Squared Error	Valid: Average Squared Error
41	0.20300	0.20324
42	0.20298	0.20321
43	0.20293	0.20314
44	0.20291	0.20328
45	0.20287	0.20316
46	0.20285	0.20308
47	0.20282	0.20303
48	0.20280	0.20303
49	0.20278	0.20299
50	0.20272	0.20301
51	0.20268	0.20304
52	0.20267	0.20308
53	0.20263	0.20314
54	0.20260	0.20334
55	0.20258	0.20325
56	0.20258	0.20320
57	0.20255	0.20322
58	0.20253	0.20330
59	0.20252	0.20332
60	0.20248	0.20326

You can print the variables _ITER_, _ASE_, and _VASE_ by using the SAS code shown in Display 5.10.

Display 5.10

```
Training Code

proc print data=&em_lib..neural_plotds noobs label;
  var _ITER_ _ASE_ _VASE_ ;
  where 41 le _ITER_ le 60;
run;
```

5.4.2 Assessing the Predictive Performance of the Estimated Model

In order to assess the predictive performance of the neural network model, run the **Model Comparison** node and open the Results window. In the Results window, the Score Rankings Overlay shows the Lift charts for the Training, Validation, and Test data sets. These are shown in Display 5.11.

Display 5.11

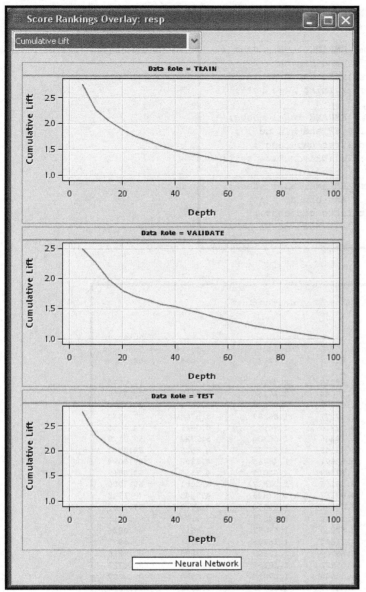

Click the arrow in the box at the top left corner of the Score Ranking Overlay window to see a list of available charts.

SAS Enterprise Miner saves the Score Rankings table as EMWS.MdlComp_EMRANK. Tables 5.2, 5.3, and 5.4 are created from the saved data set using the simple SAS code shown in Display 5.12.

Display 5.12

```
Training Code
  options center ;
  proc print data=&EM_LIB..MdlComp_EMRANK label noobs ;
   where upcase(datarole) = "TRAIN" and bin ne . ;
   var bin decile resp respc lift liftc cap capc ;
   title "Lift and Capture Rates: Training Data set";
  run;
  proc print data=&EM_LIB..MdlComp_EMRANK label noobs;
   where upcase(datarole) = "VALIDATE" and bin ne . ;
   var bin decile resp respc lift liftc cap capc ;
   title "Lift and Capture Rates: Validation Data set";
  run;
  proc print data=&EM_LIB..MdlComp_EMRANK label noobs;
   where upcase(datarole) = "TEST" and bin ne . ;
   var bin decile resp respc lift liftc cap capc ;
   title "Lift and Capture Rates: Test Data set";
  run;
```

Table 5.2

Lift and Capture Rates: Training Data set

Bin	Depth	% Response	Cumulative % Response	Lift	Cumulative Lift	% Captured Response	Cumulative % Captured Response
1	5	8.31775	8.31775	2.77258	2.77258	13.8642	13.8642
2	10	5.31131	6.81464	1.77044	2.27155	8.8518	22.7160
3	15	4.84188	6.15707	1.61396	2.05236	8.0697	30.7856
4	20	4.12431	5.64861	1.37477	1.88287	6.8788	37.6644
5	25	3.83131	5.28540	1.27710	1.76180	6.3811	44.0455
6	30	3.58335	5.00173	1.19445	1.66724	5.9723	50.0178
7	35	2.91134	4.70309	0.97045	1.56770	4.8525	54.8702
8	40	2.78073	4.46259	0.92691	1.48753	4.6392	59.5094
9	45	2.91268	4.29046	0.97089	1.43015	4.8525	64.3619
10	50	2.64467	4.12588	0.88156	1.37529	4.4081	68.7700
11	55	2.61352	3.98844	0.87117	1.32948	4.3548	73.1248
12	60	2.46397	3.86142	0.82132	1.28714	4.1059	77.2307
13	65	2.34569	3.74481	0.78190	1.24827	3.9104	81.1411
14	70	1.71661	3.59991	0.57220	1.19997	2.8617	84.0028
15	75	2.15451	3.50356	0.71817	1.16785	3.5905	87.5933
16	80	1.97366	3.40798	0.65789	1.13599	3.2883	90.8816
17	85	1.67467	3.30604	0.55822	1.10201	2.7906	93.6722
18	90	1.58819	3.21056	0.52940	1.07019	2.6484	96.3207
19	95	1.34421	3.11236	0.44807	1.03745	2.2396	98.5603
20	100	0.86420	3.00000	0.28807	1.00000	1.4397	100.000

Table 5.3

							Cumulative
		%	Cumulative		Cumulative	% Captured	% Captured
Bin	Depth	Response	% Response	Lift	Lift	Response	Response
1	5	7.49586	7.49586	2.49862	2.49862	12.5089	12.5089
2	10	6.05928	6.77825	2.01976	2.25942	10.0924	22.6013
3	15	4.24246	5.93308	1.41415	1.97769	7.0718	29.6731
4	20	3.84086	5.41043	1.28029	1.80348	6.3966	36.0697
5	25	3.85782	5.09981	1.28594	1.69994	6.4321	42.5018
6	30	4.00842	4.91792	1.33614	1.63931	6.6809	49.1827
7	35	3.42772	4.70477	1.14257	1.56826	5.7214	54.9041
8	40	3.65246	4.57347	1.21749	1.52449	6.0768	60.9808
9	45	3.11138	4.41094	1.03713	1.47031	5.1883	66.1692
10	50	2.78928	4.24858	0.92976	1.41619	4.6553	70.8244
11	55	2.26190	4.06814	0.75397	1.35605	3.7669	74.5913
12	60	2.28275	3.91945	0.76092	1.30648	3.8024	78.3937
13	65	2.08817	3.77850	0.69606	1.25950	3.4826	81.8763
14	70	1.90004	3.64450	0.63335	1.21483	3.1628	85.0391
15	75	2.04643	3.53794	0.68214	1.17931	3.4115	88.4506
16	80	1.59942	3.41680	0.53314	1.13893	2.6652	91.1158
17	85	1.61828	3.31087	0.53943	1.10362	2.7008	93.8166
18	90	1.57989	3.21483	0.52663	1.07161	2.6297	96.4463
19	95	1.32169	3.11518	0.44056	1.03839	2.2033	98.6496
20	100	0.81060	3.00000	0.27020	1.00000	1.3504	100.000

Lift and Capture Rates: Validation Data set

Table 5.4

							Cumulative
		%	Cumulative		Cumulative	% Captured	% Captured
Bin	Depth	Response	% Response	Lift	Lift	Response	Response
1	5	8.36796	8.36796	2.78932	2.78932	13.9510	13.9510
2	10	5.49570	6.93212	1.83190	2.31071	9.1587	23.1097
3	15	5.02337	6.29385	1.67446	2.09795	8.4132	31.5229
4	20	4.47671	5.84044	1.49224	1.94681	7.4547	38.9776
5	25	4.17028	5.50778	1.39009	1.83593	6.9223	45.8999
6	30	3.50818	5.17403	1.16939	1.72468	5.8573	51.7572
7	35	3.12520	4.88096	1.04173	1.62699	5.2183	56.9755
8	40	3.01327	4.64829	1.00442	1.54943	5.0053	61.9808
9	45	2.54180	4.41310	0.84727	1.47103	4.2599	66.2407
10	50	2.30884	4.20349	0.76961	1.40116	3.8339	70.0745
11	55	2.17732	4.01970	0.72577	1.33990	3.6209	73.6954
12	60	3.00147	3.93481	1.00049	1.31160	5.0053	78.7007
13	65	2.03820	3.78849	0.67940	1.26283	3.4079	82.1086
14	70	2.17732	3.67367	0.72577	1.22456	3.6209	85.7295
15	75	1.52626	3.52989	0.50875	1.17663	2.5559	88.2854
16	80	1.66239	3.41329	0.55413	1.13776	2.7689	91.0543
17	85	1.86309	3.32260	0.62103	1.10753	3.0884	94.1427
18	90	1.85065	3.24072	0.61688	1.08024	3.0884	97.2311
19	95	1.14908	3.13055	0.38303	1.04352	1.9169	99.1480
20	100	0.51256	3.00000	0.17085	1.00000	0.8520	100.000

Lift and Capture Rates: Test Data set

The lift and capture rates calculated from the Test data set (shown in Table 5.4) should be used for evaluating the models or comparing the models because the Test data set is not used in training or fine-tuning the model.

To calculate the lift and capture rates, SAS Enterprise Miner first calculates the predicted probability of response for each record in the Test data. Then it sorts the records in descending order of the predicted probabilities (also called the scores) and divides the data set into 20 groups of equal size. In Table 5.4, the column Bin shows the ranking of these groups. If the model is accurate, the table should show the highest actual response rate in the first bin, the second highest in the next bin, and so on. From the column %Response, it is clear that the average response rate for observations in the first bin is 8.36796%. The average response rate for the entire test data set is 3%. Hence the lift for Bin 1, which is the ratio of the response rate in Bin 1 to the overall response rate, is 2.7893. The lift for each bin is calculated in the same way. The first row of the column Cumulative %Response shows the response rate for the first bin. The second row shows the response rate for bins 1 and 2 combined, and so on.

The capture rate of a bin shows the percentage of likely responders that it is reasonable to expect to be captured in the bin. From the column Captured Response, you can see that 13.951% of all responders are in Bin 1.

From the Cumulative % Captured Response column of Table 5.3, you can be seen that, by sending mail to customers in the first four bins, or the top 20% of the target population, it is reasonable to expect to capture 39% of all potential responders from the target population. This assumes that the modeling sample represents the target population.

5.4.3 Receiver Operating Characteristic (ROC) Charts

Display 5.13, taken from the Results window of the **Model Comparison** node, displays ROC curves for the Training, Validation, and Test data sets. An ROC curve shows the values of the *true positive fraction* and the *false positive fraction* at different *cut-off values*, which can be denoted by P_c. In the case of the response model, if the estimated probability of response for a customer record were above a cut-off value P_c, then you would classify the customer as a responder; otherwise, you would classify the customer as a non-responder.

Display 5.13

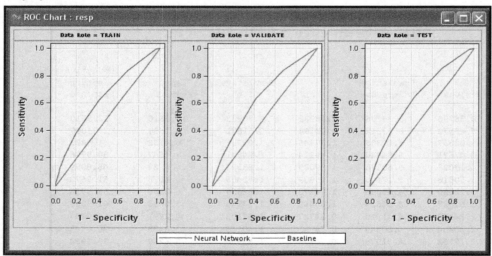

In the ROC chart, the true positive fraction is shown on the vertical axis, and the false positive fraction is on the horizontal axis for each cut-off value (P_c).

If the calculated probability of response (P_resp1) is greater than equal to the cut-off value, then the customer (observation) is classified as a responder. Otherwise, the customer is classified as non-responder.

True positive fraction is the proportion of responders correctly classified as responders. The false positive fraction is the proportion of non-responders incorrectly classified as responders. The true positive fraction is also called *sensitivity,* and *specificity* is the proportion of non-responders correctly classified as non-responders. Hence, the false positive fraction is 1-specificity. An ROC curve reflects the tradeoff between sensitivity and specificity.

The straight diagonal lines in Display 5.13 that are labeled Baseline are the ROC charts of a model that assigns customers at random to the responder group and the non-responder group, and hence has no predictive power. On these lines, sensitivity = 1- specificity at all cut-off points. The larger the area between the ROC curve of the model being evaluated and the diagonal line, the better the model. The area under the ROC curve is a measure of the predictive accuracy of the model and can be used for comparing different models.

Table 5.5 shows sensitivity and 1-specificity at various cut-off points in the validation data.

Table 5.5

```
            ROC Table: Validation Data


   Cutoff     Sensitivity     1-Specificity

  1.00000      0.00000         0.00000
  0.26583      0.00036         0.00000
  0.22075      0.00071         0.00000
  0.15334      0.00071         0.00016
  0.14645      0.00142         0.00016
  0.13954      0.00284         0.00049
  0.12783      0.00391         0.00097
  0.11956      0.00640         0.00162
  0.10987      0.00995         0.00325
  0.09991      0.01741         0.00503
  0.08999      0.02665         0.01007
  0.07991      0.04655         0.01689
  0.06999      0.07285         0.02598
  0.05995      0.12296         0.04644
  0.04999      0.20220         0.08834
  0.03998      0.36283         0.19584
  0.03000      0.63184         0.41410
  0.02000      0.84115         0.68139
  0.01000      0.98969         0.96963
  0.00000      1.00000         1.00000
```

From Table 5.5, you can see that at a cut-off probability (P_c) of 0.02, for example, the sensitivity is 0.84115. That is, at this cut-off point, you will correctly classify 84.1% of responders as responders, but you will also *incorrectly* classify 68.1% of non-responders as responders, since 1-specificity at this point is 0.68139. If instead you chose a much higher cut-off point of $P_c = 0.13954$, you would classify 0.284% of true responders as responders and 0.049% of non-responders as responders. In this case, by increasing the cut-off probability beyond which you would classify an individual as a responder, you would be reducing the fraction of false positive decisions made, while, at the same time, also reducing the fraction of true positive decisions made. These pairings of a true positive fraction with a false positive fraction are plotted as the ROC curve for the VALIDATE case in Display 5.13.

The SAS macro in Display 5.14 demonstrates the calculation of the true positive rate (TPR) and the false positive rate (FPR) at the cut-off probability of 0.02.

Display 5.14

```
%macro roccalc(PC=);
ods html file ="C:/TheBook/EM12.1/Reports/Chapter5/RespRateV.html";
title "Validation Data Set";
proc freq data=&EM_LIB..MdlComp_Validate;
 table resp / noperc nocumperc out=tab1(keep= resp count rename=(count=N));
run;
ods html close ;
ods html file ="C:/TheBook/EM12.1/Reports/Chapter5/RespRateV_cutoff.html";
Title "Cases with Predicted Probability (P_resp1) GE &PC";
title2 "Validation Data Set";
proc freq data=&EM_LIB..MdlComp_Validate;
   table resp / noperc nocumperc out=tab2(keep= resp count rename=(count=NC));
   where P_resp1 ge &PC;
run;
ods html close;
data temp;
 merge tab1 tab2 ;
 by resp ;
 if resp=0 then TYPE='FPR' ; else if resp=1 then TYPE='TPR'; Rate= NC/N;
 cutoff=&pc;
 run;
ods html file ="C:/TheBook/EM12.1/Reports/Chapter5/TPR_FPR_Cutoff.html";
 Title "True Positive Rate (TPR) and False Positive Rate (FPR)";
 title2 "at cut-off = &PC";
proc print data=temp label noobs;
 var RESP N NC TYPE RATE ;
 label N = "Number of Observations in the sample";
 label NC = "Number of Observations classified as responders ";
 label RESP = "Actual RESPONSE ";
 label TYPE = "ROC Coordinate   ";
 label Rate = "ROC Coordinate Value";
 run;
ods html close;
%mend roccalc;
%roccalc(PC=0.02);
```

Tables 5.6, 5.7, and 5.8, generated by the macro shown in Display 5.14, show the sequence of steps for calculating TPR and FPR for a given cut-off probability.

Table 5.6

Validation Data Set

The FREQ Procedure

resp	Frequency
0	6158
1	2814

Table 5.7

Cases with Predicted Probability (P_resp1) GE 0.02 Validation Data Set

The FREQ Procedure

resp	Frequency
0	4196
1	2367

Table 5.8

True Positive Rate (TPR) and False Positive Rate (FPR) at cut-off = 0.02				
Actual RESPONSE	Number of Observations in the sample	Number of Observations classified as responders	ROC Coordinate	ROC Coordinate Value
0	6158	4196	FPR	0.68139
1	2814	2367	TPR	0.84115

For more information about ROC curves, see the textbook by A. A. Afifi and Virginia Clark (2004).[4]

Display 5.15 shows the SAS code that generated Table 5.5.

Display 5.15

```
proc print data=&EM_LIB..mdlcomp_emroc noobs label;
 var cutoff sensitivity oneminusspecificity ;
 where upcase(datarole) = 'VALIDATE' and upcase(model)="NEURAL" and cutoff ne .;
 Title "ROC Table: Validation Data" ;
 label cutoff = "Cutoff" sensitivity = "Sensitivity"
       oneminusspecificity="1-Specificity";
run;
```

5.4.4 How Did the Neural Network Node Pick the Optimum Weights for This Model?

In Section 5.3, I described how the optimum weights are found in a neural network model. I described the two-step procedure of estimating and selecting the weights. In this section, I show the results of these two steps with reference to the neural network model discussed in Sections 5.4.1 and 5.4.2.

The weights such as those shown in Equations 5.1, 5.3, 5.5, 5.7, 5.9, 5.11, and 5.13 are shown in the Results window of the **Neural Network** node. You can see the estimated weights created at each iteration by opening the results window and selecting **View→Model→Weights-History**. Display 5.16 shows a partial view of the Weights-History window.

Display 5.16

ITER	AGE -> HL1	CRED -> HL1	DELINQ -> HL1	MILEAGE -> HL1	NUMTR -> HL1	AGE -> HL2	CRED -> HL2	DELINQ -> HL2	MILEAGE -> HL2	NUMTR -> HL2	AGE -> HL3
30	0.051634	0.013581	-0.04942	0.037994	-0.09108	0.021394	0.015272	-0.01862	0.032955	0.072774	0.03261
31	0.05182	0.014726	-0.05917	0.039808	-0.09145	0.021723	0.013705	-0.02283	0.035434	0.080229	0.033779
32	0.04926	0.014514	-0.06559	0.035954	-0.09289	0.020158	0.013317	-0.01744	0.035331	0.083941	0.035385
33	0.049163	0.015336	-0.07253	0.036421	-0.09141	0.019859	0.012407	-0.01959	0.036852	0.086744	0.036088
34	0.049216	0.015535	-0.07133	0.035055	-0.08937	0.019039	0.012593	-0.01825	0.036613	0.084751	0.036105
35	0.047965	0.016614	-0.08615	0.035396	-0.09089	0.019267	0.010752	-0.02111	0.039862	0.09669	0.037988
36	0.046969	0.015786	-0.08104	0.033397	-0.09163	0.019507	0.011716	-0.01786	0.038681	0.093822	0.036828
37	0.045379	0.017111	-0.09285	0.031359	-0.09064	0.019137	0.010233	-0.01878	0.041725	0.101743	0.03723
38	0.043214	0.017856	-0.10395	0.029164	-0.09141	0.01911	0.009036	-0.01862	0.044226	0.110682	0.037984
39	0.041534	0.018032	-0.10769	0.026898	-0.09245	0.019452	0.008694	-0.01763	0.045423	0.115703	0.037584
40	0.039495	0.019087	-0.11634	0.023946	-0.09319	0.019732	0.007452	-0.01835	0.048368	0.124959	0.036921
41	0.041232	0.019089	-0.10932	0.025544	-0.09307	0.020193	0.007861	-0.02079	0.047792	0.12189	0.035542
42	0.042101	0.019884	-0.11178	0.026301	-0.09267	0.019756	0.007088	-0.02322	0.048779	0.12454	0.0359
43	0.044041	0.020461	-0.10627	0.025526	-0.09514	0.016812	0.006401	-0.02138	0.047231	0.122084	0.036323
44	0.043902	0.022069	-0.1106	0.022244	-0.09944	0.014736	0.004598	-0.02197	0.049496	0.131921	0.035225
45	0.044047	0.021174	-0.1079	0.023537	-0.09872	0.015295	0.005568	-0.02131	0.048206	0.127659	0.035611
46	0.044491	0.021269	-0.10816	0.023826	-0.09965	0.01462	0.005453	-0.02128	0.047965	0.127914	0.036005
47	0.045297	0.021834	-0.10742	0.022637	-0.10376	0.01277	0.004953	-0.0206	0.048183	0.130482	0.03537
48	0.048549	0.022946	-0.10315	0.023189	-0.11333	0.010181	0.004433	-0.02328	0.049119	0.135763	0.032797
49	0.046004	0.022182	-0.10612	0.022121	-0.10898	0.011822	0.004946	-0.02065	0.048699	0.13283	0.033759
50	0.047727	0.022403	-0.10271	0.023855	-0.11699	0.01153	0.005256	-0.02343	0.049256	0.135583	0.031076
51	0.048555	0.022612	-0.09836	0.023541	-0.12846	0.010614	0.005728	-0.02371	0.049642	0.1387	0.027236

The second column in Display 5.16 shows the weight of the variable AGE in hidden unit 1 at each iteration. The seventh column shows the weight of AGE in hidden unit 2 at each iteration. The twelfth column shows the weight of AGE in the third hidden unit. Similarly, you can trace through the weights of other variables. You can save the Weights-History table as a SAS data set.

To see the final weights, open the Results window. Select **View→Model→Weights_Final**. Then, click the **Table** icon. Selected rows of the final weights_table are shown in Display 5.17.

Display 5.17

LABEL	FROM	TO	WEIGHT
AGE -> HL1	AGE	HL1	0.04600
AGE -> HL2	AGE	HL2	0.01182
AGE -> HL3	AGE	HL3	0.03376
BIAS -> HL1	BIAS	HL1	0.21515
BIAS -> HL2	BIAS	HL2	-0.03256
BIAS -> HL3	BIAS	HL3	-0.09328
BIAS -> resp0	BIAS	resp0	1.02894
BIAS -> resp1	BIAS	resp1	-1.02894
CRED -> HL1	CRED	HL1	0.02218
CRED -> HL2	CRED	HL2	0.00495
CRED -> HL3	CRED	HL3	0.03979
HL1 -> resp0	HL1	resp0	1.29172
HL1 -> resp1	HL1	resp1	-1.29172
HL2 -> resp0	HL2	resp0	-1.59773
HL2 -> resp1	HL2	resp1	1.59773
HL3 -> resp0	HL3	resp0	1.56982
HL3 -> resp1	HL3	resp1	-1.56982

Outputs of the hidden units become inputs to the target layer. In the target layer, these inputs are combined using the weights estimated by the **Neural Network** node.

In the model I have developed, the weights generated at the 49th iteration are the optimal weights, because the Average Squared Error computed from the Validation data set reaches its minimum at the 49th iteration. This is shown in Display 5.18.

Display 5.18

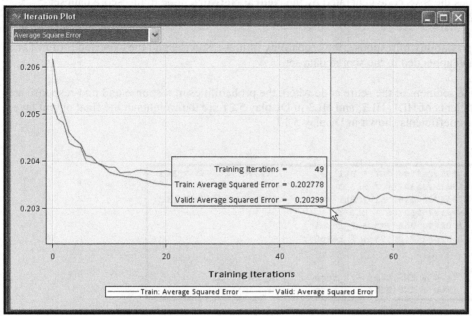

5.4.5 Scoring a Data Set Using the Neural Network Model

You can use the SAS code generated by the **Neural Network** node to score a data set within SAS Enterprise Miner or outside. This example scores a data set inside SAS Enterprise Miner.

The process flow diagram with a scoring data set is shown in Display 5.19.

Display 5.19

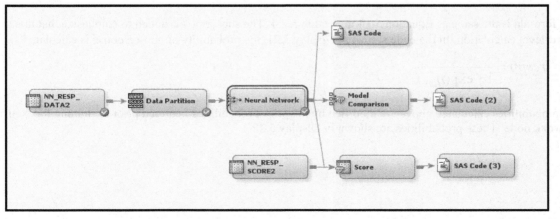

Set the **Role** property of the data set to be scored to Score, as shown in Display 5.20.

Display 5.20

Train	
Output Type	View
Role	Score
Rerun	No
Summarize	No
Drop Map Variables	No

The Score Node applies the SAS code generated by the **Neural Network** node to the Score data set NN_RESP_SCORE2, shown in Display 5.19.

For each record, the probability of response, the probability of non-response, and the expected profit of each record is calculated and appended to the scored data set.

Display 5.21 shows the segment of the score code where the probabilities of response and non-response are calculated. The coefficients of HL1, HL2, and HL3 in Display 5.21 are the weights in the final output layer. These are same as the coefficients shown in Display 5.17.

Display 5.21

```
P_resp1 =      -1.29172482403562 * HL1  +      1.5977312209735 * HL2
        +      -1.56981973510786 * HL3 ;
P_resp0 =       1.29172482403562 * HL1  +     -1.5977312209735 * HL2
        +       1.56981973510786 * HL3 ;
P_resp1 =      -1.0289412689995 + P_resp1 ;
P_resp0 =       1.0289412689995 + P_resp0 ;
DROP _EXP_BAR;
_EXP_BAR=50;
P_resp1 = 1.0 / (1.0 + EXP(MIN( - P_resp1 , _EXP_BAR)));
P_resp0 = 1.0 / (1.0 + EXP(MIN( - P_resp0 , _EXP_BAR)));
```

The code segment given in Display 5.21 calculates the probability of response using the formula $P\_resp1_i = \dfrac{1}{1 + \exp(-\eta_{i21})}$,

where $\mu_{i21} =$ -1.29172481842873 * HL1 +   1.59773122184585 * HL2

+   -1.56981973539319 * HL3 -1.0289412689995;

This formula is the same as Equation 5.14 in Section 5.2.4. The subscript i is added to emphasize that this is a record-level calculation. In the code shown in Display 5.21, the probability of non-response is calculated as $p\_resp0 = \dfrac{1}{1 + \exp(\eta_{i21})}$.

The probabilities calculated above are modified by the prior probabilities I entered prior to running the **Neural Network** node. These probabilities are shown in Display 5.22.

Display 5.22

You can enter the prior probabilities when you create the **Data Source**. Prior probabilities are entered because the responders are overrepresented in the modeling sample, which is extracted from a larger sample. In the larger sample, the proportion of responders is only 3%. In the modeling sample, the proportion of responders is 31.36%. Hence, the probabilities should be adjusted before expected profits are computed. The SAS code generated by the **Neural Network** node and passed on to the **Score** node includes statements for making this adjustment. Display 5.23 shows these statements.

Display 5.23

```
*** Update Posterior Probabilities;
P_resp1 = P_resp1 * 0.03 / 0.31368937998772;
P_resp0 = P_resp0 * 0.97 / 0.68631062001227;
drop _sum; _sum = P_resp1 + P_resp0 ;
if _sum > 4.135903E-25 then do;
   P_resp1 = P_resp1 / _sum;
   P_resp0 = P_resp0 / _sum;
end;
```

Display 5.24 shows the profit matrix used in the decision-making process.

Display 5.24

Given the above profit matrix, calculation of expected profit under the alternative decisions of classifying an individual as responder or non-responder proceeds as follows. Using the neural network model, the scoring algorithm first calculates the individual's probability of response and non-response. Suppose the calculated probability of response for an individual is 0.3, and probability of non-response is 0.7. The expected profit if the individual is classified as responder is 0.3x$5 + 0.7x (–$1.0) = $0.8. The expected profit if the individual is classified as non-responder is 0.3x ($0) + 0.7x ($0) = $0. Hence classifying the individual as responder (Decision1) yields a higher profit than if the individual is classified as non-responder (Decision2). An additional field is added to the record in the scored data set indicating the decision to classify the individual as a responder.

These calculations are shown in the score code segment shown in Display 5.25.

Display 5.25

```
*** Decision Processing;
label D_RESP = 'Decision: resp' ;
label EP_RESP = 'Expected Profit: resp' ;

length D_RESP $ 9;

D_RESP = ' ';
EP_RESP = .;

*** Compute Expected Consequences and Choose Decision;
_decnum = 1; drop _decnum;

D_RESP = '1' ;
EP_RESP = P_resp1 * 5 + P_resp0 * -1;
drop _sum;
_sum = P_resp1 * 0 + P_resp0 * 0;
if _sum > EP_RESP + 2.273737E-12 then do;
   EP_RESP = _sum; _decnum = 2;
   D_RESP = '0' ;
end;

*** End Decision Processing ;
```

5.4.6 Score Code

The score code is automatically saved by the **Score** node in the sub-directory \Workspaces\EMWSn\Score within the project directory.

For example, in my computer, the Score code is saved by the **Score** node in the folder C:\TheBook\EM12.1\EMProjects\Chapter5\Workspaces\EMWS3\Score. Alternatively, you can save the score code in some other directory. To do so, run the Score node, and then click Results. Select either the Optimized SAS Code window or the SAS Code window. Click **File→Save As**, and enter the directory and name for saving the score code.

5.5 A Neural Network Model to Predict Loss Frequency in Auto Insurance

The premium that an insurance company charges a customer is based on the degree of risk of monetary loss to which the customer exposes the insurance company. The higher the risk, the higher the premium the company charges. For proper rate setting, it is essential to predict the degree of risk associated with each current or prospective customer. Neural networks can be used to develop models to predict the risk associated with each individual.

In this example, I develop a neural network model to predict loss frequency at the customer level. I use the target variable LOSSFRQ that is a discrete version of loss frequency. The definition of LOSSFRQ is presented in the beginning of this chapter (Section 5.1.1). If the target is a discrete form of a continuous variable with more than two levels, it should be treated as an ordinal variable, as I do in this example.

The goal of the model developed here is to estimate the conditional probabilities $\Pr(LOSSFRQ = 0 \mid X)$, $\Pr(LOSSFRQ = 1 \mid X)$, and $\Pr(LOSSFRQ = 2 \mid X)$, where X is a set of inputs or explanatory variables.

I have already discussed the general framework for neural network models in Section 5.2. There I gave an example of a neural network model with two hidden layers. I also pointed out that the outputs of the final hidden layer can be considered as complex transformations of the original inputs. These are given in Equations 5.1 through 5.12.

In the following example, there is only one hidden layer, and it has three units. The outputs of the hidden layer are H_{L1}, H_{L2}, and H_{L3}. Since these are nothing but transformations of the original inputs, I can write the desired conditional probabilities as

$$\Pr(LOSSFRQ = 0 \mid H_{L1}, H_{L2}, and\ H_{L3}), \Pr(LOSSFRQ = 1 \mid H_{L1}, H_{L2}, and\ H_{L3}),$$
and $\Pr(LOSSFRQ = 2 \mid H_{L1}, H_{L2}, and\ H_{L3})$.

5.5.1 Loss Frequency as an Ordinal Target

In order for the **Neural Network** node to treat the target variable LOSSFRQ as an ordinal variable, we should set its measurement level to Ordinal in the data source.

Display 5.26 shows the process flow diagram for developing a neural network model.

Display 5.26

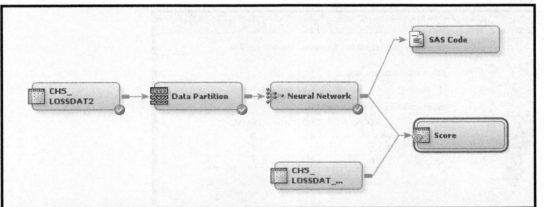

The data source for the **Input Data** node is created from the data set Ch5_LOSSDAT2 using the Advanced Metadata Advisor Options with the **Class Levels Count Threshold** property set to 8, which is the same setting shown in Display 5.2.

As pointed out in Section 5.1.1, loss frequency is a continuous variable. The target variable LOSSFRQ, used in the Neural Network model developed here, is a discrete version of loss frequency. It is created as follows:

$$LOSSFRQ = 0 \ if\ loss\ frequency\ =\ 0,$$
$$LOSSFRQ = 1 \ if\ 0\ <\ loss\ frequency\ <\ 1.5,\ and$$
$$LOSSFRQ = 2 \ if\ loss\ frequency\ \geq 1.5.$$

Display 5.27 shows a partial list of the variables contained in the data set Ch5_LOSSDAT2. The target variable is LOSSFRQ. I changed its measurement level from Nominal to Ordinal.

Display 5.27

For illustration purposes, I enter a profit matrix, which is shown in Display 5.28. A profit matrix can be used for classifying customers into different risk groups such as low, medium, and high.

Display 5.28

The SAS code generated by the **Neural Network** node, shown in Display 5.29, illustrates how the profit matrix shown in Display 5.28 is used to classify customers into different risk groups.

Display 5.29

```
*** Compute Expected Consequences and Choose Decision;
_decnum = 1; drop _decnum;


D_LOSSFRQ = '2' ;
EP_LOSSFRQ = P_LOSSFRQ2 * 5 + P_LOSSFRQ1 * -1 + P_LOSSFRQ0 * -2;
drop _sum;
_sum = P_LOSSFRQ2 * -1 + P_LOSSFRQ1 * 4 + P_LOSSFRQ0 * -1;
if _sum > EP_LOSSFRQ + 2.273737E-12 then do;
   EP_LOSSFRQ = _sum; _decnum = 2;
   D_LOSSFRQ = '1' ;
end;
_sum = P_LOSSFRQ2 * -2 + P_LOSSFRQ1 * -1 + P_LOSSFRQ0 * 3;
if _sum > EP_LOSSFRQ + 2.273737E-12 then do;
   EP_LOSSFRQ = _sum; _decnum = 3;
   D_LOSSFRQ = '0' ;
end;


*** End Decision Processing ;
```

In the SAS code shown in Display 5.29, P_LOSSFRQ2, P_LOSSFRQ1, and P_LOSSFRQ0 are the estimated probabilities of LOSSFRQ= 2, 1, and 0, respectively, for each customer.

The property settings of the **Data Partition** node for allocating records for Training, Validation, and Test are 60, 30, and 10, respectively. These settings are the same as in Display 5.4.

To define the Neural Network, click ▣ located to the right of the **Network** property of the **Neural Network** node. The Network window opens, as shown in Display 30.

Display 5.30

Property	Value
Architecture	User
Direct Connection	No
Number of Hidden Units	3
Randomization Distribution	Normal
Randomization Center	0.0
Randomization Scale	0.1
Input Standardization	Standard Deviation
Hidden Layer Combination Function	Linear
Hidden Layer Activation Function	Hyperbolic Tangent
Hidden Bias	Yes
Target Layer Combination Function	Linear
Target Layer Activation Function	Logistic
Target Layer Error Function	MBernoulli

The general formulas of Combination Functions and Activation Functions of the hidden layers are given by equations 5.1-5.12 of section 5.2.2.

In the Neural Network model developed in this section, there is only one hidden layer and three hidden units, each hidden unit producing an output. The outputs produced by the hidden units are HL1, HL2, and HL3. These are special variables that are constructed from inputs contained in the dataset Ch5_LOSSDAT2. These inputs are standardized prior to being used by the hidden layer.

5.5.1.1 Target Layer Combination and Activation Functions

As you can see from Display 5.30, I set the **Target Layer Combination Function** property to Linear and the Target **Layer Activation Function** property to Logistic. These settings result in the following calculations.

Since we set the **Target Layer Combination** Function property to be linear, the **Neural Network** node calculates linear predictors for each observation from the hidden layer outputs HL1, HL2, and HL3. Since the target variable LOSSFRQ is ordinal, taking the values 0, 1 and 2, the **Neural Network** node calculates two linear predictors.

The first linear predictor for the ith customer/record is calculated as:

$$\mu_{i1} = w_{10} + w_{11}HL1_i + w_{12}HL2_i + w_{13}HL3_i \qquad (5.18)$$

μ_{i1} is the linear predictor used in the calculation of $P(LOSSFRQ = 2 / H_i)$ for the ith customer/record.

$$where \ H_i = \begin{pmatrix} 1 \\ HL1_i \\ HL2_i \\ HL3_i \end{pmatrix},$$

$HL1_i$ = Output of the first hidden unit for the ith customer,

$HL2_i$ = Output of the second hidden unit for the ith customer,

$HL3_i$ = Output of the third hidden unit for the ith customer,

w_{11}, w_{12} and w_{13} are the weights of the outputs of the hidden units, and w_{10} is the bias.

The estimated values of the bias and weights are:

$w_{10} = -3.328952$

$w_{11} = 1.489494$

$w_{12} = 1.759599$

$w_{13} = 1.067301$

The second linear predictor for the ith customer/record is calculated as:

$$\mu_{i2} = w_{20} + w_{21}HL1_i + w_{22}HL2_i + w_{23}HL3_i \qquad (5.19)$$

The estimated values of the bias and weights are:

$w_{20} = -2.23114$

$w_{21} = 1.11754$

$w_{22} = 0.96459$

$w_{23} = 0.02544$

μ_{i2} is the linear predictor used in the calculation of $P(LOSSFRQ = 2 / H_i) + P(LOSSFRQ = 1 / H_i)$ for the ith customer/record.

μ_{i1} and μ_{i2} are treated as logits. That is,

$$\mu_{i1} = \log\left[\frac{P(LOSSFRQ = 2 \mid H_i)}{1 - P(LOSSFRQ = 2 \mid H_i)}\right] \tag{5.20}$$

$$\mu_{i2} = \log\left[\frac{P(LOSSFRQ = 2 \mid H_i) + P(LOSSFRQ = 1 \mid H_i)}{1 - \{P(LOSSFRQ = 2 \mid H_i) + P(LOSSFRQ = 1 \mid H_i)\}}\right] \tag{5.21}$$

Since μ_{i1} is based on $P(LOSSFRQ = 2 \mid H_i)$ and μ_{i2} is based on $P(LOSSFRQ = 2 \mid H_i) + P(LOSSFRQ = 1 \mid H_i)$, μ_{i1} and μ_{i2} can be called cumulative logits, and the model estimated by this procedure can be called a Cumulative Logits model. Given μ_{i1} and μ_{i2}, we can calculate the probabilities $P(LOSSFRQ = 2 \mid H_i)$, $P(LOSSFRQ = 1 \mid H_i)$ and $P(LOSSFRQ = 0 \mid H_i)$ in the following way:

$$P(LOSSFRQ = 2 \mid H_i) = \frac{1}{1 + \exp(-\mu_{i1})} \tag{5.22}$$

$$P(LOSSFRQ = 2 \mid H_i) + P(LOSSFRQ = 1 \mid H_i) = \frac{1}{1 + \exp(-\mu_{i2})} \tag{5.23}$$

$$P(LOSSFRQ = 0 \mid H_i) = 1 - \left[P(LOSSFRQ = 2 \mid H_i) + P(LOSSFRQ = 1 \mid H_i)\right] \tag{5.24}$$

$$\begin{aligned} &P(LOSSFRQ = 1 \mid H_i) \\ &= \left[P(LOSSFRQ = 2 \mid H_i) + P(LOSSFRQ = 1 \mid H_i)\right] - P(LOSSFRQ = 2 \mid H_i) \end{aligned} \tag{5.25}$$

To verify the calculations shown in equations 5.18 – 5.25, run the **Neural Network** node shown in Display 5.26 and open the Results window. In the Results window, click **View→Scoring→SAS Code** and scroll down until you see "Writing the Node LOSSFRQ" as shown in Display 5.31.

Display 5.31

```
*** ************************;
*** Writing the Node LOSSFRQ ;
*** ************************;
IF _DM_BAD EQ 0 THEN DO;
   P_LOSSFRQ2  =      1.48949434173539 * HL1  +     1.75959884200533 * HL2
         +      1.06730113426007 * HL3 ;
   P_LOSSFRQ1  =      1.11754131452423 * HL1  +     0.96458843025431 * HL2
         +      0.0254374332168 * HL3 ;
   P_LOSSFRQ2  =     -3.32895202992659 + P_LOSSFRQ2 ;
   P_LOSSFRQ1  =     -2.23114143594446 + P_LOSSFRQ1 ;
   DROP _EXP_BAR;
   _EXP_BAR=50;
   P_LOSSFRQ2  = 1.0 / (1.0 + EXP(MIN( - P_LOSSFRQ2 , _EXP_BAR)));
   P_LOSSFRQ1  = 1.0 / (1.0 + EXP(MIN( - P_LOSSFRQ1 , _EXP_BAR)));
   P_LOSSFRQ0  = 1. -  P_LOSSFRQ1 ;
   P_LOSSFRQ1  = P_LOSSFRQ1  - P_LOSSFRQ2 ;

END;
```

Scroll down further to see how each customer/record is assigned to a risk level using the profit matrix shown in Display 5.28. The steps in assigning a risk level to a customer/record are shown in Display 5.32.

Display 5.32

```
*** Decision Processing;
label D_LOSSFRQ = 'Decision: LOSSFRQ' ;
label EP_LOSSFRQ = 'Expected Profit: LOSSFRQ' ;

length D_LOSSFRQ $ 9;

D_LOSSFRQ = ' ';
EP_LOSSFRQ = .;

*** Compute Expected Consequences and Choose Decision;
_decnum = 1; drop _decnum;

D_LOSSFRQ = '2' ;
EP_LOSSFRQ = P_LOSSFRQ2 * 5 + P_LOSSFRQ1 * -1 + P_LOSSFRQ0 * -2;
drop _sum;
_sum = P_LOSSFRQ2 * -1 + P_LOSSFRQ1 * 4 + P_LOSSFRQ0 * -1;
if _sum > EP_LOSSFRQ + 2.273737E-12 then do;
   EP_LOSSFRQ = _sum; _decnum = 2;
   D_LOSSFRQ = '1' ;
end;
_sum = P_LOSSFRQ2 * -2 + P_LOSSFRQ1 * -1 + P_LOSSFRQ0 * 3;
if _sum > EP_LOSSFRQ + 2.273737E-12 then do;
   EP_LOSSFRQ = _sum; _decnum = 3;
   D_LOSSFRQ = '0' ;
end;


*** End Decision Processing ;
```

5.5.1.2 Target Layer Error Function Property

I set this **Target Layer Error Function** property to MBernoulli. Equation 5.16 is a Bernoulli error function for a binary target. If the target variable (y) takes the values 1 and 0, then the Bernoulli error function is:

$$E = -2\sum_{i=1}^{n}\left\{ y_i \ln \frac{\pi(W,X_i)}{y_i} + (1-y_i)\ln \frac{1-\pi(W,X_i)}{1-y_i}\right\}$$

If the target has c mutually exclusive levels (classes), you can create c dummy variables $y_1, y_2, y_3, ..., y_c$ for each record. In this example, the observed loss frequency has the three levels: 0, 1, and 2. These three levels (classes) are mutually exclusive. Hence, I can create three dummy variables $y_1, y_2,$ and y_3. If the observed loss frequency[5] is 0 for the ith observation, then $y_{i1} = 1$, and both y_{i2} and y_{i3} are 0. Similarly, if the observed loss frequency is 1, then $y_{i2} = 1$, and both y_{i1} and y_{i3} equals 0. If the observed loss frequency is 2, then $y_{i3} = 1$, and the other two dummy variables take on values of 0. This way of representing the target variable by dummy variables is called 1-of-c coding. Using these indicator (dummy) variables, the error function can be written as:

$$E = -2\sum_{i=1}^{n}\sum_{j=1}^{c}\left\{ y_{ij} \ln \pi(W_j,X_i)\right\}$$

where n is the number of observations in the Training data set and c is the number of mutually exclusive target levels.[6]

The weights W_j for the jth level of the target variable are chosen so as to minimize the error function. The minimization is achieved by means of an iterative process. At each iteration, a set of weights is generated, each set defining a candidate model. The iteration stops when the error is minimized or the threshold number of iterations is reached. The next task in the model development process is to select one of these models. This is done by applying the **Model Selection Criterion** property to the Validation data set.

5.5.1.3 Model Selection Criterion Property

I set the **Model Selection Criterion** property to Average Error. During the training process, the **Neural Network** node creates a number of candidate models, one at each iteration. If we set **Model Selection Criterion** to Average Error, the **Neural Network** node selects the model that has the smallest error calculated using the Validation data set.

5.5.1.4 Score Ranks in the Results Window

Open the Results window of the **Neural Network** node to see the details of the selected model.

The Results window is shown in Display 5.33.

Display 5.33

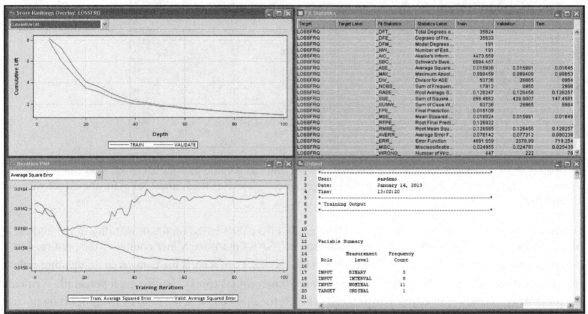

The score rankings are shown in the Score Rankings window. To see how lift and cumulative lift are presented for a model with an ordinal target, you can open the table corresponding to the graph by clicking on the Score Rankings window, and then clicking on the **Table** icon on the task bar. Selected columns from the table behind the Score Rankings graph are shown in Display 5.34.

Display 5.34

Target Label	Data Role	Event	Cumulative % Response	% Captured Response	Cumulative % Captured Response	Depth	Lift	Cumulative Lift
	TRAIN	2	1.004464	40.90909	40.90909	5	8.178165	8.178165
	TRAIN	2	0.892857	31.81818	72.72727	10	6.360795	7.26948
	TRAIN	2	0.669892	9.090909	81.81818	15	1.819401	5.454139
	TRAIN	2	0.502372	0	81.81818	20	0	4.090224
	TRAIN	2	0.446628	9.090909	90.90909	25	1.819401	3.636364
	TRAIN	2	0.372162	0	90.90909	30	0	3.030077
	TRAIN	2	0.318979	0	90.90909	35	0	2.597071
	TRAIN	2	0.279135	0	90.90909	40	0	2.272664
	TRAIN	2	0.260514	4.545455	95.45455	45	0.908685	2.121054
	TRAIN	2	0.23448	0	95.45455	50	0	1.909091
	TRAIN	2	0.213155	0	95.45455	55	0	1.735467
	TRAIN	2	0.195385	0	95.45455	60	0	1.590791
	TRAIN	2	0.188955	4.545455	100	65	0.9097	1.538435
	TRAIN	2	0.175453	0	100	70	0	1.428503
	TRAIN	2	0.163764	0	100	75	0	1.333333
	TRAIN	2	0.153524	0	100	80	0	1.249965
	TRAIN	2	0.14449	0	100	85	0	1.176409
	TRAIN	2	0.136468	0	100	90	0	1.111097
	TRAIN	2	0.129282	0	100	95	0	1.052594
	TRAIN	2	0.122823	0	100	100	0	1
	VALIDATE	2	0.892857	40	40	5	7.995536	7.995536
	VALIDATE	2	0.669643	20	60	10	3.997768	5.996652
	VALIDATE	2	0.520833	10	70	15	1.998884	4.664062
	VALIDATE	2	0.390843	0	70	20	0	3.5
	VALIDATE	2	0.357302	10	80	25	1.998884	3.199643
	VALIDATE	2	0.29773	0	80	30	0	2.66617
	VALIDATE	2	0.255183	0	80	35	0	2.285167
	VALIDATE	2	0.223339	0	80	40	0	2
	VALIDATE	2	0.223325	10	90	45	1.998884	1.999876
	VALIDATE	2	0.200983	0	90	50	0	1.799799
	VALIDATE	2	0.203004	10	100	55	1.998884	1.817905
	VALIDATE	2	0.186116	0	100	60	0	1.666667
	VALIDATE	2	0.171792	0	100	65	0	1.538395
	VALIDATE	2	0.159515	0	100	70	0	1.428457
	VALIDATE	2	0.148876	0	100	75	0	1.333184
	VALIDATE	2	0.139587	0	100	80	0	1.25
	VALIDATE	2	0.131372	0	100	85	0	1.176432
	VALIDATE	2	0.124069	0	100	90	0	1.111042
	VALIDATE	2	0.117536	0	100	95	0	1.052539
	VALIDATE	2	0.111669	0	100	100	0	1

Table: Score Rankings Overlay: LOSSFRQ

In Display 5.34, column 3 has the title **Event**. This column has a value of 2 for all rows. You can infer that SAS Enterprise Miner created the lift charts for the target level 2, which is the highest level for the target variable LOSSFRQ.

When the target is ordinal, SAS Enterprise Miner creates lift charts based on the probability of the highest level of the target variable. For each record in the test data set, SAS Enterprise Miner computes the predicted or posterior probability $Pr(lossfrq = 2 | X_i)$ from the model. Then it sorts the data set in descending order of the predicted probability and divides the data set into 20 percentiles (bins).[7] Within each percentile it calculates the proportion of cases with actual $lossfrq = 2$. The lift for a percentile is the ratio of the proportion of cases with $lossfrq = 2$ in the percentile to the proportion of cases with $lossfrq = 2$ in the entire data set.

To assess model performance, the insurance company must rank its customers by the expected value of $LOSSFRQ$, instead of the highest level of the target variable. To demonstrate how to construct a lift table based on expected value of the target, I construct the expected value for each record in the data set by using the following formula.

$$E(lossfrq | X_i) = Pr(lossfrq = 0 | X_i) * 0 + Pr(lossfrq = 1 | X_i) * 1 + Pr(lossfrq = 2 | X_i) * 2$$

Then I sorted the data set in descending order of $E(lossfrq | X_i)$, and divided the data set into 20 percentiles. Within each percentile, I calculated the mean of the *actual* or observed value of the target variable $lossfrq$.

The ratio of the mean of actual *lossfrq* in a percentile to the overall mean of actual *lossfrq* for the data set is the lift of the percentile.

These calculations were done in the **SAS Code** node, which is connected to the **Neural Network** node as shown Display 5.26.

The SAS program used in the **SAS Code** node is shown in Displays 5.88, 5.89, and 5.90 in the appendix to this chapter.

The lift tables based on $E(lossfrq \mid X_i)$ are shown in Tables 5.9, 5.10, and 5.11.

Table 5.9

		Loss Frequency: TRAIN	
Percentile	Lossfrq Mean	Lossfrq Cumulative Mean	Cumulative Lift
5	0.10726	0.10726	4.09656
10	0.08036	0.09380	3.58249
15	0.05475	0.08079	3.08550
20	0.02567	0.06700	2.55892
25	0.02905	0.05941	2.26916
30	0.01563	0.05211	1.99027
35	0.02121	0.04770	1.82156
40	0.01229	0.04327	1.65264
45	0.02344	0.04107	1.56843
50	0.02011	0.03897	1.48844
55	0.02121	0.03736	1.42672
60	0.01339	0.03536	1.35042
65	0.01341	0.03367	1.28597
70	0.01339	0.03222	1.23062
75	0.00670	0.03052	1.16569
80	0.01674	0.02966	1.13278
85	0.01563	0.02883	1.10123
90	0.01453	0.02804	1.07089
95	0.01004	0.02709	1.03470
100	0.00893	0.02618	1.00000

Table 5.10

		Loss Frequency: VALIDATE	
Percentile	Lossfrq Mean	Lossfrq Cumulative Mean	Cumulative Lift
5	0.093960	0.093960	3.62676
10	0.055804	0.074860	2.88954
15	0.049107	0.066270	2.55795
20	0.017897	0.054190	2.09169
25	0.044643	0.052279	2.01792
30	0.020089	0.046910	1.81068
35	0.024554	0.043714	1.68733
40	0.042506	0.043563	1.68150
45	0.020089	0.040953	1.58075
50	0.024554	0.039312	1.51741
55	0.017857	0.037360	1.44208
60	0.015660	0.035555	1.37238
65	0.013393	0.033849	1.30653
70	0.013393	0.032387	1.25010
75	0.000000	0.030226	1.16671
80	0.017897	0.029457	1.13701
85	0.017857	0.028774	1.11066
90	0.015625	0.028043	1.08244
95	0.008929	0.027037	1.04359
100	0.004464	0.025907	1.00000

Table 5.11

		Loss Frequency: TEST	
Percentile	Lossfrq Mean	Lossfrq Cumulative Mean	Cumulative Lift
5	0.10067	0.10067	3.76007
10	0.03356	0.06711	2.50671
15	0.04667	0.06027	2.25100
20	0.04027	0.05528	2.06457
25	0.04027	0.05228	1.95261
30	0.01333	0.04576	1.70910
35	0.04698	0.04593	1.71560
40	0.02667	0.04351	1.62527
45	0.04027	0.04315	1.61183
50	0.00671	0.03952	1.47599
55	0.00000	0.03591	1.34124
60	0.02013	0.03460	1.29224
65	0.00667	0.03244	1.21166
70	0.00671	0.03061	1.14319
75	0.03356	0.03080	1.15051
80	0.01333	0.02971	1.10956
85	0.01342	0.02875	1.07387
90	0.00667	0.02752	1.02785
95	0.01342	0.02678	1.00021
100	0.02667	0.02677	1.00000

5.5.2 Scoring a New Dataset with the Model

The Neural Network model we developed can be used for scoring a data set where the value of the target is not known. The **Score** node uses the model developed by the **Neural Network** node to predict the target levels and their probabilities for the Score data set.

Display 5.26 shows the process flow for scoring. The process flow consists of an **Input Data** node called Ch5_LOSSDAT_SCORE2, which reads the data set to be scored. The **Score** node in the process flow takes the model score code generated by the **Neural Network** node and applies it to the data set to be scored.

The output generated by the **Score** node includes the predicted (assigned) levels of the target and their probabilities. These calculations reflect the solution of Equations 5.18 through 5.25.

Display 5.35 shows the programming statements used by the **Score** node to calculate the predicted probabilities of the target levels.

Display 5.35

```
P_LOSSFRQ2  =       1.48949434173539 * HL1  +      1.75959884200533 * HL2
     +         1.06730113426007 * HL3 ;
P_LOSSFRQ1  =       1.11754131452423 * HL1  +      0.96458843025431 * HL2
     +         0.0254374332168 * HL3 ;
P_LOSSFRQ2  =      -3.32895202992659 + P_LOSSFRQ2 ;
P_LOSSFRQ1  =      -2.23114143594446 + P_LOSSFRQ1 ;
DROP _EXP_BAR;
_EXP_BAR=50;
P_LOSSFRQ2  = 1.0 / (1.0 + EXP(MIN( - P_LOSSFRQ2 , _EXP_BAR)));
P_LOSSFRQ1  = 1.0 / (1.0 + EXP(MIN( - P_LOSSFRQ1 , _EXP_BAR)));
P_LOSSFRQ0  = 1. -  P_LOSSFRQ1 ;
P_LOSSFRQ1  = P_LOSSFRQ1  - P_LOSSFRQ2 ;
```

Display 5.36 shows the assignment of the target levels to individual records using the profit matrix.

Display 5.36

```
*** Decision Processing;
label D_LOSSFRQ = 'Decision: LOSSFRQ' ;
label EP_LOSSFRQ = 'Expected Profit: LOSSFRQ' ;

length D_LOSSFRQ $ 9;

D_LOSSFRQ = ' ';
EP_LOSSFRQ = .;

*** Compute Expected Consequences and Choose Decision;
_decnum = 1; drop _decnum;

D_LOSSFRQ = '2' ;
EP_LOSSFRQ = P_LOSSFRQ2 * 5 + P_LOSSFRQ1 * -1 + P_LOSSFRQ0 * -2;
drop _sum;
_sum = P_LOSSFRQ2 * -1 + P_LOSSFRQ1 * 4 + P_LOSSFRQ0 * -1;
if _sum > EP_LOSSFRQ + 2.273737E-12 then do;
   EP_LOSSFRQ = _sum; _decnum = 2;
   D_LOSSFRQ = '1' ;
end;
_sum = P_LOSSFRQ2 * -2 + P_LOSSFRQ1 * -1 + P_LOSSFRQ0 * 3;
if _sum > EP_LOSSFRQ + 2.273737E-12 then do;
   EP_LOSSFRQ = _sum; _decnum = 3;
   D_LOSSFRQ = '0' ;
end;


*** End Decision Processing ;
```

The following code segments create additional variables. Display 5.37 shows the target level assignment process. The variable I_LOSSFRQ is created according to the posterior probability found for each record.

Display 5.37

```
*** ************************;
*** Writing the I_LOSSFRQ  AND U_LOSSFRQ ;
*** ************************;
_MAXP_  = P_LOSSFRQ2 ;
I_LOSSFRQ  = "2            " ;
U_LOSSFRQ  =                   2;
IF( _MAXP_ LT P_LOSSFRQ1  ) THEN DO;
   _MAXP_  = P_LOSSFRQ1 ;
   I_LOSSFRQ  = "1            " ;
   U_LOSSFRQ  =                   1;
END;
IF( _MAXP_ LT P_LOSSFRQ0  ) THEN DO;
   _MAXP_  = P_LOSSFRQ0 ;
   I_LOSSFRQ  = "0            " ;
   U_LOSSFRQ  =                   0;
END;
```

In Display 5.38, the variable EM_CLASSIFICATION represents predicted values based on maximum posterior probability. EM_CLASSIFICATION is the same as I_LOSSFRQ shown in Display 5.37.

Display 5.38

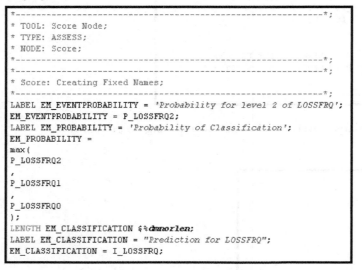

```
*-------------------------------------------------------*;
* TOOL: Score Node;
* TYPE: ASSESS;
* NODE: Score;
*-------------------------------------------------------*;
*-------------------------------------------------------*;
* Score: Creating Fixed Names;
*-------------------------------------------------------*;
LABEL EM_EVENTPROBABILITY = 'Probability for level 2 of LOSSFRQ';
EM_EVENTPROBABILITY = P_LOSSFRQ2;
LABEL EM_PROBABILITY = 'Probability of Classification';
EM_PROBABILITY =
max(
P_LOSSFRQ2
,
P_LOSSFRQ1
,
P_LOSSFRQ0
);
LENGTH EM_CLASSIFICATION $%dmnorlen;
LABEL EM_CLASSIFICATION = "Prediction for LOSSFRQ";
EM_CLASSIFICATION = I_LOSSFRQ;
```

Display 5.39 shows the list of variables created by the **Score** node using the Neural Network model.

Display 5.39

Variable Name	Creator	Variable Label	Function	Type
D_LOSSFRQ	Neural	Decision: LOSSFRQ	DECISION	C
EM_CLASSIFICATION	Score	Prediction for LOSSF...	CLASSIFICATION	C
EM_EVENTPROBABI...	Score	Probability for level 2 ...	PREDICT	N
EM_PROBABILITY	Score	Probability of Classifi...	PREDICT	N
EP_LOSSFRQ	Neural	Expected Profit: LOS...	ASSESS	N
I_LOSSFRQ	Neural	Into: LOSSFRQ	CLASSIFICATION	C
P_LOSSFRQ0	Neural	Predicted: LOSSFRQ...	PREDICT	N
P_LOSSFRQ1	Neural	Predicted: LOSSFRQ...	PREDICT	N
P_LOSSFRQ2	Neural	Predicted: LOSSFRQ...	PREDICT	N
U_LOSSFRQ	Neural	Unnormalized Into: L...	CLASSIFICATION	N
WARN	Neural	Warnings	ASSESS	C

5.5.3 Classification of Risks for Rate Setting in Auto Insurance with Predicted Probabilities

The probabilities generated by the neural network model can be used to classify risk for two purposes:

- to select new customers
- to determine premiums to be charged according to the predicted risk

For each record in the data set, you can compute the expected LOSSFRQ as:

$$E(lossfrq \mid X_i) = Pr(lossfrq = 0 \mid X_i) * 0 + Pr(lossfrq = 1 \mid X_i) * 1 + Pr(lossfrq = 2 \mid X_i) * 2.$$

Customers can be ranked by this expected frequency and assigned to different risk groups.

5.6 Alternative Specifications of the Neural Networks

The general Neural Network model presented in section 5.2 consists of linear combinations functions (Equations 5.1, 5.3, 5.5, 5.7, 5.9, and 5.11) and hyperbolic tangent activation functions (Equations 5.2, 5.4, 5.6, 5.8, 5.10, and 5.12) in the hidden layers, and a linear combination function (Equation 5.13) and a logistic activation function (Equation 5.14) in the output layer. Networks of the type presented in Equations 5.1-5.14 are called multilayer perceptrons and use linear combination functions and sigmoid activation functions in the hidden layers. Sigmoid activation functions are S-shaped, and they are shown in Displays 5.0B, 5.0C, 5.0D and 5.0E.

In this section, I introduce you to other types of networks known as Radial Basis Function (RBF) networks, which have different types of combination activation functions. You can build both Multilayer Perceptron (MLP) networks and RBF networks using the **Neural Network** node.

5.6.1 A Multilayer Perceptron (MLP) Neural Network

The neural network presented in Equations 5.1-5.14 have two hidden layers. The examples considered here have only one hidden layer with three hidden units. The **Neural Network** node allows 1–64 hidden units. You can compare the models produced by setting the **Number of Hidden Units** property to different values and pick the optimum values.

In an MLP neural network the hidden layer combination functions are linear, as shown in Equations 5.26, 5.28, and 5.30. The hidden Layer Activations Functions are sigmoid functions, as shown in Equations 5.27, 5.29, and 5.31.

Hidden Layer Combination Function, Unit 1:

The following equation is the weighted sum of inputs:

$$\eta_{i1} = b_1 + w_{11}x_{i1} + w_{21}x_{i2} + \ldots + w_{p1}x_{ip} \tag{5.26}$$

$w_{11}x_{i1} + w_{21}x_{i2} + \ldots + w_{p1}x_{ip}$ is the inner product of the weight and input vectors, where $w_{11}, w_{21}, \ldots \ldots w_{p1}$ are the weights to be estimated by the iterative algorithm of the **Neural Network** node. The coefficient b_1 (called *bias*) is also estimated by the algorithm. Equation 5.26 shows that the combination functions η_{i1} is a weighted sum of the inputs plus bias. The neural network algorithm calculates η_{i1} for the ith person or record at each iteration.

Hidden Layer Activation Function, Unit 1:

The Hidden Layer Activation Function in an MLP network is a sigmoid function such as the Hyperbolic Tangent, Arc Tangent, Elliot, or Logistic. The sigmoid functions are S-shaped, as shown in Displays 5.0B – 5.0E.

Equation 5.27 shows a sigmoid function known as the tanh function.

$$H_{i1} = \tanh(\eta_{i1}) = \frac{\exp(\eta_{i1}) - \exp(-\eta_{i1})}{\exp(\eta_{i1}) + \exp(-\eta_{i1})} \tag{5.27}$$

The neural network algorithm calculates H_{i1} for the ith case or record at each iteration.

Hidden Layer Combination Function, Unit 2:

$$\eta_{i2} = b_2 + w_{12}x_{i1} + w_{22}x_{i2} + \ldots + w_{p2}x_{ip} \tag{5.28}$$

Hidden Layer Activation Function, Unit 2:

$$H_{i2} = \tanh(\eta_{i2}) = \frac{\exp(\eta_{i2}) - \exp(-\eta_{i2})}{\exp(\eta_{i2}) + \exp(-\eta_{i2})} \tag{5.29}$$

Hidden Layer Combination Function, Unit 3:

$$\eta_{i3} = b_3 + w_{13}x_{i1} + w_{23}x_{i2} + \ldots + w_{p3}x_{ip} \tag{5.30}$$

Hidden Layer Activation Function, Unit 3:

$$H_{i3} = \tanh(\eta_{i3}) = \frac{\exp(\eta_{i3}) - \exp(-\eta_{i3})}{\exp(\eta_{i3}) + \exp(-\eta_{i3})} \tag{5.31}$$

Target Layer Combination Function:

$$\eta_i = b + w_1 H_{i1} + w_2 H_{i2} + w_3 H_{i3} \tag{5.32}$$

Target Layer Activation Function:

$$\frac{1}{1 + EXP(-\eta_i)}$$

If the target is binary, then the target layer output is

$$P(\text{Target}=1) = \frac{1}{1 + EXP(-\eta_i)} \quad \text{and} \quad P(\text{Target}=0) = \frac{1}{1 + EXP(\eta_i)}$$

If you want to use a built-in MLP network, click ⊡ located at the right of the **Network** property of the **Neural Network** node. Set the **Architecture** property to Multilayer Perceptron in the Network Properties window, as shown in Display 5.40.

Display 5.40

Property	Value
Architecture	Multilayer Perceptron
Direct Connection	No
Number of Hidden Units	5
Randomization Distribution	Normal
Randomization Center	0.0
Randomization Scale	0.1
Input Standardization	Standard Deviation
Hidden Layer Combination Function	Default
Hidden Layer Activation Function	Default
Hidden Bias	Yes
Target Layer Combination Function	Default
Target Layer Activation Function	Logistic
Target Layer Error Function	Bernoulli

When you set the **Architecture** property to Multilayer Perceptron, the **Neural Network** node uses the default values Linear and Tanh for the **Hidden Layer Combination Function** and the **Hidden Layer Activation Function** properties. The default values for the **Target Layer Combination Function** and **Target Layer Activation Function** properties are Linear and Exponential. You can change the **Target Layer Combination Function**, **Target Layer Activation Function**, and **Target Layer Error Function** properties. I changed the **Target Layer Activation Function** property to Logistic, as shown in Display 5.40. If the Target is Binary and the **Target Layer Activation Function** is Logistic, then you can set the **Target Layer Error Function** property to Bernoulli to generate a Logistic Regression-type model.

5.6.2 A Radial Basis Function (RBF) Neural Network

The Radial Basis Function neural network can be represented by the following equation:

$$y_i = \sum_{k=1}^{M} w_k \phi_k(X_i) + w_0 \tag{5.33}$$

where y_i is the target variable for i^{th} case (record or observation), w_k is the weight for the kth basis function, w_0 is the bias, ϕ_k is the kth basis function, $X_i = (x_{i1}, x_{i2}, \ldots\ldots, x_{ip})$ is a vector of inputs for the ith case, M is the number of basis functions, p is the number of inputs, and N is the number of cases (observations) in the Training data set. The weight multiplied by the basis function plus the bias can be thought of as the output of a hidden unit in a single layer network.

An example of a basis function is:

$$\phi_k(X_i) = \exp\left(-\frac{\|X_i - \mu_k\|}{2\sigma_k}\right), \text{ where } X_i = (x_{i1}, x_{i2}, \ldots\ldots, x_{ip}), \ \mu_k = (\mu_{1k}, \mu_{2k}, \ldots\ldots, \mu_{pk}),$$

i stands for the ith observation and k stands for the kth basis function. $\|X_i - \mu_k\|$ is the Squared Euclidean distance between the vectors X_i and μ_k. μ_k and σ_k are the center and width of the kth basis function. The values of the center and width are determined during the training process.

5.6.2.1 Radial Basis Function Neural Networks in the Neural Network Node

An alternative way of representing a Radial Basis neural network is by using the equation:

$$y_i = \sum_{k=1}^{M} w_k H_{ik} + w_0 \tag{5.33A}$$

where H_{ik} is the output of the kth hidden unit for the ith observation (case), M is the number of hidden units, w_k is the weight for the kth hidden unit, w_0 is the bias, and y_i is the target value for the ith observation (case).

The following examples show how the outputs of the hidden units are calculated for different types of Basis Functions, and how the target values are computed. In each case the formulas are followed by SAS code. The output of the target layer is y_i. The number of output units in the target layer can be more than one, depending on your target variable. For example, if the target is binary, then the target layer calculates two outputs, namely P_resp1 and P_resp0, for each observation, where P_resp1= $\Pr(resp = 1)$ and P_resp0 = $\Pr(resp = 0)$.

As shown by Equations 5.26, 5.28, and 5.30, the hidden layer combination function in a MLP neural network is the weighted sum or inner product of the vector of inputs and vector of corresponding weights *plus* a bias coefficient.

In contrast to the Hidden Layer Combination function in an MLP neural network, the hidden layer combination function in a RBF neural network is the Squared Euclidian Distance between the vector of inputs and the vector of corresponding weights (center points), multiplied by squared bias.

In the **Neural Network** node, the Basis functions are defined in terms of the combination and activation functions of the hidden units.

An example of a combination function used in some RBF networks is

$$\eta_{ik} = -b_k^2 \sum_{j=1}^{p} (w_{jk} - x_{ij})^2 \tag{5.34}$$

where η_{ik} = the estimated value of the combination function for the kth hidden unit

b_k = bias,

w_{jk} = weight of the jth input

x_{ij} = j^{th} input

i = i^{th} record or observation in the data set

p = the number of inputs

The RBF neural network which uses the combination function given by equation 5.34 is called the Ordinary Radial with Unequal Widths or ORBFUN.

The activation function of the hidden units of an ORBFUN network is an exponential function. Hence the output of kth hidden unit in ORBFUN network is:

$$H_{ik} = Exp\left\{-b_k{}^2 \sum_j^p (w_{jk} - x_{ij})^2\right\} \qquad (5.35)$$

As Equation 5.35 shows, the bias is not the same in all the units in the hidden layer. Since the *width* is inversely related to the bias b_k ($b_k = \dfrac{1}{\sigma_k}$), it is also different for different units. Hence the term Unequal Widths in ORBFUN.

The network with the combination function given in Equation 5.34 and the activation function given in Equation 5.35 is called the Ordinary Radial with Unequal Widths or ORBFUN.

To build an ORBFUN neural network model, click [...] located to the right of the **Network** property of the **Neural Network** node. In the Network window that opens, set the **Architecture** Property to Ordinary Radial-Unequal Width, as shown in Display 5.41.

Display 5.41

Property	Value
Architecture	Ordinary Radial - Unequal Width
Direct Connection	No
Number of Hidden Units	5
Randomization Distribution	Normal
Randomization Center	0.0
Randomization Scale	0.1
Input Standardization	Standard Deviation
Hidden Layer Combination Function	Default
Hidden Layer Activation Function	Default
Hidden Bias	Yes
Target Layer Combination Function	Default
Target Layer Activation Function	Default
Target Layer Error Function	Default

As you can see in Display 5.41, the rows corresponding to **Hidden Layer Combination Function** and **Hidden Layer Activation Function** are shaded, indicating that their values are fixed for this network. Hence this is called a built-in network in SAS Enterprise Miner terminology. However, the rows corresponding to the **Target Layer Combination Function**, **Target Layer Activation Function**, and **Target Layer Error Function** are not shaded, so you can change their values. For example, to set the **Target Layer Activation Function** to Logistic, click on the value column corresponding to **Target Layer Activation Function** property and select Logistic. Similarly you can change the **Target Layer Error** Function to Bernoulli.

As pointed out earlier in this chapter, the error function you select depends on the type of model you want to build.

Display 5.42 shows a list of RBF neural networks available in SAS Enterprise Miner. You can see all the network architectures If you click on the down-arrow in the box to the right of the **Architecture** property in the Network window.

Display 5.42

```
Ordinary Radial - Equal Width
Ordinary Radial - Unequal Width
Normalized Radial - Equal Height
Normalized Radial - Equal Volumes
Normalized Radial - Equal Width
Normalized Radial - Equal Width and Height
Normalized Radial - Unequal Width and Height
```

For a complete description of these network architectures, see *SAS Enterprise Miner: Reference Help*, which is available in the SAS Enterprise Miner Help. In the following pages I will review some of the RBF network architectures.

The general form of the Hidden Layer Combination Function of a RBF network is:

$$\eta_k = f * \log(abs(a_k)) - b_k^2 \sum_j^p (w_{jk} - x_j)^2 \qquad (5.36)$$

The Hidden Layer Activation Function is

$$H_k = \exp(\eta_k) \qquad (5.37)$$

and where x_j is the jth standardized input, the w_{jk}'s are weights representing the center (origin) of the basis function, a_k is the *height*, or altitude, b_k is a parameter sometimes referred to as the *bias*, p is the number of inputs, f is the number of connections to the unit and k refers to the kth hidden unit. Another measure that is related to *bias* is the *width*. The bias b_k is the inverse width of the kth hidden unit. (see the section "Definitions for Built-in Architectures" in *SAS Enterprise Miner: Reference Help*, which is available in the SAS Enterprise Miner Help). The coefficient a_k, called the height parameter, measures the maximum height of the bell-shaped curve presented in Displays 5.43 through 5.45. The subscript k is added to indicate that the radial basis function is used to calculate the output of the kth hidden unit. Although Equations 5.36 and 5.37 are evaluated for each observation during the training of the neural network, I have dropped the subscript *i*, which I used in Equations 5.34 and 5.35 to denote the *i*th record.

When there are p inputs, there are p weights. These weights are estimated by the **Neural Network** node. The vector of inputs for any record or observation may be thought of as a point in p-dimensional space in which the vector of weights is the center. Given the height and width, the value of the radial basis function depends on the distance between the inputs and the center.

Now I plot the radial basis functions with different heights and widths for a single input that ranges between − 10 and 10. Since there is only input, there is only one weight w_{1k}, which I arbitrarily set to 0.

Display 5.43 shows a graph of the radial basis function given in Equations 5.36 and 5.37 with height = 1 and width = 1.

Display 5.43

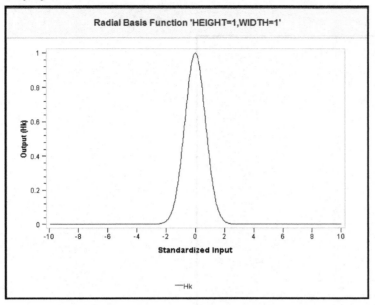

Display 5.44 shows a graph of the radial basis function given in Equation 5.36 with height = 1 and width = 4.

Display 5.44

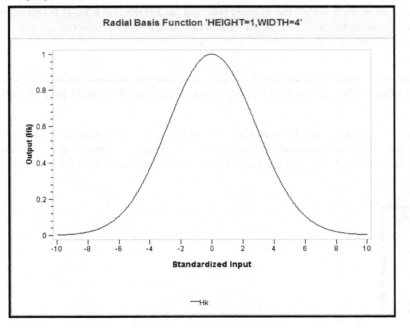

Display 5.45 shows a graph of the radial basis function given in Equation 5.36 with height = 5 and width = 4.

Display 5.45

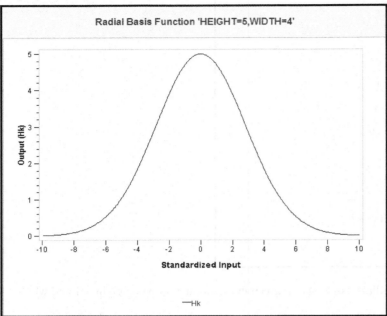

If there are three units in a hidden layer, each having a radial basis function, and if the radial basis functions are constrained to sum to 1, then they are called Normalized Radial Basis Functions in SAS Enterprise Miner. Without this constraint, they are called Ordinary Radial Basis Functions.

5.7 Comparison of Alternative Built-in Architectures of the Neural Network Node

You can produce a number of neural network models by specifying alternative architectures in the **Neural Network** node. You can then make a selection from among these models, based on lift or other measures, using the **Model Comparison** node. In this section, I compare the built-in architectures listed in the Display 5.46.

Display 5.46

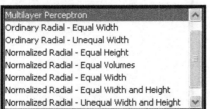

Display 5.47 shows the process flow diagram for comparing the Neural Network models produced by different architectural specifications. In all the eight models compared in this section, I set the **Assessment** property to Average Error and the **Number of Hidden Units** property to 5.

Display 5.47

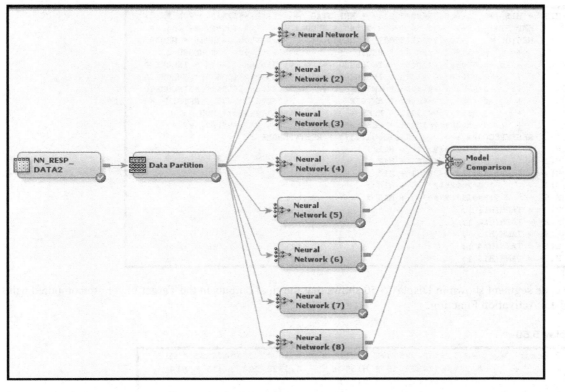

5.7.1 Multilayer Perceptron (MLP) Network

The top **Neural Network** node in Display 5.47 is a MLP network, and its property settings are shown in Display 5.48.

Display 5.48

Property	Value
Architecture	Multilayer Perceptron
Direct Connection	No
Number of Hidden Units	5
Randomization Distribution	Normal
Randomization Center	0.0
Randomization Scale	0.1
Input Standardization	Standard Deviation
Hidden Layer Combination Function	Default
Hidden Layer Activation Function	Default
Hidden Bias	Yes
Target Layer Combination Function	Default
Target Layer Activation Function	Logistic
Target Layer Error Function	Bernoulli

The default value of the **Hidden Layer Combination Function** property is Linear and the default value of the **Hidden Layer Activation Functions** property is Tanh. After running this node, you can open the Results window and view the SAS code generated by the **Neural Network** node. Display 5.49 shows a segment of the SAS code generated by the Neural Network mode with the Network settings shown in Display 5.48.

Display 5.49

```
H15   = H15  +      -0.06066092430142 * EMP_STA0  +      -0.19618565779228 *
        EMP_STA1_2  +       0.09372854722503 * HEQ0_1  +       0.09505732724066 *
        HEQ10  +     -0.41799111020572 * HEQ30  +       0.04899087475045 * HEQ50
        +       0.20196142298612 * HEQ70  +     -0.27896795409954 * HEQ90
        +      -0.57008770190488 * INCOME20  +       0.09533035076376 * INCOME30
        +       0.03682163124048 * INCOME40  +       0.16160729452535 * INCOME50
        +      -0.06542738211563 * INCOME60  +       0.00456490359612 * INCOME70
        +      -0.08992201568762 * MRTGIN  +       0.10752845325655 * MRTGIU
        +      -0.02526735023317 * MSM  +       0.23140461113131 * MSU
        +       0.04448164872264 * RESTYPECONDO  +      -0.21762258678691 *
        RESTYPECOOP  +       0.48301305184379 * RESTYPEHOME ;
H11   =      -0.46851838899142 + H11 ;
H12   =      -0.95439753413981 + H12 ;
H13   =      -0.46412258738771 + H13 ;
H14   =       1.0823290829572 + H14 ;
H15   =       0.88511260799725 + H15 ;
H11   = TANH(H11 );
H12   = TANH(H12 );
H13   = TANH(H13 );
H14   = TANH(H14 );
H15   = TANH(H15 );
```

The code segment shown in Display 5.50 shows that the final outputs in the Target Layer are computed using a Logistic Activation Function.

Display 5.50

```
P_resp1  =      -0.43903268035898 * H11  +       0.4779504876191 * H12
        +       0.23584328373816 * H13  +       0.33232292758025 * H14
        +      -0.4267541872682 * H15 ;
P_resp0  =       0.43903268022256 * H11  +      -0.47795048770249 * H12
        +      -0.23584328383112 * H13  +      -0.33232292740344 * H14
        +       0.42675418737984 * H15 ;
P_resp1  =      -0.83766717267926 + P_resp1 ;
P_resp0  =       0.83766717290929 + P_resp0 ;
DROP _EXP_BAR;
_EXP_BAR=50;
P_resp1  = 1.0 / (1.0 + EXP(MIN( - P_resp1 , _EXP_BAR)));
P_resp0  = 1.0 / (1.0 + EXP(MIN( - P_resp0 , _EXP_BAR)));
```

5.7.2 Ordinary Radial Basis Function with Equal Heights and Widths (ORBFEQ)

The combination function for the kth Hidden Unit in this case is $\eta_k = -b^2 \sum_{j=1}^{p} (w_{jk} - x_j)^2$ and the Activation function for the kth hidden unit is given by $H_k = \exp(\eta_k)$. Note that the parameter b in the Combination function above is constant for all the hidden units. Display 5.51 shows an example of the outputs of five Hidden Units of an ORBFEQ with a single input.

Display 5.51

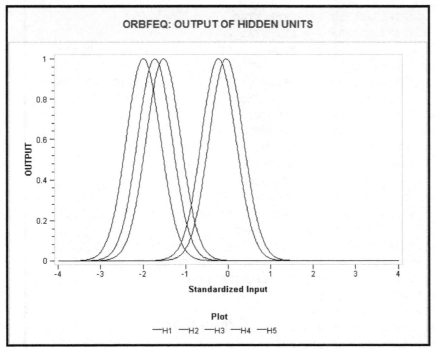

In the second **Neural Network** node in Display 5.47, the **Architecture** property is set to Ordinary Radial – Equal width as shown in Display 5.52.

Display 5.52

Property	Value
Architecture	Ordinary Radial - Equal Width
Direct Connection	No
Number of Hidden Units	5
Randomization Distribution	Normal
Randomization Center	0.0
Randomization Scale	0.1
Input Standardization	Standard Deviation
Hidden Layer Combination Function	Default
Hidden Layer Activation Function	Default
Hidden Bias	Yes
Target Layer Combination Function	Default
Target Layer Activation Function	Default
Target Layer Error Function	Default

Display 5.53 shows a segment of the SAS code, which shows the computation of the outputs of the hidden units of the ORBFEQ neural network specified in Display 5.52.

Display 5.53

```
  H15   = H15  + (    -0.83802366998253 - EMP_STA0
        )**2 + (     0.84317324480859 - EMP_STA1_2
        )**2 + (    -0.17564073673323 - HEQ0_1 )**2 + (     0.88544781052024 -
  HEQ10 )**2 + (     -0.4110278897524 - HEQ30
        )**2 + (    -0.46729890575235 - HEQ50 )**2 + (    -1.40524752389733 -
  HEQ70 )**2 + (    -0.73967069125904 - HEQ90
        )**2 + (    -0.26591219389074 - INCOME20
        )**2 + (    -0.28971325241902 - INCOME30
        )**2 + (     1.24550176575527 - INCOME40
        )**2 + (    -1.80655741684727 - INCOME50
        )**2 + (     1.44413699656461 - INCOME60
        )**2 + (    -0.66724666660526 - INCOME70
        )**2 + (     0.92764871683685 - MRTGIN )**2 + (     1.7131600121147 -
  MRTGIU )**2 + (     -0.235765622469 - MSM
        )**2 + (    -1.14750335245139 - MSU )**2 + (    -1.96868838752226 -
  RESTYPECONDO )**2 + (     0.94878849329388 - RESTYPECOOP
        )**2 + (     1.62874579746874 - RESTYPEHOME )**2;
  H11   = -      0.02620723147721 * H11 ;
  H12   = -      0.02620723147721 * H12 ;
  H13   = -      0.02620723147721 * H13 ;
  H14   = -      0.02620723147721 * H14 ;
  H15   = -      0.02620723147721 * H15 ;
  DROP _EXP_BAR;
  _EXP_BAR=50;
  H11   = EXP(MIN(H11 , _EXP_BAR));
  H12   = EXP(MIN(H12 , _EXP_BAR));
  H13   = EXP(MIN(H13 , _EXP_BAR));
  H14   = EXP(MIN(H14 , _EXP_BAR));
  H15   = EXP(MIN(H15 , _EXP_BAR));
```

> Equal Widths (annotation bracketing the five H11–H15 equations)

Display 5.54 shows the computation of the outputs of the target layer of the ORBFEQ neural network specified in Display 5.52.

Display 5.54

```
  P_resp1  =     -2.29043997245064 * H11 +      4.69131355398006 * H12
           +     -6.39803294538255 * H13 +      2.66173017560067 * H14
           +     -0.06038445318379 * H15 ;
  P_resp1  =     -0.38960261652833 + P_resp1 ;
  P_resp0  = 0;
  _MAX_  = MAX (P_resp1 , P_resp0 );
  _SUM_  = 0.;
  P_resp1  = EXP(P_resp1  - _MAX_);
  _SUM_  = _SUM_ + P_resp1 ;
  P_resp0  = EXP(P_resp0  - _MAX_);
  _SUM_  = _SUM_ + P_resp0 ;
  P_resp1  = P_resp1  / _SUM_ ;
  P_resp0  = P_resp0  / _SUM_ ;
```

5.7.3 Ordinary Radial Basis Function with Equal Heights and Unequal Widths (ORBFUN)

In this network, the combination function for the kth unit is $\eta_k = -b_k^2 \sum_{j=1}^{p} (w_{jk} - x_j)^2$, where w_{jk} are the weights iteratively calculated by the **Neural Network** node, x_j is the jth input (standardized), and b_k is the reciprocal of the width for the kth hidden unit. Unlike the ORBFEQ example in which a single value for b is calculated to be used with all the hidden units, the constant b_k in this case can differ from one hidden unit to another. The output of the kth hidden unit is given by $H_k = \exp(\eta_k)$.

Display 5.55shows an example of the output of the five hidden units $H_k (\text{k} = 1, 2, 3, 4 \text{ and } 5)$ with a single standardized input and arbitrary values for w_{jk} and b_k.

Display 5.55

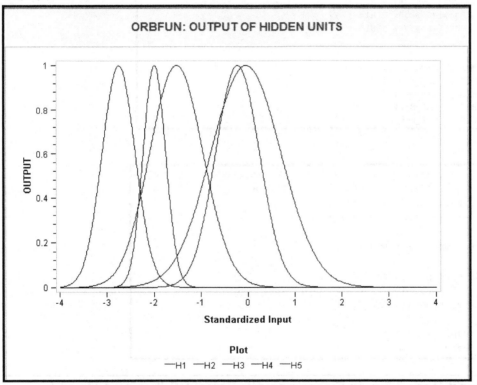

The Architecture property of the ORBFUN **Neural Network** node (the third node from the top in Display 5.46) is set to Ordinary Radial – Unequal Width. The remaining Network properties are the same as in Display 5.52.

Displays 5.56 and 5.57 show the SAS code segment showing the calculation of the outputs of the hidden and target layers of this ORBFUN network.

Display 5.56

```
H15  = H15  + (    1.32592021388042 - EMP_STA0
       )**2 + (   -0.64464187887107 - EMP_STA1_2
       )**2 + (    1.07356758777762 - HEQ0_1 )**2 + (     1.7196396490955 -
     HEQ10 )**2 + (   -0.30921462328772 - HEQ30
       )**2 + (   -1.33934705394679 - HEQ50 )**2 + (   -2.10874917414115 -
     HEQ70 )**2 + (    0.87510405522492 - HEQ90
       )**2 + (    1.0503456010738 - INCOME20
       )**2 + (    0.31262764549167 - INCOME30
       )**2 + (   -0.82039001075217 - INCOME40
       )**2 + (    0.55664571989613 - INCOME50
       )**2 + (    0.48938964411422 - INCOME60
       )**2 + (   -0.94838905847271 - INCOME70
       )**2 + (   -0.74533431647581 - MRTGIN )**2 + (    -0.97026843400431 -
     MRTGIU )**2 + (   -1.82028941778904 - MSM
       )**2 + (    0.73439057243203 - MSU )**2 + (     0.02966634436638 -
     RESTYPECONDO )**2 + (    0.24213947846912 - RESTYPECOOP
       )**2 + (    1.57951421070838 - RESTYPEHOME )**2;
H11  = -     0.00450500128911 * H11 ;
H12  = -     2.12354480036853 * H12 ;
H13  = -     0.45081420848381 * H13 ;         } Unequal Widths
H14  = -     1.09370490008504 * H14 ;
H15  = -     0.40319640385244 * H15 ;
DROP _EXP_BAR;
_EXP_BAR=50;
H11  = EXP(MIN(H11 , _EXP_BAR));
H12  = EXP(MIN(H12 , _EXP_BAR));
H13  = EXP(MIN(H13 , _EXP_BAR));
H14  = EXP(MIN(H14 , _EXP_BAR));
H15  = EXP(MIN(H15 , _EXP_BAR));
```

Display 5.57

```
P_resp1  =      11.391143788901 * H11  + -2.7296115329023E-18 * H12
         +     -0.00037899955484 * H13  +  2.8030266851505E-12 * H14
         +      0.00113611240964 * H15 ;
P_resp1  =     -3.22054437607751 + P_resp1 ;
P_resp0  = 0;
_MAX_ = MAX (P_resp1 , P_resp0 );
_SUM_ = 0.;
P_resp1  = EXP(P_resp1  - _MAX_);
_SUM_ = _SUM_ + P_resp1 ;
P_resp0  = EXP(P_resp0  - _MAX_);
_SUM_ = _SUM_ + P_resp0 ;
P_resp1  = P_resp1  / _SUM_;
P_resp0  = P_resp0  / _SUM_;
```

5.7.4 Normalized Radial Basis Function with Equal Widths and Heights (NRBFEQ)

In this case, the combination function for the kth hidden unit is $\eta_k = -b^2 \sum_{j=1}^{p} (w_{jk} - x_j)^2$, where w_{jk} are the weights iteratively calculated by the **Neural Network** node, x_j is the jth input (standardized), and b is the square root of the reciprocal of the width. The height k^{th} hidden unit is = $a_k = 1$. Since $\log(1) = 0$, it does not appear in the combination function shown above. Here it is implied that $f = 1$ (See equation 5.36).

The output of the kth hidden unit is given by $H_k = \dfrac{\exp(\eta_k)}{\displaystyle\sum_{m=1}^{5} \exp(\eta_m)}$. This is called the

Softmax Activation Function. Display 5.58 shows the output of the five hidden units $H_k (k = 1, 2, 3, 4 \text{ and } 5)$ with a single standardized input, and arbitrary values for w_{jk} and b.

In this case, the height $a_k = 1$ for all the hidden units. Hence $\log(a_k) = 0$.

Display 5.58

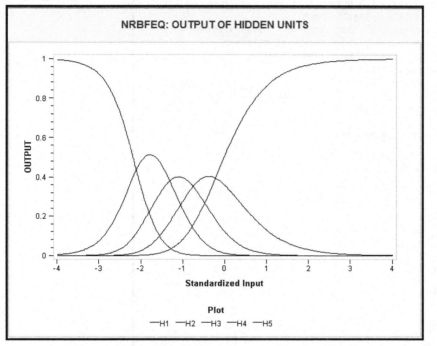

For a NRBFEQ neural network, the **Architecture** property is set to Normalized Radial – Equal Width and Height. The remaining Network settings are the same as in Display 5.52.

Displays 5.59 and 5.60 show segments from the SAS code generated by the **Neural Network** node. Display 5.59 shows the computation of the outputs of the hidden units and Display 5.60 shows the calculation of the target layer outputs.

Display 5.59

```
H15 = H15 + (    -0.16384398225545 - EMP_STA0
     )**2 + (    -1.24262668728065 - EMP_STA1_2
     )**2 + (     0.78008139319127 - HEQ0_1 )**2 + (     1.22894706685711 -
     HEQ10 )**2 + (     0.04258684848552 - HEQ30
     )**2 + (     2.20108898341708 - HEQ50 )**2 + (     0.07832975481428 -
     HEQ70 )**2 + (     0.62093931237862 - HEQ90
     )**2 + (     1.54866961587827 - INCOME20
     )**2 + (     1.48212682673191 - INCOME30
     )**2 + (     1.69863574918261 - INCOME40
     )**2 + (     0.59258166307803 - INCOME50
     )**2 + (    -1.06725049591633 - INCOME60
     )**2 + (     0.46843145619781 - INCOME70
     )**2 + (     0.18859020447816 - MRTGIN )**2 + (    -1.61560946026487 -
     MRTGIU )**2 + (     0.4771574194171 - MSM
     )**2 + (     1.12506013693023 - MSU )**2 + (     0.00682801905887 -
     RESTYPECONDO )**2 + (    -0.47621826615509 - RESTYPECOOP
     )**2 + (    -1.85594552394525 - RESTYPEHOME )**2;
H11 = -      0.10959230264485 * H11 ;
H12 = -      0.10959230264485 * H12 ;
H13 = -      0.10959230264485 * H13 ;          Equal Widths
H14 = -      0.10959230264485 * H14 ;
H15 = -      0.10959230264485 * H15 ;
_MAX_ = MAX (H11 , H12 , H13 , H14 , H15 );
_SUM_ = 0.;
H11 = EXP(H11 - _MAX_);
_SUM_ = _SUM_ + H11 ;
H12 = EXP(H12 - _MAX_);
_SUM_ = _SUM_ + H12 ;
H13 = EXP(H13 - _MAX_);
_SUM_ = _SUM_ + H13 ;
H14 = EXP(H14 - _MAX_);
_SUM_ = _SUM_ + H14 ;
H15 = EXP(H15 - _MAX_);
_SUM_ = _SUM_ + H15 ;
H11 = H11 / _SUM_;
H12 = H12 / _SUM_;
H13 = H13 / _SUM_;          Normalized
H14 = H14 / _SUM_;
H15 = H15 / _SUM_;
```

Display 5.60

```
P_resp1 =    -3.37966494396031 * H11 +      0.4682110521405 * H12
        +    -1.32282625739425 * H13 +     -3.6395603689828 * H14
        +    -1.09143517831771 * H15 ;
P_resp0 = 0;
_MAX_ = MAX (P_resp1 , P_resp0 );
_SUM_ = 0.;
P_resp1 = EXP(P_resp1 - _MAX_);
_SUM_ = _SUM_ + P_resp1 ;
P_resp0 = EXP(P_resp0 - _MAX_);
_SUM_ = _SUM_ + P_resp0 ;
P_resp1 = P_resp1 / _SUM_;
P_resp0 = P_resp0 / _SUM_;
```

5.7.5 Normalized Radial Basis Function with Equal Heights and Unequal Widths (NRBFEH)

The combination function here is $\eta_k = -b_k^{\ 2} \sum_{j=1}^{p} (w_{jk} - x_j)^2$, where w_{jk} the weights are iteratively calculated

by the **Neural Network** node, x_j is the jth input (standardized), and b_k is the reciprocal of the width for the

k^{th} hidden unit. The output of the kth hidden unit is given by $H_k = \dfrac{\exp(\eta_k)}{\sum\limits_{m=1}^{5} \exp(\eta_m)}$.

Display 5.61 shows the output of the five hidden units $H_k (k = 1, 2, 3, 4 \text{ and } 5)$ with a single standardized input AGE (p =1), and arbitrary values for w_{jk} and b_k. The height a_k =1 for all the hidden units. Hence $\log(a_k) = 0$.

Display 5.61

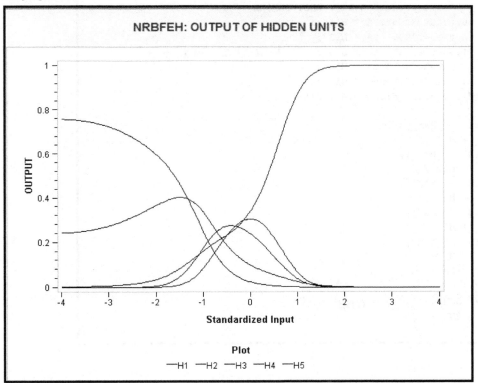

The **Architecture** property for the NRBFEH neural network is set to Normalized Radial – Equal Height. The remaining Network settings are the same as in Display 5.52.

Display 5.62 shows a segment of the SAS code used to compute the hidden units in this NRBEH network.

Display 5.62

```
H15  = H15  + (    -0.20498702809535 - EMP_STA0
      )**2 + (    -1.53715597362906 - EMP_STA1_2
      )**2 + (    -0.25737452771557 - HEQ0_1 )**2 + (    -0.09395982108002 -
HEQ10 )**2 + (     0.29222313572515 - HEQ30
      )**2 + (     1.10102074230184 - HEQ50 )**2 + (     1.05433012749055 -
HEQ70 )**2 + (    -0.72594057830564 - HEQ90
      )**2 + (     0.27574917372838 - INCOME20
      )**2 + (     0.35324662108168 - INCOME30
      )**2 + (     1.24219476320619 - INCOME40
      )**2 + (    -0.27812796446945 - INCOME50
      )**2 + (     0.52852349458741 - INCOME60
      )**2 + (    -1.54666369447877 - INCOME70
      )**2 + (     0.16408226063963 - MRTGIN )**2 + (    -1.83549049739692 -
MRTGIU )**2 + (     0.06099704426224 - MSM
      )**2 + (     1.30362850487163 - MSU )**2 + (    -0.21052916751609 -
RESTYPECONDO )**2 + (     1.16143034666871 - RESTYPECOOP
      )**2 + (    -0.67885145250911 - RESTYPEHOME )**2;
H11  = -      0.01152752559772 * H11 ;  ⎫
H12  = -      0.00906317743923 * H12 ;  ⎪
H13  = -      1.68851628457561 * H13 ;  ⎬ Unequal Widths
H14  = -      1.38295403933468 * H14 ;  ⎪
H15  = -      1.80149903990254 * H15 ;  ⎭
_MAX_ = MAX (H11 , H12 , H13 , H14 , H15 );
_SUM_ = 0.;
H11  = EXP(H11  - _MAX_);
_SUM_ = _SUM_ + H11 ;
H12  = EXP(H12  - _MAX_);
_SUM_ = _SUM_ + H12 ;
H13  = EXP(H13  - _MAX_);
_SUM_ = _SUM_ + H13 ;
H14  = EXP(H14  - _MAX_);
_SUM_ = _SUM_ + H14 ;
H15  = EXP(H15  - _MAX_);
_SUM_ = _SUM_ + H15 ;
H11  = H11  / _SUM_;  ⎫
H12  = H12  / _SUM_;  ⎪
H13  = H13  / _SUM_;  ⎬ Normalized
H14  = H14  / _SUM_;  ⎪
H15  = H15  / _SUM_;  ⎭
```

Display 5.63 shows the SAS code used to compute the target layer outputs.

Display 5.63

```
P_resp1  =    -7.28172300395717 * H11  +       2.31985635543082 * H12
         +  2.4923106566143E-21 * H13  +      -0.809431544435 * H14
         +  3.3120882586631E-13 * H15 ;
P_resp0  = 0;
_MAX_  = MAX (P_resp1 , P_resp0 );
_SUM_  = 0.;
P_resp1  = EXP(P_resp1  - _MAX_);
_SUM_  = _SUM_ + P_resp1 ;
P_resp0  = EXP(P_resp0  - _MAX_);
_SUM_  = _SUM_ + P_resp0 ;
P_resp1  = P_resp1 / _SUM_;
P_resp0  = P_resp0 / _SUM_;
```

5.7.6 Normalized Radial Basis Function with Equal Widths and Unequal Heights (NRBFEW)

The combination function in this case is:

$\eta_k = f * \log(abs(a_k)) - b^2 \sum_{j}^{p} (w_{jk} - x_j)^2$, where a_k is an altitude parameter representing the maximum

height, f is the number of connections to the unit, and b is the square root of reciprocal of the width.

The output of the kth hidden unit is given by $H_k = \dfrac{\exp(\eta_k)}{\sum_{m=1}^{5} \exp(\eta_m)}$.

Display 5.64 shows the output of the five hidden units $H_k (k = 1, 2, 3, 4 \text{ and } 5)$ with a single standardized input and arbitrary values for w_{jk}, a_k, and b.

Display 5.64

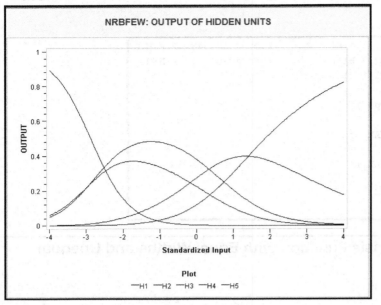

The **Architecture** property of the estimated NRBFEW network is set to Normalized Radial – Equal Width. The remaining Network settings are the same as in Display 5.52.

Display 5.65 shows the code segment showing the calculation of the output of the hidden units in the estimated NRBFEW network.

Display 5.65

```
H15   = H15   + (    -2.35858567500572  - EMP_STA0
      )**2 + (     2.40518623330855  - EMP_STA1_2
      )**2 + (     0.91927859442274  - HEQ0_1 )**2 + (    -0.45148651869942  -
   HEQ10 )**2 + (    -0.32072103277871  - HEQ30
      )**2 + (    -1.06226285807783  - HEQ50 )**2 + (     0.0603090728379  -
   HEQ70 )**2 + (     0.28333390637084  - HEQ90
      )**2 + (     3.26876479340412  - INCOME20
      )**2 + (     2.30191760870978  - INCOME30
      )**2 + (     1.77312115888372  - INCOME40
      )**2 + (     0.98619742489415  - INCOME50
      )**2 + (    -0.72106504210076  - INCOME60
      )**2 + (    -1.13941338986417  - INCOME70
      )**2 + (     0.27032103311026  - MRTGIN )**2 + (    -1.01616930524604  -
   MRTGIU )**2 + (    -0.35879811387852  - MSM
      )**2 + (    -0.53946915745049  - MSU )**2 + (     0.05262748845409  -
   RESTYPECONDO )**2 + (    -0.01959558001093  - RESTYPECOOP
      )**2 + (    -4.73871955675152  - RESTYPEHOME )**2;
H11   = 30 * LOG(ABS(     0.05471300284154)) -     0.01629029958408 * H11 ;
H12   = 30 * LOG(ABS(     0.02468890023713)) -     0.01629029958408 * H12 ;
H13   = 30 * LOG(ABS(     1.78754153859408)) -     0.01629029958408 * H13 ;
H14   = 30 * LOG(ABS(    -0.19255087809684)) -     0.01629029958408 * H14 ;
H15   = 30 * LOG(ABS(     1.75736846888498)) -     0.01629029958408 * H15 ;
_MAX_ = MAX (H11 , H12 , H13 , H14 , H15 );
_SUM_ = 0.;
H11   = EXP(H11  - _MAX_);
_SUM_ = _SUM_ + H11 ;
H12   = EXP(H12  - _MAX_);
_SUM_ = _SUM_ + H12 ;
H13   = EXP(H13  - _MAX_);
_SUM_ = _SUM_ + H13 ;
H14   = EXP(H14  - _MAX_);
_SUM_ = _SUM_ + H14 ;
H15   = EXP(H15  - _MAX_);
_SUM_ = _SUM_ + H15 ;
H11   = H11   / _SUM_;
H12   = H12   / _SUM_;
H13   = H13   / _SUM_;
H14   = H14   / _SUM_;
H15   = H15   / _SUM_;
```

Unequal Heights

Equal Widths

Normalized

Display 5.66 shows the calculation of the outputs of the target layer.

Display 5.66

```
P_resp1  =     -0.16134105457285 * H11  + -2.7066232071686E-12 * H12
        +     -2.87470911871013 * H13  +    -0.61862989003733 * H14
        +      3.8236889805613 * H15 ;
P_resp0  = 0;
_MAX_ = MAX (P_resp1 , P_resp0 );
_SUM_ = 0.;
P_resp1  = EXP(P_resp1  - _MAX_);
_SUM_ = _SUM_ + P_resp1 ;
P_resp0  = EXP(P_resp0  - _MAX_);
_SUM_ = _SUM_ + P_resp0 ;
P_resp1  = P_resp1  / _SUM_;
P_resp0  = P_resp0  / _SUM_;
```

5.7.7 Normalized Radial Basis Function with Equal Volumes (NRBFEV)

In this case, the combination function is:

$$\eta_k = f * \log(abs(b_k)) - b_k^2 \sum_j^p (w_{jk} - x_j)^2$$

The output of the jth hidden unit is given by $H_k = \dfrac{\exp(\eta_k)}{\sum\limits_{m=1}^{5} \exp(\eta_m)}$.

Display 5.67 shows the output of the five hidden units $H_k (k = 1, 2, 3, 4 \text{ and } 5)$ with a single standardized input, and arbitrary values for w_{jk} and b_k .

Display 5.67

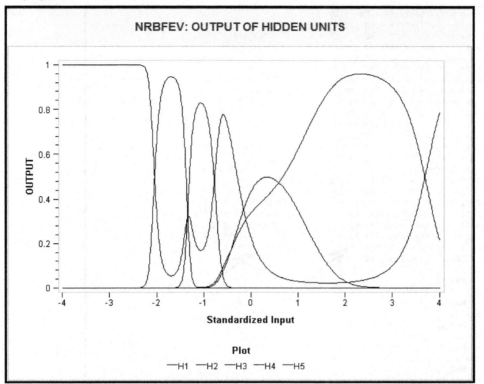

The **Architecture** property of the NRBFEV architecture used in this example is set to Normalized Radial –
Equal Volumes. The remaining Network settings are the same as in Display 5.52.

Display 5.68 shows a segment of the SAS code generated by the **Neural Network** node for the NRBFEV
network.

Display 5.68

```
H15  = H15 + (    -0.40056247815467 - EMP_STA0
       )**2 + (   0.32729497150828 - EMP_STA1_2
       )**2 + (   0.19936397828165 - HEQ0_1 )**2 + (    0.54549764323708 -
       HEQ10 )**2 + (   -1.39804195099283 - HEQ30
       )**2 + (   -0.4438101616772 - HEQ50 )**2 + (    0.35650965821771 -
       HEQ70 )**2 + (    0.92525871861177 - HEQ90
       )**2 + (   -1.85240554070194 - INCOME20
       )**2 + (   1.83633083327827 - INCOME30
       )**2 + (   0.40649052867803 - INCOME40
       )**2 + (   -1.29392141537124 - INCOME50
       )**2 + (   -0.33869220429378 - INCOME60
       )**2 + (   1.06004722323373 - INCOME70
       )**2 + (   -0.36265728712526 - MRTGIN )**2 + (    -1.05222898968267 -
       MRTGIU )**2 + (   -1.31499410501194 - MSM
       )**2 + (   0.62252950954715 - MSU )**2 + (    0.97959339029623 -
       RESTYPECONDO )**2 + (   -1.80294662082152 - RESTYPECOOP
       )**2 + (   -0.06653581595028 - RESTYPEHOME )**2;
H11  = 30 * LOG(ABS(    1.89509986131827)) -    3.59140348436854 * H11 ;
H12  = 30 * LOG(ABS(    0.20294501095193)) -    0.04118667747027 * H12 ;
H13  = 30 * LOG(ABS(    0.19170606368155)) -    0.03675121485227 * H13 ;
H14  = 30 * LOG(ABS(    0.20923990034599)) -    0.0437813358968 * H14 ;
H15  = 30 * LOG(ABS(   -1.44807685508502)) -    2.09692657823292 * H15 ;
_MAX_ = MAX (H11 , H12 , H13 , H14 , H15 );
_SUM_ = 0.;
H11  = EXP(H11 - _MAX_);
_SUM_ = _SUM_ + H11 ;
H12  = EXP(H12 - _MAX_);
_SUM_ = _SUM_ + H12 ;
H13  = EXP(H13 - _MAX_);
_SUM_ = _SUM_ + H13 ;
H14  = EXP(H14 - _MAX_);
_SUM_ = _SUM_ + H14 ;
H15  = EXP(H15 - _MAX_);
_SUM_ = _SUM_ + H15 ;
H11  = H11 / _SUM_;
H12  = H12 / _SUM_;
H13  = H13 / _SUM_;
H14  = H14 / _SUM_;
H15  = H15 / _SUM_;
```

b   b**2

Display 5.69 shows the computation of the outputs of the target layer in the estimated NRBFEV network.

Display 5.69

```
P_resp1  =    -0.70263501920271 * H11  +    0.44649332033907 * H12
         +    4.03874078281119 * H13  +   -2.09048998863334 * H14
         +    -0.00022347123334 * H15 ;
P_resp0  = 0;
_MAX_ = MAX (P_resp1 , P_resp0 );
_SUM_ = 0.;
P_resp1  = EXP(P_resp1 - _MAX_);
_SUM_ = _SUM_ + P_resp1 ;
P_resp0  = EXP(P_resp0 - _MAX_);
_SUM_ = _SUM_ + P_resp0 ;
P_resp1  = P_resp1 / _SUM_;
P_resp0  = P_resp0 / _SUM_;
```

5.7.8 Normalized Radial Basis Function with Unequal Widths and Heights (NRBFUN)

Here, the combination function for the kth hidden unit is:

$$\eta_k = f * \log(abs(a_k)) - b_k{}^2 \sum_i^p (w_{jk} - x_j)^2,$$ where a_k is the altitude and b_k is the reciprocal of the width.

The output of the kth hidden unit is given by $H_k = \dfrac{\exp(\eta_k)}{\displaystyle\sum_{m=1}^{5} \exp(\eta_m)}$.

Display 5.70 shows the output of the five hidden units $H_k (\text{k} = 1, 2, 3, 4 \text{ and } 5)$ with a single standardized input and arbitrary values for w_{jk} , a_k , and b_k .

Display 5.70

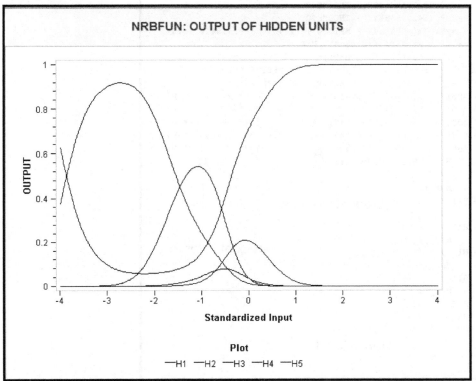

The **Architecture** property for the NRBFUN network is set to Normalized Radial – Unequal Width and Height. The remaining Network settings are the same as in Display 5.52.

Display 5.71 shows the segment of the SAS code used for calculating the outputs of the hidden units.

Display 5.71

```
H15  = H15  + (      0.787301277859 - EMP_STA0
      )**2 + (    1.81318270630709 - EMP_STA1_2
      )**2 + (    0.25477008690416 - HEQ0_1 )**2 + (   -0.15870990655028 -
    HEQ10 )**2 + (    0.56996376518127 - HEQ30
      )**2 + (    0.30377183316445 - HEQ50 )**2 + (   -0.32452628630545 -
    HEQ70 )**2 + (    0.66742331777625 - HEQ90
      )**2 + (    0.57509430491254 - INCOME20
      )**2 + (     0.9516977499014 - INCOME30
      )**2 + (   -0.97567775285454 - INCOME40
      )**2 + (   -0.67866386017564 - INCOME50
      )**2 + (    0.22920301070227 - INCOME60
      )**2 + (   -0.96456209633617 - INCOME70
      )**2 + (    0.94654469019615 - MRTGIN )**2 + (   -0.91012827678969 -
    MRTGIU )**2 + (    -0.4129064910113 - MSM
      )**2 + (   -0.74646120172093 - MSU )**2 + (    -1.75019109310648 -
    RESTYPECONDO )**2 + (   -0.55910129596964 - RESTYPECOOP
      )**2 + (    -3.6087905911793 - RESTYPEHOME )**2;
H11  = 30 * LOG(ABS(    1.52033241169643)) -     2.95872863341104 * H11 ;
H12  = 30 * LOG(ABS(    0.16077737790526)) -     0.21805094559163 * H12 ;
H13  = 30 * LOG(ABS(     0.7926221090244)) -     0.65440850633632 * H13 ;
H14  = 30 * LOG(ABS(    1.9823076609088)) -      0.01474387848664 * H14 ;
H15  = 30 * LOG(ABS(    1.83312804392597)) -     0.00927077558606 * H15 ;
_MAX_ = MAX (H11 , H12 , H13 , H14 , H15 );
_SUM_ = 0.;
H11  = EXP(H11  - _MAX_);
_SUM_ = _SUM_ + H11 ;                Unequal Heights
H12  = EXP(H12  - _MAX_);
_SUM_ = _SUM_ + H12 ;
H13  = EXP(H13  - _MAX_);            Unequal Widths
_SUM_ = _SUM_ + H13 ;
H14  = EXP(H14  - _MAX_);
_SUM_ = _SUM_ + H14 ;
H15  = EXP(H15  - _MAX_);
_SUM_ = _SUM_ + H15 ;
H11  = H11  / _SUM_;
H12  = H12  / _SUM_;
H13  = H13  / _SUM_;                 Normalized
H14  = H14  / _SUM_;
H15  = H15  / _SUM_;
```

Display 5.72 shows the SAS code needed for calculating the outputs of the target layer of the NRBFUN network.

Display 5.72

```
P_resp1  =   2.7453665467764E-26 * H11  + -1.3062764423596E-26 * H12
      +       0.05432132044973 * H13  +     -3.03951150686288 * H14
      +       3.84688447357827 * H15 ;
P_resp0  = 0;
_MAX_  = MAX (P_resp1 , P_resp0 );
_SUM_  = 0.;
P_resp1  = EXP(P_resp1  - _MAX_);
_SUM_  = _SUM_ + P_resp1 ;
P_resp0  = EXP(P_resp0  - _MAX_);
_SUM_  = _SUM_ + P_resp0 ;
P_resp1  = P_resp1  / _SUM_;
P_resp0  = P_resp0  / _SUM_;
```

5.7.9 User-Specified Architectures

With the user-specified **Architecture** setting, you can pair different Combination Functions and Activation Functions. Each pair generates a different Neural Network model, and all possible pairs can produce a large number of Neural Network models.

You can select the following settings for the Hidden Layer Activation Function:

- ArcTan
- Elliott
- Hyperbolic Tangent
- Logistic
- Gauss
- Sine
- Cosine
- Exponential
- Square
- Reciprocal
- Softmax

You can select the following settings for the Hidden Layer Combination Function:

- Add
- Linear
- EQSlopes
- EQRadial
- EHRadial
- EWRadial
- EVRadial
- XRadial

5.7.9.1 Linear Combination Function with Different Activation Functions for the Hidden Layer

Begin by setting the **Architecture** property to User, the **Target Layer Combination Function** property to Default, and the **Target Layer Activation Function** property to Default. In addition, set the **Hidden Layer Combination Function** property to Linear, the **Input Standardization** property to Standard Deviation, and the **Hidden Layer Activation Function** property to one of the available values.

If you set the **Hidden Layer Combination Function** property to Linear, the output of the kth hidden layer is estimated by Equation 5.38 for the kth hidden unit.

$$\eta_k = w_{0k} + \sum_{j=1}^{p} w_{jk} x_j \qquad (5.38)$$

where $w_{1k}, w_{2k}, ...w_{pk}$ are the weights, $x_1, x_2, ...x_p$ are the standardized inputs,[8] and w_{0k} is the bias coefficient. You can use the Hidden Layer Combination Function given by equation 5.38 with each of the following Hidden Layer Activation Functions.

Arc Tan $\qquad\qquad H_k = (2 / \pi) * \tan^{-1}(\eta_k)$

Elliot $\qquad\qquad\quad H_k = \dfrac{\eta_k}{1 + |\eta_k|}$

Hyperbolic Tangent $\quad H_k = \tanh(\eta_k)$

Logistic $\qquad\qquad\; H_k = \dfrac{1}{1 + \exp(-\eta_k)}$

Gauss $\qquad\qquad\; H_k = \exp(-0.5 * \eta^2_k)$

Sine $\qquad\qquad\quad\; H_k = \sin(\eta_k)$

Cosine $\qquad\qquad\; H_k = \cos(\eta_k)$

Exponential $\qquad\; H_k = \exp(\eta_k)$

Square $\qquad\qquad\; H_k = \eta_k^{\,2}$

Reciprocal $\qquad\quad H_k = \dfrac{1}{\eta_k}$

Softmax $\qquad\qquad H_k = \dfrac{\exp(\eta_k)}{\displaystyle\sum_{j=1}^{M} \exp(\eta_j)}$

where $k = 1, 2, 3, ...M$ and M is the number of hidden units.

The above specifications of the Hidden Layer Activation Function with the Hidden Layer Combination Function shown in equation 5.38 will result in 11 different neural network models.

5.7.9.2 Default Activation Function with Different Combination Functions for the Hidden Layer

You can set the **Hidden Layer Combination Function** property to the following values. For each item, I show the formula for the combination function for the kth hidden unit.

Add

$$\eta_k = \sum_{j=1}^{p} x_j, \text{where } x_1, x_2, ...x_p \text{ are the standardized inputs.}$$

Linear

$$\eta_k = w_{0k} + \sum_{j=1}^{p} w_{jk} x_j, \text{where } w_{jk} \text{ is the weight of } j^{th} \text{ input in the } k^{th} \text{ hidden unit and } w_{0k} \text{ is the bias}$$

coefficient.

EQSlopes

$\eta_k = w_{0k} + \sum_{j=1}^{p} w_j x_j$, where the w_j weights (coefficients) on the inputs do not differ from one hidden unit to

the next, though the bias coefficients, given by the w_{0k}, may differ.

EQRadial

$\eta_k = -b^2 \sum_{j=1}^{p} (w_{jk} - x_j)^2$, where x_j is the jth standardized input, and w_{jk} and b are calculated iteratively by

the **Neural Network** node. The coefficient b does not differ from one hidden unit to the next.

EHRadial

$\eta_k = -b_k^2 \sum_{j=1}^{p} (w_{jk} - x_j)^2$, where x_j is the jth standardized input, and w_{jk} and b_k are calculated iteratively

by the **Neural Network** node.

EWRadial

$\eta_k = f * \log(abs(a_k)) - b^2 \sum_{j}^{p} (w_{jk} - x_j)^2$, where x_j is the jth standardized input, w_{jk}, a_k, and b are

calculated iteratively by the **Neural Network** node, and f is the number of connections to the unit. In SAS Enterprise Miner, the constant f is called fan-in. The fan-in of a unit is the number of other units, from the preceding layer, feeding into that unit.

EVRadial

$\eta_k = f * \log(abs(b_k)) - b_k^2 \sum_{k}^{p} (w_{jk} - x_j)^2$, where w_{ij} and b_j are calculated iteratively by the **Neural**

Network node, and f is the number of other units feeding in to the unit. .

XRadial

$\eta_k = f * \log(abs(a_k)) - b_k^2 \sum_{j=1}^{p} (w_{jk} - x_j)^2$, where w_{ij}, b_j, and a_j are calculated iteratively by the

Neural Network node.

5.7.9.3 List of Target Layer Combination Functions for the User-Defined Networks
Add, Linear, EQSlope, EQRadial, EHRadial, EWRadial, EVRadial, and XRadial.

5.7.9.4 List of Target Layer Activation Functions for the User-Defined Networks
Identity, Linear, Exponential, Square, Logistic, and Softmax.

5.8 AutoNeural Node

As its name suggests, the **AutoNeural** node automatically configures a Neural Network model. It uses default combination functions and error functions. The algorithm tests different activation functions and selects the one that is optimum.

Display 5.73 shows the property settings of the **AutoNeural** node used in this example. Segments of SAS code generated by **AutoNeural** node are shown in Displays 5.74 and 5.75.

Display 5.73

Property	Value
General	
Node ID	AutoNeural
Imported Data	
Exported Data	
Notes	
Train	
Variables	
Model Options	
Architecture	Single Layer
Termination	Overfitting
Train Action	Search
Target Layer Error Function	Default
Maximum Iterations	50
Number of Hidden Units	5
Tolerance	Medium
Total Time	One Hour
Increment and Search Options	
Adjust Iterations	Yes
Freeze Connections	No
Total Number of Hidden Units	30
Final Training	Yes
Final Iterations	5
Activation Functions	
Direct	Yes
Exponential	No
Identity	No
Logistic	No
Normal	Yes
Reciprocal	No
Sine	Yes
Softmax	No
Square	No
Tanh	Yes

Display 5.74 shows the computation of outputs of the hidden units in the selected model.

Display 5.74

```
H1x1_5  = H1x1_5  +     -0.08252182859441 * EMP_STA0
        +       0.11832519492891 * EMP_STA1_2  +      -0.28438460977136 * HEQ0_1
        +      -0.25307314010436 * HEQ10  +     -0.21565665066707 * HEQ30
        +      -0.32564107793331 * HEQ50  +     -0.10548576896147 * HEQ70
        +      -0.42153355060319 * HEQ90  +     -0.08413766026447 * INCOME20
        +      -0.30345811392182 * INCOME30  +      -0.14439497556527 * INCOME40
        +      -0.83538773267711 * INCOME50  +       0.22088121551623 * INCOME60
        +       0.15175184854691 * INCOME70  +       0.12875734422697 * MRTGIN
        +      -0.2749817932864 * MRTGIU  +      -0.43364449447834 * MSM
        +      -0.47196973905534 * MSU  +      0.07176563660215 * RESTYPECONDO
        +       0.02758219691167 * RESTYPECOOP  +      -0.55927980441533 *
     RESTYPEHOME ;
H1x1_1  =      1.66704498013572 + H1x1_1 ;
H1x1_2  =     -2.16873847202774 + H1x1_2 ;
H1x1_3  =      1.43142431968495 + H1x1_3 ;
H1x1_4  =     -0.00885415890161 + H1x1_4 ;
H1x1_5  =     -0.18404756989638 + H1x1_5 ;
H1x1_1  = SIN(H1x1_1 );
H1x1_2  = SIN(H1x1_2 );
H1x1_3  = SIN(H1x1_3 );
H1x1_4  = SIN(H1x1_4 );
H1x1_5  = SIN(H1x1_5 );
```

From Display 5.74 it is clear that in the selected model inputs are combined using a Linear Combination Function in the hidden layer. A Sin Activation Function is applied to calculate the outputs of the hidden units.

Display 5.75 shows the calculation of the outputs of the Target layer.

Display 5.75

```
P_resp1  =     -0.7222170358321 * H1x1_1  +      -1.3128659568079 * H1x1_2
         +       1.04372392633489 * H1x1_3  +     -0.78093988310698 * H1x1_4
         +       0.54352791117617 * H1x1_5 ;
P_resp1  =     -2.14110809104329 + P_resp1 ;
P_resp0  = 0;
_MAX_  = MAX (P_resp1 , P_resp0 );
_SUM_  = 0.;
P_resp1  = EXP(P_resp1  - _MAX_);
_SUM_  = _SUM_ + P_resp1 ;
P_resp0  = EXP(P_resp0  - _MAX_);
_SUM_  = _SUM_ + P_resp0 ;
P_resp1  = P_resp1 / _SUM_;
P_resp0  = P_resp0 / _SUM_;
```

5.9 DMNeural Node

DMNeural node fits a non linear equation using bucketed principal components as inputs. The model derives the principal components from the inputs in the training data set. As explained in Chapter 2, the principal components are weighted sums of the original inputs, the weights being the eigenvectors of the variance covariance or correlation matrix of the inputs. Since each observation has a set of inputs, you can construct a Principal Component value for each observation from the inputs of that observation. The Principal components can be viewed as new variables constructed from the original inputs.

The **DMNeural** node selects the best principal components using the R-square criterion in a linear regression of the target variable on the principal components. The selected principal components are then binned or bucketed. These bucketed variables are used in the models developed at different stages of the training process.

The model generated by the DMNeural Node is called an additive nonlinear model because it is the sum of the models generated at different stages of the training process. In other words, the output of the final model is the sum of the outputs of the models generated at different stages of the training process as shown in Display 5.80.

In the first stage a model is developed using the response variable is used as the target variable. An identity link is used if the target is interval scaled, and a logistic link function is used if the target is binary. The residuals of the model from the first stage are used as the target variable values in the second stage. The residuals of the model from the second stage are used as the target variable values in the third stage. The output of the final model is the sum of the outputs of the models generated in the first, second and third stages as can be seen in Display 5.80. The first, second and third stages of the model generation process are referred to as stage 0, stage 1 and stage 2 in the code shown in Displays 5.77 through 5.80. See *SAS Enterprise Miner 12.1 Reference Help*.

Display 5.76 shows the Properties panel of the **DMNeural** node used in this example.

Display 5.76

Property	Value
General	
Node ID	DMNeural
Imported Data	
Exported Data	
Notes	
Train	
Variables	
⊟ DMNeural Network	
├ Lower Bound R2	5.0E-5
├ Max Component	5
├ Max EigenVector	400
├ Max Function Call	500
├ Max Iteration	200
└ Max Stage	3
⊟ Convergence Criteria	
├ Absolute Gradient	5.0E-4
└ Gradient	1.0E-8
⊟ Model Criteria	
├ Selection	Default
└ Optimization	SSE
Print Option	Default
Status	
Create Time	1/19/13 8:19 AM

After running the **DMNeural** node, open the Results window and click **Scoring→SAS code**. Scroll down in the SAS Code window to view the SAS code, shown in Displays 5.77 through 5.79.

Display 5.77 shows that the response variable is used as the target only in stage 0.

Display 5.77

```
/******************************************************/
* Selected activation function at stage 0 = SIN;
/******************************************************/
;
_YHAT0=0.8202154303
+0.41235892923889*SIN(1.60679535346274*_SPRIN01*(2/ARCOS(-1)))
+1.75294093758576*SIN(-0.17380300467156*_SPRIN02*(2/ARCOS(-1)))
+1.69048613728421*SIN(-0.15185018249214*_SPRIN03*(2/ARCOS(-1)))
+0.40660479197134*SIN(1.18080159361068*_SPRIN04*(2/ARCOS(-1)))
+0.39674958357393*SIN(-0.39464021714727*_SPRIN05*(2/ARCOS(-1)))
;
/*--- Target level is binary, take a logistic link function --*/
if _YHAT0 > 0 then _YHAT0=1/(1+exp(-_YHAT0));
else _YHAT0=exp(_YHAT0)/(1+exp(_YHAT0));
;
```

The model at stage 1 is shown in Display 5.77. This code shows that the algorithm uses the residuals of the model from stage 0 (_RHAT1) as the new target variable.

Display 5.78

```
/******************************************************/
* Selected activation function at stage 1 = SQUARE;
/******************************************************/
;
_RHAT1=0.015700806
+(0.0012793540995+-0.00499568950094*_SPRIN11)*_SPRIN11
+(0.0123103473237+0.00106987123715*_SPRIN12)*_SPRIN12
+(0.01354849210796+-0.00041734334168*_SPRIN13)*_SPRIN13
+(-0.00820053653591+-0.00256956052109*_SPRIN14)*_SPRIN14
+(-0.00731019932776+-0.00294377939465*_SPRIN15)*_SPRIN15
;
```

The model at stage 2 is shown in Display 5.79. In this stage, the algorithm uses the residuals of the model from stage 1 (_RHAT2) as the new target variable. This process of building the model in one stage from the residuals of the model from the previous stage can be called an additive stage-wise process.

Display 5.79

```
/******************************************************/
* Selected activation function at stage 2 = SQUARE;
/******************************************************/
;
_RHAT2=-0.000106029
+(-0.01032137286945+-0.00177139266884*_SPRIN21)*_SPRIN21
+(0.01132075640421+-0.00317884804596*_SPRIN22)*_SPRIN22
+(0.00287934750535+0.00170161047966*_SPRIN23)*_SPRIN23
+(-0.00435703724489+0.00064561558558*_SPRIN24)*_SPRIN24
+(0.00301978885486+0.00060559115769*_SPRIN25)*_SPRIN25
;
```

You can see in Display 5.79 that the Activation Function selected in stage 2 is SQUARE. The selection of the best activation function at each stage in the training process is based on the smallest SSE (Sum of Squared Error).

In Display 5.80, you can see that the final output of the model is the sum of the outputs of the models generated in stages 0, 1, and 2. Hence the underlying model can be described as an additive nonlinear model.

Display 5.80

```
_tmpPredict
=_YHAT0
+_RHAT1
+_RHAT2
;
_tmpPredict=1-_tmpPredict;
If _tmpPredict > 1 then _tmpPredict=1;
else if _tmpPredict < 0 then _tmpPredict=0;
P_resp1=_tmpPredict;
label P_resp1 = "Predicted: resp=1";
P_resp0=1-_tmpPredict;
label P_resp0 = "Predicted: resp=0";
```

5.10 Dmine Regression Node

The **Dmine Regression** node generates a Logistic Regression for a binary target. The estimation of the Logistic Regression proceeds in three steps.

In the first step, a preliminary selection is made, based on Minimum R-Square. For the original variables, the R-Square is calculated from a regression of the target on each input; for the binned variables, it is calculated from a one-way Analysis of Variance (ANOVA).

In the second step, a sequential forward selection process is used. This process starts by selecting the input variable that has the highest correlation coefficient with the target. A regression equation (model) is estimated with the selected input. At each successive step of the sequence, an additional input variable that provides the largest incremental contribution to the Model R-Square is added to the regression. If the lower bound for the incremental contribution to the Model R-Square is reached, the selection process stops.

The **Dmine Regression** node includes the original inputs and new categorical variables called AOV16 variables which are constructed by binning the original variables. These AOV16 variables are useful in taking account of the non-linear relationships between the inputs and the target variable.

After selecting the best inputs, the algorithm computes an estimated value (prediction) of the target variable for each observation in the training data set using the selected inputs.

In the third step, the algorithm estimates a logistic regression (in the case of a binary target) with a single input, namely the estimated value or prediction calculated in the first step.

Display 5.81 shows the Properties panel of the **Dmine Regression** node with default property settings.

Display 5.81

Property	Value
General	
Node ID	DmineReg
Imported Data	
Exported Data	
Notes	
Train	
Variables	
Maximum Variable Number	3000
⊟ R-Square Options	
┆ Minimum R-Square	0.0050
┖ Stop R-Square	5.0E-4
⊟ Created Variables	
┆ Use AOV16 Variables	Yes
┆ Use Group Variables	Yes
┖ Use Interactions	No
Print Option	Default
Use SPD Engine Library	Yes
Status	
Create Time	1/15/13 10:42 AM

You should make sure that the **Use AOV16 Variables** property is set to Yes if you want to include them in the variable selection and model estimation.

Display 5.82 shows the variable selected in the first step.

Display 5.82

```
The DMINE Procedure

              R-Squares for Target Variable: resp

Effect                      DF          R-Square

Class: RESTYPE               3          0.026340
Group: RESTYPE               2          0.026337
AOV16: NUMTR                 7          0.017844
Var:   NUMTR                 1          0.017175
Class: MRTGI                 2          0.011288
Group: MRTGI                 1          0.011248
```

Display 5.83 shows the variables selected by the forward least-squares stepwise regression in the second step.

Display 5.83

```
The DMINE Procedure

                        Effects Chosen for Target: resp

Effect               DF       R-Square        F Value      p-Value

Group: RESTYPE        2       0.026337      242.595910      <.0001
AOV16: NUMTR          7       0.018029       48.325141      <.0001
```

5.11 Comparing the Models Generated by DMNeural, AutoNeural, and Dmine Regression Nodes

In order to compare the predictive performance of the models produced by the **DMNeural**, **AutoNeural**, and **Dmine Regression** nodes, I created the process flow diagram shown in Display 5.84.

Display 5.84

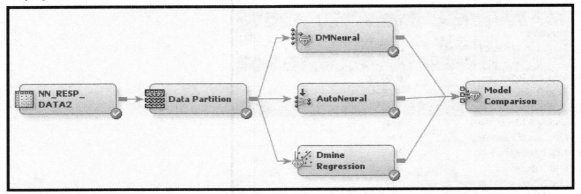

The results window of the **Model Comparison** node shows the Cumulative %Captured Response charts for the three models for the Training, Validation, and Test data sets.

Display 5.85 shows the Cumulative %Captured Response charges for the Training data set.

Display 5.85

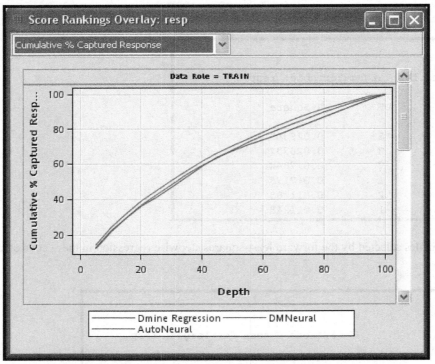

Display 5.86 shows the Cumulative %Captured Response charges for the Validation data set.

Display 5.86

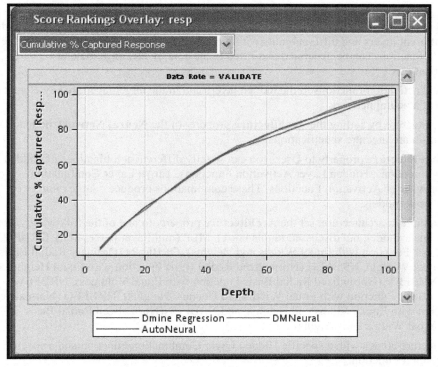

Display 5.87 shows the Cumulative %Captured Response charges for the Test data set.

Display 5.87

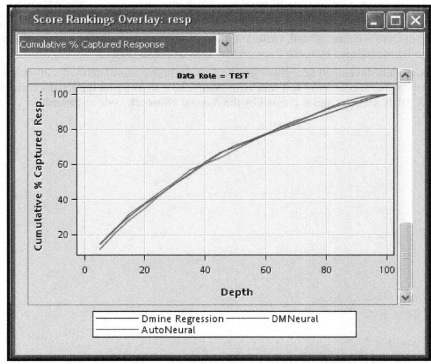

From the Cumulative %Captured Response, there does not seem to be a significant difference in the predictive performance of the three models compared.

5.12 Summary

- A neural network is essentially nothing more than a complex nonlinear function of the inputs. Dividing the network into different layers and different units within each layer makes it very flexible. A large number of nonlinear functions can be generated and fitted to the data by means of different architectural specifications.

- The combination and activation functions in the hidden layers and in the target layer are key elements of the *architecture* of a neural network.

- You specify the architecture by setting the **Architecture** property of the **Neural Network** node to User or to one of the built-in architecture specifications.

- When you set the **Architecture** property to User, you can specify different combinations of Hidden Layer Combination Functions, Hidden Layer Activation Functions, Target Layer Combination Functions, and Target Layer Activation Functions. These combinations produce a large number of potential neural network models.

- If you want to use a built-in architecture, set the **Architecture** property to one of the following values: GLM (generalized linear model, not discussed in this book), MLP (multilayer perceptron), ORBFEQ (Ordinary Radial Basis Function with Equal Widths and Heights), ORBFUN (Ordinary Radial Basis Function with Unequal Widths), NRBFEH (Normalized Radial Basis Function with Equal Heights and Unequal Widths), NRBFEV (Normalized Radial Basis Function with Equal Volumes), NRBFEW (Normalized Radial Basis Function with Equal Widths and Unequal Heights), NRBFEQ (Normalized Radial Basis Function with Equal Widths and Heights, and NRBFUN(Normalized Radial Basis Functions with Unequal Widths and Heights).

- Each built-in architecture comes with a specific Hidden Layer Combination Function and a specific Hidden Layer Activation Function.

- While the specification of the Hidden Layer Combination and Activation functions can be based on such criteria as model fit and model generalization, the selection of the Target Layer Activation Function should also be guided by theoretical considerations and the type of output you are interested in.

- In addition to the **Neural Network** node, three additional nodes, namely **DMNeural**, **AutoNeural**, and **Dmine Regression** nodes, were demonstrated. However, there is no significant difference in predictive performance of the models developed by these three nodes in the example used.

- The response and risk models developed here show you how to configure neural networks in such a way that they are consistent with economic and statistical theory, how to interpret the results correctly, and how to use the SAS data sets and tables created by the **Neural Network** node to generate customized reports.

5.13 Appendix to Chapter 5

Displays 5.88, 5.89, and 5.90 show the SAS code for calculating the lift for *LOSSFRQ*.

Display 5.88

```
libname lib2 'C:\TheBook\EM12.1\DATA\Chapter5' ;
run ;
options center ;
data lib2.Ch5_risk_NN_train ;
 set &EM_import_data;
run;
data lib2.Ch5_risk_NN_validate ;
 set &EM_import_validate ;
run;

data lib2.Ch5_risk_NN_test ;
 set &EM_import_test ;
run;

 %macro lifts(ds=);
 data &ds ;
  set lib2.ch5_risk_NN_&ds ;
  keep p_lossfrq0 p_lossfrq1 p_lossfrq2 lossfrq expected_lfrq;
  expected_lfrq = 0*p_lossfrq0 + 1*p_lossfrq1 + 2*p_lossfrq2 ;
 run ;
 proc sort data=&ds;
  by descending expected_lfrq;
 run ;
proc sql noprint;
 select count(*) into : nvl from
 work.&ds ;
quit ;
```

Display 5.89

```
data &ds ;
 retain count 0 ;
 set &ds ;
 count+1 ;
 if count < (1/20)*&nvl then dec=5; else
 if count < (2/20)*&nvl then dec=10 ; else
 if count < (3/20)*&nvl then dec=15 ; else
 if count < (4/20)*&nvl then dec=20 ; else
 if count < (5/20)*&nvl then dec=25 ; else
 if count < (6/20)*&nvl then dec=30 ; else
 if count < (7/20)*&nvl then dec=35; else
 if count < (8/20)*&nvl then dec=40 ; else
 if count < (9/20)*&nvl then dec=45 ; else
 if count < (10/20)*&nvl then dec=50 ; else
 if count < (11/20)*&nvl then dec=55 ; else
 if count < (12/20)*&nvl then dec=60 ; else
 if count < (13/20)*&nvl then dec=65 ; else
 if count < (14/20)*&nvl then dec=70 ; else
 if count < (15/20)*&nvl then dec=75 ; else
 if count < (16/20)*&nvl then dec=80 ; else
 if count < (17/20)*&nvl then dec=85 ; else
 if count < (18/20)*&nvl then dec=90 ; else
 if count < (19/20)*&nvl then dec=95 ; else
 dec = 100 ;
 run ;
proc means data=&ds noprint ;
  class dec ;
  var lossfrq ;
  output out= outsum sum(lossfrq) = sum_lossfrq mean(lossfrq)=mean_lossfrq;
run ;
```

Display 5.90

```
data Total(keep=sum_lossfrq rename=(sum_lossfrq=Tot_lossfrq)) deciles ;
  set outsum ;
  if _TYPE_ = 0 then output Total;
  else output deciles ;
run ;
data tables ;
 set deciles ;
 if _N_ = 1 then set total ;
run;
data lib2.Ch5_risk_Lift_NN_&ds ;
 set tables ;
 retain cumsum 0 nobs 0;
 cumsum + sum_lossfrq ;
 capc = cumsum/tot_lossfrq ;
 gmean = tot_lossfrq/&nvl ;
 nobs+_freq_ ;
 meanc = cumsum/nobs ;
 liftc = meanc/gmean;
 label dec='Percentile'
       Mean_lossfrq='Lossfrq Mean'
       Meanc='Lossfrq Cumulative Mean'
       liftc ='Cumulative Lift' ;
 RUN;
 proc print data=lib2.ch5_risk_Lift_NN_&DS label noobs ;
    var dec mean_lossfrq meanc liftc ;
 Title1 "             Loss Frequency: &DS" ;
 Title2 ;
RUN;
%mend lifts ;
%lifts(ds=TRAIN);
%lifts(ds=VALIDATE);
%lifts(ds=TEST) ;
```

5.14 Exercises

1. Create a data source with the SAS data set Ch5_Exdata. Use the Advanced Metadata Advisor Options to create the metadata. Customize the metadata by setting the **Class Levels Count Threshold** property to 8. Set the **Role** of the variable EVENT to Target. On the **Prior Probabilities** tab of the Decision Configuration step, set **Adjusted Prior** to 0.052 for level 1 and 0.948 for level 0.

2. Partition the data such that 60% of the records are allocated for training, 30% for Validation, and 10% for Test.

3. Attach three **Neural Network** nodes and an **Auto Neural** node.
   Set the **Model Selection Criterion** to Average Error in all the three **Neural Network** nodes.

   a. In the first **Neural Network** node, set the **Architecture** property to Multilayer Perceptron and change the **Target Layer Activation Function** to Logistic. Set the **Number of Hidden Units** property to 5. Open the **Optimization** property and set the **Maximum Iterations** property to 100. Use default values for all other Network properties.

   b. Open the Results window and examine the Cumulative Lift and Cumulative %Captured Response charts.

   c. In the first **Neural Network** node, change the **Target Layer Error Function** property to Bernoulli, while keeping all other settings same as in (a).

   d. Open the Results window and examine the Cumulative Lift and Cumulative %Captured Response charts. Did the model improved by changing the value of the **Target Layer Error Function** property to Bernoulli?

   e. Open the Results window and the Score Code window by clicking **File→Score→SAS Code**. Verify that the formulas used in calculating the outputs of the hidden units and the estimated probabilities of the event are what you expected.

   f. In the second **Neural Network** node, set the **Architecture** property to Ordinary Radial-Equal Width and the **Number of Hidden Units** property to 5. Use default values for all other Network properties.

   g. In the third **Neural Network** node, set the **Architecture** Property to Normalized Radial-Equal Width. Set the **Number of Hidden Units** property to 5. Use default values for all other Network properties.

   h. Set the **AutoNeural** node options as shown in Display 5.91.

Display 5.91

Architecture	Single Layer
Termination	Overfitting
Train Action	Search
Target Layer Error Function	Logistic
Maximum Iterations	50
Number of Hidden Units	5
Tolerance	Medium
Total Time	One Hour

i. Use the default values for the properties of the **Dmine Regression** node.

Attach a **Model Comparison** node and run all models and compare the results. Which model is the best?

Display 5.92 shows the suggested process diagram for this exercise.

Display 5.92

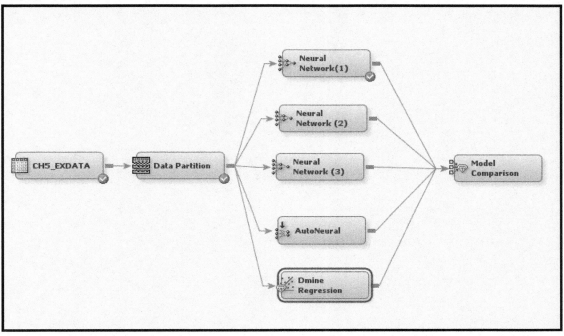

Notes

1. Although the general discussion here considers two hidden layers, the two neural network models developed later in this book are based on a single hidden layer.
2. The terms *observation*, *record*, *case*, and *person* are used interchangeably.
3. Bishop, C.M. (1995) Neural Networks for Pattern Recognition. New York: Oxford University Press
4. Afifi, A.A and Clark, Virginia, *Computer Aided Multivariate Analysis*, CRC Press, 2004.
5. This is the target variable *LOSSFRQ*.
6. For the derivation of the error functions and the underlying probability distributions, see Bishop, C.M (1995) Neural Networks for Pattern Recognition, New York: Oxford University Press.
7. SAS Enterprise Miner uses the terms *bin*, *percentile*, and *depth*.
8. To transform an input to standardized units, the mean and standard deviation of the input across all records are computed, the mean is subtracted from the original input values, and the result is then divided by the standard deviation, for each individual record.

Chapter 6: Regression Models

6.1 Introduction

This chapter explores the **Regression** node in detail using two practical business applications—one requiring a model to predict a binary target and the other requiring a model to predict a continuous target. Before developing these two models, I present an overview of the types of models that can be developed in the **Regression** node, the theory and rationale behind each type of model, and an explanation of how to set various properties of the **Regression** node to get the desired models.

6.2 What Types of Models Can Be Developed Using the Regression Node?

6.2.1 Models with a Binary Target

When the target is binary, either numeric or character, the **Regression** node estimates a logistic regression. The **Regression** node produces SAS code to calculate the probability of the event (such as response or attrition). The computation of the probability of the event is done through a *link function*. A *link function* shows the relation between the probability of the event and a linear predictor, which is a linear combination of the inputs (explanatory variables).

The linear predictor can be written as $\beta'x$, where x is a vector of inputs and β is the vector of coefficients estimated by the **Regression** node. For a response model, where the target takes on a value of 1 or 0 (1 stands for response and 0 for non-response), the link function is

$$\log\left(\frac{\Pr(y=1\mid x)}{1-\Pr(y=1\mid x)}\right) = \beta'x \tag{6.1}$$

where y is the target and $\Pr(y=1\mid x)$ is the probability of the target variable taking the value 1 for a customer (or a record in the data set), given the values of the inputs or explanatory variables for that customer. The left-hand side of Equation 6.1 is the logarithm of odds or log-odds of the event "response". Since Equation 6.1 shows the relation between the log-odds of the event and the linear predictor $\beta'x$, it is called a *logit* link.

If you solve the link function given in Equation 6.1 for $\Pr(y=1\mid x)$, you get:

$$\Pr(y=1\mid x) = \frac{\exp(\beta x)}{1+\exp(\beta x)} \tag{6.2}$$

Equation 6.2 is called the *inverse link* function. In Chapter 5, where we discussed neural networks, we called equations of this type *activation functions*.

From Equation 6.2, you can see that

$$1-\Pr(y=1\mid x) = 1-\frac{\exp(\beta x)}{1+\exp(\beta x)} = \frac{1}{1+\exp(\beta x)} \tag{6.3}$$

Dividing Equation 6.3 by Equation 6.2 and taking logs, I once again obtain the logit link given in Equation 6.1.

The important feature of this function for our purposes is that it produces values that lie only between 0 and 1, just as the as the probability of response and non-response do.

In addition to the logit link, the **Regression** node provides probit and complementary log-log link functions. (These are discussed in Section 6.3.2.) All of these link functions keep the predicted probabilities between 0 and 1. In cases where it is not appropriate to constrain the predicted values to lie between 0 and 1, no link function is needed.

There are also some well-accepted theoretical rationales behind some of these link functions. For readers who are interested in pursuing this line of reasoning further, a discussion of the theory is included in Section 6.3.2.

Display 6.1 shows the process flow diagram for a regression with a binary target.

Display 6.1

Display 6.2 shows a partial list of the variables in the input data set used in this process flow. You can see from Display 6.2 that the target variable RESP is binary.

Display 6.2

Name	Role	Level	Report
NVAR82	Input	Interval	No
NVAR83	Input	Interval	No
NVAR84	Input	Interval	No
NVAR85	Input	Interval	No
NVAR86	Input	Interval	No
NVAR87	Input	Interval	No
NVAR88	Input	Interval	No
NVAR89	Input	Interval	No
NVAR9	Input	Nominal	No
NVAR90	Input	Interval	No
NVAR91	Input	Interval	No
NVAR92	Input	Interval	No
NVAR93	Input	Interval	No
NVAR94	Input	Interval	No
NVAR95	Input	Interval	No
NVAR96	Input	Nominal	No
NVAR97	Input	Interval	No
NVAR98	Input	Interval	No
NVAR99	Input	Interval	No
TYPE	Rejected	Unary	No
matchkey	Rejected	Nominal	No
nvar001	Input	Interval	No
resp	Target	Binary	No

Display 6.3 shows the properties of the **Regression** node that is in the process flow.

Display 6.3

Property	Value
General	
Node ID	Reg
Imported Data	
Exported Data	
Notes	
Train	
Variables	
☐Equation	
Main Effects	Yes
Two-Factor Interactions	No
Polynomial Terms	No
Polynomial Degree	2
User Terms	No
Term Editor	
☐Class Targets	
Regression Type	Logistic Regression
Link Function	Logit
☐Model Options	
Suppress Intercept	No
Input Coding	Deviation
☐Model Selection	
Selection Model	Stepwise
Selection Criterion	None
Use Selection Defaults	Yes
Selection Options	

After running the **Regression** node, you can open the SAS code from the Results window by clicking **View→Scoring→SAS code**.

Displays 6.4A and 6.4B show the SAS code used for calculating the linear predictor and the probabilities of response and no response for each record in a Score data set.

Display 6.4A

```
*** Effect: NVAR103 ;
_TEMP = NVAR103 ;
_LP0 = _LP0 + (    0.00003694562876 * _TEMP);

*** Effect: NVAR253 ;
_TEMP = NVAR253 ;
_LP0 = _LP0 + (    0.16797834307168 * _TEMP);

*** Naive Posterior Probabilities;
drop _MAXP _IY _P0 _P1;
_TEMP =    -4.54086566603301 + _LP0;
if (_TEMP < 0) then do;
   _TEMP = exp(_TEMP);
   _P0 = _TEMP / (1 + _TEMP);
end;
else _P0 = 1 / (1 + exp(-_TEMP));
_P1 = 1.0 - _P0;
```

Display 6.4B

```
221    *** Posterior Probabilities and Predicted Level;
222    label P_resp1 = 'Predicted: resp=1' ;
223    label P_resp0 = 'Predicted: resp=0' ;
224    P_resp1 = _P0;
225    _MAXP = _P0;
226    _IY = 1;
227    P_resp0 = _P1;
228    if (_P1 > _MAXP + 1E-8) then do;
229       _MAXP = _P1;
230       _IY = 2;
231    end;
```

In Display 6.4A, _LP0 is the linear combination of the inputs and _TEMP = _LP0 + the estimated intercept (-4.54086566603301). _TEMP is the linear predictor $\beta'x$ shown in Equation 6.1.

You learn more about modeling binary targets in Section 6.4, where a model with a binary target is developed for a marketing application.

6.2.2 Models with an Ordinal Target

Ordinal targets, also referred to as *ordered polychotomous* targets, take on more than two discrete, ordered values. An example of an ordinal target with three levels (0, 1, and 2) is the discretized version of the loss frequency variable discussed in Chapter 5.

If you want your target to be treated as an ordinal variable, you must set its measurement scale to Ordinal. However, if the target variable has fewer than the number of levels specified in the Advanced Advisor Metadata options, SAS Enterprise Miner sets its measurement scale to Nominal (see Section 2.5, Displays 2.10 and 2.11). If the measurement scale of your target variable is currently set to Nominal but you want to model it as an ordinal target, then you must change its measurement scale setting to Ordinal. In the Appendix to Chapter 3, I showed how to change the measurement scale of a variable in a data source.

When the target is ordinal, then, by default, the **Regression** node produces a Proportional Odds (or Cumulative Logits) model, as discussed in Chapter 5, Section 5.5.1.1. (The following example illustrates the proportional odds model as well.) When the target is nominal, on the other hand, it produces a Generalized Logits model, which is presented in Section 6.2.3. Although the default value of the **Regression Type** property is Logistic Regression and the default value of the **Link Function** property (link functions are discussed in detail in Section 6.3.2) is Logit for both ordinal targets and nominal targets, the models produced are different for targets with different measurement scales. Hence, the setting of the measurement scale determines which type of model you get. Using some simple examples, I will demonstrate how different types of models are produced for different measurement scales even though the **Regression** node properties are all set to the same values.

Display 6.5 shows the process flow for a regression with an ordinal target.

Display 6.5

Display 6.6 shows the list of variables and their measurement scales, or levels. In the second and third columns, you can see that the role of the variable $LOSSFRQ$ is a *target* and its measurement scale (level) is ordinal.

Display 6.6

Name	Role	Level	Report	Type	Number of Levels
AGE	Input	Interval	No	Numeric	.
ANTHFT	Input	Binary	No	Character	2
CRED	Input	Interval	No	Numeric	.
DELINQ	Input	Interval	No	Numeric	.
DEPC	Input	Binary	No	Character	2
EMP_STA	Input	Nominal	No	Character	3
GENDER	Input	Binary	No	Character	2
HEQ	Input	Nominal	No	Numeric	7
INCOME	Input	Nominal	No	Numeric	7
LOSSFRQ	Target	Ordinal	No	Numeric	3
MFDU	Input	Binary	No	Numeric	2
MILEAGE	Input	Interval	No	Numeric	.
MOB	Input	Binary	No	Character	2
MRTGI	Input	Nominal	No	Character	3
MS	Input	Nominal	No	Character	3

Display 6.7 shows the Properties panel of the **Regression** node in the process flow shown in Display 6.5.

Display 6.7

Property	Value
General	
Node ID	Reg
Imported Data	
Exported Data	
Notes	
Train	
Variables	
⊟Equation	
├Main Effects	Yes
├Two-Factor Interactions	No
├Polynomial Terms	No
├Polynomial Degree	2
├User Terms	No
└Term Editor	
⊟Class Targets	
├Regression Type	Logistic Regression
└Link Function	Logit
⊟Model Options	
├Suppress Intercept	No
└Input Coding	Deviation
⊟Model Selection	
├Selection Model	Stepwise
├Selection Criterion	None
├Use Selection Defaults	Yes
└Selection Options	

Note the settings in the Class Targets section. These settings are the default values for class targets (categorical targets), including binary, ordinal, and nominal targets. Since the target variable $LOSSFRQ$ is ordinal, taking on the values 0, 1, or 2 for any given record of the data set, the **Regression** node uses a cumulative logits link.

The expression $\dfrac{\Pr(lossfrq = 2)}{1 - \Pr(lossfrq = 2)}$ is called the *odds ratio*. It is the odds that the event $LOSSFRQ = 2$ will occur. In other words, the odds ratio represents the odds of the loss frequency equaling 2. Similarly, the expression $\dfrac{\Pr(lossfrq = 1) + \Pr(lossfrq = 2)}{1 - \{\Pr(lossfrq = 1) + \Pr(lossfrq = 2)\}}$ represents the odds of the event $LOSSFRQ = 1$ or 2. When I take the logarithms of the above expressions, they are called *log-odds* or *logits*.

Thus the logit for the event $LOSSFRQ = 2$ is $\log\left(\dfrac{\Pr(lossfrq = 2)}{1 - \Pr(lossfrq = 2)}\right)$, and the logit for the event $LOSSFRQ = 1$ or 2 is $\log\left(\dfrac{\Pr(lossfrq = 1) + \Pr(lossfrq = 2)}{1 - \{\Pr(lossfrq = 1) + \Pr(lossfrq = 2)\}}\right)$.

The logits shown above are called *cumulative logits*. When a cumulative logits link is used, the logits of the event $LOSSFRQ = 1$ or 2 are larger than the logits for the event $LOSSFRQ = 2$ by a positive amount. In a Cumulative Logits model, the logits of the event $LOSSFRQ = 2$ and the event $LOSSFRQ = 1$ or 2 differ by a constant amount for all observations (all customers). To illustrate this phenomenon, I run the **Regression** node with an ordinal target ($LOSSFRQ$). Display 6.8, which is taken from the Results window, shows the equations estimated by the **Regression** node.

Display 6.8

Analysis of Maximum Likelihood Estimates							
Parameter	DF	Estimate	Standard Error	Wald Chi-Square	Pr > ChiSq	Standardized Estimate	Exp(Est)
Intercept 2	1	-2.8840	0.4145	48.42	<.0001		0.056
Intercept 1	1	0.2133	0.3236	0.43	0.5097		1.238
AGE	1	-0.0231	0.00413	31.27	<.0001	-0.1998	0.977
CRED	1	-0.00498	0.000527	89.39	<.0001	-0.2563	0.995
NPRVIO	1	0.2252	0.0665	11.47	0.0007	0.0830	1.253

From the coefficients given in Display 6.8, I can write the following two equations:[1]

$$\log\left(\frac{\Pr(lossfrq = 2)}{1 - \Pr(lossfrq = 2)}\right) = -2.8840 - 0.0231 * AGE - 0.00498 * CRED \tag{6.4}$$
$$+ 0.2252 * NPRVIO$$

$$\log\left(\frac{\Pr(lossfrq = 2) + \Pr(lossfrq = 1)}{1 - \{\Pr(lossfrq = 2) + \Pr(lossfrq = 1)\}}\right) = 0.2133 - 0.0231 * AGE - 0.00498 * CRED \tag{6.5}$$
$$+ 0.2252 * NPRVIO$$

To interpret the results, take the example of a customer (Customer A) who is 50 years of age (AGE=50), has a credit score of 670(CRED=670), and has two prior traffic violations (NPRVIO=2).

From Equation 6.4, the log-odds (logit) of the event $LOSSFRQ = 2$ for Customer A are:

$$\log\left(\frac{\Pr(lossfrq = 2)}{1 - \Pr(lossfrq = 2)}\right) = -2.8840 - 0.0231*50 - 0.00498*670 \qquad (6.4\text{A})$$
$$+ 0.2252*2$$

Therefore, the odds of the event $LOSSFRQ = 2$ for Customer A are

$$\left(\frac{\Pr(lossfrq = 2)}{1 - \Pr(lossfrq = 2)}\right) = e^{-2.8840 - 0.0231*50 - 0.00498*670 + 0.2252*2} \qquad (6.4\text{B})$$

From Equation 6.5, the log-odds (logit) of the event $LOSSFRQ = 1$ or 2 for Customer A are:

$$\log\left(\frac{\Pr(lossfrq = 2) + \Pr(lossfrq = 1)}{1 - \{\Pr(lossfrq = 2) + \Pr(lossfrq = 1)\}}\right) = 0.2133 - 0.0231*50 - 0.00498*670 \qquad (6.5\text{A})$$
$$+ 0.2252*2$$

The odds of the event $LOSSFRQ = 1$ or 2 for Customer A is

$$\left(\frac{\Pr(lossfrq = 2) + \Pr(lossfrq = 1)}{1 - \{\Pr(lossfrq = 2) + \Pr(lossfrq = 1)\}}\right) = e^{0.2133 - 0.0231*50 - 0.00498*670 + 0.2252*2} \qquad (6.5\text{B})$$

The difference in the logits (not the odds, but the log-odds) for the events $LOSSFRQ = 1$ or 2 and $LOSSFRQ = 2$ is equal to 0.2133–(–2.8840) =3.0973, which is obtained by subtracting Equation 6.4A from Equation 6.5A. Irrespective of the values of the inputs AGE, CRED, and NPRVIO, the difference in the logits (or logit difference) for the events $LOSSFRQ = 1$ or 2 and $LOSSFRQ = 2$ is the same constant 3.0973. Now turning to the odds of the event, rather than the log-odds, if I divide Equation 6.5B by 6.4B, I get the ratio of the odds of the event $LOSSFRQ = 1$ or 2 to the odds of the event $LOSSFRQ = 2$. This ratio is equal to

$$\frac{e^{0.2133}}{e^{-2.8840}} = e^{0.2133 + 2.8840} = e^{3.0973} = 22.1381.$$

By this result, I can see that the odds of the event $LOSSFRQ = 1$ or 2 is proportional to the odds of the event $LOSSFRQ = 2$, and that the proportionality factor is 22.1381. This proportionality factor is the same for any values of the inputs (AGE, CRED, and NPRVIO in this example). Hence the name is *Proportional Odds model*.

Now I compare these results to those for Customer B, who is the same age (50) and has the same credit score (670) as Customer A, but who has three prior violations (NPRVIO=3) instead of two. In other words, all of the inputs except for NPRVIO are the same for Customer B and Customer A. I show how the log-odds and the odds for the events $LOSSFRQ = 2$ and $LOSSFRQ = 1$ or 2 for Customer B differ from those for Customer A.

From Equation 6.4, the log-odds (logit) of the event $LOSSFRQ = 2$ for Customer B are:

$$\log\left(\frac{\Pr(lossfrq = 2)}{1 - \Pr(lossfrq = 2)}\right) = -2.8840 - 0.0231*50 - 0.00498*670 \qquad (6.4\text{C})$$
$$+ 0.2252*3$$

Therefore, the odds of the event $LOSSFRQ = 2$ for Customer B are:

$$\left(\frac{\Pr(lossfrq = 2)}{1 - \Pr(lossfrq = 2)} \right) = e^{-2.8840 - 0.0231*50 - 0.00498*670 + 0.2252*3} \tag{6.4D}$$

From Equation 6.5, the log-odds (logit) of the event $LOSSFRQ = 1$ or 2 for Customer B are:

$$\log \left(\frac{\Pr(lossfrq = 2) + \Pr(lossfrq = 1)}{1 - \{\Pr(lossfrq = 2) + \Pr(lossfrq = 1)\}} \right) = 0.2133 - 0.0231*50 - 0.00498*670 \\ + 0.2252*3 \tag{6.5C}$$

The odds of the event $LOSSFRQ = 1$ or 2 for Customer B are:

$$\left(\frac{\Pr(lossfrq = 2) + \Pr(lossfrq = 1)}{1 - \{\Pr(lossfrq = 2) + \Pr(lossfrq = 1)\}} \right) = e^{0.2133 - 0.0231*50 - 0.00498*670 + 0.2252*3} \tag{6.5D}$$

It is clear that the difference between log-odds of the event $LOSSFRQ = 2$ for Customer B and Customer A is 0.2252, which is obtained by subtracting 6.4A from 6.4C.

It is also clear that the difference between log-odds of the event $LOSSFRQ = 1$ or 2 for Customer B and Customer A is also 0.2252, which is obtained by subtracting 6.5A from 6.5C.

The difference in the log-odds (0.2252) is simply the coefficient of the variable NPRVIO, whose value differs by 1 for customers A and B. Generalizing this result, we now see that the coefficient of any of the inputs in the log-odds Equations 6.4 and 6.5 indicates the change in log-odds per unit of change in the input.

The odds of the event $LOSSFRQ = 2$ for Customer B relative to Customer A is $e^{0.2252} = 1.252573$, which is obtained by dividing Equation 6.4D by Equation 6.4B. In this example, an increase of one prior violation indicates a 25.2% increase in the odds of the event $LOSSFRQ = 2$. I leave it to you to verify that an increase of one prior violation has the same impact, a 25.2% increase in the odds of the event $LOSSFRQ = 1$ or 2 as well. (Hint: Divide Equation 6.5D by Equation 6.5B.)

Equations 6.4 and 6.5 and the identity $\Pr(lossfrq = 0) + \Pr(lossfrq = 1) + \Pr(lossfrq = 2) = 1$ are used by the **Regression** node to calculate the three probabilities $\Pr(lossfrq = 0)$, $\Pr(lossfrq = 1)$, and $\Pr(lossfrq = 2)$. This is shown in the SAS code generated by the **Regression** node, shown in Displays 6.9A and 6.9B.

Display 6.9A

```
*** Compute Linear Predictor;
drop _TEMP;
drop _LP0;
_LP0 = 0;

*** Effect: AGE ;
_TEMP = AGE ;
_LP0 = _LP0 + (   -0.02307201027961 * _TEMP);

*** Effect: CRED ;
_TEMP = CRED ;
_LP0 = _LP0 + (   -0.00498225050304 * _TEMP);

*** Effect: NPRVIO ;
_TEMP = NPRVIO ;
_LP0 = _LP0 + (    0.22524384228134 * _TEMP);

*** Naive Posterior Probabilities;
drop _MAXP _IY _P0 _P1 _P2;
_TEMP =    -2.88399896897927 + _LP0;
if (_TEMP < 0) then do;
   _TEMP = exp(_TEMP);
   _P0 = _TEMP / (1 + _TEMP);
end;
else _P0 = 1 / (1 + exp(-_TEMP));
_TEMP =     0.21333987443107 + _LP0;
if (_TEMP < 0) then do;
   _TEMP = exp(_TEMP);
   _P1 = _TEMP / (1 + _TEMP);
end;
else _P1 = 1 / (1 + exp(-_TEMP));
_P2 = 1.0 - _P1;
_P1 = _P1 - _P0;
```

Display 6.9B

```
*** Posterior Probabilities and Predicted Level;
label P_LOSSFRQ2 = 'Predicted: LOSSFRQ=2' ;
label P_LOSSFRQ1 = 'Predicted: LOSSFRQ=1' ;
label P_LOSSFRQ0 = 'Predicted: LOSSFRQ=0' ;
P_LOSSFRQ2 = _P0;
_MAXP = _P0;
_IY = 1;
P_LOSSFRQ1 = _P1;
if (_P1 > _MAXP + 1E-8) then do;
   _MAXP = _P1;
   _IY = 2;
end;
P_LOSSFRQ0 = _P2;
if (_P2 > _MAXP + 1E-8) then do;
   _MAXP = _P2;
   _IY = 3;
end;
```

By setting Age=50, CRED=670, and NPRVIO=2, and by using the SAS statements from Displays 6.9A and 6.9B, I generated the probabilities that $LOSSFRQ$ =0, 1, and 2, as shown in Display 6.10. The SAS code executed for generating these probabilities is shown in Displays 6.77 and 6.78 in the appendix to this chapter.

Display 6.10

ORDINAL TARGET AGE=50, CRED=670 and NPRVIO=2		
Predicted: LOSSFRQ=2	Predicted: LOSSFRQ=1	Predicted: LOSSFRQ=0
.000981723	0.020311	0.97871

6.2.3 Models with a Nominal (Unordered) Target

Unordered polychotomous variables are nominal-scaled categorical variables where there is no particular order in which to rank the category labels. That is, you cannot assume that one category is higher or better than another. Targets of this type arise in market research where marketers are modeling customer preferences. In consumer preference studies and also in this chapter letters like A, B, C, and D may represent different products or different brands of the same product.

When the target is nominal categorical, the default values of the **Regression** node properties produce a model of the following type:

$$P(T = D \mid X) = \frac{\exp(\alpha_0 + \beta_0 X)}{1 + \exp(\alpha_0 + \beta_0 X) + \exp(\alpha_1 + \beta_1 X) + \exp(\alpha_2 + \beta_2 X)} \tag{6.6}$$

$$P(T = C \mid X) = \frac{\exp(\alpha_1 + \beta_1 X)}{1 + \exp(\alpha_0 + \beta_0 X) + \exp(\alpha_1 + \beta_1 X) + \exp(\alpha_2 + \beta_2 X)} \tag{6.7}$$

$$P(T = B \mid X) = \frac{\exp(\alpha_2 + \beta_2 X)}{1 + \exp(\alpha_0 + \beta_0 X) + \exp(\alpha_1 + \beta_1 X) + \exp(\alpha_2 + \beta_2 X)} \tag{6.8}$$

$$P(T = A \mid X) = \frac{1}{1 + \exp(\alpha_0 + \beta_0 X) + \exp(\alpha_1 + \beta_1 X) + \exp(\alpha_2 + \beta_2 X)} \tag{6.9}$$

where T is the target variable; A, B, C, and D represent levels of the target variable T; X is the vector of input variables; and $\alpha_0, \alpha_1, \alpha_2, \beta_0, \beta_1$, and β_2 are the coefficients estimated by the **Regression** node. Note that β_0, β_1, and β_2 are row vectors, and α_o, α_1, and α_2 are scalars. A model of the type presented in Equations 6.6, 6.7, 6.8, and 6.9 is called a *Generalized Logits* model or a logistic regression with a *generalized logits* link.

In the following example, I show that the measurement scale of the target variable determines which model is produced by the **Regression** node. I use the same data that I used in Section 6.2.2, except that I have changed the measurement scale of the target variable $LOSSFRQ$ to Nominal. Since the variable $LOSSFRQ$ is a discrete version of the continuous variable Loss Frequency, it should be treated as ordinal. I have changed its measurement scale to Nominal only to highlight the fact that the model produced depends on the measurement scale of the target variable.

Display 6.11 shows the flow chart for a regression with a nominal target.

Display 6.11

The variables in the data set are set to values identical to the list shown in Display 6.6 with the exception that the measurement scale (level) of the target variable $LOSSFRQ$ is set to Nominal. Display 6.12 shows the coefficients of the model estimated by the **Regression** node.

Display 6.12

```
                        Analysis of Maximum Likelihood Estimates

                                   Standard      Wald                Standardized
Parameter      LOSSFRQ   DF   Estimate    Error   Chi-Square   Pr > ChiSq     Estimate   Exp(Est)

Intercept         2       1    1.5946    1.2965      1.51       0.2187                     4.927
Intercept         1       1   -0.1186    0.3367      0.12       0.7248                     0.888
AGE               2       1   -0.0324    0.0204      2.53       0.1115        -0.2805      0.968
AGE               1       1   -0.0228    0.00421    29.42       <.0001        -0.1976      0.977
CRED              2       1   -0.0134    0.00241    30.75       <.0001        -0.6871      0.987
CRED              1       1   -0.00449   0.000548   67.08       <.0001        -0.2309      0.996
NPRVIO            2       1    0.2186    0.2848      0.59       0.4426         0.0805      1.244
NPRVIO            1       1    0.2240    0.0680     10.86       0.0010         0.0825      1.251
```

Display 6.12 shows that the **Regression** node has produced two sets of coefficients corresponding to the two equations for calculating the probabilities of the events $LOSSFRQ=2$ and $LOSSFRQ=1$. In contrast to the coefficients shown in Display 6.8, the two equations shown in Display 6.12 have different intercepts *and* different slopes (i.e., the coefficients of the variables AGE, $CRED$, and $NPRVIO$ are also different).

The **Regression** node gives the following formulas for calculating the probabilities of the events $LOSSFRQ=2$, $LOSSFRQ=1$, and $LOSSFRQ=0$.

$$\Pr(LOSSFRQ=2) = \frac{\exp(1.5946 - 0.0324*AGE - 0.0134*CRED + 0.2186*NPRVIO)}{1 + \exp(1.5946 - 0.0324*AGE - 0.0134*CRED + 0.2186*NPRVIO) + \exp(-0.1186 - 0.0228*AGE - 0.00449*CRED + 0.2240*NPRVIO)} \tag{6.10}$$

$$\Pr(LOSSFRQ=1) = \frac{\exp(-0.1186 - 0.0228*AGE - 0.00449*CRED + 0.2240*NPRVIO)}{1 + \exp(1.5946 - 0.0324*AGE - 0.0134*CRED + 0.2186*NPRVIO) + \exp(-0.1186 - 0.0228*AGE - 0.00449*CRED + 0.2240*NPRVIO)} \tag{6.11}$$

$$\Pr(LOSSFRQ=0) = \frac{1}{1 + \exp(1.5946 - 0.0324*AGE - 0.0134*CRED + 0.2186*NPRVIO) + \exp(-0.1186 - 0.0228*AGE - 0.00449*CRED + 0.2240*NPRVIO)} \tag{6.12}$$

Equation 6.12 implies that

$$\Pr(LOSSFRQ=0) = 1 - \Pr(LOSSFRQ=1) - \Pr(LOSSFRQ=2) \tag{6.13}$$

You can check the SAS code produced by the **Regression** node to verify that the model produced by the **Regression** node is the same as Equations 6.10, 6.11, 6.12, and 6.13.

In the results window, click on View→ Scoring→SAS Code to see the SAS code generated by the Regression Node. Displays 6.13A and 6.13B show segments of this code.

Display 6.13A

```
*** Compute Linear Predictor;
drop _TEMP;
drop _LP0 _LP1;
_LP0 = 0;
_LP1 = 0;

*** Effect: AGE ;
_TEMP = AGE ;
_LP0 = _LP0 + (    -0.0323912354634 * _TEMP);
_LP1 = _LP1 + (    -0.0228181055783 * _TEMP);

*** Effect: CRED ;
_TEMP = CRED ;
_LP0 = _LP0 + (   -0.01335715577514 * _TEMP);
_LP1 = _LP1 + (   -0.00448783417963 * _TEMP);

*** Effect: NPRVIO ;
_TEMP = NPRVIO ;
_LP0 = _LP0 + (    0.21864487608312 * _TEMP);
_LP1 = _LP1 + (    0.22396412290576 * _TEMP);

*** Naive Posterior Probabilities;
drop _MAXP _IY _P0 _P1 _P2;
drop _LPMAX;
_LPMAX= 0;
_LP0 =     1.59463711057944 + _LP0;
if _LPMAX < _LP0 then _LPMAX = _LP0;
_LP1 =    -0.11855123381699 + _LP1;
if _LPMAX < _LP1 then _LPMAX = _LP1;
_LP0 = exp(_LP0 - _LPMAX);
_LP1 = exp(_LP1 - _LPMAX);
_LPMAX = exp(-_LPMAX);
_P2 = 1 / (_LPMAX + _LP0 + _LP1);
_P0 = _LP0 * _P2;
_P1 = _LP1 * _P2;
_P2 = _LPMAX * _P2;
```

Display 6.13B

```
*** Posterior Probabilities and Predicted Level;
label P_LOSSFRQ2 = 'Predicted: LOSSFRQ=2' ;
label P_LOSSFRQ1 = 'Predicted: LOSSFRQ=1' ;
label P_LOSSFRQ0 = 'Predicted: LOSSFRQ=0' ;
P_LOSSFRQ2 = _P0;
_MAXP = _P0;
_IY = 1;
P_LOSSFRQ1 = _P1;
if (_P1 >  _MAXP + 1E-8) then do;
   _MAXP = _P1;
   _IY = 2;
end;
P_LOSSFRQ0 = _P2;
if (_P2 >  _MAXP + 1E-8) then do;
   _MAXP = _P2;
   _IY = 3;
end;
```

By setting Age=50, CRED=670, and NPRVIO=2, and by using the statements shown in Displays 6.13A and 6.13B, I generated the probabilities that $LOSSFRQ$ =0, 1, and 2, as shown in Display 6.14. The SAS code executed for generating these probabilities is given in Displays 6.79 and 6.80 in the appendix to this chapter.

Display 6.14

NOMINAL TAGET AGE=50,CRED=670 and NPRVIO=2		
Predicted: LOSSFRQ=2	Predicted: LOSSFRQ=1	Predicted: LOSSFRQ=0
.000191840	0.021487	0.97832

6.2.4 Models with Continuous Targets

When the target is continuous, the **Regression** node estimates a linear equation (linear in parameters). The score function is $E(Y / X) = \beta' X$, where Y is the target variable (such as expenditure, income, savings, etc.), X is a vector of inputs (explanatory variables), and β is the vector of coefficients estimated by the **Regression** node.

6.3 An Overview of Some Properties of the Regression Node

A clear understanding of **Regression Type, Link Function, Selection Model**, and **Selection Model Criterion** properties is essential for developing predictive models.

In this section, I present a description of these properties in detail.

6.3.1 Regression Type Property

The choices available for this property are Logistic Regression and Linear Regression.

6.3.1.1 Logistic Regression

If your target is categorical (binary, ordinal, or nominal), Logistic Regression is the default regression type. If the target is binary (either numeric or character), the **Regression** node gives you a Logistic Regression with logit link by default. If the target is categorical with more than two categories, and if its measurement scale (referred to as *level* in the Variables table of the **Input Data Source** node) is declared as ordinal, the **Regression** node by default gives you a model based on a cumulative logits link, as shown in Section 6.2.2. If the target has more than two categories, and if its measurement scale is set to Nominal, then the **Regression** node gives, by default, a model with a generalized logits link, as shown in Section 6.2.3.

6.3.1.2 Linear Regression

If your target is continuous or if its measurement scale is interval, then the **Regression** node gives you an ordinary least squares regression by default. A unit link function is used.

6.3.2 Link Function Property

In this section, I discuss various values to which you can set the **Link Function** property and how the **Regression** node calculates the predictions of the target for each value of the **Link Function** property. I start with an explanation of the theory behind the logit and probit link functions for a binary target. You may skip the theoretical explanation and go directly to the formula that the **Regression** node uses for calculating the target for each specified value of the **Link Function** property.

I start with the example of a response model, where the target variable takes one of the two values, response and non-response (1 and 0, respectively).

Assume that there is a latent variable y^* for each record or customer in the data set. Now assume that if the value of $y^* > 0$, then the customer responds. The value of y^* depends on the customer's characteristics as represented by the inputs or explanatory variables, which I can denote by the vector x. Assume that

$$y^* = \beta' x + U \tag{6.14}$$

where β is the vector of coefficients and U is a random variable. Different assumptions about the distribution of the random variable U give rise to different link functions. I will show how this is so.

The probability of response is

$$\Pr(y=1\,|\,x) = \Pr(y^* > 0\,|\,x) = \Pr(\beta'x + U > 0) = \Pr(U > -\beta'x) \qquad (6.15)$$

where y is the target variable.

Therefore,

$$\Pr(y=1\,|\,x) = \Pr(U > -\beta'x) = 1 - F(-\beta'x) \qquad (6.16)$$

where $F(.)$ is the Cumulative Distribution Function (CDF) of the random variable U.

6.3.2.1 Logit Link

If I choose the Logit value for the **Link Function** property, then the CDF above takes on the following value:

$$F(-\beta'x) = \frac{1}{1 + e^{\beta'x}} \quad \text{and} \quad 1 - F(-\beta'x) = 1 - \frac{1}{1 + e^{\beta'x}} = \frac{1}{1 + e^{-\beta'x}} \qquad (6.17)$$

Hence, the probability of response is calculated as

$$\Pr(y=1\,|\,x) = \frac{1}{1 + e^{-\beta'x}} = \frac{e^{\beta'x}}{1 + e^{\beta'x}} \qquad (6.18)$$

From Equation 6.18 you can see that the link function is

$$\log\left(\frac{\Pr(y=1\,|\,x)}{1 - \Pr(y=1\,|\,x)}\right) = \beta'x. \qquad (6.19)$$

$\beta'x$ is called a *linear predictor* since it is a linear combination of the inputs.

The **Regression** node estimates the linear predictor and then applies Equation 6.18 to get the probability of response.

6.3.2.2 Probit Link

If I choose the Probit value for the **Link Function** property, then it is assumed that the random variable U in Equation 6.14 has a normal distribution with mean=0 and standard deviation=1. In this case I have

$$\Pr(y=1\,|\,x) = 1 - F(-\beta'x) = F(\beta'x) \qquad (6.20)$$

$$\text{where } F(\beta'x) = \int_{-\infty}^{\beta'x} \frac{1}{\sqrt{2\pi}} e^{-u^2}\, du \qquad (6.21).$$

To estimate the probability of response, the **Regression** node uses the *probnorm* function to calculate the probability in Equation 6.20. Thus, it calculates the probability that y=1 as

$$\Pr(y=1\,|\,x) = probnorm(\beta'x), \text{ where the } probnorm \text{ function gives } F(\beta'x).$$

6.3.2.3 Complementary Log-Log Link (Cloglog)

In this case, the **Regression** node calculates the probability of response as

$$\Pr(y = 1 \mid x) = 1 - \exp(-\exp(\beta' x)) \tag{6.22}$$

which is the cumulative extreme value distribution.

6.3.2.4 Identity Link

When the target is interval-scaled, the **Regression** node uses an identity link. This means that the predicted value of the target variable is equal to the linear predictor $\beta' x$. As a result, the predicted value of the target variable is not restricted to the range 0 to 1.

6.3.3 Selection Model Property

The value of this property determines the model selection method used for selecting the variables for inclusion in a regression, whether it is a logistic or linear regression. You can set the value of this property to None, Backward, Forward, or Stepwise.

If you set the value to None, all inputs (sometimes referred to as *effects*) are included in the model, and there is only one step in the model selection process.

6.3.3.1 Backward Elimination Method

To use this method, set the **Selection Model** property to Backward, set the **Use Selection Defaults** property to **No**, and set the **Stay Significance Level** property to a desired threshold level of significance (such as 0.025, 0.05, 0.10, etc.) by opening the Selection Options window. Also, set the Maximum Number of Steps to a value such as 100 in the Selection Options window..

In this method, the process of model fitting starts with Step 0, where all inputs are included in the model. For each parameter estimated, a Wald Chi-Square test statistic is computed. In Step 1, the input whose coefficient is least significant and that also does not meet the threshold set by **Stay Significance Level** property in the Selection Options group (i.e., the p-value of its Chi-Square test statistic is greater than the value to which the **Stay Significance Level** property is set) is removed. Also, from the remaining variables, a new model is estimated and a Wald Chi-Square test statistic is computed for the coefficient of each input in the estimated model. In Step 2, the input whose coefficient is least significant (and that does not meet the **Stay Significance Level** threshold) is removed, and a new model is built. This process is repeated until no more variables fail to meet the threshold specified. Once a variable is removed at any step, it does not enter the equation again.

If this selection process takes ten steps (including Step 0) before it comes to a stop, then ten models are available when the process stops. From these ten models, the **Regression** node selects one model. The criterion used for the selection of a final model is determined by the value to which the **Selection Criterion** property is set. If the **Selection Criterion** property is set to None, then the final model selected is the one from the last step of the Backward Elimination process. If the **Selection Criterion** property is set to any value other than None, then the **Regression** node uses the criterion specified by the value of the **Selection Criterion** property to select the final model. Alternative values of the **Selection Criterion** property and a description of the procedures the **Regression** node follows to select the final model according to each of these criteria are discussed in Section 6.3.4.

6.3.3.1A Backward Elimination Method When the Target Is Binary

To demonstrate the backward selection method, I use a data set with only ten inputs and a binary target. Display 6.15 shows the process flow diagram for this demonstration.

Display 6.15

To demonstrate the backward selection method I used a data set with only ten inputs and a binary target shown in Display 6.16

Display 6.16

Name	Role	Level	Type	Number of Levels
CVAR001	Input	Nominal	Character	10
NVAR103	Input	Interval	Numeric	.
NVAR179	Input	Interval	Numeric	.
NVAR2	Input	Interval	Numeric	.
NVAR253	Input	Interval	Numeric	.
NVAR27	Input	Interval	Numeric	.
NVAR305	Input	Interval	Numeric	.
NVAR5	Input	Interval	Numeric	.
NVAR7	Input	Interval	Numeric	.
NVAR8	Input	Interval	Numeric	.
resp	Target	Binary	Numeric	2

Display 6.17 shows the property settings of the **Regression** node in the process flow shown in Display 6.15.

Display 6.17

Class Targets	
Regression Type	Logistic Regression
Link Function	Logit
Model Options	
Suppress Intercept	No
Input Coding	Deviation
Model Selection	
Selection Model	Backward
Selection Criterion	None
Use Selection Defaults	Yes
Selection Options	...

By setting the **Use Selection Defaults** properties to Yes, the value of the **Stay Significance Level** property is set to 0.05 by default. When the **Stay Significance Level** property is set to 0.05, the **Regression** node eliminates variables with a *p*-value greater than 0.05 in the backward elimination process. You can set the **Stay Significance Level** property to a value different from the default value of 0.05 by setting the **Use Selection Defaults** property to No, and then opening the Selection Options window by clicking ... next to the **Selection Options** property.

Display 6.18 shows the Selection Options window.

Display 6.18

Property	Value
Sequential Order	No
Entry Significance Level	0.05
Stay Significance Level	0.05
Start Variable Number	0
Stop Variable Number	0
Force Candidate Effects	0
Hierarchy Effects	Class
Moving Effect Rule	None
Maximum Number of Steps	100

Displays 6.19A, 6.19B, and 6.19C illustrate the backward selection method.

Display 6.19A shows that at Step 0, all of the ten inputs are included in the model.

Display 6.19A

```
Backward Elimination Procedure

Step 0: The following effects were entered.

Intercept CVAR001 NVAR103 NVAR179 NVAR2 NVAR253 NVAR27 NVAR305 NVAR5 NVAR7 NVAR8
```

Display 6.19B shows that in Step 1, the variable NVAR5 was removed because its Wald Chi-Square Statistic had the highest p-value (0.8540) (which is above the threshold significance level of 0.05 specified in the **Stay Significance Level** property). Next, a model with only nine inputs (all except NVAR5) was estimated, and a Wald Chi-Square statistic was computed for the coefficient of each input. Of the remaining variables, NVAR8 had the highest p-value for the Wald Test statistic, and so this input was removed in Step 2 and a logistic regression was estimated with the remaining eight inputs. This process of elimination continued until two more variables, NVAR2 and NVAR179, were removed at Steps 3 and 4 respectively.

Display 6.19B

```
NOTE: No (additional) effects met the 0.05 significance level for removal from the model.


             Summary of Backward Elimination

          Effect            Number      Wald
   Step    Removed    DF      In      Chi-Square   Pr > ChiSq

    1      NVAR5      1       9        0.0338       0.8540
    2      NVAR8      1       8        0.0750       0.7842
    3      NVAR2      1       7        0.1001       0.7517
    4      NVAR179    1       6        2.5626       0.1094
```

The Wald Test statistic was then computed for the coefficient of each input included in the model. At this stage, none of the p-values were above the threshold value of 0.05. Hence, the final model is the model developed at the last step, namely Step 4. Display 6.19C shows the variables included in the final model and shows that the p-values for all the included inputs are below the **Stay Significance Level** property value of 0.05.

Display 6.19C

```
The selected model is the model trained in the last step (Step 4).
It consists of the following effects:

Intercept CVAR001 NVAR103 NVAR253 NVAR27 NVAR305 NVAR7

    Likelihood Ratio Test for Global Null Hypothesis: BETA=0

   -2 Log Likelihood          Likelihood
   Intercept    Intercept &     Ratio
     Only       Covariates    Chi-Square    DF    Pr > ChiSq

   2163.746      1719.165      444.5805     14      <.0001

         Type 3 Analysis of Effects

                      Wald
   Effect     DF    Chi-Square   Pr > ChiSq

   CVAR001     9     91.5441      <.0001
   NVAR103     1    176.8076      <.0001
   NVAR253     1     50.4588      <.0001
   NVAR27      1     13.3377      0.0003
   NVAR305     1     18.5836      <.0001
   NVAR7       1      9.3743      0.0022
```

6.3.3.1B Backward Elimination Method When the Target Is Continuous

The backward elimination method for a continuous target is similar to that for a binary target. However, when the target is continuous, an *F* statistic is used instead of Wald Chi-Square statistic for calculating the *p*-values.

To demonstrate the backward elimination property for a continuous target, I used the process flow shown Display 6.20.

Display 6.20

As shown in Display 6.21, there are sixteen inputs in the modeling data set used in this example.

Display 6.21

Name	Role	Level	Type	Number of Levels
CVAR05	Input	Nominal	Character	6
CVAR14	Input	Binary	Character	2
CVR13	Input	Binary	Character	2
LDELBAL	Target	Interval	Numeric	.
NVAR009	Input	Interval	Numeric	.
NVAR027	Input	Interval	Numeric	.
NVAR048	Input	Interval	Numeric	.
NVAR051	Input	Interval	Numeric	.
NVAR054	Input	Interval	Numeric	.
NVAR077	Input	Interval	Numeric	.
NVAR084	Input	Nominal	Numeric	5
NVAR085	Input	Interval	Numeric	.
NVAR106	Input	Interval	Numeric	.
NVAR112	Input	Interval	Numeric	.
NVAR148	Input	Interval	Numeric	.
NVAR174	Input	Nominal	Numeric	5
NVAR190	Input	Interval	Numeric	.

Display 6.22 shows the properties of the **Regression** node in the process flow shown in Display 6.20.

Display 6.22

Open the Selection Options window (shown in Display 2.23) by clicking 	⬜.

Display 6.23

Property	Value
Sequential Order	No
Entry Significance Level	0.05
Stay Significance Level	0.025
Start Variable Number	0
Stop Variable Number	0
Force Candidate Effects	0
Hierarchy Effects	Class
Moving Effect Rule	None
Maximum Number of Steps	100

In Step 0 (See Display 6.24A), all 16 inputs were included in the model.

Display 6.24A

```
Backward Elimination Procedure

Step 0: The following effects were entered.

Intercept  CVAR05   CVAR14   CVR13    NVAR009   NVAR027   NVAR048
NVAR051    NVAR054  NVAR077  NVAR084  NVAR085   NVAR106
NVAR112    NVAR148  NVAR174  NVAR190
```

In the backward elimination process, four inputs were removed, one at each step. These results are shown in Display 6.24B.

Display 6.24B

```
NOTE: No (additional) effects met the 0.025 significance level
for removal from the model.

              Summary of Backward Elimination

             Effect              Number
     Step    Removed      DF        In     F Value    Pr > F

       1     NVAR112      1         15       2.23     0.1354
       2     NVAR077      1         14       2.59     0.1081
       3     NVAR106      1         13       3.91     0.0482
       4     CVR13        1         12       3.98     0.0465
```

Display 6.24C shows inputs included in the final model.

Display 6.24C

```
The selected model is the model trained in the last step (Step 4).
It consists of the following effects:

Intercept  CVAR05  CVAR14  NVAR009  NVAR027  NVAR048  NVAR051
           NVAR054  NVAR084  NVAR085  NVAR148  NVAR174  NVAR190

                          Analysis of Variance

                                    Sum of
Source               DF             Squares     Mean Square   F Value   Pr > F

Model                19           783.721209      41.248485     21.20   <.0001
Error               936          1821.252197       1.945782
Corrected Total     955          2604.973407

               Model Fit Statistics

R-Square      0.3009     Adj R-Sq       0.2867
AIC         656.1627     BIC          658.6388
SBC         753.4179     C(p)          28.7482

            Type 3 Analysis of Effects

                    Sum of
Effect       DF     Squares    F Value   Pr > F

CVAR05        4    24.3879       3.13    0.0142
CVAR14        1    73.8882      37.97    <.0001
NVAR009       1    54.0651      27.79    <.0001
NVAR027       1    12.8831       6.62    0.0102
NVAR048       1    42.9301      22.06    <.0001
NVAR051       1    10.3940       5.34    0.0210
NVAR054       1    27.0675      13.91    0.0002
NVAR084       3    23.1783       3.97    0.0079
NVAR085       1    10.1905       5.24    0.0223
NVAR148       1    10.6235       5.46    0.0197
NVAR174       3    19.4993       3.34    0.0188
NVAR190       1    33.1522      17.04    <.0001
```

Display 6.24C shows the variables included in the final model, along with the F statistics and p-values for the variable coefficients. The F statistics are partial F's since they are adjusted for the other effects already included in the model. In other words, the F statistic shown for each variable is based on a partial sum of squares, which measures the increase in the model sum of squares due to the addition of the variable to a model that already contains all the other main effects (11 in this case) plus any interaction terms (0 in this case). This type of analysis, where the contribution of each variable is adjusted for all other effects including the interaction effects, is called a Type 3 analysis.

At this point, it may be worth mentioning Type 2 sum of squares. According to Littell, Freund, and Spector, "Type 2 sum of squares for a particular variable is the increase in Model sum of square due to adding the variable to a model that already contains all other variables. If the model contains only main effects, then type 3 and type 2 analyses are the same."[2] For a discussion of partial sums of squares and different types of analysis, see *SAS System for Linear Models.*[3]

6.3.3.2 Forward Selection Method

To use this method, set the **Selection Model** property to Forward, set the **Use Selection Default** property to No, and set the **Entry Significance Level** property to a desired threshold level of significance, such as 0.025 0.05, etc.

6.3.3.2A Forward Selection Method When the Target Is Binary

At Step 0 of the forward selection process, the model consists of only the intercept and no other inputs. At the beginning of Step 1, the **Regression** node calculates the score Chi-Square statistic[4] for each variable not included in the model and selects the variable with the largest score statistic, provided that it also meets the **Entry Significance Level** threshold value. At the end of Step 1, the model consists of the intercept plus the first variable selected. In Step 2, the **Regression** node again calculates the score Chi-Square statistic for each variable not included in the model and similarly selects the best variable. This process is repeated until none of the remaining variables meets the **Entry Significance Level** threshold. Once a variable is entered into the model at any step, it is never removed.

To demonstrate the Forward Selection method, I used a data set with a binary target. The process flow diagram for this demonstration is shown in Display 6.25.

Display 6.25

In the **Regression** node, I set the **Selection Model** property to Forward, the **Entry Significance Level** property to 0.025, and the **Selection Criterion** property to None. Displays 6.26A, 6.26B, and 6.26C show how the Forward Selection method works when the target is binary.

Display 6.26A

```
Forward Selection Procedure


Step 0: Intercept entered.
```

Display 6.26B

```
NOTE: No (additional) effects met the 0.025 significance level
      for entry into the model.

                  Summary of Forward Selection

              Effect              Number         Score
      Step    Entered      DF       In        Chi-Square    Pr > ChiSq

        1     NVAR103       1        1         140.8246       <.0001
        2     CVAR001       9        2         154.9416       <.0001
        3     IMP_NVAR253   1        3          99.9514       <.0001
        4     IMP_NVAR305   1        4          44.0484       <.0001
        5     IMP_NVAR215   1        5          13.5667       0.0002
        6     IMP_NVAR100   1        6           9.3321       0.0023
        7     IMP_NVAR6     1        7           9.8942       0.0017
        8     IMP_NVAR179   1        8           8.7175       0.0032
        9     NVAR27        1        9          10.5958       0.0011
       10     IMP_NVAR306   1       10           7.9773       0.0047
       11     NVAR304       6       11          18.1012       0.0060
       12     IMP_NVAR286   1       12           6.5607       0.0104
       13     IMP_NVAR278   1       13           6.9671       0.0083
       14     IMP_CVAR6     1       14           6.1874       0.0129
       15     IMP_CVAR9     1       15           6.1878       0.0129
       16     IMP_NVAR177   1       16           5.0289       0.0249
```

Display 6.26C

```
The selected model is the model trained in the last step (Step 16).
It consists of the following effects:

Intercept  CVAR001  IMP_CVAR6  IMP_CVAR9  IMP_NVAR100  IMP_NVAR177
IMP_NVAR179  IMP_NVAR215  IMP_NVAR253  IMP_NVAR278  IMP_NVAR286
IMP_NVAR305  IMP_NVAR306  IMP_NVAR6  NVAR103  NVAR27  NVAR304

      Likelihood Ratio Test for Global Null Hypothesis: BETA=0

    -2 Log Likelihood           Likelihood
   Intercept    Intercept &       Ratio
      Only      Covariates     Chi-Square     DF     Pr > ChiSq

    2789.110     2268.000       521.1099       29      <.0001

         Type 3 Analysis of Effects

                          Wald
   Effect       DF    Chi-Square    Pr > ChiSq

   CVAR001       9     126.0418       <.0001
   IMP_CVAR6     1       5.8594       0.0155
   IMP_CVAR9     1       6.1195       0.0134
   IMP_NVAR100   1       6.8563       0.0088
   IMP_NVAR177   1       4.7098       0.0300
   IMP_NVAR179   1      11.8901       0.0006
   IMP_NVAR215   1       2.5535       0.1100
   IMP_NVAR253   1      67.4345       <.0001
   IMP_NVAR278   1       6.4460       0.0111
   IMP_NVAR286   1       9.7380       0.0018
   IMP_NVAR305   1       1.0760       0.2996
   IMP_NVAR306   1       8.0894       0.0045
   IMP_NVAR6     1      14.3459       0.0002
   NVAR103       1     134.5030       <.0001
   NVAR27        1      14.3750       0.0001
   NVAR304       6      11.1109       0.0850
```

Display 6.26C shows that twelve of the sixteen variables in the final model have *p*-values below the threshold level of 0.025. These variables were significant when they entered the equation (see Display 6.26B), and remained significant even after other variables entered the equation at the subsequent steps. However, the variables IMP_NVAR177, IMP_NVAR215, IMP_NVAR305, and IMP_NVAR304 were significant when they entered the equation (see Display 6.26B), but they became insignificant (their *p*-values are greater than 0.025) in the final model. You should examine these variables and determine whether they are collinear with other variables, have a highly skewed distribution, or both.

6.3.3.2B Forward Selection Method When the Target Is Continuous

Once again, at Step 0 of the forward selection process, the model consists only of the intercept. At Step 1, the **Regression** node calculates, for each variable not included in the model, the reduction in the sum of squared residuals that would result if that variable were included in the model. It then selects the variable that, when added to the model, results in the largest reduction in the sum of squared residuals, provided that the *p*-value of the *F* statistic for the variable is less than or equal to the **Entry Significance Level** threshold. At Step 2, the variable selected in Step 1 is added to the model, and the **Regression** node again calculates, for each variable not included in the model, the reduction in the sum of squared residuals that would result if that variable were included in the model. Again, the variable whose inclusion results in the largest reduction in the sum of squared residuals is added to the model, and so on, until none of the remaining variables meets the **Entry Significance Level** threshold value. As with the binary target, once a variable is entered into the model at any step, it is never removed.

Display 6.27 shows the process flow created for demonstrating the Forward Selection method when the target is continuous.

Display 6.27

The target variable in the **Input Data** shown in Display 6.27 is continuous. In the **Regression** node, I set the **Regression Type** property to Linear Regression, the **Selection Model** property to Forward, the **Selection Criterion** property to None, and the **Use Selection Defaults property** to Yes. The **Entry Significance Level** property is set to its default value of 0.05.

Displays 6.28A, 6.28B and 6.28C show how the Forward Selection method works when the target is continuous.

Display 6.28A

```
Forward Selection Procedure


Step 0: Intercept entered.
```

Display 6.28B

```
NOTE: No (additional) effects met the 0.05 significance level
      for entry into the model.

              Summary of Forward Selection

           Effect            Number
   Step    Entered      DF      In    F Value   Pr > F

     1     CVAR14       1       1     122.40    <.0001
     2     NVAR011      1       2      26.48    <.0001
     3     NVAR176      2       3      11.38    <.0001
     4     NVAR048      1       4       8.25    0.0042
     5     NVAR224      1       5       6.33    0.0121
     6     NVAR094      1       6       7.77    0.0055
     7     CVR13        1       7       6.33    0.0121
     8     NVAR188      3       8       2.83    0.0379
     9     NVAR152      1       9       4.17    0.0416
    10     NVAR193      1      10       3.86    0.0500
```

Display 6.28C

```
The selected model is the model trained in the last step (Step 10).
It consists of the following effects:

Intercept  CVAR14  CVR13  NVAR011  NVAR048  NVAR094  NVAR152  NVAR176
           NVAR188  NVAR193  NVAR224

                        Analysis of Variance

                            Sum of
Source              DF      Squares    Mean Square   F Value   Pr > F

Model               13    453.743241    34.903326     18.44    <.0001
Error              593   1122.585457     1.893061
Corrected Total    606   1576.328698

           Model Fit Statistics

R-Square      0.2878    Adj R-Sq      0.2722
AIC         401.2206    BIC         403.9726
SBC         462.9400    C(p)         12.0947

           Type 3 Analysis of Effect

                    Sum of
Effect      DF      Squares   F Value   Pr > F

CVAR14       1     34.1080    18.02    <.0001
CVR13        1     14.7836     7.81    0.0054
NVAR011      1     48.9905    25.88    <.0001
NVAR048      1     15.4707     8.17    0.0044
NVAR094      1     14.4990     7.66    0.0058
NVAR152      1      7.8196     4.13    0.0426
NVAR176      2     23.0487     6.09    0.0024
NVAR188      3     17.5634     3.09    0.0266
NVAR193      1      7.3049     3.86    0.0500
NVAR224      1     17.8924     9.45    0.0022
```

6.3.3.3 Stepwise Selection Method

To use this method, set the **Selection Model** property to Stepwise, the **Selection Default** property to No, the **Entry Significance Level** property to a desired threshold level of significance (such as 0.025, 0.05, 0.1, etc.) and the **Stay Significance Level** to a desired value (such as 0.025, 0.05, etc.).

The stepwise selection procedure consists of both forward and backward steps, although the backward steps may or may not yield any change in the forward path. The variable selection is similar to the forward selection method except that the variables already in the model do not necessarily remain in the model. When a new variable is entered at any step, some or all variables that are already included in the model may become insignificant. Such variables may be removed by the same procedures used in the backward method. Thus, variables are entered by forward steps and removed by backward steps. A forward step may be followed by more than one backward step.

6.3.3.3A Stepwise Selection Method When the Target Is Binary

Display 6.29 shows the process flow used for demonstrating the Stepwise Selection method.

Display 6.29

Display 6.30 shows a partial list of the variables in the input data set RESP_SMPL_CLEAN4_C.

Display 6.30

Name	Role	Level	Type	Number of Levels
NVAR92	Input	Interval	Numeric	.
NVAR93	Input	Interval	Numeric	.
NVAR94	Input	Interval	Numeric	.
NVAR95	Input	Interval	Numeric	.
NVAR96	Input	Nominal	Numeric	6
NVAR97	Input	Interval	Numeric	.
NVAR98	Input	Interval	Numeric	.
NVAR99	Input	Interval	Numeric	.
TYPE	Rejected	Unary	Numeric	1
matchkey	Rejected	Nominal	Character	21
nvar001	Input	Interval	Numeric	.
resp	Target	Binary	Numeric	2

As shown in Display 6.30, the target variable RESP is binary. Display 6.31 shows the property settings of the **Regression** node.

Display 6.31

Class Targets	
Regression Type	Logistic Regression
Link Function	Logit
Model Options	
Suppress Intercept	No
Input Coding	Deviation
Model Selection	
Selection Model	Stepwise
Selection Criterion	None
Use Selection Defaults	No
Selection Options	...
Optimization Options	

Display 6.32 shows the **Selection Options** properties.

Display 6.32

Property	Value
Sequential Order	No
Entry Significance Level	0.1
Stay Significance Level	0.025
Start Variable Number	0
Stop Variable Number	0
Force Candidate Effects	0
Hierarchy Effects	Class
Moving Effect Rule	None
Maximum Number of Steps	100

The results are shown in Displays 6.33A, 6.33B, and 6.33C.

Display 6.33A

```
Stepwise Selection Procedure


Step 0: Intercept entered.
```

Display 6.33B

```
NOTE: Model building terminates because the last effect entered is removed by the Wald test criterion.

                        Summary of Stepwise Selection

                  Effect                     Number        Score        Wald
      Step    Entered      Removed     DF       In      Chi-Square   Chi-Square    Pr > ChiSq

        1     NVAR103                   1        1        140.8246                   <.0001
        2     CVAR001                   9        2        154.9416                   <.0001
        3     IMP_NVAR253               1        3         99.9514                   <.0001
        4     IMP_NVAR305               1        4         44.0484                   <.0001
        5     IMP_NVAR215               1        5         13.5667                    0.0002
        6               IMP_NVAR215     1        4                      2.3657        0.1240
```

Display 6.33C

```
The selected model is the model trained in the last step (Step 6).
It consists of the following effects:

Intercept  CVAR001  IMP_NVAR253  IMP_NVAR305  NVAR103

    Likelihood Ratio Test for Global Null Hypothesis: BETA=0

    -2 Log Likelihood           Likelihood
  Intercept    Intercept &        Ratio
    Only        Covariates     Chi-Square      DF      Pr > ChiSq

  2789.110      2371.898        417.2117        12        <.0001

        Type 3 Analysis of Effects

                            Wald
  Effect        DF     Chi-Square     Pr > ChiSq

  CVAR001        9      124.1487       <.0001
  IMP_NVAR253    1       69.9897       <.0001
  IMP_NVAR305    1       35.9927       <.0001
  NVAR103        1      129.0491       <.0001
```

Comparing the variables shown in Display 6.33C with those shown in Display 6.26C, you can see that the stepwise selection method produced a slightly more parsimonious (and therefore better) solution than the forward selection method. The forward selection method selected 16 variables (Display 6.26C), while the stepwise selection method selected 4 variables.

6.3.3.3B Stepwise Selection Method When the Target Is Continuous

The stepwise procedure in the case of a continuous target is similar to the stepwise procedure used when the target is binary. In the forward step, the R-square, along with the F statistic, is used to evaluate the input variables. In the backward step, partial F statistics are used.

Display 6.34 shows the process flow for demonstrating the Stepwise Selection method when the target is continuous.

Display 6.34

Display 6.35 shows a partial list of variables included in the Input data set in the process flow diagram shown in Display 6.34.

Display 6.35

Name	Role	Level	Type
CVR08	Input	Nominal	Character
CVR13	Input	Binary	Character
LDELBAL	Target	Interval	Numeric
MATCHKEY	Rejected	Nominal	Character
NVAR001	Input	Interval	Numeric
NVAR002	Input	Interval	Numeric
NVAR003	Input	Interval	Numeric
NVAR004	Input	Interval	Numeric
NVAR005	Input	Interval	Numeric

Display 6.36 shows the property settings of the **Regression** node used in this demonstration.

Display 6.36

Class Targets	
Regression Type	Linear Regression
Link Function	Logit
Model Options	
Suppress Intercept	No
Input Coding	Deviation
Model Selection	
Selection Model	Stepwise
Selection Criterion	None
Use Selection Defaults	No
Selection Options	
Optimization Options	

In the Selection Options window, I set the **Entry Significance Level** property to 0.1, the **Stay Significance Level** property to 0.025, and the **Maximum Number of Steps** property to 100.

The results are shown in Displays 6.37A, 6.37B, and 6.37C.

Display 6.37A

```
Stepwise Selection Procedure

Step 0: Intercept entered.
                        Analysis of Variance

                            Sum of
Source              DF      Squares    Mean Square   F Value    Pr > F

Model                0            0                      .         .
Error              606    1576.328698   2.601202
Corrected Total    606    1576.328698

          Model Fit Statistics

R-Square      0.0000    Adj R-Sq      0.0000
AIC         581.2753    BIC         582.5935
SBC         585.6838    C(p)        225.0122

        Analysis of Maximum Likelihood Estimates

                             Standard
Parameter     DF    Estimate    Error    t Value    Pr > |t|

Intercept      1     9.0938     0.0655    138.92     <.0001

Step 1: Effect CVAR14 entered.
```

Display 6.37B

```
NOTE: Model building terminates because the last effect entered
      is removed by the Wald test criterion.

               Summary of Stepwise Selection

                  Effect                Number
       Step    Entered   Removed    DF    In    F Value    Pr > F

         1     CVAR14                1     1     122.40    <.0001
         2     NVAR011               1     2      26.48    <.0001
         3     NVAR176               2     3      11.38    <.0001
         4     NVAR048               1     4       8.25    0.0042
         5     NVAR224               1     5       6.33    0.0121
         6     NVAR094               1     6       7.77    0.0055
         7     CVR13                 1     7       6.33    0.0121
         8     NVAR188               3     8       2.83    0.0379
         9               NVAR188     3     7       2.83    0.0379
```

Display 6.37C

```
The selected model is the model trained in the last step (Step 9).
It consists of the following effects:

Intercept CVAR14 CVR13 NVAR011 NVAR048 NVAR094 NVAR176 NVAR224

                    Analysis of Variance

                         Sum of
Source              DF    Squares    Mean Square   F Value    Pr > F

Model                8   422.282026   52.785253    27.35    <.0001
Error              598  1154.046672    1.929844
Corrected Total    606  1576.328698

          Model Fit Statistics

R-Square      0.2679    Adj R-Sq      0.2581
AIC         407.9982    BIC         409.9786
SBC         447.6749    C(p)         18.6606

          Type 3 Analysis of Effects

                    Sum of
Effect        DF    Squares    F Value    Pr > F

CVAR14         1    42.2974     21.92     <.0001
CVR13          1    12.2252      6.33     0.0121
NVAR011        1    44.9943     23.31     <.0001
NVAR048        1    12.2179      6.33     0.0121
NVAR094        1    15.3324      7.94     0.0050
NVAR176        2    33.6566      8.72     0.0002
NVAR224        1    22.1318     11.47     0.0008
```

6.3.4 Selection Criterion Property[5]

In Section 6.3.3, I pointed out that during the backward, forward, and stepwise selection methods, the **Regression** node creates one model at each step. When you use any of these selection methods, the **Regression** node selects the final model based on the value to which the **Selection Criterion** property is set. If the **Selection Criterion** property is set to None, then the final model is simply the one created in the last step of the selection process. But any value other than None requires the **Regression** node to use the criterion specified by the value of the **Selection Criterion** property to select the final model.

You can set the value of the **Selection Criterion** property to any one of the following options:

- Akaike Information Criterion (AIC)
- Schwarz Bayesian Criterion (SBC)
- Validation Error
- Validation Misclassification
- Cross Validation Error
- Cross Validation Misclassification
- Validation Profit/Loss
- Profit/Loss
- Cross Validation Profit/Loss
- None

In this section, I demonstrate how these criteria work by using a small data set TestSelection, as shown in Display 6.38.

Display 6.38

For all of the examples shown in this section, in the Properties panel, I set the **Regression Type** property to Logistic Regression, the **Link Function** property to Logit, the **Selection Model** property to Backward, the **Stay Significance Level** property to 0.05, and the **Selection Criterion** property to one of the values listed above.

The Selection Options properties are shown in Display 6.39.

Display 6.39

Property	Value
Sequential Order	No
Entry Significance Level	0.05
Stay Significance Level	0.05
Start Variable Number	0
Stop Variable Number	0
Force Candidate Effects	0
Hierarchy Effects	Class
Moving Effect Rule	None
Maximum Number of Steps	100

To understand the Akaike Information Criterion and the Schwarz Bayesian Criterion, it is helpful to first review the computation of the errors in models with binary targets (such as response to a direct mail marketing campaign).

Let $\hat{\pi}(x_i)$ represent the predicted probability of response for the ith record, where x_i is the vector of inputs for the ith customer record. Let y_i represent the actual outcome for the ith record, where $y_i = 1$ if the customer responded and $y_i = 0$ otherwise. For a responder, the error is measured as the difference between the log of the actual value for y_i and the log of the predicted probability of response ($\hat{\pi}(x_i)$). Since $y_i = 1$,

$$Error =_i \log\left(\frac{y_i}{\hat{\pi}(x_i)}\right) = -\log\left(\frac{\hat{\pi}(x_i)}{y_i}\right) = -y_i \log\left(\frac{\hat{\pi}(x_i)}{y_i}\right).$$

For a non-responder the error is represented as

$$Error_i = -(1 - y_i)\log\left(\frac{1 - \hat{\pi}(x_i)}{1 - y_i}\right).$$

Combining both the response and non-response possible outcomes, the error for the ith customer record can be represented by

$$Error_i = -\left[y_i \log\left(\frac{\hat{\pi}_i(x_i)}{y_i}\right) + (1 - y_i)\log\left(\frac{1 - \hat{\pi}_i(x_i)}{1 - y_i}\right)\right].$$

Notice that the second term in the brackets goes to zero in the case of a responder. Hence, the first term gives the value of the error in that case. And in the case of a non-responder, the first term goes to zero and the error is given by the second term alone.

Summing the errors for all the records in the data set, I get

$$ErrorSum = -\sum_{i=1}^{n}\left[y_i \log\left(\frac{\hat{\pi}_i(x_i)}{y_i}\right) + (1 - y_i)\log\left(\frac{1 - \hat{\pi}_i(x_i)}{1 - y_i}\right)\right] \tag{6.23}$$

where n is the number of records in the data set.

In general, Equation 6.23 is used as the measure of model error for models with a binary target. By taking the anti-logs of Equation 6.23, we get

$$-\prod_{i=1}^{n}\frac{(\hat{\pi}_i(x_i))^{y_i}}{y_i^{y_i}}\frac{(1 - \hat{\pi}_i(x_i))^{1-y_i}}{(1 - y_i)^{1-y_i}} \tag{6.24}$$

which can be rewritten as

$$-\frac{\prod_{i=1}^{n}(\hat{\pi}_i(x_i))^{y_i}(1 - \hat{\pi}_i(x_i))^{1-y_i}}{\prod_{i=1}^{n}y_i^{y_i}(1 - y_i)^{1-y_i}} \tag{6.25}$$

The denominator of Equation 6.25 is 1, and the numerator is simply the likelihood of the model, often represented by the symbol L, so I can write

$$-\prod_{i=1}^{n} \left(\hat{\pi}_i(x_i)\right)^{y_i} \left(1-\hat{\pi}_i(x_i)\right)^{1-y_i} = -L \qquad (6.26)$$

Taking logs of the terms on both sides in Equation 6.27, I get

$$-LogL = -\sum_{i=1}^{n} \left[y_i \log\left(\hat{\pi}_i(x_i)\right) + (1-y_i)\log\left(1-\hat{\pi}_i(x_i)\right) \right] \qquad (6.27)$$

The expression given in Equation 6.27 is multiplied by 2 to obtain a quantity with a known distribution, giving the expression

$$-2LogL = -2\sum_{i=1}^{n} \left[y_i \log\left(\hat{\pi}_i(x_i)\right) + (1-y_i)\log\left(1-\hat{\pi}_i(x_i)\right) \right] \qquad (6.28)$$

The quantity shown in Equation 6.28 is used as a measure of model error for comparing models when the target is binary.

In the case of a continuous target, the model error is the sum of the squares of the errors of all the individual records of the data set, where the errors are the simple differences between the actual and predicted values.

6.3.4.1 Akaike Information Criterion (AIC)

When used for comparing different models, the Akaike Information Criterion (AIC) adjusts the quantity shown in Equation 6.28 in the following way.

The Akaike Information Criterion (AIC) statistic is

$$AIC = -2LogL + 2(k+s) \qquad (6.29)$$

where k is the number of response levels minus 1, and s is the number of explanatory variables in the model. The AIC statistic has two parts. The first part is $-2LogL$, which is a measure of model error, and it tends to decrease as we add more explanatory variables to the model. The second part $2(k+s)$ can be considered as a penalty term that depends on the number of explanatory variables included in the model. It increases with the number of explanatory variables added to the model. Inclusion of the penalty term is to reduce the chances of including too many explanatory variables, which may lead to over- fitting.

When I performed a model selection on my example data set (using backward elimination) and set the **Selection Criterion** property to **Akaike Information Criterion**, there were four steps in the model selection process. The model from Step 1 was selected since the Akaike Information Criterion statistic was the lowest at step1. Display 6.40 shows the Akaike Information Criterion Statistic at each step of the iteration.

Display 6.40

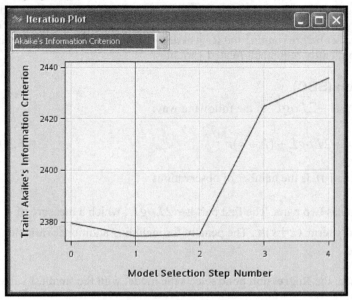

Displays 6.41 shows the Backward Elimination process and Display 6.42 shows the variables in the selected model.

Display 6.41

```
NOTE: No (additional) effects met the 0.05 significance level for removal from the model.

                          Summary of Backward Elimination

                                                                      Akaike
               Effect                 Number        Wald           Information
       Step    Removed       DF        In       Chi-Square   Pr > ChiSq   Criterion

        1      NVAR5         1          9          0.0150       0.9025      2374.8
        2      NVAR2         1          8          0.0639       0.8004      2375.6
        3      NVAR8         1          7          0.1023       0.7491      2424.9
        4      OPT_NVAR179   1          6          3.8337       0.0502      2435.9
```

Display 6.42

```
The selected model, based on the Akaike information criterion,
is the model trained in Step 1. It consists of the following effects:

Intercept  CVAR001  NVAR103  NVAR2  NVAR253  NVAR27  NVAR305  NVAR7
           NVAR8  OPT_NVAR179

        Likelihood Ratio Test for Global Null Hypothesis: BETA=0

     -2 Log Likelihood            Likelihood
     Intercept    Intercept &       Ratio
       Only       Covariates     Chi-Square      DF      Pr > ChiSq

     2163.746      1715.708       448.0378        17        <.0001

             Type 3 Analysis of Effects
                           Wald
     Effect        DF    Chi-Square    Pr > ChiSq

     CVAR001        9      92.6156       <.0001
     NVAR103        1     177.8976       <.0001
     NVAR2          1       0.0639       0.8004
     NVAR253        1      51.1330       <.0001
     NVAR27         1      14.0992       0.0002
     NVAR305        1      16.8802       <.0001
     NVAR7          1       9.1586       0.0025
     NVAR8          1       0.0820       0.7746
     OPT_NVAR179    1       3.8324       0.0503
```

You can see in Display 6.41 that the model created in Step 1 has the smallest value of AIC statistic. However, as you can see from Display 6.42, the selected model (created in Step 1) includes two variables (NVAR2 and NVAR8) that are not significant or have a high *p*-value. You can avoid this type of situation by eliminating (or combining) collinear variables from the data set before running the **Regression** node. You can use the **Variable Clustering** node, discussed in Chapter 2, to identify redundant/related variables.

6.3.4.2 Schwarz Bayesian Criterion (SBC)

The Schwarz Bayesian Criterion (SBC) adjusts $-2LogL$ in the following way:

$$SBC = -2LogL + (k + s)n \qquad (6.30)$$

where k and s are same as in Equation 6.29 and n is the number of observations.

Like the AIC statistic, the SBC statistic also has two parts. The first part is $-2LogL$, which a measure of model error, and the second part is the penalty term $(k+s)n$. The penalty for including additional variables is higher in SBC statistic than in AIC statistic.

When the Schwarz Bayesian Criterion is used, the **Regression** node selects the model with the smallest value for the Schwarz Bayesian Criterion statistic.

As Displays 6.43 and 6.44 show, the model selection process again took only four steps, but this time the model from Step 2 is selected.

Display 6.43 shows a plot of the SBC statistic at each step in the backward elimination process.

Display 6.43

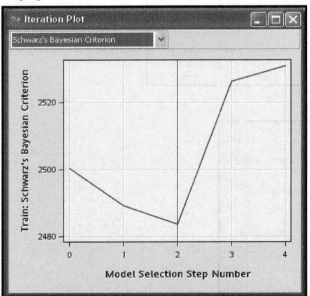

Display 6.44 shows the variables removed at successive steps of backward elimination process.

Display 6.44

```
NOTE: No (additional) effects met the 0.05 significance level for removal from the model.

                    Summary of Backward Elimination

                                                            Schwarz
            Effect              Number      Wald            Bayesian
    Step    Removed       DF      In      Chi-Square   Pr > ChiSq   Criterion

      1     NVAR5         1        9        0.0150       0.9025      2489.1
      2     NVAR2         1        8        0.0639       0.8004      2483.5
      3     NVAR8         1        7        0.1023       0.7491      2526.5
      4     OPT_NVAR179   1        6        3.8337       0.0502      2531.1
```

Display 6.45 shows the variables included in the selected model.

Display 6.45

```
The selected model, based on the Schwarz Bayesian criterion,
is the model trained in Step 2. It consists of the following effects:

Intercept  CVAR001  NVAR103  NVAR253  NVAR27  NVAR305  NVAR7  NVAR8
           OPT_NVAR179

    Likelihood Ratio Test for Global Null Hypothesis: BETA=0

    -2 Log Likelihood            Likelihood
  Intercept    Intercept &         Ratio
    Only       Covariates       Chi-Square     DF     Pr > ChiSq

  2163.746     1715.772          447.9731       16       <.0001

        Type 3 Analysis of Effects

                        Wald
  Effect     DF     Chi-Square   Pr > ChiSq

  CVAR001     9       92.5686      <.0001
  NVAR103     1      177.8362      <.0001
  NVAR253     1       51.1887      <.0001
  NVAR27      1       14.0552       0.0002
  NVAR305     1       16.8653      <.0001
  NVAR7       1        9.1576       0.0025
  NVAR8       1        0.1023       0.7491
  OPT_NVAR179 1        3.8386       0.0501
```

6.3.4.3 Validation Error

In the case of a binary target, validation error is the error calculated by the negative log likelihood expression shown in Equation 6.28. The error is calculated from the Validation data set for each of the models generated at the various steps of the selection process. The parameters of each model are used to calculate the predicted probability of response for each record in the data set. In Equation 6.28 this is shown as $\hat{\pi}(x_i)$, where x_i is the vector of inputs for the ith record in the Validation data set. (In the case of a continuous target, the validation error is the sum of the squares of the errors calculated from the Validation data set, where the errors are the simple differences between the actual and predicted values for the continuous target.) The model with the smallest validation error is selected, as shown in Displays 6.46 and 6.47.

Display 6.46

```
NOTE: No (additional) effects met the 0.05 significance level for removal from the model.

                    Summary of Backward Elimination

            Effect              Number      Wald            Validation
    Step    Removed       DF      In      Chi-Square   Pr > ChiSq   Error Rate

      1     NVAR5         1        9        0.0150       0.9025      1782.5
      2     NVAR2         1        8        0.0639       0.8004      1782.6
      3     NVAR8         1        7        0.1023       0.7491      1761.1
      4     OPT_NVAR179   1        6        3.8337       0.0502      1759.7
```

Display 6.47

```
The selected model, based on the error rate for the validation data,
is the model trained in Step 4. It consists of the following effects:

Intercept CVAR001 NVAR103 NVAR253 NVAR27 NVAR305 NVAR7

    Likelihood Ratio Test for Global Null Hypothesis: BETA=0

  -2 Log Likelihood           Likelihood
 Intercept     Intercept &       Ratio
    Only        Covariates     Chi-Square      DF      Pr > ChiSq

   2163.746       1719.165       444.5805       14        <.0001

        Type 3 Analysis of Effects

                            Wald
 Effect          DF     Chi-Square    Pr > ChiSq

 CVAR001          9       91.5441       <.0001
 NVAR103          1      176.8076       <.0001
 NVAR253          1       50.4588       <.0001
 NVAR27           1       13.3377       0.0003
 NVAR305          1       18.5836       <.0001
 NVAR7            1        9.3743       0.0022
```

6.3.4.4 Validation Misclassification

For analyses with a binary target such as response, a record in the Validation data set is classified as a responder if the posterior probability of response is greater than the posterior probability of non-response. Similarly, if the posterior probability of non-response is greater than the posterior probability of response, the record is classified as a non-responder. These posterior probabilities are calculated for each customer in the Validation data set using the parameters estimated for each model generated at each step of the model-selection process. Since the validation data includes the actual customer response for each record, you can check whether the classification is correct or not. If a responder is correctly classified as a responder, or a non-responder is correctly classified as a non-responder, then there is no misclassification; otherwise, there is a misclassification. The error rate for each model can be calculated by dividing the total number of misclassifications in the validation data set by the total number of records. The validation misclassification rate is calculated for each model generated at the various steps of the selection process, and the model with the smallest validation misclassification rate is selected.

Display 6.48 shows the model selected by the **Validation Misclassification** criterion.

Display 6.48

```
NOTE: No (additional) effects met the 0.05 significance level for removal from the model.

                    Summary of Backward Elimination
                                                            Validation
            Effect              Number      Wald          Misclassification
 Step      Removed      DF       In     Chi-Square  Pr > ChiSq     Rate
   1       NVAR5         1        9       0.0150      0.9025       0.1003
   2       NVAR2         1        8       0.0639      0.8004       0.1003
   3       NVAR8         1        7       0.1023      0.7491       0.1000
   4       OPT_NVAR179   1        6       3.8337      0.0502       0.0991

The selected model, based on the misclassification rate for the validation data,
is the model trained in Step 4. It consists of the following effects:
    Intercept CVAR001 NVAR103 NVAR253 NVAR27 NVAR305 NVAR7

    Likelihood Ratio Test for Global Null Hypothesis: BETA=0

  -2 Log Likelihood           Likelihood
 Intercept     Intercept &       Ratio
    Only        Covariates     Chi-Square      DF      Pr > ChiSq

   2163.746       1719.165       444.5805       14        <.0001

        Type 3 Analysis of Effects

                            Wald
 Effect          DF     Chi-Square    Pr > ChiSq

 CVAR001          9       91.5441       <.0001
 NVAR103          1      176.8076       <.0001
 NVAR253          1       50.4588       <.0001
 NVAR27           1       13.3377       0.0003
 NVAR305          1       18.5836       <.0001
 NVAR7            1        9.3743       0.0022
```

6.3.4.5 Cross Validation Error

In SAS Enterprise Miner, cross validation errors are calculated from the Training data set. One observation is omitted at a time, and the model is re-estimated with the remaining observations. The re-estimated model is then used to predict the value of the target for the omitted observation, and then the error of the prediction, in the case of a binary target, is calculated using the formula given in Equation 6.28 with $n=1$. (In the case of a continuous target, the error is the squared difference between the actual and predicted values.) Next, another observation is omitted, the model is re-estimated and used to predict the omitted observation, and this new prediction error is calculated. This process is repeated for each observation in the Training data set, and the aggregate error is found by summing the individual errors.

If you do not have enough observations for training and validation, then you can select the **Cross Validation Error** criterion for model selection. The data sets I used have several thousand observations. Therefore, in my example, it may take a long time to calculate the cross validation errors, especially since there are several models. Hence this criterion is not tested here.

6.3.4.6 Cross Validation Misclassification Rate

The cross validation misclassification rate is also calculated from the Training data set. As in the case described in Section 6.3.4.5, one observation is omitted at a time, and the model is re-estimated every time. The re-estimated model is used to calculate the posterior probability of response for the omitted observation. The observation is then assigned a target level (such as response or non-response) based on maximum posterior probability. SAS Enterprise Miner then checks the actual response of the omitted customer to see if this re-estimated model misclassifies the omitted customer. This process is repeated for each of the observations, after which the misclassifications are counted, and the misclassification rate is calculated. For the same reason cited in Section 6.3.4.5, I have not tested this criterion.

6.3.4.7 The Validation Profit/Loss Criterion

In order to use the **Validation Profit/Loss** criterion, you must provide SAS Enterprise Miner with a profit matrix. To demonstrate this method, I have provided the profit matrix shown in Display 6.49.

Display 6.49

When calculating the profit/loss associated with a given model, the Validation data set is used. From each model produced by the backward elimination process, a formula for calculating posterior probabilities of response and non-response is generated in this example. Using these posterior probabilities, the expected profits under Decision1 (assigning the record to the responder category) and Decision2 (assigning the record to the non-responder category) are calculated. If the expected profit under Decision1 is greater than the expected profit under Decision2, then the record is assigned the target level responder; otherwise, it is set to non-responder. Thus, each record in the validation data set is assigned to a target level.

The following equations show how the expected profit is calculated, given the posterior probabilities calculated from a model and the profit matrix shown in Display 6.49.

The expected profit under Decision1 is

$$E_{1i} = \$10 * \hat{\pi}(x_i) + (-\$1)\left(1 - \hat{\pi}(x_i)\right) \qquad (6.31)$$

where E_{1i} is the expected profit under Decision1 for the ith record, $\hat{\pi}(x_i)$ is the posterior probability of response of the ith record, and x_i is the vector of inputs for the ith record in the validation data set.

The expected profit under Decision2 (assigning the record to the target level non-responder) is

$$E_{2i} = \$0 * \hat{\pi}(x_i) + \$0\left(1 - \hat{\pi}(x_i)\right) \qquad (6.32)$$

which is always zero in this case, given the profit matrix I have chosen.

The rule for classifying customers is:

If $E_{1i} > E_{2i}$, then assign the record to the target level responder. Otherwise, assign the record to the target level non-responder.

Having assigned a target category such as responder or non-responder to each record in the Validation data set, you can examine whether the decision taken is correct by comparing the assignment to the actual value. If a customer record is assigned the target class of responder and the customer is truly a responder, then the company has made a profit of $10. This is the actual profit that I would have made—not the expected profit calculated before. If, on the other hand, we assigned the target class responder and the customer turned out to be a non-responder, then I incur a loss of $1. According to the profit matrix given in Display 6.49, there is no profit or loss if I assigned the target level non-responder to a true responder or a target level of non-responder to a true non-responder. Thus, based on the decision and the actual outcome for each record, I can calculate the actual profit for each customer in the Validation data set. And by summing the profits across all of the observations, I obtain the validation profit/loss. These calculations are made for each model generated at various iterations of the model selection process. The model that yields the highest profit is selected.

Display 6.50 shows the model selected by this criterion for my example problem.

Display 6.50

```
                    Summary of Backward Elimination

          Effect                Number        Wald              Validation
   Step    Removed       DF        In    Chi-Square   Pr > ChiSq    Profit
    1      NVAR5          1         9       0.0150       0.9025      1393.0
    2      NVAR2          1         8       0.0639       0.8004      1417.0
    3      NVAR8          1         7       0.1023       0.7491      1487.0
    4      OPT_NVAR179    1         6       3.8337       0.0502      1464.0

The selected model, based on the total profit for the validation data,
is the model trained in Step 3. It consists of the following effects:

Intercept  CVAR001  NVAR103  NVAR253  NVAR27  NVAR305  NVAR7  OPT_NVAR179

    Likelihood Ratio Test for Global Null Hypothesis: BETA=0

   -2 Log Likelihood           Likelihood
   Intercept    Intercept &      Ratio
     Only       Covariates    Chi-Square      DF     Pr > ChiSq

   2163.746      1715.876      447.8696        15       <.0001

        Type 3 Analysis of Effects

                         Wald
   Effect        DF   Chi-Square   Pr > ChiSq

   CVAR001        9     92.4868      <.0001
   NVAR103        1    177.9873      <.0001
   NVAR253        1     51.2445      <.0001
   NVAR27         1     13.9899      0.0002
   NVAR305        1     16.8519      <.0001
   NVAR7          1      9.1248      0.0025
   OPT_NVAR179    1      3.8337      0.0502
```

6.3.4.8 Profit/Loss Criterion

When this criterion is selected, the profit/loss is calculated in the same way as the validation profit/lLoss above, but it is done using the training data set instead of the validation data. Again, you resort to this option only when there is insufficient data to create separate Training and Validation data sets. Display 6.51 shows the results of applying this option to my example data.

Display 6.51

```
                   Summary of Backward Elimination

             Effect              Number      Wald
    Step     Removed       DF      In     Chi-Square    Pr > ChiSq    Profit

      1      NVAR5         1       9       0.0150        0.9025       1968.0
      2      NVAR2         1       8       0.0639        0.8004       1961.0
      3      NVAR8         1       7       0.1023        0.7491       1900.0
      4      OPT_NVAR179   1       6       3.8337        0.0502       1879.0

The selected model, based on the total profit for the training data,
is the model trained in Step 1. It consists of the following effects:

Intercept  CVAR001  NVAR103  NVAR2  NVAR253  NVAR27  NVAR305  NVAR7  NVAR8
           OPT_NVAR179

       Likelihood Ratio Test for Global Null Hypothesis: BETA=0

    -2 Log Likelihood          Likelihood
    Intercept   Intercept &      Ratio
      Only      Covariates    Chi-Square     DF     Pr > ChiSq

    2163.746    1715.708      448.0378       17      <.0001

            Type 3 Analysis of Effects

                          Wald
    Effect       DF    Chi-Square    Pr > ChiSq

    CVAR001      9      92.6156       <.0001
    NVAR103      1     177.8976       <.0001
    NVAR2        1       0.0639       0.8004
    NVAR253      1      51.1330       <.0001
    NVAR27       1      14.0992       0.0002
    NVAR305      1      16.8802       <.0001
    NVAR7        1       9.1586       0.0025
    NVAR8        1       0.0820       0.7746
    OPT_NVAR179  1       3.8324       0.0503
```

6.3.4.9 The Cross Validation Profit/Loss Criterion

The process by which the **Cross Validation Profit/Loss** criterion is calculated is similar to the cross validation misclassification rate method described in Section 6.3.4.6. At each step, one observation is omitted, the model is re-estimated, and the re-estimated model is used to calculate posterior probabilities of the target levels (such as response or no-response) for the omitted observation. Based on the posterior probability and the profit matrix that is specified, the omitted observation is assigned a target level in such a way that the expected profit is maximized for that observation, as described in Section 6.3.4.7. Actual profit for the omitted record is calculated by comparing the actual target level and the assigned target level. This process is repeated for all the observations in the data set. By summing the profits across all of the observations, I obtain the cross validation profit/loss. These calculations are made for each model generated at various steps of the model selection process. The model that yields the highest profit is selected. The selected model is shown in Display 6.52

Display 6.52

```
                    Summary of Backward Elimination
                                                              Cross-
              Effect              Number      Wald           Validation
      Step    Removed       DF      In    Chi-Square  Pr > ChiSq  Profit

        1     NVAR5          1       9      0.0150     0.9025     1755.4
        2     NVAR2          1       8      0.0639     0.8004     1758.8
        3     NVAR8          1       7      0.1023     0.7491     1708.3
        4     OPT_NVAR179    1       6      3.8337     0.0502     1701.7
```

The selected model, based on the cross-validation total profit,
is the model trained in Step 2. It consists of the following effects:

Intercept CVAR001 NVAR103 NVAR253 NVAR27 NVAR305 NVAR7 NVAR8 OPT_NVAR179

```
        Likelihood Ratio Test for Global Null Hypothesis: BETA=0

       -2 Log Likelihood          Likelihood
   Intercept      Intercept &        Ratio
      Only        Covariates      Chi-Square     DF    Pr > ChiSq

    2163.746       1715.772        447.9731      16      <.0001
```

```
            Type 3 Analysis of Effects

                          Wald
   Effect       DF    Chi-Square    Pr > ChiSq

   CVAR001       9      92.5686      <.0001
   NVAR103       1     177.8362      <.0001
   NVAR253       1      51.1887      <.0001
   NVAR27        1      14.0552      0.0002
   NVAR305       1      16.8653      <.0001
   NVAR7         1       9.1576      0.0025
   NVAR8         1       0.1023      0.7491
   OPT_NVAR179   1       3.8386      0.0501
```

6.4 Business Applications

The purpose of this section is to illustrate the use of the **Regression** node with two business applications, one requiring a binary target, and the other a continuous target.

The first application is concerned with identifying the current customers of a hypothetical bank who are most likely to respond to an invitation to sign up for Internet banking that carries a monetary reward with it. A pilot study was already conducted in which a sample of customers were sent the invitation. The task of the modeler is to use the results of the pilot study to develop a predictive model that can be used to score all customers and send mail to those customers who are most likely to respond (a *binary* target).

The second application is concerned with identifying customers of a hypothetical bank who are likely to increase their savings deposits by the largest dollar amounts if the bank increases the interest paid to them by a preset number of basis points. A sample of customers was tested first. Based on the results of the test, the task of the modeler is to develop a model to use for predicting the amount each customer would increase his savings deposits (a *continuous* target), given the increase in the interest paid. (Had the test also included several points representing decrease, increase, and no change from the current rate, then the estimated model could be used to test the price sensitivity in any direction. However, in the current study, the bank is interested only in the consequences of a rate increase to customer's savings deposits.)

Before using the **Regression** node for modeling, you need to clean the data. This very important step includes purging some variables, imputing missing values, transforming the inputs, and checking for spurious correlation between the inputs and the target. Spurious correlations arise when input variables are derived in part from the target variables, generated from target variable behavior, or are closely related to the target variable. Post-event inputs (inputs recorded after the event being modeled had occurred) might also cause spurious correlations.

I performed a prior examination of the variables using the **StatExplore** and **MultiPlot** nodes, and excluded some variables, such as the one shown Displays 6.53, from the data set.

Display 6.53

The FREQ Procedure			
Frequency **Row Pct**	**Table of CVAR12 by resp**		
		resp	
CVAR12	0	1	Total
N	1212	0	1212
	100.00	0.00	
Y	8278	1079	9357
	88.47	11.53	
Total	9490	1079	10569

In Display 6.53, the variable CVAR12 is a categorical input, taking the values Y and N. A look at the distribution of the variable shown in the 2x2 table suggests that it might be derived from the target variable or connected to it somehow, or it might be an input with an uneven spread across the target levels. Variables such as these can produce undesirable results, and they should be excluded from the model if their correlation with the target is truly spurious. However, there is also the possibility that this variable has valuable information for predicting the target. Clearly, only a modeler who knows where his data came from, how the data were constructed, and what each variable represents in terms of the characteristics and behaviors of customers will be able to detect spurious correlations. As a precaution before applying the modeling tool, you should examine every variable included in the final model (at a minimum) for this type of problem.

In order to demonstrate alternative types of transformations, I present three models for each type of target that is discussed. The first model is based on untransformed inputs. This type of model can be used to make a preliminary identification of important inputs. For the second model, the transformations are done by the **Transform Variables** node, and for the third model transformations are done by the **Decision Tree** node.

The **Transform Variables** node has a variety of transformations (discussed in detail in Section 2.9.6 of Chapter 2 and in Section 3.5 of Chapter 3). I use some of these transformation methods here for demonstration purposes.

The transformations produced by the **Decision Tree** node yield a new categorical variable, and this happens in the following way. First, a Decision Tree model is built based on the inputs in order to predict the target (or classify records). Next, a categorical variable is created, where the value (category) of the variable for a given record is simply the leaf to which the record belongs. The **Regression** node then uses this categorical variable as a class input. The **Regression** node creates a dummy variable for each category and uses it in the regression equation. These dummy variables capture the interactions between different inputs included in the tree, which are potentially very important.

If there are a large number of variables, you can use the **Variable Selection** node to screen important variables prior to using the **Regression** node. However, since the **Regression** node itself has several methods of variable selection (as shown in Sections 6.3.3 and 6.3.4), and since a moderate number of inputs is contained in each of my data sets, I did not use the **Variable Selection** node prior to using the **Regression** node in these examples.

Two data sets are used in this chapter: one for modeling the binary target and the other for a modeling the continuous target. Both of these are simulated. The results may therefore seem too good to be true at times, and not so good at other times.

6.4.1 Logistic Regression for Predicting Response to a Mail Campaign

For predicting response, I explore three different ways of developing a logistic regression model. The models developed are Model_B1, Model_B2, and Model_B3. The process flow for each model is shown in Display 6.54.

Display 6.54

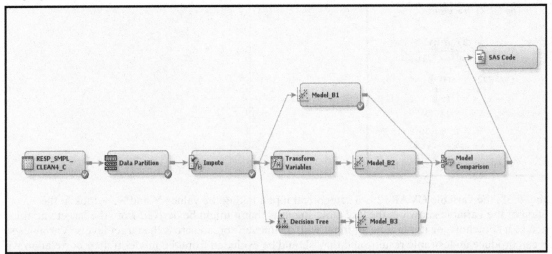

In the process flow, the first node is the **Input Data Source** node that reads the modeling data. From this modeling data set, the **Data Partition** node creates three data sets Part_Train, Part_Validate, and Part_Test according to the **Training**, **Validation,** and **Test** properties, which I set at 40%, 30%, and 30%, respectively. These data sets, with the roles of Train, Validate and Test, are passed to the **Impute** node. In the **Impute** node, I set the **Default Input Method** property to Median for the interval variables, and I set it to **Count** for the class variables. The **Impute** node creates three data sets Impt_Train, Impt_Validate, and Impt_Test including the imputed variables, and passes them to the next node. The last node in the process flow shown in Display 6.54 is the **Model Comparison** node. This node generates various measures such as lift, capture rate, etc., from the training, validation, and test data sets.

6.4.1.1 Model_B1

The process flow for Model_B1 is at the top segment of Display 6.54. The **Regression** node uses the data set Impt_Train for effect selection. Each step of the effects selection method generates a model. The Validation data set is used by the **Regression** node to select the best model from those generated at several steps of the selection process. Finally, the data set Impt_Test is used to get an independent assessment of the model. This is done in the **Model Comparison** node at the end of the process flow.

Display 6.55 shows the settings of the **Regression** node properties for Model_B1.

Display 6.55

Class Targets	
Regression Type	Logistic Regression
Link Function	Logit
Model Options	
Suppress Intercept	No
Input Coding	Deviation
Model Selection	
Selection Model	Stepwise
Selection Criterion	Validation Error
Use Selection Defaults	No
Selection Options	...

By opening the Results window and scrolling through the output, you can see the models created at different steps of the effect selection process and the final model selected from these steps. In this example, the **Regression** node took six steps to complete the effect selection. The model created at Step 4 is selected as the final model since it minimized the validation error.

Display 6.56 shows the variables included in the final model.

Display 6.56

```
The selected model, based on the error rate for the validation data,
is the model trained in Step 4. It consists of the following effects:

Intercept  CVAR001  IMP_NVAR253  IMP_NVAR305  NVAR103

      Likelihood Ratio Test for Global Null Hypothesis: BETA=0

     -2 Log Likelihood             Likelihood
     Intercept      Intercept &       Ratio
        Only        Covariates     Chi-Square      DF      Pr > ChiSq

     2789.110        2371.898        417.2117       12        <.0001

            Type 3 Analysis of Effects

                            Wald
     Effect         DF   Chi-Square   Pr > ChiSq

     CVAR001         9    124.1487      <.0001
     IMP_NVAR253     1     69.9897      <.0001
     IMP_NVAR305     1     35.9927      <.0001
     NVAR103         1    129.0491      <.0001
```

Display 6.56 shows a partial view of the output in the Results window. Display 6.57 shows the lift charts for the final Model_B1. These charts are displayed under **Score Rankings Overlay** in the Results window.

Display 6.57

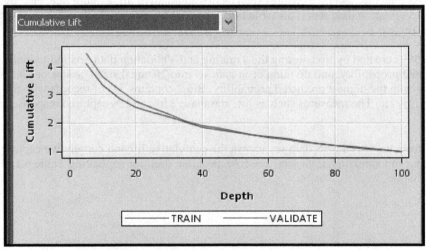

If you click anywhere under **Score Rankings Overlay** in the Results window and select the **Table** icon, you will see the table shown in Display 6.58.

Display 6.58

Data Role	Event	Bin	Depth	% Response	Cumulative % Response	% Captured Response	Cumulative % Captured Response	Lift	Cumulative Lift	Gain
TRAIN	1	1	5	41.98113	41.98113	20.60185	20.60185	4.108709	4.108709	310.8709
TRAIN	1	2	10	26.06835	34.04255	12.73148	33.33333	2.551123	3.331757	233.1757
TRAIN	1	3	15	19.33962	29.13386	9.490741	42.82407	1.892776	2.851341	185.1341
TRAIN	1	4	20	17.06161	26.12293	8.333333	51.15741	1.669826	2.556661	155.6661
TRAIN	1	5	25	16.11374	24.12488	7.87037	59.02778	1.577058	2.361111	136.1111
TRAIN	1	6	30	14.15094	22.45863	6.944444	65.97222	1.384958	2.198034	119.8034
TRAIN	1	7	35	9.952607	20.67568	4.861111	70.83333	0.974065	2.023536	102.3536
TRAIN	1	8	40	7.075472	18.97163	3.472222	74.30556	0.692479	1.856761	85.67606
TRAIN	1	9	45	9.952607	17.97162	4.861111	79.16667	0.974065	1.758889	75.88895
TRAIN	1	10	50	6.161137	16.79281	3.009259	82.17593	0.602993	1.643519	64.35185
TRAIN	1	11	55	9.433962	16.1221	4.62963	86.80556	0.923305	1.577876	57.78757
TRAIN	1	12	60	7.582938	15.4119	3.703704	90.50926	0.742145	1.508369	50.83687
TRAIN	1	13	65	5.660377	14.65988	2.777778	93.28704	0.553983	1.434768	43.47675
TRAIN	1	14	70	3.317536	13.85135	1.62037	94.90741	0.324688	1.355637	35.56369
TRAIN	1	15	75	1.421801	13.02428	0.694444	95.60185	0.139152	1.274691	27.46914
TRAIN	1	16	80	1.415094	12.29678	0.694444	96.2963	0.138496	1.20349	20.34902
TRAIN	1	17	85	2.369668	11.71397	1.157407	97.4537	0.23192	1.14645	14.64504
TRAIN	1	18	90	0.943396	11.11403	0.462963	97.91667	0.092331	1.087734	8.773428
TRAIN	1	19	95	2.369668	10.65472	1.157407	99.07407	0.23192	1.042781	4.278114
TRAIN	1	20	100	1.895735	10.2176	0.925926	100	0.185536	1	0
VALIDATE	1	1	5	45.91195	45.91195	22.53086	22.53086	4.492002	4.492002	349.2002
VALIDATE	1	2	10	28.48101	37.22397	13.88889	36.41975	2.786568	3.641975	264.1975
VALIDATE	1	3	15	22.01258	32.14286	10.80247	47.22222	2.1537	3.144841	214.4841

The table shown in Display 6.58 contains several measures such as lift, cumulative lift, capture rate, etc., by percentile for the Training and Validation data sets. This table is saved is a SAS data set, and you can use it for making custom tables such as Table 6.1.

The table shown in Display 6.58 is created by first sorting the Training and Validation data sets individually in descending order of the predicted probability, and dividing each data set into 20 equal sized groups or bins, such that Bin 1 contains 5% records with the highest predicted probability, Bin 2 contains 5% of records with the next highest predicted probability, etc. The measures such as lift, cumulative lift, and % captured response, etc. are calculated for each bin.

Table 6.1, which was created from the saved SAS data set, shows the cumulative lift and cumulative capture rate by bin/depth for the training and validation data sets. The SAS program used for generating Table 6.1 is given in the appendix to this chapter.

Table 6.1

			Model_B1		
Bin	Percentile	Cumulative Lift Training	Cumulative Lift Validation	Cumulative % Captured Response Training	Cumulative % Captured Response Validation
1	5	4.10871	4.49200	20.602	22.531
2	10	3.33176	3.64198	33.333	36.420
3	15	2.85134	3.14484	42.824	47.222
4	20	2.55666	2.74691	51.157	54.938
5	25	2.36111	2.50459	59.028	62.654
6	30	2.19803	2.29424	65.972	68.827
7	35	2.02354	2.05375	70.833	71.914
8	40	1.85676	1.91358	74.306	76.543
9	45	1.75889	1.81007	79.167	81.481
10	50	1.64352	1.69753	82.176	84.877
11	55	1.57788	1.57082	86.806	86.420
12	60	1.50837	1.48663	90.509	89.198
13	65	1.43477	1.41466	93.287	91.975
14	70	1.35564	1.33598	94.907	93.519
15	75	1.27469	1.26311	95.602	94.753
16	80	1.20349	1.21142	96.296	96.914
17	85	1.14645	1.16173	97.454	98.765
18	90	1.08773	1.11111	97.917	100.000
19	95	1.04278	1.05246	99.074	100.000
20	100	1.00000	1.00000	100.000	100.000

The lift for any percentile (bin) is the ratio of the observed response rate in the percentile to the overall observed response rate in the data set. The capture rate (% captured response) for any percentile shows the ratio of the number of observed responders within that percentile to the total number of observed responders in the data set. The cumulative lift and the lift are the same for the first percentile. However, the cumulative lift for the second percentile is the ratio of the observed response rate in the first two percentiles combined to the overall observed response rate. The observed response rate for the first two percentiles is calculated as the ratio of the total number of observed responders in the first two percentiles to the total number of records in the first two percentiles, and so on. The cumulative capture rate for the second percentile is the ratio of the number of observed responders in the first two percentiles to the total number of observed responders in the sample, and so on for the remaining percentiles.

For example, the last column for 40th percentile shows that 76.543% of all responders are in the top 40% of the ranked data set. The ranking is done by the scores (predicted probability).

6.4.1.2 Model_B2

The middle section of the Display 6.54 shows the process flow for Model_B2. This model is built by using transformed variables. All interval variables are transformed by the Maximum Normal transformation. Display 6.59 shows the settings of the Transform Variables node for this model.

To include the original as well as transformed variables for consideration in the model, you can reset the **Hide** and **Reject** properties to No. However, in this example, I left these properties at the default value of Yes, so that Model_B2 is built solely on the transformed variables.

Display 6.59

Default Methods	
Interval Inputs	Maximum Normal
Interval Targets	None
Class Inputs	Group rare levels
Class Targets	None
Treat Missing as Level	Yes
Sample Properties	
Method	Random
Size	Default
Random Seed	12345
Optimal Binning	
Number of Bins	4
Missing Values	Use in Search
Grouping Method	
Cutoff Value	0.5
Group Missing	Yes
Number of Bins	4
Add Minimum Value to Offset Value	Yes
Offset Value	1
Score	
Use Meta Transformation	Yes
Hide	Yes
Reject	Yes

The value of the **Interval Inputs** property is set to Maximum Normal. This means that for each input X, the transformations $\log(X)$, $X^{1/4}$, $sqrt(X)$, X^2, X^4, and e^X are evaluated. The transformation that yields sample quantiles that are closest to the theoretical quantiles of a normal distribution is selected. Suppose, by using one of these transformations, you get the transformed variable Y. This means that the 0.75_quantile for Y from the sample is that value of Y such that 75% of the observations are at or below that value of Y. The 0.75_quantile for a standard normal distribution is 0.6745 given by $P(Z \le 0.6745) = 0.75$, where Z is a normal random variable with mean 0 and standard deviation 1. The 0.75_quantile for Y is compared with 0.6745; likewise, the other quantiles are compared with the corresponding quantiles of the standard normal distribution. To compare the sample quantiles with the quantiles of a standard normal distribution, you must first standardize the variables. Display 6.60 shows the standardization of the variable and the transformation that yielded maximum normality for each variable. In the code in this display, the variables Trans_SCALEVAR_100 and Trans_SCALEVAR_101 are the standardized variables.

Display 6.60

```
*---------------------------------------------------------*;
* TRANSFORM: IMP_NVAR205 , (max(IMP_NVAR205-0, 0.0)/216151.33)**0.25;
*---------------------------------------------------------*;
drop Trans_SCALEVAR_100;
label PWR_IMP_NVAR205 = 'Transformed: Imputed NVAR205';
if IMP_NVAR205 eq . then PWR_IMP_NVAR205 = .;
else do;
Trans_SCALEVAR_100 = max(IMP_NVAR205-0, 0.0)/216151.33;
PWR_IMP_NVAR205 = (Trans_SCALEVAR_100)**0.25;
end;
*---------------------------------------------------------*;
* TRANSFORM: IMP_NVAR21 , (max(IMP_NVAR21-0, 0.0)/17946.92)**2;
*---------------------------------------------------------*;
drop Trans_SCALEVAR_101;
label SQR_IMP_NVAR21 = 'Transformed: Imputed NVAR21';
if IMP_NVAR21 eq . then SQR_IMP_NVAR21 = .;
else do;
Trans_SCALEVAR_101 = max(IMP_NVAR21-0, 0.0)/17946.92;
SQR_IMP_NVAR21 = (Trans_SCALEVAR_101)**2;
end;
```

The variables are first scaled and then transformed, and the transformation that yielded quantiles closest to the theoretical normal quantiles is selected.

The **Regression** node properties settings are the same as those for Model_B1.

Display 6.61 shows the variables selected in the final model.

Display 6.61

```
The selected model, based on the error rate for the validation data,
is the model trained in Step 5. It consists of the following effects:

Intercept  LOG_IMP_NVAR253  LOG_IMP_NVAR305  PWR_IMP_NVAR205  SQRT_IMP_NVAR286
           SQRT_NVAR103

    Likelihood Ratio Test for Global Null Hypothesis: BETA=0

    -2 Log Likelihood             Likelihood
  Intercept      Intercept &        Ratio
    Only         Covariates       Chi-Square       DF     Pr > ChiSq

   2789.110       2501.154         287.9555         5        <.0001

              Type 3 Analysis of Effects

                                  Wald
Effect               DF      Chi-Square     Pr > ChiSq

LOG_IMP_NVAR253       1        73.2588        <.0001
LOG_IMP_NVAR305       1        44.2287        <.0001
PWR_IMP_NVAR205       1        16.6450        <.0001
SQRT_IMP_NVAR286      1        17.2357        <.0001
SQRT_NVAR103          1       120.7532        <.0001
```

Display 6.62 shows the cumulative lift for the Training and Validation data sets.

Display 6.62

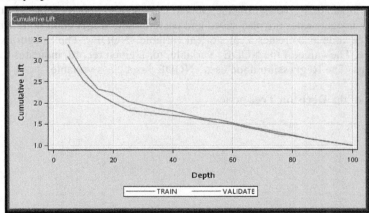

Display 6.62 shows that the model seems to make a better prediction for the Validation sample than it does for the Training data set. This does happen occasionally, and is no cause for alarm. It may be caused by the fact that the final model selection is based on the minimization of the validation error, which is calculated from the validation data set, since I set the **Selection Criterion** property to Validation Error.

When I allocated the data between the Training, Validation, and Test data sets, I used the default method, which is stratified partitioning with the target variable serving as the stratification variable. Partitioning done this way usually avoids the situations where the model performs better on the Validation data set than on the Training data set because of the differences in outcome concentrations.

The lift table corresponding to Display 6.62 and the cumulative capture rate are shown in Table 6.2.

Table 6.2

			Model_B2		
Bin	Percentile	Cumulative Lift Training	Cumulative Lift Validation	Cumulative % Captured Response Training	Cumulative % Captured Response Validation
1	5	3.09307	3.38439	15.509	16.975
2	10	2.54509	2.74691	25.463	27.469
3	15	2.21942	2.32266	33.333	34.877
4	20	2.01294	2.23765	40.278	44.753
5	25	1.83333	2.03575	45.833	50.926
6	30	1.78928	1.95473	53.704	58.642
7	35	1.75241	1.86865	61.343	65.432
8	40	1.71215	1.82099	68.519	72.840
9	45	1.67146	1.72779	75.231	77.778
10	50	1.60648	1.63580	80.324	81.790
11	55	1.54842	1.60448	85.185	88.272
12	60	1.50451	1.53292	90.278	91.975
13	65	1.42053	1.44314	92.361	93.827
14	70	1.34572	1.37566	94.213	96.296
15	75	1.27469	1.30425	95.602	97.840
16	80	1.22085	1.23457	97.685	98.765
17	85	1.16279	1.16173	98.843	98.765
18	90	1.10059	1.10768	99.074	99.691
19	95	1.04765	1.05246	99.537	100.000
20	100	1.00000	1.00000	100.000	100.000

6.4.1.3 Model_B3

The bottom segment of Display 6.54 shows the process flow for Model_B3. In this case, the **Decision Tree** node is used for selecting the inputs. The inputs that give significant splits in growing the tree are selected. In addition, inputs selected for creating the surrogate rules are also saved and passed to the **Regression** node with their roles set to **Input**. The **Decision Tree** node also creates a categorical variable _NODE_. The categories of this variable identify the leaves of the tree. The value of the NODE_ variable, for a given record, indicates the leaf of the tree to which the record belongs. The **Regression** node uses _NODE_ as a class variable.

Display 6.63 shows the property settings of the **Decision Tree** node.

Display 6.63

Property	Value
Splitting Rule	
Interval Criterion	ProbF
Nominal Criterion	ProbChisq
Ordinal Criterion	Entropy
Significance Level	0.05
Missing Values	Use in search
Use Input Once	No
Maximum Branch	2
Maximum Depth	6
Minimum Categorical Size	5
Split Precision	4
Node	
Leaf Size	5
Number of Rules	5
Number of Surrogate Rules	0
Split Size	.
Split Search	
Subtree	
Method	Largest
Number of Leaves	4
Assessment Measure	Decision
Assessment Fraction	0.25
Cross Validation	
Observation Based Importance	
P-Value Adjustment	
Output Variables	
Leaf Variable	Yes
Performance	Disk
Score	
Variable Selection	Yes
Leaf Role	Input

The **Significance Level**, **Leaf Size,** and the **Maximum Depth** properties control the growth of the tree, as discussed in as in Chapter 4. No pruning is done, since the Subtree **Method** property is set to Largest. Therefore, the Validation data set is not used in the tree development. As a result of the settings in the Output Variables and Score groups, the **Decision Tree** node creates a class variable _NODE_, which indicates the leaf node to which a record is assigned. The variable _NODE_ will be passed to the next node, the **Regression** node, as a categorical input, along with the other selected variables.

In the Model Selection group of the **Regression** node, the **Selection Model** property is set to Stepwise, and the **Selection Criterion** property to Validation Error. These are the same settings used earlier for Model_B1 and Model_B2.

The variables selected by the Regression Node are given in Display 6.64.

Display 6.64

```
The selected model, based on the error rate for the validation data,
is the model trained in Step 6. It consists of the following effects:

Intercept CVAR001 IMP_NVAR133 NVAR103 _NODE_

     Likelihood Ratio Test for Global Null Hypothesis: BETA=0

     -2 Log Likelihood              Likelihood
   Intercept      Intercept &          Ratio
     Only         Covariates        Chi-Square        DF      Pr > ChiSq

   2789.110        1861.426          927.6832          29       <.0001

         Type 3 Analysis of Effects

                             Wald
   Effect        DF      Chi-Square     Pr > ChiSq

   CVAR001        9       21.6363        0.0101
   IMP_NVAR133    1        3.8968        0.0484
   NVAR103        1       15.0113        0.0001
   _NODE_        18      136.1133        <.0001
```

Display 6.65 shows the lift chart for the Training and Validation data sets. I saved the table behind these charts by clicking on the **Table** icon, clicking **File→Save As**, and then typing a data set name for saving the table. Table 6.3 is created from the saved SAS data set.

Display 6.65

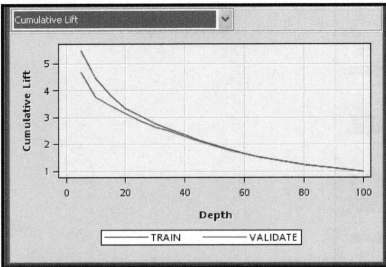

Table 6.3

				Model_B3	
Bin	Percentile	Cumulative Lift Training	Cumulative Lift Validation	Cumulative % Captured Response Training	Cumulative % Captured Response Validation
1	5	5.49367	4.67661	27.546	23.457
2	10	4.46548	3.76543	44.676	37.654
3	15	3.82234	3.43260	57.407	51.543
4	20	3.35489	3.14815	67.130	62.963
5	25	3.06481	2.87473	76.620	71.914
6	30	2.78418	2.65432	83.565	79.630
7	35	2.53273	2.46802	88.657	86.420
8	40	2.35421	2.27623	94.213	91.049
9	45	2.13947	2.11174	96.296	95.062
10	50	1.97222	1.95062	98.611	97.531
11	55	1.80509	1.79522	99.306	98.765
12	60	1.66654	1.64609	100.000	98.765
13	65	1.53801	1.53334	100.000	99.691
14	70	1.42838	1.42857	100.000	100.000
15	75	1.33333	1.33305	100.000	100.000
16	80	1.24978	1.25000	100.000	100.000
17	85	1.17641	1.17625	100.000	100.000
18	90	1.11088	1.11111	100.000	100.000
19	95	1.05253	1.05246	100.000	100.000
20	100	1.00000	1.00000	100.000	100.000

The three models are compared by the **Model Comparison** node.

Display 6.66 shows the lift charts for all three models together for the Training, Validation, and Test data sets.

Display 6.66

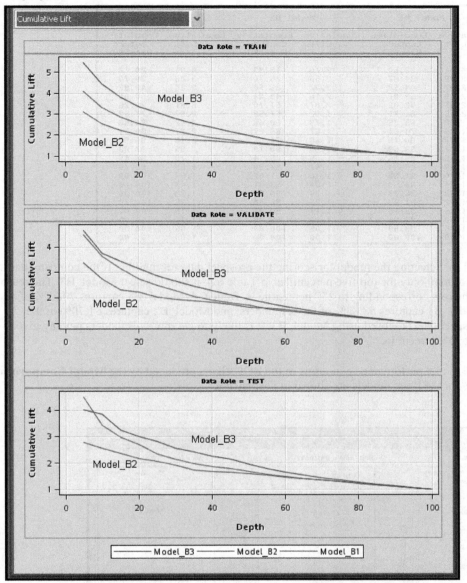

Click the **Table** icon located below the menu bar in the Results window to see the data behind the charts. SAS Enterprise Miner saves this data as a SAS data set. In this example, the data set is named &EM_LIB..MdlComp_EMRANK. Table 6.4 is created from this data set. It shows a comparison of lift charts and capture rates for the Test data set.

Table 6.4

Percentile	Model_B1		Model_B2		Model_B3	
	Cumulative Lift	Cumulative Capture Rate(%)	Cumulative Lift	Cumulative Capture Rate(%)	Cumulative Lift	Cumulative Capture Rate(%)
5	4.51	22.60	2.78	13.93	4.01	20.12
10	3.46	34.67	2.56	25.70	3.86	38.70
15	2.97	44.58	2.37	35.60	3.26	48.92
20	2.69	53.87	2.16	43.34	3.00	60.06
25	2.35	58.82	2.07	51.70	2.81	70.28
30	2.16	64.71	1.92	57.59	2.57	77.09
35	1.98	69.35	1.73	60.68	2.42	84.83
40	1.86	74.61	1.67	66.87	2.27	90.71
45	1.78	80.19	1.66	74.61	2.13	95.98
50	1.67	83.59	1.60	80.19	1.96	97.83
55	1.56	85.76	1.55	85.14	1.79	98.45
60	1.48	88.54	1.47	87.93	1.65	98.76
65	1.40	91.33	1.40	91.02	1.54	100.00
70	1.34	93.50	1.34	93.81	1.43	100.00
75	1.28	96.28	1.28	96.28	1.33	100.00
80	1.22	97.83	1.21	96.90	1.25	100.00
85	1.15	98.14	1.17	99.07	1.18	100.00
90	1.10	99.07	1.10	99.38	1.11	100.00
95	1.05	99.69	1.05	100.00	1.05	100.00
100	1.00	100.00	1.00	100.00	1.00	100.00

This table can be used for selecting the model for scoring the prospect data for mailing. If the company has a budget that permits it to mail only the top five percentiles in Table 6.4, it should select Model_B3. The reason for this is that, from the row corresponding to 25th percentile in Table 6.4, Model_B3 captures 70.28% of all responders, while Model_B1 captures 58.82% of the responders, and Model_B2 captures 51.70% of the responders. If the prospect data is scored using Model_B3, it is likely to get more responders per any given number of mailings at 25th percentile.

You can also compare model performance e in terms of the ROC charts produced by the **Model Comparison** node. The ROC charts for the three models are shown in Display 6.67.

Display 6.67

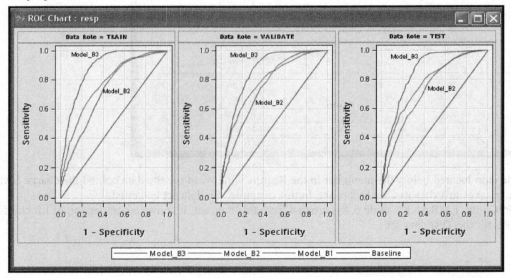

An ROC curve is a plot of Sensitivity (True Positive Rate) versus 1-Specifity (False Positive Rate) at different cut-off values.

To draw an ROC curve of a model for a given data set, you have to classify each customer (record) in the data set based on a cutoff value and the predicted probability. If the predicted probability of response for a customer is equal to or greater than the cutoff value, then you classify the customer as responder (1) If not, you classify the customer as non-responder (0).

For given cutoff value, you may classify a true responder as responder or a true non-responder as responder. If a true responder is classified as responder, then it is called true positive, and the total number of true positives in the data set is denoted by *TP*. If a true non-responder classified as a responder, then it is called a false positive, and the total number of false positives is denoted by *FP*. If a true responder is classified as non-responder, it is called a false negative, and the number of false negatives is denoted by *FN*. If a true non-responder is classified as non-responder, then it is called a true negative, and the number of true negatives is denoted by *TN*.

The total number of true responders in a data set is *TP* + *FN* = *P*, and the total number of true non-responders is *FP*+*T N* = *N*.

For a given cutoff value, you can calculate Sensitivity and Specificity as follows:

$$Sensitivity = \frac{TP}{P} = TPR$$

$$Specificiy = \frac{TN}{N}$$

$$1 - Specificiy = \frac{N - TN}{N} = \frac{FP}{N} = FPR$$

An ROC curve is obtained by calculating sensitivity and specificity for different cutoff values and plotting them as shown in Display 6.67. See Chapter 5, Section 5.4.3 for a detailed description of how to draw ROC charts.

The size of the area below an ROC curve is an indicator of the accuracy of a model. From Display 6.67 it is apparent that the area below the ROC curve of Model_B3 is the largest for the Training, Validation, and Test data sets.

6.4.2 Regression for a Continuous Target

Now I turn to the second business application discussed earlier, which involves developing a model with a continuous target. As I did in the case of binary targets, in this example I build three versions of the model described above: Model_C1, Model_C2, and Model_C3. All three predict the increase in customers' savings deposits in response to a rate increase, given a customer's demographic and lifestyle characteristics, assets owned, and past banking transactions. My primary goal is to demonstrate the use of the **Regression** node when the target is continuous. In order to do this, I use simulated data of a price test conducted by a hypothetical bank, as explained at the beginning of Section 6.4.

The basic model is $\log(\Delta Deposits_{it}) = \beta' X_i$, where $\Delta Deposits_{it}$ is the change in deposits in the ith account during the time interval $(t, t + \Delta t)$. An increase in interest rate is offered at time t. β is the vector of coefficients to be estimated using the **Regression** node. X_i is the vector of inputs for the ith account. This vector includes customers' transactions prior to the rate change and other customer characteristics.

The target variable LDELBAL is $\log(\Delta Deposits_{it})$ for the models developed here. My objective is to find answers to this question: Of those who responded to the rate increase, which customers are likely to increase their savings the most? The change in deposits is positive by definition, but it is much skewed in the sample created. Hence, I made a log transformation of the dependent variable. When the model is created, you can score the prospects using either the estimated value of $\log(\Delta Deposits_{it})$ or $\Delta Deposits_{it}$.

Display 6.68 shows the process flow for the three models to be tested.

Display 6.68

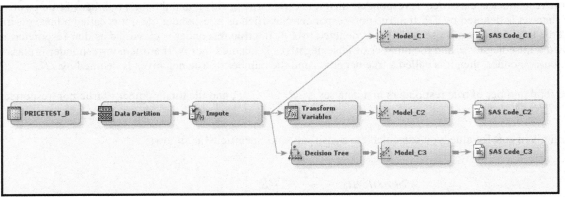

The SAS data set containing the price test results is read by the **Input Data Source** node. From this data set, the **Data Partition** node creates the three data sets **Part_Train**, **Part_Validate**, and **Part_Test** with the roles Train, Validate, and Test, respectively. 40% of the records are randomly allocated for Training, 30% for Validation, and 30% for Test.

Display 6.69 shows the property settings for the **Regression** node. It is the same for all three models.

Display 6.69

Class Targets	
Regression Type	Linear Regression
Link Function	Logit
Model Options	
Suppress Intercept	No
Input Coding	Deviation
Model Selection	
Selection Model	Stepwise
Selection Criterion	Validation Error
Use Selection Defaults	No
Selection Options	[...]

In addition to the above settings, the **Entry Significance Level** property is set to 0.1, and the **Stay Significance Level** property is set to 0.05.

6.4.2.1 Model_C1

The top section of Display 6.68 shows the process flow for Model_C1. In this model, the inputs are used without any transformations. The target is continuous; therefore, the coefficients of the regression are obtained by minimizing the error sums of squares.

From the Iteration Plot shown in Display 6.70 it can be seen that the stepwise selection process took 21 steps to complete the effect selection. After the 21th step, no additional inputs met the 0.1 significance level for entry into the model. The selected model is the one trained at Step 5 since it has the smallest validation error. Display 6.71 shows the selected model.

Display 6.70

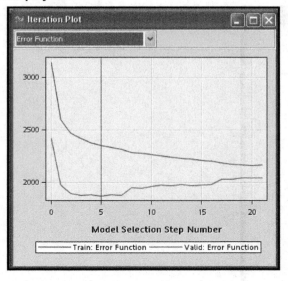

Display 6.71

```
The selected model, based on the error rate for the validation data,
is the model trained in Step 5. It consists of the following effects:

Intercept  CVAR14  IMP_NVAR048  IMP_NVAR054  NVAR009  NVAR174

                      Analysis of Variance

                             Sum of
Source              DF       Squares      Mean Square    F Value    Pr > F

Model                7     791.282729     113.040390      55.78     <.0001
Error             1160    2350.923422       2.026658
Corrected Total   1167    3142.206151

                  Model Fit Statistics

R-Square      0.2518    Adj R-Sq        0.2473
AIC         833.0339    BIC           834.4815
SBC         873.5383    C(p)           57.5791

                  Type 3 Analysis of Effects

                           Sum of
Effect              DF     Squares    F Value    Pr > F

CVAR14               1    140.2208     69.19     <.0001
IMP_NVAR048          1     37.6777     18.59     <.0001
IMP_NVAR054          1     22.6154     11.16     0.0009
NVAR009              1     41.5166     20.49     <.0001
NVAR174              3     92.1909     15.16     <.0001
```

Display 6.72 shows the plot of the predicted mean and actual mean of the target variable by percentile.

Display 6.72

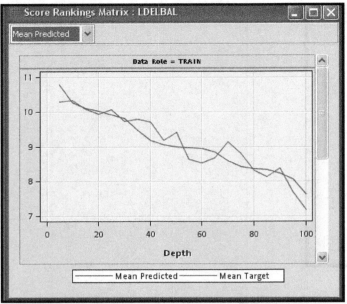

The **Regression** node appends the predicted and actual values of the target variable for each record in the Training, Validation, and Test data sets. These data sets are passed to the **SAS Code** node.

Table 6.5 is constructed from the Validation data set. When you run the **Regression** node, it calculates the predicted values of the target variable for each record in the Training, Validation, and Test data sets. The predicted value of the target is P_LDELBAL in this case. This new variable is added to the three data sets.

In the **SAS Code** node, I use the Validation data set's predicted values in P_LDELBAL to create the decile, lift, and capture rates for the deciles. The program for these tasks is shown in Displays 6.73A, 6.73B, and 6.73C.

Table 6.5 includes special lift and capture rates that are suitable for continuous variables in general, and for banking applications in particular.

Table 6.5

```
Model_C1
                           INCREASE IN DEPOSITS
'                           Cumulative  Cumulative  Cumulative  Cumulative
 Percentile     Sum      Mean    Sum        Mean       Lift     Capture
                                                                Rate(%)
      5      $933,328  $21,705   $933,328  $21,705     1.90      9.3
     10      $916,227  $20,823  $1,849,555 $21,259     1.86     18.5
     15    $1,042,700  $23,698  $2,892,256 $22,078     1.93     28.9
     20      $483,248  $10,983  $3,375,504 $19,289     1.69     33.8
     25    $1,084,943  $25,231  $4,460,447 $20,461     1.79     44.6
     30      $558,922  $12,703  $5,019,368 $19,158     1.68     50.2
     35    $1,252,661  $28,470  $6,272,029 $20,497     1.80     62.7
     40      $889,215  $20,209  $7,161,244 $20,461     1.79     71.6
     45      $203,863   $4,633  $7,365,107 $18,693     1.64     73.7
     50      $261,953   $6,092  $7,627,059 $17,453     1.53     76.3
     55      $313,393   $7,123  $7,940,453 $16,508     1.45     79.4
     60      $364,940   $8,294  $8,305,392 $15,820     1.39     83.1
     65      $352,320   $8,007  $8,657,712 $15,216     1.33     86.6
     70      $398,065   $9,047  $9,055,777 $14,773     1.29     90.6
     75      $191,026   $4,442  $9,246,803 $14,096     1.23     92.5
     80      $147,518   $3,353  $9,394,321 $13,420     1.18     93.9
     85      $210,877   $4,793  $9,605,198 $12,910     1.13     96.1
     90      $150,201   $3,414  $9,755,399 $12,380     1.08     97.6
     95      $138,214   $3,141  $9,893,613 $11,891     1.04     98.9
    100      $106,387   $2,418 $10,000,000 $11,416     1.00    100.0
```

The capture rate shown in the last column of Table 6.5 indicates that if the rate incentive generated an increase of $10 million in deposits, 9.3% of this increase would be coming from the accounts in the top percentile, 18.5% would be coming from the top two percentiles, and so on. Therefore, by targeting only the top 50%, the bank can get 76.3% of the increase in deposits.

By looking at the last row of the Cumulative Mean column, you can see that the average increase in deposits for the entire data set is $11,416. However, the average increase for the top demi-decile is $21,705 – almost double the overall mean. Hence, the lift for the top percentile is 1.90. The cumulative lift for the top two percentiles is 1.86, and so on.

Display 6.73A

```
 libname lib2 "c:\TheBook\EM12.1\Reports\Chapter6";
 run;
 %let model= Model_C1;

data Reg_Validate;
  set &EM_LIB..Reg_Validate;
  delbal=exp(ldelbal);
  run;
proc sort data=Reg_Validate out=validate;
  by descending p_ldelbal ;
  run ;
proc sql noprint;
  select count(*) into : nv1 from
  work.validate ;
 quit ;
 %let scale = 10000000;
data validate ;
  retain count 0 ;
  set validate ;
  count+1 ;
  if count < (1/20)*&nv1  then dec=5; else
  if count < (2/20)*&nv1  then dec=10 ; else
  if count < (3/20)*&nv1  then dec=15 ; else
  if count < (4/20)*&nv1  then dec=20 ; else
  if count < (5/20)*&nv1  then dec=25 ; else
  if count < (6/20)*&nv1  then dec=30 ; else
  if count < (7/20)*&nv1  then dec=35; else
  if count < (8/20)*&nv1  then dec=40 ; else
  if count < (9/20)*&nv1  then dec=45 ; else
  if count < (10/20)*&nv1 then dec=50 ; else
  if count < (11/20)*&nv1 then dec=55 ; else
  if count < (12/20)*&nv1 then dec=60 ; else
  if count < (13/20)*&nv1 then dec=65 ; else
  if count < (14/20)*&nv1 then dec=70 ; else
  if count < (15/20)*&nv1 then dec=75 ; else
  if count < (16/20)*&nv1 then dec=80 ; else
  if count < (17/20)*&nv1 then dec=85 ; else
  if count < (18/20)*&nv1 then dec=90 ; else
  if count < (19/20)*&nv1 then dec=95 ; else
  dec = 100 ;
  run ;
```

Display 6.73B (*continued from Display 6.73A*)

```
proc means data=validate noprint ;
   class dec ;
   var delbal ;
   output out= outsum sum(delbal) = sum_delbal mean(delbal)=mean_delbal;
run ;
data Total(keep=sum_delbal rename=(sum_delbal=Tot_delbal)) deciles ;
   set outsum ;
   if _TYPE_ = 0 then output Total;
   else output deciles ;
run ;
data total ;
   set total ;
   scale = &scale/tot_delbal ;
run ;
data tables ;
   set deciles ;
   if _N_ = 1 then set total ;
run;
data lib2.Lift_&model ;
   set tables ;
   retain cumsum 0 nobs 0;
   sum_delbal= sum_delbal*scale ;
   tot_delbal= tot_delbal*scale ;
   mean_delbal = mean_delbal*scale;
   cumsum + sum_delbal ;
   capc = 100*cumsum/tot_delbal ;
   gmean = tot_delbal/&nv1 ;
   nobs+_freq_ ;
   meanc = cumsum/nobs ;
   liftc = meanc/gmean;
run ;
```

Display 6.73C (*continued from Display 6.73B*)

```
data _NULL_ ;

   file print ;
   set lib2.Lift_&model ;
   if _N_ = 1 then do ;
   put @1 "&model" ;
   put ' ' ;
   put @35 'INCREASE IN DEPOSITS' ;
   put ' ' ;
   put @1 '' @40 'Cumulative' @52 'Cumulative' @64 'Cumulative' @76 'Cumulative' ;
   put @1 'Percentile ' @20 'Sum' @30 'Mean' @44 'Sum' @57 'Mean' @66 'Lift' @76'Capture' ;
   put @1 '  ' @76 'Rate(%)'  ;
   put ' ' ;
   end ;
   put @2 dec 4. @15 sum_delbal dollar10.0 @25 mean_delbal dollar10.0
       @38 cumsum dollar12.0 @50 meanc dollar11.0 @64 liftc 6.2 @75 capc 6.1 ;
run ;
```

6.4.2.2 Model_C2

The process flow for Model_C2 is given in Display 6.68. The **Transform Variables** node creates three data sets: Trans_Train, Trans_Validate, and Trans_Test. These data sets include the transformed variables. The interval inputs are transformed by the Maximum Normal method, which is described in Section 2.9.6 of Chapter 2. In addition, I set the **Class Inputs** property to Dummy indicators and **Treat Missing as Level** property to Yes in the **Transformation** node. In the **Regression** node, the effect selection is done by using the Stepwise method, and the **Selection Criterion** property is set to Validation Error.

Display 6.74 shows the Score Rankings in the **Regression** node's Results window.

Display 6.74

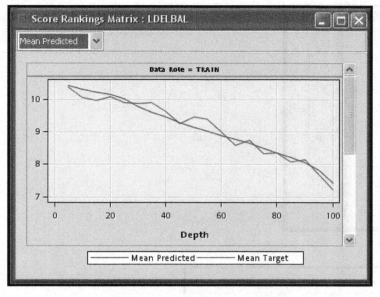

Table 6.6 shows the lift and capture rates of Model_C2.

Table 6.6

```
Model_C2
                           INCREASE IN DEPOSITS
                                  Cumulative  Cumulative  Cumulative  Cumulative
Percentile      Sum       Mean       Sum         Mean        Lift      Capture
                                                                       Rate(%)
     5      $1,107,696   $25,760  $1,107,696    $25,760      2.26        11.1
    10        $699,991   $15,909  $1,807,687    $20,778      1.82        18.1
    15        $753,927   $17,135  $2,561,615    $19,554      1.71        25.6
    20        $810,457   $18,419  $3,372,071    $19,269      1.69        33.7
    25        $562,237   $13,075  $3,934,308    $18,047      1.58        39.3
    30      $1,214,688   $27,607  $5,148,996    $19,653      1.72        51.5
    35      $1,031,405   $23,441  $6,180,400    $20,197      1.77        61.8
    40        $397,971    $9,045  $6,578,372    $18,795      1.65        65.8
    45        $839,096   $19,070  $7,417,468    $18,826      1.65        74.2
    50        $366,506    $8,523  $7,783,974    $17,812      1.56        77.8
    55        $318,393    $7,236  $8,102,367    $16,845      1.48        81.0
    60        $226,383    $5,145  $8,328,750    $15,864      1.39        83.3
    65        $237,563    $5,399  $8,566,313    $15,055      1.32        85.7
    70        $176,108    $4,002  $8,742,421    $14,262      1.25        87.4
    75        $246,380    $5,730  $8,988,801    $13,702      1.20        89.9
    80        $153,698    $3,493  $9,142,499    $13,061      1.14        91.4
    85        $278,275    $6,324  $9,420,774    $12,662      1.11        94.2
    90        $292,620    $6,650  $9,713,394    $12,327      1.08        97.1
    95        $188,874    $4,293  $9,902,268    $11,902      1.04        99.0
   100         $97,732    $2,221 $10,000,000    $11,416      1.00       100.0
```

6.4.2.3 Model_C3

The process flow for Model_C3 is given in the lower region of Display 6.68. After the missing values were imputed, the data was passed to the **Decision Tree** node, which selected some important variables and created a class variable called _NODE_. This variable indicates the node to which a record belongs. The selected variables and special class variables were then passed to the **Regression** node with their roles set to **Input**. The **Regression** node made further selections from these variables, and the final equation consisted of only the _NODE_ variable. In the **Decision Tree** node, I set the **Subtree Method** property to Largest, and in the Score group, I set the **Variable Selection** property to Yes and the **Leaf Role** property to Input.

Display 6.75 shows the **Score Rankings** in the **Regression** node's Results window.

Display 6.75

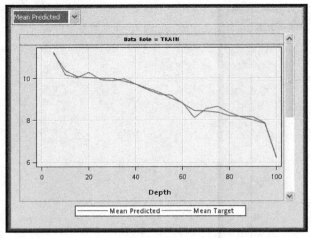

Table 6.7

Model_C3						
			INCREASE IN DEPOSITS			
Percentile	Sum	Mean	Cumulative Sum	Cumulative Mean	Cumulative Lift	Cumulative Capture Rate(%)
5	$906,279	$21,076	$906,279	$21,076	1.85	9.1
10	$819,478	$18,624	$1,725,756	$19,836	1.74	17.3
15	$966,695	$21,970	$2,692,452	$20,553	1.80	26.9
20	$643,361	$14,622	$3,335,813	$19,062	1.67	33.4
25	$697,183	$16,214	$4,032,995	$18,500	1.62	40.3
30	$983,159	$22,345	$5,016,154	$19,146	1.68	50.2
35	$419,151	$9,526	$5,435,305	$17,762	1.56	54.4
40	$572,750	$13,017	$6,008,055	$17,166	1.50	60.1
45	$315,819	$7,178	$6,323,874	$16,050	1.41	63.2
50	$261,236	$6,075	$6,585,110	$15,069	1.32	65.9
55	$587,368	$13,349	$7,172,478	$14,912	1.31	71.7
60	$1,085,251	$24,665	$8,257,729	$15,729	1.38	82.6
65	$289,298	$6,575	$8,547,028	$15,021	1.32	85.5
70	$263,372	$5,986	$8,810,400	$14,373	1.26	88.1
75	$303,270	$7,053	$9,113,670	$13,893	1.22	91.1
80	$154,117	$3,503	$9,267,787	$13,240	1.16	92.7
85	$119,588	$2,718	$9,387,375	$12,617	1.11	93.9
90	$138,339	$3,144	$9,525,714	$12,088	1.06	95.3
95	$231,511	$5,262	$9,757,224	$11,727	1.03	97.6
100	$242,776	$5,518	$10,000,000	$11,416	1.00	100.0

A comparison of Model_C1, Model_C2, and Model_C3 is shown in Table 6.8.

Table 6.8

				The SAS System					
Percentile	Cumulative Mean			Cumulative Lift			Cumulative Capture Rate		
	Model_C1	Model_C2	Model_C3	Model_C1	Model_C2	Model_C3	Model_C1	Model_C2	Model_C3
5	$21,705	$25,760	$21,076	1.90	2.26	1.85	9.33	11.08	9.06
10	$21,259	$20,778	$19,836	1.86	1.82	1.74	18.50	18.08	17.26
15	$22,078	$19,554	$20,553	1.93	1.71	1.80	28.92	25.62	26.92
20	$19,289	$19,269	$19,062	1.69	1.69	1.67	33.76	33.72	33.36
25	$20,461	$18,047	$18,500	1.79	1.58	1.62	44.60	39.34	40.33
30	$19,158	$19,653	$19,146	1.68	1.72	1.68	50.19	51.49	50.16
35	$20,497	$20,197	$17,762	1.80	1.77	1.56	62.72	61.80	54.35
40	$20,461	$18,795	$17,166	1.79	1.65	1.50	71.61	65.78	60.08
45	$18,693	$18,826	$16,050	1.64	1.65	1.41	73.65	74.17	63.24
50	$17,453	$17,812	$15,069	1.53	1.56	1.32	76.27	77.84	65.85
55	$16,508	$16,845	$14,912	1.45	1.48	1.31	79.40	81.02	71.72
60	$15,820	$15,864	$15,729	1.39	1.39	1.38	83.05	83.29	82.58
65	$15,216	$15,055	$15,021	1.33	1.32	1.32	86.58	85.66	85.47
70	$14,773	$14,262	$14,373	1.29	1.25	1.26	90.56	87.42	88.10
75	$14,096	$13,702	$13,893	1.23	1.20	1.22	92.47	89.89	91.14
80	$13,420	$13,061	$13,240	1.18	1.14	1.16	93.94	91.42	92.68
85	$12,910	$12,662	$12,617	1.13	1.11	1.11	96.05	94.21	93.87
90	$12,380	$12,327	$12,088	1.08	1.08	1.06	97.55	97.13	95.26
95	$11,891	$11,902	$11,727	1.04	1.04	1.03	98.94	99.02	97.57
100	$11,416	$11,416	$11,416	1.00	1.00	1.00	100.0	100.0	100.0

The cumulative capture rate at 50th percentile is 77.84 for Model_C2, Therefore, by targeting the top 50% given by Model_C2, the bank can get 77.84% of the increase in deposits. The cumulative capture rates for Model_C1 and Model_C3 are smaller than for Model_C3 at the 50th percentile.

6.5 Summary

- Four types of models are demonstrated using the Regression Node. These are models with binary, ordinal, nominal (unordered) and continuous targets.
- For each model, the underlying theory is explained with equations which are verified from the SAS code produced by the Regression Node.
- The Regression Type, Link Function, Selection Model, and Selection Options properties are demonstrated in detail.
- A number of examples are given to demonstrate alternative methods of model selection and model assessment by setting different values to the Selection Model and Selection Criterion properties. In each case, the produced output is analyzed.
- Various tables produced by the Regression Node and the Model Comparison Node are saved and other tables are produced from them.
- Two business applications are demonstrated - one requiring a model with binary target, and the other a model with continuous target.

    1) The model with binary target is for identifying the current customers of a hypothetical bank who are most likely to respond to a monetarily rewarded invitation to sign up for Internet banking.

    2) The model with continuous target is for identifying customers of a hypothetical bank who are likely to increase their savings deposits by the largest dollar amounts if the bank increases the interest paid to them by a preset number of basis points.

    3) Three models are developed for each application and the best model is identified.

6.6 Appendix to Chapter 6

SAS code used to generate Table 6.1.

Display 6.76

```
libname mylib "C:\TheBook\Em12.1\Data\Chapter6";
%let table=ScoreRankings_Model_B1;
data train(rename=(liftc=liftc_train capc=capc_train))
     validate(rename=(liftc=liftc_valid capc=capc_valid)) ;
     set mylib.&table (keep=Decile Bin datarole liftc capc);

  if upcase(dataRole)='TRAIN'  then output train; else
  if upcase(dataRole) = 'VALIDATE' then output validate;
  if decile = 0 then delete ;
  drop datarole ;
run;


data both ;
   *retain decile bin liftc_train  liftc_valid capc_train capc_valid ;
   merge train validate ;
   by decile ;
    if decile = 0 then delete ;
   label liftc_train = "Cumulative Lift Training"
        capc_train = "Cumulative % Captured Response Training"
        liftc_valid = "Cumulative Lift Validation"
        capc_valid = "Cumulative % Captured Response Validation"
        Decile = "Percentile";

run ;
title "Model_B1";
proc print data=both noobs label split="/"  ;
   var bin decile liftc_train  liftc_valid capc_train capc_valid ;
run ;
```

Display 6.77

```
%let AGE = 50;
%let NPRVIO = 2;
%let CRED = 670;

*** Compute Linear Predictor;
Data ordinal;
 AGE=&AGE;
 NPRVIO = &NPRVIO;
 CRED=&CRED;
drop _TEMP;
drop _LP0;
_LP0 = 0;

*** Effect: AGE ;
_TEMP = AGE ;
_LP0 = _LP0 + (  -0.02307201027961 * _TEMP);

*** Effect: CRED ;
_TEMP = CRED ;
_LP0 = _LP0 + (  -0.00498225050304 * _TEMP);

*** Effect: NPRVIO ;
_TEMP = NPRVIO ;
_LP0 = _LP0 + (   0.22524384228134 * _TEMP);

*** Naive Posterior Probabilities;
drop _MAXP _IY _P0 _P1 _P2;
_TEMP =     -2.88399896897927 + _LP0;
if (_TEMP < 0) then do;
   _TEMP = exp(_TEMP);
   _P0 = _TEMP / (1 + _TEMP);
end;
else _P0 = 1 / (1 + exp(-_TEMP));
_TEMP =     0.21333987443107 + _LP0;
if (_TEMP < 0) then do;
   _TEMP = exp(_TEMP);
   _P1 = _TEMP / (1 + _TEMP);
end;
else _P1 = 1 / (1 + exp(-_TEMP));
_P2 = 1.0 - _P1;
_P1 = _P1 - _P0;
```

Display 6.78

```
*** Posterior Probabilities and Predicted Level;
label P_LOSSFRQ2 = 'Predicted: LOSSFRQ=2' ;
label P_LOSSFRQ1 = 'Predicted: LOSSFRQ=1' ;
label P_LOSSFRQ0 = 'Predicted: LOSSFRQ=0' ;
P_LOSSFRQ2 = _P0;
_MAXP = _P0;
_IY = 1;
P_LOSSFRQ1 = _P1;
if (_P1 >  _MAXP + 1E-8) then do;
   _MAXP = _P1;
   _IY = 2;
end;
P_LOSSFRQ0 = _P2;
if (_P2 >  _MAXP + 1E-8) then do;
   _MAXP = _P2;
   _IY = 3;
end;
run;

ods listing;
ods html file="C:\TheBook\EM12.1\Reports\Chapter6\Ordinal.html";
proc print data=ORDINAL label noobs;
title "ORDINAL TARGET" ;
title2 "AGE=50, CRED=670 and NPRVIO=2";
var P_LOSSFRQ2 P_LOSSFRQ1 P_LOSSFRQ0;
run;
```

Display 6.79

```
%let AGE= 50;
%let NPRVIO = 2;
%let CRED = 670;
DATA NOMINAL;
   *** Compute Linear Predictor;
drop _TEMP;
drop _LP0 _LP1;
_LP0 = 0;
_LP1 = 0;
AGE = &AGE;
NPRVIO = &NPRVIO;
CRED = &CRED;
*** Effect: AGE ;
_TEMP = AGE ;
_LP0 = _LP0 + (    -0.0323912354634 * _TEMP);
_LP1 = _LP1 + (    -0.0228181055783 * _TEMP);
*** Effect: CRED ;
_TEMP = CRED ;
_LP0 = _LP0 + (    -0.01335715577514 * _TEMP);
_LP1 = _LP1 + (    -0.00448783417963 * _TEMP);
*** Effect: NPRVIO ;
_TEMP = NPRVIO ;
_LP0 = _LP0 + (    0.21864487608312 * _TEMP);
_LP1 = _LP1 + (    0.22396412290576 * _TEMP);
*** Naive Posterior Probabilities;
drop _MAXP _IY _P0 _P1 _P2;
drop _LPMAX;
_LPMAX = 0;
_LP0 =      1.59463711057944 + _LP0;
if _LPMAX < _LP0 then _LPMAX = _LP0;
_LP1 =     -0.11855123381699 + _LP1;
if _LPMAX < _LP1 then _LPMAX = _LP1;
_LP0 = exp(_LP0 - _LPMAX);
_LP1 = exp(_LP1 - _LPMAX);
_LPMAX = exp(-_LPMAX);
_P2 = 1 / (_LPMAX + _LP0 + _LP1);
_P0 = _LP0 * _P2;
_P1 = _LP1 * _P2;
_P2 = _LPMAX * _P2;
```

Display 6.80

```
*** Posterior Probabilities and Predicted Level;
label P_LOSSFRQ2 = 'Predicted: LOSSFRQ=2' ;
label P_LOSSFRQ1 = 'Predicted: LOSSFRQ=1' ;
label P_LOSSFRQ0 = 'Predicted: LOSSFRQ=0' ;
P_LOSSFRQ2 = _P0;
_MAXP = _P0;
_IY = 1;
P_LOSSFRQ1 = _P1;
if (_P1 > _MAXP + 1E-8) then do;
   _MAXP = _P1;
   _IY = 2;
end;
P_LOSSFRQ0 = _P2;
if (_P2 > _MAXP + 1E-8) then do;
   _MAXP = _P2;
   _IY = 3;
end;
run;
ods html file= "C:\TheBook\EM12.1\REPORTS\Chapter6\Nominal.html";
proc print data=NOMINAL label noobs;
   title "NOMINAL TAGET";
   title2 "AGE=50,CRED=670 and NPRVIO=2";
   var P_LOSSFRQ2 P_LOSSFRQ1 P_LOSSFRQ0 ;
run;
ods html close;
```

6.6 Exercises

1. Create a data source using the dataset TestSelection.
   a. Customize the metadata using the Advanced Metadata Advisor Options. Set the **Class Levels Threshold** property to 8.
   b. Set the role of the variable RESP to Target.
   c. Enter the following Decision Weights:

Level		DECISION1	DECISION2
1	...	10.0	0.0
0	...	-1.0	0.0

2. Add a **Data Partition** node to the process flow and allocate 40% of the records for Training, 30% for Validation and 30% for the Test data set.
3. Add a **Regression** node to the process flow and set the following properties:
   a. **Selection Model** property to Forward
   b. **Selection Criterion** to Schwarz Bayesian Criterion
   c. **Use Selection Defaults** property to Yes
4. Run the **Regression** node.
5. Plot the Schwarz's Bayesian Criterion at each step of the iteration.
6. Change the **Selection Criterion** to Akaike Information Criterion.
7. Repeat steps 4 and 5.
8. Compare the iteration plots and the selected models.

Notes

1. Despite the fact that Intercept 2 is not statistically significant, I show the calculations to illustrate the proportional odds model.
2. Ramon C. Littell, Rudolf J. Freund, and, Philip C. Spector. *SAS System for Linear Models* (Cary, NC: SAS Institute Inc., 1991), 21, 22, 156.
3. Littell, Freund, and Spector, *SAS System for Linear Models*.
4. For the computation of the score Chi-Square statistic, see "The LOGISTIC Procedure" in *SAS/STAT® User's Guide*.
5. My explanations of the procedures described here are not written from the actual computational algorithms used inside SAS Enterprise Miner. I have no access to those. There may be some variance between my descriptions and the actual computations in some places. Nonetheless, these explanations should provide a clear view of how the various criteria of model selection are applied in general, and can most certainly be applied using SAS Enterprise Miner.

Chapter 7: Comparison and Combination of Different Models

7.1 Introduction

This chapter compares the output and performance of three modeling tools examined in the previous three chapters—**Decision Tree**, **Regression**, and **Neural Network**—by using all three tools to develop two types of models—one with a binary target and one with an ordinal target. I hope this chapter will help you decide what approach to take.

The model with a binary target is developed for a hypothetical bank that wants to predict the likelihood of a customer's attrition so that it can take suitable action to prevent the attrition if necessary. The model with an ordinal target is developed for a hypothetical insurance company that wants to predict the probability of a given frequency of insurance claims for each of its existing customers, and then use the resulting model to further profile customers who are most likely to have accidents.

The **Stochastic Boosting** and **Ensemble** nodes are demonstrated for combining models.

7.2 Models for Binary Targets: An Example of Predicting Attrition

Attrition can be modeled as either a binary or a continuous target. When you model attrition as a binary target, you predict the probability of a customer "attriting" during the next few days or months, given the customer's general characteristics and change in the pattern of the customer's transactions. With a continuous target, you predict the expected time of attrition, or the residual lifetime, of a customer. For example, if a bank or credit card company wants to identify customers who are likely to terminate their accounts at *any* point within a predefined interval of time in the future, the company can model attrition as a binary target. If, on the other hand, they are interested in predicting the *specific* time at which the customer is likely to attrit, then the company should model attrition as a continuous target, and use techniques such as survival analysis.

In this chapter, I present an example of a hypothetical bank that wants to develop an early warning system to identify customers who are most likely to close their investment accounts in the next three months. To meet this business objective, an attrition model with a binary target is developed.

The customer record in the modeling data set for developing an attrition model consists of three types of variables (fields):

- Variables indicating the customer's past transactions (such as deposits, withdrawals, purchase of equities, etc.) by month for several months
- Customer characteristics (demographic and socio-economic) such as age, income, lifestyle, etc.
- Target variable indicating whether a customer attrited during a pre-specified interval

Assume that the model was developed during December 2006, and that it was used to predict attrition for the period January1, 2007 through March 31, 2007.

Display 7.1 shows a chronological view of a data record in the data set used for developing an attrition model.

Display 7.1

Here are some key input design definitions that are labeled in Display 7.1:

- inputs/explanatory variables window:
  This window refers to the period from which the transaction data is collected.
- operational lag:
  The model excludes any data for the month of July 2006 to allow for the operational lag that comes into play when the model is used for forecasting in real time. This lag may represent the period between the time at which a customer's transaction takes place and the time at which it is recorded in the data base and becomes available to be used in the model.

- performance window:
  This is the time interval in which the customers in the sample data set are observed for attrition. If a customer attrited during this interval, then the target variable ATTR takes the value 1; if not, it takes the value 0.

The model should exclude all data from any of the inputs that have been captured during the performance window.

If the model is developed using the data shown in Display 7.1 and used for predicting, or scoring, attrition propensity for the period January 1, 2007 through March 2007, then the inputs window, the operational lag, and the prediction window look like what is shown in Display 7.2 when the scoring or prediction is done.

Display 7.2

```
┌─────────────────────────────────────────────────────────────────────────────────┐
│  Inputs/Explanatory variables window                    Performance Window        │
│                                                                                   │
│  • Customer transactions by month                       Event (Target Variable)   │
│    June 1, 2006 – Nov 30, 2006      Operational lag          Attrition = 1         │
│                                        = 1 month            No Attrition = 0       │
│  • Customer characteristics observed                                              │
│    at a time point                                                                │
│    on or before Nov 30, 2006.                                                     │
│                                                                                   │
│  June 1, 2006              Nov 30, 2006             Jan 1, 2007    Mar 31, 2007    │
└─────────────────────────────────────────────────────────────────────────────────┘
```

At the time of prediction (say on Dec 31, 2006), all data in the inputs/explanatory variables window is available. In this hypothetical example, however, the inputs are not available for the month of December 2006 because of the time lag in collecting the input data. In addition, no data is available for the prediction window, since it resides in the future (see Display 7.2).

As mentioned earlier, we assume that the model was developed during December 2006. The modeling data set was created by observing a sample of current customers with investment accounts during a time interval of three months starting August 1, 2006 through October 31, 2006. This is the performance window, as shown in Display 7.1. The fictitious bank is interested in predicting the probability of attrition, specifically among their customers' investment accounts. Accordingly, if an investment account is closed during the performance window, an indicator value of 1 is entered on the customer's record; otherwise, a value of 0 is given. Thus, for the target variable ATTR, 1 represents attrition and 0 represents no attrition.

Each record in the modeling data set includes the following details:

- monthly transactions
- balances
- monthly rates of change of customer balances
- specially constructed trend variables

These trend variables indicate patterns in customer transaction behavior, etc., for all accounts held by the customer during the Inputs/ Explanatory variables window, which spans the period of January 1, 2006, through June 30, 2006, as shown in Display 7.1. The customer's records also show how long the customer has held an investment account with the bank (investment account tenure). In addition, each record is appended with the customer's age, marital status, household income, home equity, life-stage indicators, and number of months the customer has been on the books (customer tenure), etc. All of these details are candidates for inclusion in the model.

In the process flow shown in Display 7.3, the data source is created first from the dataset ATTRITION2_C. The **Input Data** node reads this data. The data is then partitioned such that 45% of the records are allocated to training, 35% to validation, and 25% for testing. I allocated more records for training and validation than for testing so that the models are estimated as accurately as possible. The value of the **Partitioning Method** property is set to Default, which performs stratification with respect to the target variable if the target variable is categorical. I also imputed missing observations using the **Impute** node. Display 7.3 shows the process flow for modeling the binary target.

Display 7.3

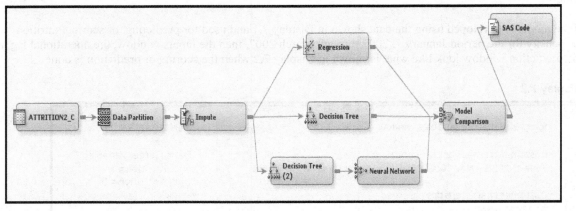

7.2.1 Logistic Regression for Predicting Attrition

The top segment of Display 7.3 shows the process flow for a logistic regression model of attrition. In the **Regression** node, I set the **Model Selection** property to Stepwise, as this method was found to produce the most parsimonious model. For the stepwise selection, I set the **Entry Significance Level** property to 0.05 and the **Stay Significance Level** property to 0.05. I set the **Use Selection Defaults** property to No and the Maximum **Number of Steps** property to 100. Before making a final choice of value for the **Selection Criterion** property, I tested the model using different values. In this example, the **Validation Error** criterion produced a model that made the most business sense. Display 7.4 shows the model that is estimated by the **Regression** node with these property settings.

Display 7.4

Analysis of Maximum Likelihood Estimates					
Parameter	DF	Estimate	Standard Error	Wald Chi-Square	Pr > ChiSq
Intercept	1	-1.3223	0.1336	97.93	<.0001
btrend	1	-0.1485	0.00864	295.45	<.0001
duration	1	-0.0299	0.00717	17.37	<.0001

The variables that are selected by the **Regression** node are btrend and duration. The variable btrend measures the trend in the customer's balances during the six-month period prior to the operational lag period and the operational lag period. If there is a *downward trend* in the balances over the period, e.g., a *decline* of 1 percent, then the odds of attrition increase by $100 * \left\{ e^{-0.1485(-1)} - 1 \right\} = 16.0\%$. This may seem a bit high, but the direction of the result does make sense. (Also, keep in mind that these estimates are based on simulated data. It is best not to consider them as general results.)

The variable duration measures the customer's investment account tenure. Longer tenure corresponds to lower probability of attrition. This can be interpreted as the positive effect of customer loyalty.

Display 7.5 shows the Cumulative Lift chart from the Results window of the **Regression** node.

Display 7.5

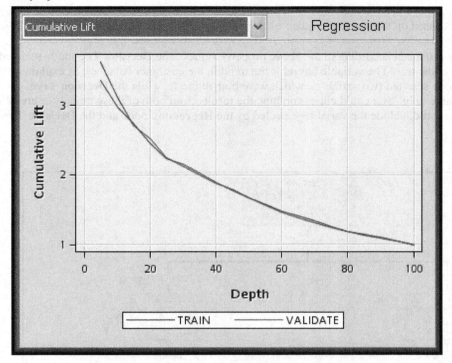

Table 7.1 shows the cumulative lift and capture rates for the Train, Validation, and Test data sets.

Table 7.1

REG		TRAIN		VALIDATE		TEST	
Bin	Percentile	Cumulative Lift	Cumulative Capture Rate(%)	Cumulative Lift	Cumulative Capture Rate(%)	Cumulative Lift	Cumulative Capture Rate(%)
1	5	3.37	16.89	3.63	18.26	3.53	17.65
2	10	2.98	29.82	3.11	31.18	2.90	29.02
3	15	2.75	41.23	2.71	40.73	2.85	42.75
4	20	2.46	49.12	2.53	50.56	2.69	53.73
5	25	2.24	55.92	2.25	56.18	2.52	63.14
6	30	2.16	64.69	2.12	63.76	2.37	70.98
7	35	2.02	70.83	2.00	69.94	2.13	74.51
8	40	1.89	75.44	1.88	75.28	2.01	80.39
9	45	1.77	79.82	1.80	80.90	1.86	83.53
10	50	1.67	83.33	1.66	83.15	1.70	85.10
11	55	1.57	86.18	1.56	85.67	1.59	87.45
12	60	1.48	88.60	1.46	87.64	1.48	88.63
13	65	1.40	91.23	1.38	89.61	1.41	91.37
14	70	1.33	93.42	1.32	92.13	1.34	93.73
15	75	1.26	94.52	1.24	92.98	1.28	95.69
16	80	1.19	95.39	1.19	95.22	1.22	97.65
17	85	1.14	97.15	1.13	96.35	1.16	98.82
18	90	1.10	98.90	1.09	98.03	1.11	100.00
19	95	1.05	99.34	1.05	99.44	1.05	100.00
20	100	1.00	100.00	1.00	100.00	1.00	100.00

7.2.2 Decision Tree Model for Predicting Attrition

The middle section of Display 7.3 shows the process flow for the Decision Tree model of attrition.

I set the **Splitting Rule Criterion** property to ProbChisq, the **Subtree Method** property to Assessment, and the **Assessment Measure** property to Average Square Error. My choice of these property values is somewhat arbitrary, but it resulted in a tree that can serve as an illustration. You can set alternative values for these properties and examine the trees produced.

According to the property values I set, the nodes are split on the basis of *p*-values of the Pearson Chi-Square, and the sub-tree selected is based on the Average Square Error calculated from the Validation data set.

Display 7.6 shows the tree produced according to the above property values. The **Decision Tree** node selected only one variable, btrend, in the tree. The variable btrend is the trend in the customer balances, as explained earlier. The **Regression** node selected two variables, which were both numeric, while the **Decision Tree** selected one numeric variable only. You could either combine the results from both of these models or create a larger set of variables that would include the variables selected by the **Regression** node and the **Decision Tree** node.

Display 7.6

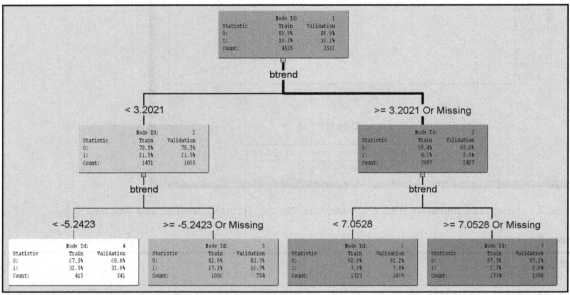

Display 7.7 shows the cumulative lift of the Decision Tree model.

Display 7.7

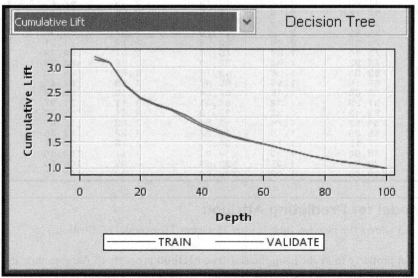

The cumulative lift and cumulative capture rates for Train, Validation, and Test data are shown in Table 7.2.

Table 7.2

TREE		TRAIN		VALIDATE		TEST	
Bin	Percentile	Cumulative Lift	Cumulative Capture Rate(%)	Cumulative Lift	Cumulative Capture Rate(%)	Cumulative Lift	Cumulative Capture Rate(%)
1	5	3.23	16.19	3.16	15.89	3.01	15.07
2	10	3.10	31.03	3.11	31.18	2.94	29.46
3	15	2.63	39.57	2.63	39.44	2.60	39.08
4	20	2.40	48.06	2.38	47.70	2.43	48.71
5	25	2.26	56.56	2.24	55.97	2.33	58.34
6	30	2.17	65.09	2.14	64.23	2.26	67.96
7	35	2.03	71.07	1.97	69.02	2.11	74.01
8	40	1.86	74.61	1.82	72.86	1.92	76.79
9	45	1.74	78.13	1.70	76.70	1.77	79.57
10	50	1.63	81.65	1.61	80.54	1.65	82.35
11	55	1.55	85.19	1.53	84.40	1.55	85.13
12	60	1.48	88.71	1.47	88.24	1.46	87.91
13	65	1.40	90.78	1.39	90.32	1.39	90.25
14	70	1.32	92.10	1.31	91.70	1.31	91.64
15	75	1.25	93.41	1.24	93.08	1.24	93.04
16	80	1.18	94.74	1.18	94.47	1.18	94.43
17	85	1.13	96.05	1.13	95.85	1.13	95.83
18	90	1.08	97.37	1.08	97.23	1.08	97.22
19	95	1.04	98.69	1.04	98.62	1.04	98.62
20	100	1.00	100.00	1.00	100.00	1.00	100.00

7.2.3 A Neural Network Model for Predicting Attrition

The bottom segment of Display 7.3 shows the process flow for the Neural Network model of attrition.

In developing a Neural Network model of attrition, I tested the following two approaches:

- Use all the inputs in the data set, set the **Architecture** property to MLP (multi/layer perceptron), the **Model Selection Criterion** property to Average Error, and the **Number of Hidden Units** property to 10.

- Use the same property settings as above, but use only selected inputs. The included variables are: _NODE_, a special variable created by the Decision Tree Node that preceded the Neural Network Node; btrend, a trend variable selected by the Decision Tree Node and duration which was manually included by setting the "use" property to "Yes" in the variables table of the Neural Network Node.

As expected, the first approach resulted in an over-fitted model. There was considerable deterioration in the lift calculated from the Validation data set relative to that calculated from the Training data set.

The second approach yielded a much more robust model. With this approach, I compared the cumulative lift of the different models generated by setting the **Number of Hidden Units** property to 5, 10, 15, and 20. The model that has the best lift for the Validation data is the one generated by setting the **Number of Hidden Units** property to 10.

Using only a selected number of inputs enables you to quickly test a variety of architectural specifications. For purposes of illustration, I present below the results from a model with the following property settings:

- **Architecture** property: Multilayer Perceptron
- **Model Selection Criterion** property: Average Error
- **Number of Hidden Units** property: 10

The bottom segment of Display 7.3 shows the process flow for the neural network model of attrition. In this process flow, I used the **Decision Tree** node with the **Splitting Rule Criterion** property set to ProbChisq, **Assessment Measure** property set to Average Square Error, **Leaf Variable property** to Yes, **Variable**

Selection property to Yes, and **Leaf Role property** to Input to select the inputs for use in the **Neural Network** node.

The lift charts for this model are shown in Display 7.8.

Display 7.8

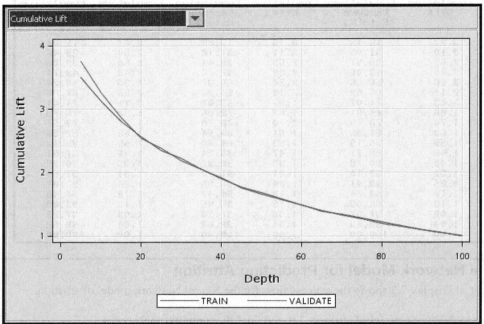

Table 7.3 shows the cumulative lift and cumulative capture rates for the Train, Validation, and Test data sets.

Table 7.3

NEURAL		TRAIN		VALIDATE		TEST	
Bin	Percentile	Cumulative Lift	Cumulative Capture Rate(%)	Cumulative Lift	Cumulative Capture Rate(%)	Cumulative Lift	Cumulative Capture Rate(%)
1	5	3.50	17.54	3.74	18.82	3.45	17.25
2	10	3.11	31.14	3.25	32.58	2.98	29.80
3	15	2.80	42.11	2.86	42.98	2.77	41.57
4	20	2.55	51.10	2.53	50.56	2.55	50.98
5	25	2.33	58.33	2.37	59.27	2.43	60.78
6	30	2.21	66.45	2.17	65.17	2.30	69.02
7	35	2.05	71.71	2.05	71.63	2.15	75.29
8	40	1.90	75.88	1.90	76.12	2.00	80.00
9	45	1.76	79.17	1.75	78.93	1.84	82.75
10	50	1.67	83.55	1.65	82.58	1.69	84.71
11	55	1.58	87.06	1.57	86.52	1.58	87.06
12	60	1.49	89.25	1.48	89.04	1.48	88.63
13	65	1.39	90.35	1.40	91.01	1.42	92.16
14	70	1.33	93.20	1.32	92.42	1.34	93.73
15	75	1.27	95.18	1.26	94.38	1.26	94.90
16	80	1.21	96.71	1.19	95.22	1.19	95.29
17	85	1.15	97.81	1.14	96.91	1.13	96.08
18	90	1.09	98.46	1.10	98.60	1.08	97.25
19	95	1.04	98.90	1.04	99.16	1.04	98.82
20	100	1.00	100.00	1.00	100.00	1.00	100.00

Display 7.9 shows the lift charts for all models from the **Model Comparison** node.

Display 7.9

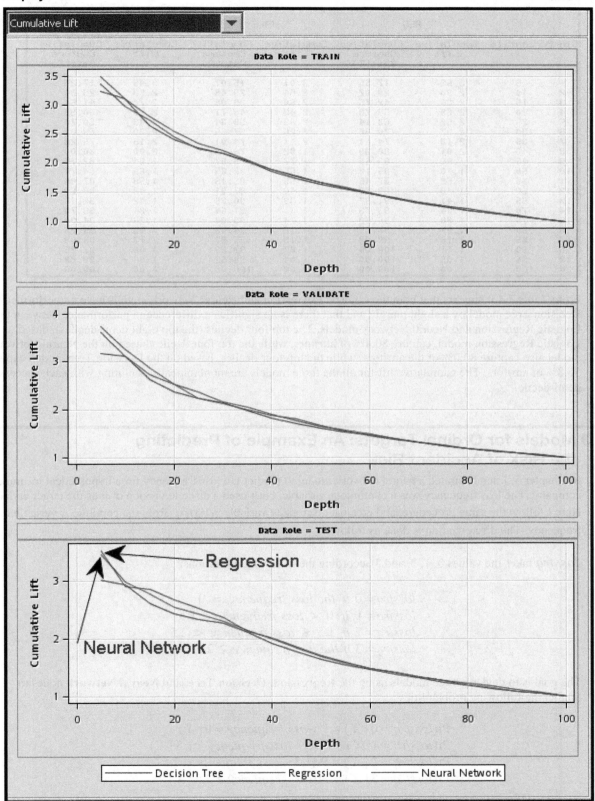

Table 7.4 shows a comparison of the three models using Test data.

Table 7.4

		REG		TREE		NEURAL	
Bin	Percentile	Cumulative Lift	Cumulative Capture Rate(%)	Cumulative Lift	Cumulative Capture Rate(%)	Cumulative Lift	Cumulative Capture Rate(%)
1	5	3.53	17.65	3.01	15.07	3.45	17.25
2	10	2.90	29.02	2.94	29.46	2.98	29.80
3	15	2.85	42.75	2.60	39.08	2.77	41.57
4	20	2.69	53.73	2.43	48.71	2.55	50.98
5	25	2.52	63.14	2.33	58.34	2.43	60.78
6	30	2.37	70.98	2.26	67.96	2.30	69.02
7	35	2.13	74.51	2.11	74.01	2.15	75.29
8	40	2.01	80.39	1.92	76.79	2.00	80.00
9	45	1.86	83.53	1.77	79.57	1.84	82.75
10	50	1.70	85.10	1.65	82.35	1.69	84.71
11	55	1.59	87.45	1.55	85.13	1.58	87.06
12	60	1.48	88.63	1.46	87.91	1.48	88.63
13	65	1.41	91.37	1.39	90.25	1.42	92.16
14	70	1.34	93.73	1.31	91.64	1.34	93.73
15	75	1.28	95.69	1.24	93.04	1.26	94.90
16	80	1.22	97.65	1.18	94.43	1.19	95.29
17	85	1.16	98.82	1.13	95.83	1.13	96.08
18	90	1.11	100.00	1.08	97.22	1.08	97.25
19	95	1.05	100.00	1.04	98.62	1.04	98.82
20	100	1.00	100.00	1.00	100.00	1.00	100.00

From Table 7.4 it appears that both the Logistic Regression and Neural Network models have outperformed the Decision Tree model by a slight margin and that there is no significant difference in performance between the Logistic Regression and Neural Network models. The top four deciles (the top eight demi-deciles), based on the Logistic Regression model, capture 80.4% of attritors, while the top four deciles based on the Neural Network model also capture 80.0% of the attritors, while the top four deciles, based on the Decision Tree model, capture 76.8% of attritors. The cumulative lift for all the three models are monotonically declining with each successive demi-decile.

7.3 Models for Ordinal Targets: An Example of Predicting the Risk of Accident Risk

In Chapter 5, I demonstrated a neural network model to predict the loss frequency for a hypothetical insurance company. The loss frequency was a continuous variable, but I used a discrete version of it as the target variable. Here I follow the same procedure, and create a discretized variable *lossfrq* from the continuous variable loss frequency. The discretization is done as follows:

lossfrq takes the values 0, 1, 2, and 3 according the following definitions:

$$lossfrq = 0 \text{ if the loss frequency } = 0$$
$$lossfrq = 1 \text{ if } 0 < \text{loss frequency} < 1.5$$
$$lossfrq = 2 \text{ if } 1.5 \leq \text{loss frequency} < 2.5$$
$$lossfrq = 3 \text{ if the loss frequency } \geq 2.5.$$

The goal is to develop three models using the **Regression**, **Decision Tree**, and **Neural Network** nodes to predict the following probabilities:

$$\Pr(lossfrq = 0 \mid X) = \Pr(loss \ frequency = 0 \mid X)$$
$$\Pr(lossfrq = 1 \mid X) = \Pr(0 < loss \ frequency < 1.5 \mid X)$$
$$\Pr(lossfrq = 2 \mid X) = \Pr(1.5 \leq loss \ frequency < 2.5 \mid X)$$
$$\Pr(lossfrq = 3 \mid X) = \Pr(loss \ frequency \geq 2.5 \mid X)$$

where X is a vector of inputs or explanatory variables.

When you are using a target variable such as *lossfrq,* which is a discrete version of a continuous variable, the variable becomes ordinal if it has more than two levels.

The first model I develop for predicting the above probabilities is a Proportional Odds model (or logistic regression with cumulative logits link) using the **Regression** node. As explained in Section 6.2.2 of Chapter 6, the **Regression** node produces such a model if the measurement level of the target is set to Ordinal.

The second model I develop is a Decision Tree model using the **Decision Tree** node. The **Decision Tree** node does not produce any equations, but it does provide certain rules for portioning the data set into disjoint groups (leaf nodes). The rules are stated in terms of input ranges (or definitions). For each group, the **Decision Tree** node gives the predicted probabilities:

$$Pr(lossfrq = 0 \mid X), Pr(lossfrq = 1 \mid X), Pr(lossfrq = 2 \mid X) \text{ and } Pr(lossfrq = 3 \mid X).$$

The **Decision Tree** node also assigns a target level such as 0, 1, 2, and 3 to each group, and hence to all the records belonging to that group.

The third model is developed by using the **Neural Network** node, which produces a Logit type model, provided you set the **Target Activation Function** property to Logistic. With this setting, you get a model similar to the one shown in Section 5.5.1.1 of Chapter 5.

Display 7.10 shows the process flow for comparing the ordinal target.

Display 7.10

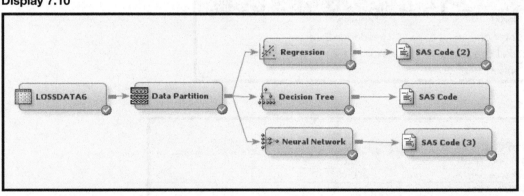

7.3.1 Lift Charts and Capture Rates for Models with Ordinal Targets

7.3.1.1 Method Used by SAS Enterprise Miner

When the target is ordinal, SAS Enterprise Miner creates lift charts based on the probability of the highest level of the target variable. The highest level in this example is 3. For each record in the Test data set, SAS Enterprise Miner computes the predicted, or posterior, probability $Pr(lossfrq = 3 \mid X_i)$ from the model. Then it sorts the data set in descending order of the predicted probability and divides the data set into 20 deciles, which I refer to alternatively as demi-deciles. Within each decile, SAS Enterprise Miner calculates the proportion of cases, as well as the total number of cases, with *lossfrq* = 3. The lift for a decile is the ratio of the proportion of cases where *lossfrq* =3 in the decile to the proportion of cases with *lossfrq* =3 in the entire data set. The capture rate is the ratio of the total number of cases with *lossfrq* =3 within the decile to the total number of cases with *lossfrq* =3 in the entire data set.

7.3.1.2 An Alternative Approach Using Expected Lossfrq

As pointed out earlier in Section 5.5.1.4, you sometimes need to calculate lift and capture rate based on the expected value of the target variable. For each model, I present lift tables and capture rates based on the expected value of the target variable.

For each record in the Test data set, I calculate the expected *lossfrq* as

$$E(lossfrq \mid X_i) = Pr(lossfrq = 0 \mid X_i) * 0 + Pr(lossfrq = 1 \mid X_i) * 1$$
$$+ Pr(lossfrq = 2 \mid X_i) * 2 + Pr(lossfrq = 3 \mid X_i) * 3.$$

Next, I sort the records of the data set in descending order of $E(lossfrq \mid X_i)$, divide the data set into 20 demi-deciles (bins), and within each demi-decile, I calculate the mean of the actual or observed value of the target variable *lossfrq*. The ratio of the mean of actual *lossfrq* in a demi-decile to the overall mean of actual *lossfrq* for the data set is the lift for the demi-decile. Likewise, the capture rate for a demi-decile is the ratio of the sum of the actual *lossfrq* for all the records within the demi-decile to the sum of the actual *lossfrq* of all the records in the entire data set.

7.3.2 Logistic Regression with Proportional Odds for Predicting Risk in Auto Insurance

Given that the measurement level of the target variable is set to Ordinal, a Proportional Odds model is estimated by the **Regression** node using the property settings shown in Displays 7.11 and 7.12.

Display 7.11

Model Selection	
Selection Model	Stepwise
Selection Criterion	Validation Error
Use Selection Defaults	No
Selection Options	[...]

Display 7.12

Property	Value
Sequential Order	No
Entry Significance Level	0.05
Stay Significance Level	0.05
Start Variable Number	0
Stop Variable Number	0
Force Candidate Effects	0
Hierarchy Effects	Class
Moving Effect Rule	None
Maximum Number of Steps	100

The reasons for the settings of the **Selection Model** property and the **Selection Criterion** property are cited in previous sections. The choice of Validation Error is also prompted by the lack of an appropriate profit matrix.

Display 7.13 shows the estimated equations for the Proportional Odds model.

Display 7.13

```
              Analysis of Maximum Likelihood Estimates

                            Standard        Wald
Parameter        DF  Estimate    Error  Chi-Square  Pr > ChiSq

Intercept 3       1   -2.0400   0.3486       34.25      <.0001
Intercept 2       1   -0.0786   0.2629        0.09      0.7650
Intercept 1       1    1.6415   0.2509       42.79      <.0001
AGE               1   -0.0343   0.00332     106.53      <.0001
CRED              1   -0.00593  0.000407    212.44      <.0001
NPRVIO            1    0.5624   0.0407      190.88      <.0001
```

Display 7.13 shows that the estimated model is Logistic with Proportional Odds. Accordingly, the slopes (coefficients of the explanatory variables) for the three logit equations are the same, but the intercepts are different. There are only three equations in Display 7.13, while there are four events represented by the four levels (0, 1, 2, and 3) of the target variable *lossfrq* . The three equations given in Display 7.13 plus the fourth equation, given by the identity

$$\Pr(lossfrq = 0) + \Pr(lossfrq = 1) + \Pr(lossfrq = 2) + \Pr(lossfrq = 3) = 1$$

are sufficient to solve for the probabilities of the four events. (See Section 6.2.2 of Chapter 6 to review how these probabilities are calculated by the **Regression** node.) In the model presented in Section 6.2.2, the target variable has only three levels, while the target variable in the model presented in Display 7.13 has four levels: 0, 1, 2, and 3.

The variables selected by the **Regression** node are AGE (age of the insured), CRED (credit score of the insured), and NPRVIO (number of prior violations). The cumulative lift charts from the Results window of the **Regression** node are shown in Display 7.14.

Display 7.14

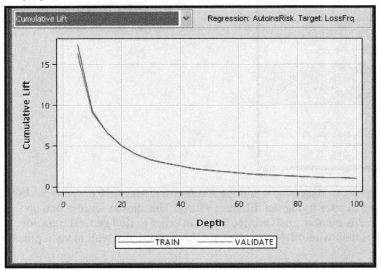

The cumulative lift charts shown in Display 7.14 are based on the highest level of the target variable, as described in Section 7.3.1. Using Test data, alternative cumulative lift, and capture rates based on $E(lossfrq)$, were computed and are shown in Table 7.5.

Table 7.5

REG			Loss Frequency					
Bin	Percentile	Number of Policies	Total	Mean	Cumulative Total	Cumulative Mean	Cumulative Lift	Cumulative Capture Rate(%)
1	5	450	126	0.280	126	0.280	5.09	25.4%
2	10	451	58	0.129	184	0.204	3.71	37.1%
3	15	451	61	0.135	245	0.181	3.29	49.4%
4	20	450	28	0.062	273	0.151	2.75	55.0%
5	25	451	43	0.095	316	0.140	2.55	63.7%
6	30	451	33	0.073	349	0.129	2.35	70.4%
7	35	451	20	0.044	369	0.117	2.13	74.4%
8	40	450	18	0.040	387	0.107	1.95	78.0%
9	45	451	18	0.040	405	0.100	1.81	81.7%
10	50	451	9	0.020	414	0.092	1.67	83.5%
11	55	451	13	0.029	427	0.086	1.57	86.1%
12	60	450	12	0.027	439	0.081	1.48	88.5%
13	65	451	10	0.022	449	0.077	1.39	90.5%
14	70	451	8	0.018	457	0.072	1.32	92.1%
15	75	451	4	0.009	461	0.068	1.24	92.9%
16	80	450	9	0.020	470	0.065	1.18	94.8%
17	85	451	11	0.024	481	0.063	1.14	97.0%
18	90	451	10	0.022	491	0.061	1.10	99.0%
19	95	451	5	0.011	496	0.058	1.05	100.0%
20	100	451	0	0.000	496	0.055	1.00	100.0%

7.3.3 Decision Tree Model for Predicting Risk in Auto Insurance

Display 7.15 shows the settings of the properties of the **Decision Tree** node.

Display 7.15

Splitting Rule	
Interval Criterion	ProbF
Nominal Criterion	ProbChisq
Ordinal Criterion	Entropy
Significance Level	0.2
Missing Values	Use in search
Use Input Once	No
Maximum Branch	2
Maximum Depth	6
Minimum Categorical Size	5
Split Precision	4
Node	
Split Search	
Subtree	
Method	Assessment
Number of Leaves	1
Assessment Measure	Average Square Error
Assessment Fraction	0.25

Note that I set the **Subtree Method** property to Assessment and the **Assessment Measure** property to Average Square Error. These choices were arrived at after trying out different values. The chosen settings end up yielding a reasonably sized tree—one that is not too small nor too large. In order to find general rules for making these choices, you need to experiment with different values for these properties with many replications.

Displays 7.16A and 7.16B show a part of the tree produced by the property settings given in Display 7.15.

Display 7.16A

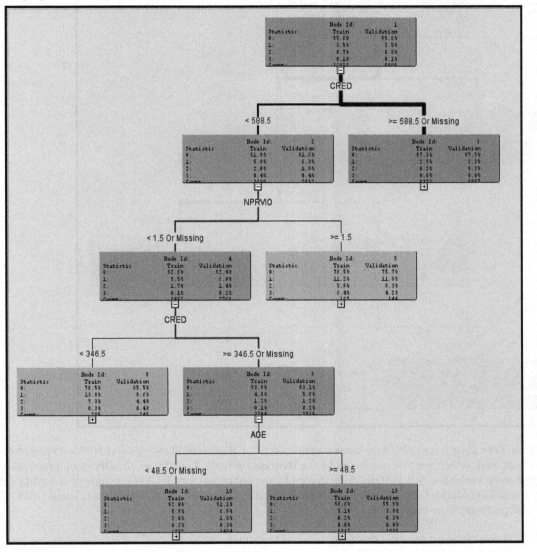

Display 7.16B

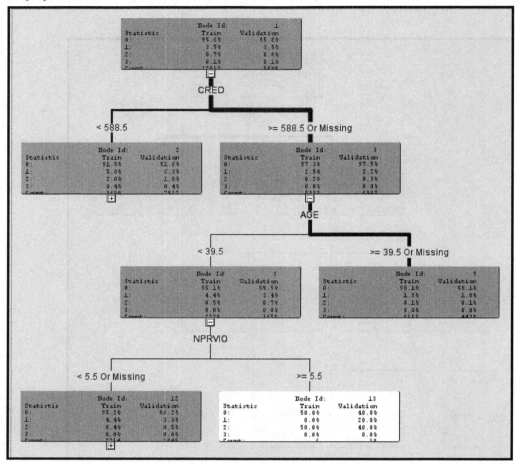

The **Decision Tree** node has selected the three variables AGE, CRED (credit rating), and NPRVIO (number of previous violations), which were also selected by the **Regression** node. In addition, the **Decision Tree** node selected five more variables: MILEAGE (miles driven by the policy holder), NUMTR (number of credit cards owned by the policyholder), DELINQ (number of delinquencies), HEQ (value of home equity), and DEPC (number of department store cards owned).

Display 7.17 shows the importance of the variables used in splitting the nodes.

Display 7.17

Variable Name	Label	Number of Splitting Rules	Importance	Validation Importance	Ratio of Validation to Training Importance
CRED		5	1	1	1
NPRVIO		3	0.8202	0.9011	1.0987
AGE		4	0.6104	0.5522	0.9047
MILEAGE		1	0.3222	0.2651	0.8228
NUMTR		1	0.2864	0.3626	1.266
DELINQ		1	0.2738	0.2384	0.8706
HEQ		1	0.261	0.2129	0.8155
DEPC		1	0.2202	0.1499	0.6809

Display 7.18 shows the lift charts for the Decision Tree model.

Display 7.18

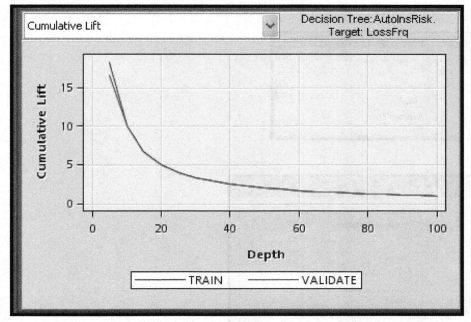

Table 7.6 shows the lift charts calculated using the expected loss frequency.

Table 7.6

Bin	Percentile	Number of Policies	Total	Mean	Cumulative Total	Cumulative Mean	Cumulative Lift	Cumulative Capture Rate(%)
1	5	450	95	0.211	95	0.211	3.84	19.2%
2	10	451	48	0.106	143	0.159	2.88	28.8%
3	15	451	92	0.204	235	0.174	3.16	47.4%
4	20	450	26	0.058	261	0.145	2.63	52.6%
5	25	451	26	0.058	287	0.127	2.32	57.9%
6	30	451	40	0.089	327	0.121	2.20	65.9%
7	35	451	13	0.029	340	0.108	1.96	68.5%
8	40	450	26	0.058	366	0.102	1.85	73.8%
9	45	451	13	0.029	379	0.093	1.70	76.4%
10	50	451	15	0.033	394	0.087	1.59	79.4%
11	55	451	15	0.033	409	0.082	1.50	82.5%
12	60	450	9	0.020	418	0.077	1.40	84.3%
13	65	451	9	0.020	427	0.073	1.32	86.1%
14	70	451	8	0.018	435	0.069	1.25	87.7%
15	75	451	8	0.018	443	0.066	1.19	89.3%
16	80	450	12	0.027	455	0.063	1.15	91.7%
17	85	451	18	0.040	473	0.062	1.12	95.4%
18	90	451	5	0.011	478	0.059	1.07	96.4%
19	95	451	3	0.007	481	0.056	1.02	97.0%
20	100	451	15	0.033	496	0.055	1.00	100.0%

TREE — Loss Frequency

7.3.4 Neural Network Model for Predicting Risk in Auto Insurance

The property settings for the **Neural Network** node are shown in Displays 7.19 and 7.20.

Display 7.19

Train	
Variables	...
Continue Training	No
Network	...
Optimization	...
Initialization Seed	12345
Model Selection Criterion	Average Error
Suppress Output	No

Display 7.20

Property	Value
Architecture	Multilayer Perceptron
Direct Connection	No
Number of Hidden Units	3
Randomization Distribution	Normal
Randomization Center	0.0
Randomization Scale	0.1
Input Standardization	Standard Deviation
Hidden Layer Combination Function	Default
Hidden Layer Activation Function	Default
Hidden Bias	Yes
Target Layer Combination Function	Default
Target Layer Activation Function	Logistic
Target Layer Error Function	Default

Because the number of inputs available in the data set is small, I passed all of them into the **Neural Network** node. I set the **Number of Hidden Units** property to its default value of 3, since any increase in this value did not improve the results significantly.

Display 7.21 shows the lift charts for the neural networks model.

Display 7.21

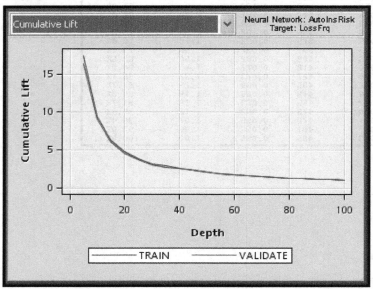

Table 7.7 shows the cumulative lift and capture rates for the Neural Network model based on the expected loss frequency using the Test data.

Table 7.7

Neural Bin	Percentile	Number of Policies	Total	Mean	Cumulative Total	Cumulative Mean	Cumulative Lift	Cumulative Capture Rate(%)
1	5	450	129	0.287	129	0.287	5.21	26.0%
2	10	451	57	0.126	186	0.206	3.75	37.5%
3	15	451	37	0.082	223	0.165	3.00	45.0%
4	20	450	54	0.120	277	0.154	2.79	55.8%
5	25	451	33	0.073	310	0.138	2.50	62.5%
6	30	451	21	0.047	331	0.122	2.22	66.7%
7	35	451	30	0.067	361	0.114	2.08	72.8%
8	40	450	19	0.042	380	0.105	1.92	76.6%
9	45	451	12	0.027	392	0.097	1.76	79.0%
10	50	451	17	0.038	409	0.091	1.65	82.5%
11	55	451	13	0.029	422	0.085	1.55	85.1%
12	60	450	18	0.040	440	0.081	1.48	88.7%
13	65	451	10	0.022	450	0.077	1.40	90.7%
14	70	451	13	0.029	463	0.073	1.33	93.3%
15	75	451	10	0.022	473	0.070	1.27	95.4%
16	80	450	3	0.007	476	0.066	1.20	96.0%
17	85	451	7	0.016	483	0.063	1.15	97.4%
18	90	451	7	0.016	490	0.060	1.10	98.8%
19	95	451	6	0.013	496	0.058	1.05	100.0%
20	100	451	0	0.000	496	0.055	1.00	100.0%

7.4 Comparison of All Three Accident Risk Models

Table 7.8 shows the lift and capture rates calculated for the Test data set ranked by $E(lossfrq)$, as outlined in Section 7.3.

Table 7.8

Bin	Percentile	Regression Cumulative Lift	Regression Cumulative Capture Rate (%)	Decision Tree Cumulative Lift	Decision Tree Cumulative Capture Rate (%)	Neural Network Cumulative Lift	Neural Network Cumulative Capture Rate(%)
1	5	5.09	25.4%	3.84	19.2%	5.21	26.0%
2	10	3.71	37.1%	2.88	28.8%	3.75	37.5%
3	15	3.29	49.4%	3.16	47.4%	3.00	45.0%
4	20	2.75	55.0%	2.63	52.6%	2.79	55.8%
5	25	2.55	63.7%	2.32	57.9%	2.50	62.5%
6	30	2.35	70.4%	2.20	65.9%	2.22	66.7%
7	35	2.13	74.4%	1.96	68.5%	2.08	72.8%
8	40	1.95	78.0%	1.85	73.8%	1.92	76.6%
9	45	1.81	81.7%	1.70	76.4%	1.76	79.0%
10	50	1.67	83.5%	1.59	79.4%	1.65	82.5%
11	55	1.57	86.1%	1.50	82.5%	1.55	85.1%
12	60	1.48	88.5%	1.40	84.3%	1.48	88.7%
13	65	1.39	90.5%	1.32	86.1%	1.40	90.7%
14	70	1.32	92.1%	1.25	87.7%	1.33	93.3%
15	75	1.24	92.9%	1.19	89.3%	1.27	95.4%
16	80	1.18	94.8%	1.15	91.7%	1.20	96.0%
17	85	1.14	97.0%	1.12	95.4%	1.15	97.4%
18	90	1.10	99.0%	1.07	96.4%	1.10	98.8%
19	95	1.05	100.0%	1.02	97.0%	1.05	100.0%
20	100	1.00	100.0%	1.00	100.0%	1.00	100.0%

From Table 7.8, it is clear that the logistic regression with proportional odds is the winner in terms of lift and capture rates. The table shows that, for the logistic regression, the cumulative lift at Bin 8 is 1.95. This means that the actual average loss frequency of the customers who are in the top 40 percentiles based on the logistic regression is 1.95 times the overall average loss frequency. The corresponding numbers for the Neural Network and Decision Tree models are 1.92 and 1.85, respectively.

7.5 Boosting and Combining Predictive Models

Sometimes you may be able to produce more accurate predictions using a combination of models instead of a single model. Gradient Boosting and Ensemble methods are two different ways of combining models. Gradient Boosting starts with an initial model and updates it by successively adding a sequence of regression trees in a step-wise manner. Each tree in the sequence is created by using the residuals (gradients of an error function) from the model in the previous step as the target. If the data set used to create each tree in the sequence of trees is a random sample from the Training data set, then the method is called Stochastic Gradient Boosting. In Stochastic Gradient Boosting, a sample is selected at each step by Random Sampling Without Replacement. In the selected sample, observations are weighted according to the errors from the model in the previous step. In SAS Enterprise Miner 12.1, Gradient Boosting is done by the **Gradient Boosting** node.

While Boosting and Stochastic Boosting combine the models created sequentially from the initial model, the Ensemble method combines a collection of models independently created for the same target. The models combined can be of different types. The **Ensemble** node can be used to combine different models created by different nodes as demonstrated in Section 7.5.4.

The description of the techniques given in sections 7.5.1 and 7.5.2 may not reflect the exact computational steps used by SAS Enterprise Miner. They are intended to give you an understanding of Gradient Boosting and Stochastic Gradient Boosting in general. To gain further insights into Gradient Boosting and Stochastic Gradient Boosting techniques, you should read the papers by Friedman.[1]

7.5.1 Gradient Boosting

The Gradient Boosting algorithm starts with an initial model represented by a mapping function, which can be written as:

$$F_0(x) = \arg\min_\gamma \sum_{i=1}^{N} \Psi(y_i, \gamma) \tag{7.1}$$

where $F_0(x)$ is the initial model, N is the number of records in the Training data, Ψ is a Loss Function, γ is a parameter that can be called the "node score" and y_i is the actual value of the target variable for the ith record. Equation 7.1 implies that the value of the parameter γ is chosen such that the Loss given by $\sum_{i=1}^{N} \Psi(y_i, \gamma)$ is minimized. Suppose we use the Loss Function $(y_i - \gamma)^2$ for the ith record. Then the aggregate loss is calculated as:

$$\sum_{i=1}^{N} \Psi(y_i, \gamma) = \sum_{i=1}^{N} (y_i - \gamma)^2 \tag{7.2}$$

Minimization of the right-hand side of the equation 7.2 leads to the estimate:

$$\gamma = \frac{\sum_{i=1}^{N} y_i}{N} = \overline{y}$$, which is the mean of the target variable in the Training data set.

Using the initial model specified by Equation 7.1, the predicted value of the target variable for each record is the overall mean. Denoting the predicted value of the target for the ith record from the initial model by $\hat{y}_{i0}$:

$$\hat{y}_{i0} = \overline{y} \tag{7.3}$$

Hence we can write the Loss Functions in terms of the predicted value of the target as:

$$\left(y_i - \hat{y}_{i0} \right)^2 \qquad (7.4)$$

If we use the Loss Function $\left| y_i - \gamma \right|$, then the value of the parameter γ that minimizes the Loss is the median denoted by $\tilde{y}$. The Loss Function $\left(y_i - \gamma \right)^2$ is called the *Least Squares Loss* and the Loss Function $\left| y_i - \gamma \right|$ is called the *Least Absolute Deviation Loss*. The Appendix to this chapter describes other Loss Functions. To make the explanations simple, I will use Least Squares Loss in all of the illustrations in this section.

A mapping function is an equation or set of definitions which enable you to calculate the predicted value of the target variable for each record from the values of inputs for that record. Thus, a mapping functions maps inputs into the target variable.

Calculation of the Residual

The residual for the ith record is defined as the negative of the derivative of the loss function *with respect to the predicted value*. The first derivative is also referred to as the gradient. In the case of the Least Squares Loss, using the Loss Function written in terms of the predicted value of the target variable (Equation 7.4), the gradient (derivative) is:

$$-2 \left(y_i - \hat{y}_{i0} \right) \qquad (7.5)$$

and the negative gradient is $2 \left(y_i - \hat{y}_{i0} \right)$.

Dropping the constant term 2, the gradient is $\left(y_i - \hat{y}_{i0} \right)$, which is nothing but the difference between the actual and predicted values of the target variable for each record.

Hence, when you use Least Squares Loss Function, the residual is calculated as:

$$\left(y_i - \hat{y}_{i0} \right) \qquad (7.6)$$

The residual calculated from the initial model becomes the target variable for the decision tree developed in the first iteration, the residual from the tree developed in the first iteration becomes the target variable in the second iteration, and so on.

Iteration 1:

The residuals from the initial model, shown in Equation 7.6 are used as the target variable in the first iteration tree.

$$y_{i1} = y_i - \hat{y}_{i0}, i = 1 \ to \ N \qquad (7.7)$$

A tree is developed using y_{i1} as the target along with the input vector x_i for $i = 1 \ to \ N$. Suppose this tree has L terminal nodes. For each terminal node, calculate a predicted value of the target variable $\gamma_{\ell 1}$ where

$\gamma_{\ell 1} = \arg\min_\gamma \sum_{i \in R_{\ell 1}} \Psi \left(y_i, F_0 \left(x_i \right) + \gamma \right)$. In the case of a Least Squares Loss function, for a record that is in

the ℓth terminal node, find the parameter γ such that $\sum_{i \in R_{\ell 1}} \left\{ \left(y_i - \left(\hat{y}_{i0} + \gamma \right) \right) \right\}^2$ is minimized. I leave it to you

to calculate the estimate of γ. The estimate of the parameter γ for the ℓth terminal node in iteration 1 can be denoted by γ_{l1}.

The predicted value of the target for the ith record in the ℓth terminal node is given by $\hat{y}_{i1} = \hat{y}_{i0} + v.\gamma_{\ell 1}$, where v $(0 < v \leq 1)$ is called the "shrinkage" parameter. All the records in a given node will have the same predicted value for the target in this iteration.

Iteration 2:

The residuals from the model in iteration 1 are used for the target variable in developing the second iteration tree. Therefore the target variable for iteration 2 is:

$$y_{i2} = y_i - \hat{y}_{i1}, i = 1 \ to \ N \tag{7.8}$$

A tree is developed using y_{i2} as the target along with the input vector x_i for $i = 1 \ to \ N$. Suppose that this tree has L terminal nodes. For each terminal node, calculate a predicted value of the target variable $\gamma_{\ell 2}$ as $\gamma_{\ell 2} = \arg\min_{\gamma} \sum_{i_i \in R_{\ell 2}} \Psi(y_i, F_1(x_i) + \gamma)$. In the case of a Least Squares Loss function, for a record that is in the ℓth terminal node, find the parameter γ such that $\sum_{i \in R_{\ell 2}} \{(y_i - (\hat{y}_{i1} + \gamma))\}^2$ is minimized. The estimate of the parameter γ for the ℓth terminal node in iteration 2 can be denoted by γ_{l2}.

The predicted value of the target for the ith record in the ℓth terminal node is given by $\hat{y}_{i2} = \hat{y}_{i1} + v.\gamma_{\ell 2}$.

The predicted values are updated for iterations 2, 3, 4 ...m-1 in the same way.

Iteration m:

The residuals from the model in iteration m-1 are used for the target variable in developing the second iteration tree. Therefore the target variable for iteration m is:

$$y_{im} = y_i - \hat{y}_{im-1}, i = 1 \ to \ N \tag{7.9}$$

A tree is developed using y_{im} as the target along with the input vector x_i for $i = 1 \ to \ N$. Suppose that this tree has L terminal nodes. For each terminal node calculate a predicted value of the target variable as $\gamma_{\ell m}$ as $\gamma_{\ell m} = \arg\min_{\gamma} \sum_{i \in R_{\ell m}} \Psi(y_i, F_{m-1}(x_i) + \gamma)$. In the case of a Least Squares Loss function, for a record that is in the ℓth terminal node, the parameter γ is estimated such that $\sum_{i \in R_{\ell m}} \{(y_i - (\hat{y}_{im-1} + \gamma))\}^2$ is minimized. .

The estimate of the parameter γ for the ℓth terminal node in iteration m can be denoted by γ_{lm}.

The predicted value of the target for the ith record in the ℓth terminal node is given by $\hat{y}_{im} = \hat{y}_{im-1} + v.\gamma_{\ell m}$.

7.5.2 Stochastic Gradient Boosting

In Stochastic Gradient Boosting, at each iteration, the tree is developed using a random sample (without replacement) from the Training data set. In the sampling at each step, weights are assigned to the observations according the prediction errors in the previous step, assigning higher weights to observations with larger errors.

7.5.3 An Illustration of Boosting Using the Gradient Boosting Node

Display 7.22 shows the process flow diagram for a comparison of models created by **Gradient Boosting** and **Decision Tree** nodes.

Display 7.22

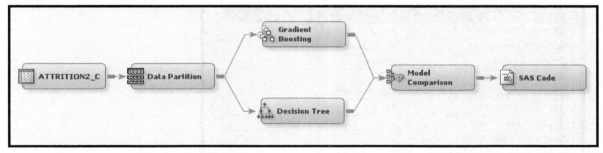

The property settings of the **Gradient Boosting** node are shown in Display 7.23.

Display 7.23

Property	Value
General	
Node ID	Boost
Imported Data	
Exported Data	
Notes	
Train	
Variables	
□ Series Options	
├ N Iterations	100
├ Seed	12345
├ Shrinkage	0.1
└ Train Proportion	60
□ Splitting Rule	
├ Huber M-Regression	No
├ Maximum Branch	2
├ Maximum Depth	2
├ Minimum Categorical Size	5
├ Re-use Variable	1
├ Categorical Bins	30
├ Interval Bins	100
├ Missing Values	Use in search
└ Performance	Disk
□ Node	
├ Leaf Fraction	0.1
├ Number of Surrogate Rules	0
└ Split Size	,
□ Split Search	
├ Exhaustive	5000
└ Node Sample	20000
□ Subtree	
└ Assessment Measure	Average Square Error

The property settings for the **Decision Tree** node are shown in Display 7.24.

Display 7.24

Splitting Rule	
Interval Criterion	ProbF
Nominal Criterion	ProbChisq
Ordinal Criterion	Entropy
Significance Level	0.2
Missing Values	Use in search
Use Input Once	No
Maximum Branch	2
Maximum Depth	6
Minimum Categorical Size	5
Split Precision	4
Node	
Split Search	
Subtree	
Method	Assessment
Number of Leaves	1
Assessment Measure	Average Square Error

Table 7.9 shows a comparison of Cumulative Lift and Cumulative Capture Rate for the two models for the Test data.

Table 7.9

TEST DATA

		Gradient Boosting		Decision Tree	
Bin	Percen-tile	Cumulative Lift	Cumulative Capture Rate(%)	Cumulative Lift	Cumulative Capture Rate(%)
1	5	3.08	15.42	3.05	15.28
2	10	3.03	30.37	3.04	30.51
3	15	2.73	41.12	2.56	38.48
4	20	2.83	56.54	2.38	47.54
5	25	2.58	64.49	2.27	56.70
6	30	2.33	70.09	2.19	65.85
7	35	2.17	76.17	2.11	73.92
8	40	1.99	79.44	1.93	77.00
9	45	1.84	82.71	1.78	80.11
10	50	1.67	83.64	1.66	83.22
11	55	1.58	86.92	1.57	86.32
12	60	1.47	88.32	1.48	88.81
13	65	1.39	90.65	1.39	90.22
14	70	1.31	92.06	1.31	91.62
15	75	1.28	95.79	1.24	93.03
16	80	1.21	97.20	1.18	94.42
17	85	1.15	98.13	1.13	95.83
18	90	1.09	98.13	1.08	97.24
19	95	1.04	99.07	1.04	98.65
20	100	1.00	100.00	1.00	100.00

By comparing the Cumulative Lift and Cumulative Capture rates for Gradient Boosting and Decision Tree models from the 15th percentile onwards, we can find that Gradient Boosting has yielded slightly better predictions than the Decision Tree model.

7.5.4 The Ensemble Node

We can use the **Ensemble** node to combine the three models' **Logistic Regression**, **Decision Tree**, and **Neural Network** nodes, as shown in Display 7.25.

Display 7.25

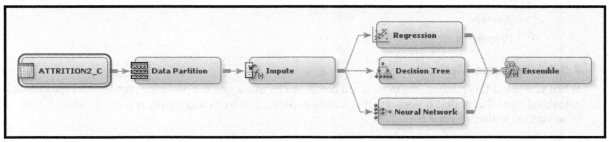

The target variable used in the **Input Data** node is ATTR, which is a binary variable. It takes the value 1 if a customer attrited during a given time interval and 0 if the customer did not attrit. When you run the **Regression**, the **Decision Tree**, the **Neural Network**, and the **Ensemble** nodes as shown in Display 7.25, the **Ensemble** node calculates the posterior probabilities and an "Into" variable for each record in the data set from each model.

The **Ensemble** node creates the variables listed Table 7.10.

Table 7.10

		Values for illustration
Reg_P_ATTR1	Probability that a customer attrits -- estimated from the Logistic Regression Model	0.6
Reg_P_ATTR0	Probability that a customer does not attrit -- estimated from the Logistic Regression Model .	0.4
Reg_I_ATTR	"Into" Variable: Indicates the target level assigned to a record based on the estimated probabilities Reg_P_ATTR1 and Reg_ATTR0 calculatd from the Logistic Regression Model.	"1"
Tree_P_ATTR1	Probability that a customer attrits -- estimated from the Decision Tree Model.	0.52
Tree_P_ATTR0	Probability that a customer does not attrit -- estimated from the Decision Tree Model.	0.48
Tree_I_ATTR	"Into" Variable: Indicates the target level assigned to a record based on the estimated probabilities Tree_P_ATTR1 and Tree_ATTR0 calculatd from the Decision Tree Model.	"1"
Neural_P_ATTR1	Probability that a customer attrits -- estimated from the Neural Network Model.	0.3
Neural_P_ATTR0	Probability that a customer does not attrit -- estimated from the Neural Network Model.	0.7
Neural_I_ATTR	"Into" Variable: Indicates the target level assigned to a record based on the estimated probabilities Neural_P_ATTR1 and Neural_ATTR0 calculatd from the Neural Network Model.	"0"

In the last column of Table 7.10, I assigned hypothetical values for each of the variables for illustrating the different combination methods of the **Ensemble** node. Note that the value of the "Into" variable from each model is based on the Maximum Posterior Probability criterion. That is, if the predicted probability from a model for level 1 is higher than the predicted probability for level 0, then the record (customer) is given the label 1. If 1 represents the event that the customer attrited and 0 represents the event that the customer did not attrit, then by assigning level 1 to the customer, the model is classifying him/her as an "attritor". You can use some other criteria such as Maximum Expected Profits by providing a weight matrix or a profit matrix.

For example, from Table 7.10 it can be seen that the probability of attrition calculated from a Logistic Regression model for the hypothetical customer is 0.6, and the probability of no attrition for the same customer is 0.4. Since 0.6 >0.4, the record is assigned level 1 if you use the Logistic Regression model for classification.

In the **Ensemble** node there are three methods for combining multiple models. They are:

- Average
- Maximum
- Voting
  - Average
  - Proportion

7.5.4.1 Average Method

When you select this method, the models are combined by averaging the posterior probabilities (probabilities calculated from the model) if your target is a class variable, and by averaging the predicted values if your target is an interval scaled variable.

To select this method, set the **Posterior Probabilities** property to Average for a class target (including binary) and set the **Predicted Values** property to Average for an interval (continuous) target.

Using the values shown in Table 7.10, the posterior probabilities for the combined model are:

$$P\_ATTR1 = (Reg\_P\_ATTR1 + Tree\_P\_ATTR1 + Neural\_P\_ATTR1)/3;$$

$$= (0.6+0.52+0.30)/3 = 0.473,$$

$$P\_ATTR0 = (Reg\_P\_ATTR0 + Tree\_P\_ATTR0 + Neural\_P\_ATTR0)/3;$$

$$= (0.4+0.48+0.7)/3 = 0.527$$

The "Into" variable for the combined model is I_ATTR = 0.

Since P_ATTR0 > P_ATTR1, the record is assigned to the target level 0, which means that the customer is labeled as non-attriter.

7.5.4.2 Maximum Method

In this method, the models are combined by taking the maximum of the posterior probabilities for class targets, and by taking the maximum of predicted values for interval targets from the collection of models.

To select this method, set the **Posterior Probabilities** property to Maximum for a class target (including binary), and set the **Predicted Values** property to Maximum for an interval (continuous) target.

Using the values shown in Table 7.10, the posterior probabilities for the combined model are:

$$P\_ATTR1 = max (Reg\_P\_ATTR1, Tree\_P\_ATTR1, Neural\_P\_ATTR1);$$

$$= max (0.6, 0.52, 0.3) = 0.6,$$

$$P\_ATTR0 = 0.4$$

The "Into" variable for the combined model is I_ATTR = 1.

Since P_ATTR1 > P_ATTR0, the record is assigned to the target level 1, which means that the customer is labeled as an attriter.

7.5.4.3 Voting _ Average Method

This method is available for class targets only. To select this method for a class target, set the **Posterior Probabilities** property to Voting and the **Voting Posterior Probabilities** property to Average. When you set these values, the posterior probabilities of the combined model are calculated by averaging the posterior properties from the models which are in the majority group.

From the example shown in Table 7.10, two models, the Logistic Regression and the Decision Tree models, have assigned the record to 1, and one model, the Neural Network model, has assigned the record to 0. That is, two models are predicting that the customer will attrit, and one model is predicting that customer will not attrit. Since majority of the models are predicting attrition, the **Ensemble** node calculates the posterior probabilities of the combined model from the two models (Logistic Regression and the Decision Tree models). Since we set the **Voting Posterior Probabilities** property to Average, the **Ensemble** node calculates the posterior probabilities from the Logistic Regression and the Decision Tree models by averaging their posterior probabilities.

Hence, from the values shown in Table 7.10, the posterior probabilities for the combined model are P_ATTR1 = (0.6+0.52)/2 = 0.560, P_ATTR0 = (0.4+0.48/2 = 0.440

The "Into" variable for the combined model I_ATTR = 1. Since P_ATTR1 > P_ATTR0, the record is assigned to the target level 1 which means that the customer is labeled as an attriter.

7.5.4.4 Voting _ Proportion Method

If you set the **Posterior Probabilities** property to Voting and the **Voting Posterior Probabilities** property to Proportion, then the posterior probability for a record is calculated as the ratio of the number of models in the majority group and the total number of models included in the **Ensemble** node for combining.

In our example, there are three models – Logistic Regression, Decision Tree, and Neural Network. From Table 7.10, we can see that two of these models (Logistic Regression and Decision Tree) are predicting that the customer is an attriter i.e. the "Into" variable for these models is 1, and one model is predicting that the customer is not an attriter for the record. Therefore, the posterior probability of level 1 for the combined model for that record is 2/3.

Using the values shown in Table 7.10, the posterior probabilities and the "Into" variable for the combined model are:

P_ATTR1 =0.667, P_ATTR0 =0.333 and I_ATTR = 1

Since P_ATTR1 > P_ATTR0, the record is assigned to the target level 1, which means that the customer is labeled as an attriter.

Table 7.11 shows a summary of the different ways the **Ensemble** node can combine multiple models.

Table 7.11

Model	P_Attr1	P_Attr0	I_Attr
Logistic Regression	0.600	0.400	1
Decisiton Tree	0.520	0.480	1
Neural Network	0.300	0.700	0
Ensemble _Average	0.473	0.527	0
Ensemble _Maximum	0.600	0.400	1
Ensemble_Voting_Average	0.560	0.440	1
Ensemble_Voting_Proportion	0.667	0.333	1

P_Attr1: Probability that a customer will attrit within a time interval

P_Attr0: Probability that a customer will not attrit within the given time interval

I_Attr ("Into" Variable) : Indicates the target level assigned to a record.
The target levels are assigned to the records based on the posterior probabilities
computed from the model, and on the "maximum posterior probability" criterion .

7.5.5 Comparing the Gradient Boosting and Ensemble Methods of Combining Models

The task of comparing the combined model from the **Ensemble** node with the individual models included in the Ensemble is left as an exercise. In this section, we compare the combined model generated by the **Ensemble** node with the one generated by the **Gradient Boosting** node. Display 7.26 shows the process flow diagram for demonstrating the **Ensemble** node and also for comparing the models generated by the **Ensemble** and **Gradient Boosting** nodes.

Display 7.26

The property settings of the **Ensemble** node are shown in Display 7.27. The property settings of the **Gradient Boosting** node are same as those shown Display 7.23.

Display 7.27

Property	Value
General	
Node ID	Ensmbl
Imported Data	
Exported Data	
Notes	
Train	
Variables	
⊟ Interval Target	
└ Predicted Values	Average
⊟ Class Target	
├ Posterior Probabilities	Average
└ Voting Posterior Probabilities	Average

Table 7.12 shows Cumulative Lift and Cumulative Capture Rate from the models developed by the **Ensemble** and **Gradient Boosting** nodes for the Test data.

Table 7.12

Bin	Percen-tile	Gradient Boosting Cumulative Lift	Gradient Boosting Cumulative Capture Rate(%)	Ensemble Cumulative Lift	Ensemble Cumulative Capture Rate(%)
1	5	3.04	15.30	3.71	18.66
2	10	2.72	27.24	3.05	30.60
3	15	2.66	39.93	2.81	42.16
4	20	2.65	52.99	2.67	53.36
5	25	2.54	63.43	2.51	62.69
6	30	2.32	69.78	2.32	69.78
7	35	2.14	75.00	2.11	73.88
8	40	1.98	79.10	1.94	77.61
9	45	1.82	81.72	1.80	80.97
10	50	1.70	85.07	1.68	83.96
11	55	1.59	87.31	1.55	85.07
12	60	1.47	88.43	1.45	87.31
13	65	1.38	89.93	1.38	89.93
14	70	1.31	91.42	1.32	92.16
15	75	1.24	93.28	1.27	95.52
16	80	1.19	95.15	1.20	95.90
17	85	1.13	96.27	1.14	96.64
18	90	1.09	98.13	1.09	97.76
19	95	1.04	98.88	1.04	98.88
20	100	1.00	100.00	1.00	100.00

From Table 7.12, it appears that the **Ensemble** method has produced better predictions than **Gradient Boosting** in percentiles 5-20.

7.6 Appendix to Chapter 7

7.6.1 Least Squares Loss

If the target variable is y and the estimated equation is $f(x)$, then the error for the ith record is $y_i - \hat{y}_i$, where x is the vector of inputs, $\hat{y}_i = f(x_i)$ is the predicted value of target variable, and y_i is the actual/observed value of the target variable for the ith record. The aggregate loss function for all the observations in the data set is:

$$\sum_{i=1}^{n}\{y_i - f(x_i)\}^2$$

7.6.2 Least Absolute Deviation Loss

$$\sum_{i=1}^{n} abs\{y_i - f(x_i)\}$$

7.6.3 Huber-M Loss

For the ith record:

$$L_i = 0.5[y_i - f(x_i)]^2 \qquad \text{if } |(y_i - f(x_i)| < \delta$$

$$L_i = \delta(|(y_i - f(x_i)| - 0.5\delta) \qquad \text{if } |(y_i - f(x_i)| \geq \delta$$

7.6.4 Logit Loss

In Equation 6.28 in Chapter 6, we have shown that the error in a logistic regression is often measured by the negative of the log-likelihood. Therefore the error for the ith record is:

$$-\log(L_i) = -\left[y_i \log\left(\hat{\pi}_i(x_i)\right) + (1-y_i)\log\left(1-\hat{\pi}_i(x_i)\right)\right] \tag{7.10}$$

If the ith record refers to a responder, then $y_i = 1$ and the likelihood for the ith record becomes

$$L_i = \Pr(y_i = 1 \mid x_i) = \hat{\pi}_i(x_i) = \frac{1}{1+\exp(-\beta' x_i)} \tag{7.11}$$

where β is the vector of estimated coefficients and x_i is the vector of inputs for the ith record.

If the ith record refers to a non-responder, then $y_i = 0$ and the likelihood for the ith record becomes

$$L_i = \Pr(y_i = 0 \mid x_i) = 1 - \hat{\pi}_i(x_i) = \frac{1}{1+\exp(\beta' x_i)} \tag{7.12}$$

By introducing the indicator variable z_i, which takes the value $+1$ when $y_i = 1$ and -1 when $y_i = 0$, we can replace the right-hand side of equations 7.11 and 7.12 with the single expression $\dfrac{1}{1+\exp(-z_i \beta' x_i)}$, and write the log-likelihood for the ith record as

$$\log(L_i) = \log\left[\frac{1}{1+\exp(-z_i \beta' x_i)}\right] = -\log\left[1+\exp(-z_i \beta' x_i)\right] \tag{7.13}$$

As shown in Chapter 6, the error is equal to the negative log-likelihood. Therefore the error for the ith record is $-\log(L_i) = \log\left[1+\exp(-z_i \beta' x_i)\right]$. The error is also referred to as Loss. The aggregate Loss for all n records in a data set is:

$$\sum_{i=1}^{n} \log\left[1+\exp(-z_i \beta' x_i)\right] \tag{7.14}$$

7.7 Exercises

1. Complete the following tasks:
   a. Create a data source using the data set Ch7_Exercise.
   b. Partition the data set such that 50% of the observations are allocated to Training, 30% to Validation, and 20% to Test
   c. Use the Advanced Metadata Advisor Options and customize by setting the **Class Level Count Threshold** property to 5.
   d. Set the Role of the variable Event to Target.
   e. In the Data Source Wizard, select **No** when prompted for Decision Processing.
   f. Develop a decision tree by setting the **Subtree Method** property to Assess and the **Assessment Measure** property to Average Square Error.
   g. Develop another model using the **Gradient Boosting** node.

      h.   Compare Cumulative Lift and Cumulative Capture Rate for the two models using the Test data.

  2. Complete the following tasks:
      a.   Create a Decision Tree model.
      b.   Create a Neural Network model.
      c.   Create a Logistic Regression model.
      d.   Combine the three models (a, b, and c above) using the **Ensemble** node.
      e.   Compare the Cumulative Lift, Cumulative Capture Rate for the models created in 1g (Gradient Boosting) and 2d (Ensemble).

Of the five models you created (-, Gradient Boosting (1g), Decision Tree (2a), Neural Network (2b), Logistic Regression (2c), and Ensemble (2d), which model would you recommend? Why?

Note

1. (1) Friedman, J. H. (1999). *Stochastic Gradient Boosting, Technical Report*, Dept. of Statistics, Stanford University. (2) Friedman, J. H. (1999). *Greedy Function Approximation: A Gradient Boosting Machine*. Technical Report, Department of Statistics, Stanford University.

Chapter 8: Customer Profitability

8.1 Introduction

This chapter presents a general framework for calculating the profitability of different groups of customers, using a simplified example. The methodology can be extended to calculate profits at the individual customer level as well.

My goal is to illustrate how costs of acquisition and the costs associated with risk factors can affect the decision to acquire new customers, using the example of a credit card company.

For example, suppose you have a population of 10,000 prospects. A response model is used to score each of these 10,000 prospects. The prospects are arranged in descending order of the predicted probability of response and divided into deciles. A risk model is then used to calculate the risk rate for each prospect. In this simplified example, the *risk rate* is the probability of a customer defaulting on payment. Assume that a default results in a net loss of $10 for the credit card company on the average. If the customer does not default, the company makes $100. (These numbers are fictitious, and I use them for demonstration purposes only.) Table 8.1 shows the estimated response and risk rates for the deciles of the prospect population.

Table 8.1

Decile	Response Rate	Risk Rate
1	0.050	0.100
2	0.045	0.092
3	0.040	0.080
4	0.030	0.067
5	0.020	0.059
6	0.010	0.042
7	0.005	0.038
8	0.004	0.029
9	0.003	0.018
10	0.002	0.010

From Table 8.1, you can see that response rate and risk rate move in the same direction. The response rate is highest in the first decile and declines with succeeding higher-numbered deciles. A similar pattern is observed for the risk rate also. One reason why this may occur is that when a credit card company solicits applications for credit cards, the groups that respond most are likely to be those who cannot get credit elsewhere because of their relatively high risk rates.

In Table 8.1, the term *risk rate* is used in a general sense. A risk rate of 10% in decile 1 in Table 8.1 means that 10% of the persons who responded and are in decile 1 tend to default on their payments. (Here I am assuming that each person who responded is issued a credit card, but this assumption can be easily relaxed without violating the logic.) The response rate in a particular decile is the proportion of individuals in the decile who are responders. Display 8.1 graphs the response and risk rates for each decile.

Display 8.1

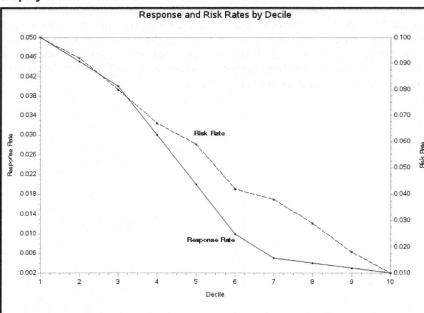

In Display 8.1, the horizontal axis shows the decile number. Decile 1 is the top decile, and decile 10 is the lowest decile.

Note: The example presented in this chapter is hypothetical and greatly simplified for the purpose of exposition; the costs and revenue figures are also quite arbitrary. The example refers to a credit card company but it can be extended to insurance and other companies as well. Details of the methodology should be modified according to the specific situation being analyzed. In many situations, for example, customers with a high response rate to a direct mailing campaign might also pose higher risks to the company that is soliciting business.

8.2 Acquisition Cost

New customers are acquired through channels such as newspaper or Internet advertisements, radio and TV broadcasting, direct mail, etc. If a company spends X dollars on a particular channel, and as a result it acquires n customers, then the average cost of acquisition per customer is X / n. For direct mail, this can be calculated easily. Suppose the response rate is r in a segment of the target population and suppose the cost of sending one mail piece is m dollars. If mail is sent to N customers from that segment, then the average cost of acquiring a customer is

$$\frac{N.m}{rN} = \frac{m}{r}.$$

This means that as the response rate decreases, the average acquisition cost increases. Display 8.2 illustrates this relationship.

Display 8.2

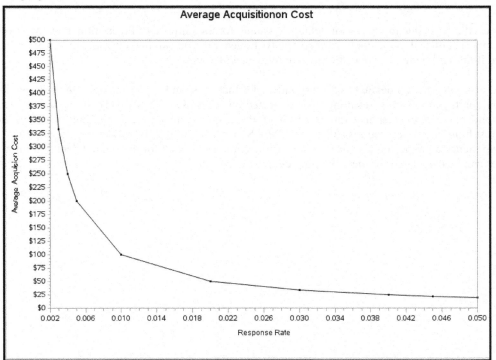

Table 8.2 shows the acquisition cost by decile. Since the response rate declines with succeeding higher-numbered deciles, the average acquisition cost increases. When the company sends mail to the 1,000 prospects in the top decile (decile 1) inviting them to purchase a product, at a cost of $1 per piece of mail, the total cost is $1,000. In return, the company acquires 50 customers, as the response rate in the top decile is 0.05. The average

cost of acquisition is therefore $1,000/50 = $20. Similarly, the average cost of acquisition for the second decile is $1,000/45 = $22.22. The last column shows the cumulative acquisition cost. If all the prospects in the top two deciles are sent a mail piece, the total cost would be $2,000. This is the cumulative acquisition cost, and it is an important item in determining the optimum cut-off point for mailing.

Table 8.2

Decile	Nummber of Customers	Cost per Mail-piece	Average Acquisition Cost	Total Acquisition Cost	Cumulative Acquisition Cost
1	1,000	$1	$20	$1,000	$1,000
2	1,000	$1	$22	$1,000	$2,000
3	1,000	$1	$25	$1,000	$3,000
4	1,000	$1	$33	$1,000	$4,000
5	1,000	$1	$50	$1,000	$5,000
6	1,000	$1	$100	$1,000	$6,000
7	1,000	$1	$200	$1,000	$7,000
8	1,000	$1	$250	$1,000	$8,000
9	1,000	$1	$333	$1,000	$9,000
10	1,000	$1	$500	$1,000	$10,000

8.3 Cost of Default

If a customer defaults on payment, his credit card company incurs a loss. The credit card company might face other types of risk, but for the purpose of illustration, only one type of risk, the risk of default, is considered here.

Table 8.3 illustrates the costs due to the risk of default. Assume, for the purpose of illustration, that each event (default) results in a cost of $10 to the credit card company. In general, you can include more events with associated probabilities and with more realistic costs in these calculations.

In this simplified example, the top decile has 50 responders. Of these responders, five people (the risk rate is 0.1) tend to default on their payments, resulting in an expected loss of $50 to the company. In the second decile there are 45 responders (since the response rate is 4.5%). Of these 45 customers, 9.2% are likely to default on their payments, resulting in an expected loss of 45 × 0.092 × $10 = $41. If the company targets the top two deciles, it will experience an expected loss of $91. This is the cumulative loss[1] for the second decile. Similarly cumulative losses can be calculated for the remaining deciles.

Table 8.3

Decile	Number of Responders	Risk Rate	Loss per Event	Loss Amount	Cumulative Loss
1	50	0.100	$10	$50	$50
2	45	0.092	$10	$41	$91
3	40	0.080	$10	$32	$123
4	30	0.067	$10	$20	$144
5	20	0.059	$10	$12	$155
6	10	0.042	$10	$4	$160
7	5	0.038	$10	$2	$161
8	4	0.029	$10	$1	$163
9	3	0.018	$10	$1	$163
10	2	0.010	$10	$0	$163

8.4 Revenue

Revenue depends on the number of customers who do not default. These revenues are shown in Table 8.4. To simplify the example, I have assumed that a customer who defaults on payment generates no revenue. This assumption can easily be relaxed without violating the logic of my argument.

In the top decile, there are 50 responders, but only 45 of them are customers in good standing at the end of the year (assuming that our analysis is based on calculations for one year). Hence the revenue generated in the top decile is 45 × $100 = $4,500. A similar calculation yields revenue of $4,086 for the second decile. Hence the cumulative revenue for the first two deciles is $8,586. Similarly, cumulative revenue calculated for all the remaining deciles is shown in Table 8.4.

Table 8.4

Decile	Number of Responders	Risk Rate	Average Revenue	Revenue	Cumulative Revenue
1	50	0.100	$100	$4,500	$4,500
2	45	0.092	$100	$4,086	$8,586
3	40	0.080	$100	$3,680	$12,266
4	30	0.067	$100	$2,799	$15,065
5	20	0.059	$100	$1,882	$16,947
6	10	0.042	$100	$958	$17,905
7	5	0.038	$100	$481	$18,386
8	4	0.029	$100	$388	$18,774
9	3	0.018	$100	$295	$19,069
10	2	0.010	$100	$198	$19,267

8.5 Profit

If, for example, the company targets the top two deciles, the expected revenue will be $8,586, the expected acquisition cost will be $2,000, and the expected losses will be $91. Therefore, the expected cumulative profit will be $8,586 – $2,000 – $91 = $6,495. Table 8.5 shows the cumulative profit for all the deciles.

Table 8.5

Decile	Cumulative Revenue	Cumulative Cost	Cumulative Proft
1	$4,500	$1,050	$3,450
2	$8,586	$2,091	$6,495
3	$12,266	$3,123	$9,143
4	$15,065	$4,144	$10,922
5	$16,947	$5,155	$11,792
6	$17,905	$6,160	$11,746
7	$18,386	$7,161	$11,225
8	$18,774	$8,163	$10,612
9	$19,069	$9,163	$9,906
10	$19,267	$10,163	$9,104

The optimum cut-off point is where the cumulative profit peaks. In our example, it occurs at decile 5. The company can maximize its profit by sending mail to only the 5,000 prospects who are in the top five deciles. This can be seen from Display 8.3.

Display 8.3

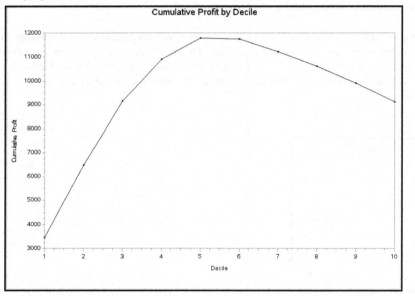

The cumulative profit peaks at the fifth decile because, beyond the fifth decile, the marginal profit earned by mailing to an additional decile is negative.

The *marginal cost* at the fifth decile is defined as the additional cost that the company would incur in acquiring the responders in the next (sixth) decile. The *marginal revenue* at the fifth decile is defined as the additional revenue the company would earn if it acquired all the responders in the next (sixth) decile. The *marginal profit* at the fifth decile is defined as the additional profit the company would make if it acquired the responders in the next (sixth) decile, which is the marginal revenue minus the marginal cost.

Beyond the fifth decile, the marginal cost outweighs the marginal revenue. Hence the marginal profit is negative. This is depicted in Display 8.4.

Display 8.4

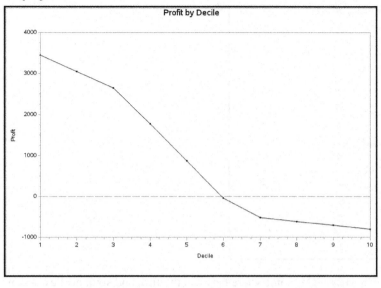

8.6 The Optimum Cut-off Point

The optimization problem can be analyzed in terms of marginal revenue and marginal cost. In this simplified example, marginal revenue (MR) at any decile is the additional revenue that is gained by mailing to the next decile, in addition to the previous deciles. Similarly, the marginal cost (MC) at any decile is the additional cost the company incurs by adding an additional decile for mailing. The marginal cost and marginal revenue, which are derived from the figures in Table 8.5, are shown in Table 8.6. In Table 8.5, you can see that if the company mails to the top five deciles, the cumulative (total) revenue is $16,947 and cumulative (total) cost is $5,155. If the company wants to mail to the top six deciles, the revenue will be $17,905 and cost will be $6,160. Hence the marginal revenue at 5 is $17,905 − $16,947 = $958 and marginal cost is $6,160 − $5,155 = $1005. Since marginal cost exceeds marginal revenue, the cumulative profits will decline by adding the sixth decile to the mailing. The marginal revenues and marginal costs at different deciles shown in Table 8.6 are plotted in Display 8.5.

Table 8.6

Decile	Marginal Revenue	MC	Profit by Decile
1	$4,500	$1,050	$3,450
2	$4,086	$1,041	$3,045
3	$3,680	$1,032	$2,648
4	$2,799	$1,020	$1,779
5	$1,882	$1,012	$870
6	$958	$1,004	$-46
7	$481	$1,002	$-521
8	$388	$1,001	$-613
9	$295	$1,001	$-706
10	$198	$1,000	$-802

Display 8.5

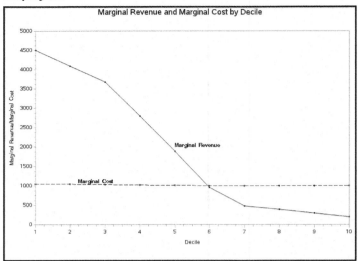

Display 8.5 shows that the point at which MR=MC is somewhere between the fifth and sixth deciles. As an approximation, you can stop mailing after the fifth decile.

In the above example, the profits are calculated for one year only for the sake of simplicity. Alternatively, you can calculate profits over a longer period of time or add other complications to the analysis to make it more pertinent to the business problem you are trying to solve. All these calculations can be made in the **SAS Code** node.

8.7 Alternative Scenarios of Response and Risk

In the example presented in Table 8.1, it is assumed that response rate and risk rate move in the same direction. The response rate is highest in the first decile, and declines with succeeding higher-numbered deciles. A similar pattern is observed for the risk rate. However, there are many situations in which response rate and risk rate may not show this type of a pattern. In such situations, you can estimate two scores for each customer—one based on probability of response, and the other based on risk. Next, the prospects can be arranged in a 10x10 matrix, each cell of the matrix consisting of the number of customers belonging to the same response-decile and risk-decile. Acquisition cost, cost of risk, revenue, and profit can be calculated for each cell, and acquisition decisions can be made for each group of customers based on the profitability of the cells.

8.8 Customer Lifetime Value

At the end of Section 8.6, I pointed out that the analysis of customer acquisition based on the marginal profit from mailing to additional prospects could be performed with a more distant horizon than the one year analysis shown in my example. Taking this to its logical conclusion, calculations of profitability at the level of an individual customer can be extended further by calculating the customer's lifetime value. For this you need answers to three questions:

- What is the expected residual lifetime of each customer?
- What is the flow of revenue that the customer is expected to generate in his "lifetime"?
- What are the future costs to the company of acquiring and retaining each customer?

Proper analysis of these elements can yield valuable insights into the type of actions that a company can take for extending the residual lifetime, the customer lifetime value, or both.

8.9 Suggestions for Extending Results

In the analysis presented in this chapter, Revenue, Cost, and Profit are calculated for each decile. Alternatively, you can do the same calculations for each percentile in your data base of prospects, rather than for each decile. This would enable you to choose the cutoff point for a mailing or other marketing promotion with greater precision. In addition, in my example, only one type of risk is identified. If, however, there is more than one type of risk, you can model competing risks by means of logistic hazard functions using either the **Regression** node or the **Neural Network** node of SAS Enterprise Miner.

Note

1. The losses are calculated using the estimated probabilities, and hence, strictly speaking, they should be called *expected losses*. But here I use the terms *losses* and *expected losses* interchangeably.

Chapter 9: Introduction to Predictive Modeling with Textual Data

9.1 Introduction

This chapter shows how you can use SAS Enterprise Miner's text mining nodes[1] to quantify unstructured textual data and create data matrices and SAS data sets that can be used for statistical analysis and predictive modeling. Some examples of textual data are: web pages, news articles, research papers, insurance reports, etc. In text mining, each web page, news article, research paper, etc. is called a document. A collection of documents, called a corpus of documents, is required for developing predictive models or classifiers. Each document is originally represented by a string of characters. In order to build predictive models from this data, you need to represent each document by a numeric vector whose elements are frequencies or some other measure of occurrence of different terms in the document. Quantifying textual information is nothing but converting the string of characters in a document to a numerical vector of frequencies of occurrences of different words or terms. The numerical vector is usually a column in the term-document matrix, whose columns represent the documents. A data matrix is the transpose of the term-document matrix. You can attach a target column with known class labels of the documents to the data matrix. Examples of class labels are: Automobile Related, Health Related, etc. You need a data matrix with a target variable for developing predictive models. This chapter shows how you can use the SAS Enterprise Miner's text miner nodes to help create a data matrix, reduce its dimensions in order to create a compact version of the data matrix, and include it in a SAS data set which can be used for statistical analysis and predictive modeling.

In order to better understand the purpose of quantifying textual data, let us take an example from marketing. Suppose an automobile seller (advertiser) wants to know if a visitor to the Web is intending to buy an automobile in the near future. A person who intends to buy an automobile in the near future can be called an "auto intender". In order to determine whether a visitor to the Web is an auto intender or not, the auto seller needs to determine if the pages the visitor views frequently are auto related. The auto seller can use a predictive model that can help him decide if a web page is auto related. Logistic regressions, neural networks, decision trees, and support vector machines can be used for assigning class labels such as Auto Related and Not Auto Related to the web pages and other documents.

9.1.1 Quantifying Textual Data: A Simplified Example

The need for quantifying textual data also arises in query processing by search engines. Suppose you send the query "Car dealers" to a search engine. The search engine compares your query with a collection of documents with pre-assigned labels. For illustration, suppose there are only three documents in the collection (in the real world there may be thousands of documents) where each document is represented by a vector of terms as shown in Table 9.1

Table 9.1

	D1	D2	D3
Term	Document1	Document 2	Document 3
bank	0	0	3
deposit	0	0	3
report	1	1	1
taxes	0	0	3
health	8	0	0
medicine	7	0	0
car	0	8	0
driver	0	6	0
gasoline	0	1	0
domestic	0	1	1
foreign	1	1	7
exchange	0	0	4
currency	0	0	5
auto	0	7	0
dealer	0	6	1

9.1.1.1 Term-Document Matrix

The rows in a term-document matrix represent the occurrence (some measure of frequency) of a term in the documents represented by the columns. The table shown in Display 9.1 is an example of a term-document matrix.

In Table 9.1, Document1 is represented by column D1, Document2 by column2, and Document3 by column 3. There are 15 terms and 3 documents. So the term-document matrix is 15x3, and the document-term matrix, which is the transpose of the term-document matrix, is 3X15. In order to compare the query with each document, we need to represent the query as a vector consistent with the document vectors.

The query is converted to the row vector $q' = (0\ 0\ 0\ 0\ 0\ 0\ 1\ 0\ 0\ 0\ 0\ 0\ 0\ 0\ 1)$, where 1 indicates the occurrence of a term in the query and 0 indicates the non-occurrence. In the query vector q the 7th and 15th elements are equal to 1, and all other elements are equal to 0 since the 7th and 15th elements of the term vector are "car" and "dealer". The term vector is the first column in Table 9.1 and we can write it in vector notation as

T' = (*bank, deposit, report, taxes, health, medicine, car, driver, gasoline, domestic, foreign, exchange, currency, auto, dealer*).

In order to facilitate the explanations, let us represent the documents by the following vectors:

$$D1' = (0\ 0\ 1\ 0\ 8\ 7\ 0\ 0\ 0\ 0\ 1\ 0\ 0\ 0\ 0)\ ,\ D2' = (0\ 0\ 1\ 0\ 0\ 0\ 8\ 6\ 1\ 1\ 1\ 0\ 0\ 7\ 6)$$

and $D3' = (3\ 3\ 1\ 3\ 0\ 0\ 0\ 0\ 0\ 1\ 7\ 4\ 5\ 0\ 1)$.

In order to determine which document is closest to the query, we need to calculate the *scores* of each document vector with the query. The score of a document can be calculated as the inner product of the query vector and the document vector.

The score for Document 1 is: $D1'q = 0$

The score for Document 2 is $D2'q = 14$

The score for Document 3 is: $D3'q = 1$

9.1.1.2 Boolean Retrieval

The above calculations show how a query can be processed or information is retrieved from a collection of documents. An alternative way of processing queries is the Boolean Retrieval method. In this method, the occurrence of a word is represented by 1, and the non-occurrence of a word is represented by 0.

$$D1' = (0\ 0\ 1\ 0\ 1\ 1\ 0\ 0\ 0\ 0\ 1\ 0\ 0\ 0\ 0),\ D2' = (0\ 0\ 1\ 0\ 0\ 0\ 1\ 1\ 1\ 1\ 1\ 0\ 0\ 1\ 1)$$

and $D3' = (1\ 1\ 1\ 1\ 0\ 0\ 0\ 0\ 0\ 1\ 1\ 1\ 1\ 0\ 1)$.

The query is represented by a vector whose components are the frequencies of the occurrences of various words. Suppose the query consists of four occurrences of the word "car" and three occurrences of the word "dealer." Then the query can be represented by the vector:

$$q' = (0\ 0\ 0\ 0\ 0\ 0\ 4\ 0\ 0\ 0\ 0\ 0\ 0\ 0\ 3)$$

The score for Document 1 is: $D1'q = 0$

The score for Document 2 is $D2'q = 7$

The score for Document 3 is: $D3'q = 3$.

Since the score is the highest for Document2, the person who submitted the query should be directed to Document2, which contains information on auto dealers.

9.1.1.3 Document-Term Matrix and the Data Matrix

A document-term matrix is the transpose of the term-document matrix, such as the one shown in Display 9.1. A data matrix is same as the document-term matrix. The data matrix can be augmented by an additional column that shows the labels given to the documents.

In the illustration given above, the columns of the term-document matrix are the vectors $D1$, $D2$ and $D3$. The rows of the document-term or the data matrix are same as the columns of the term-document matrix.

You can verify that the vectors $D1, D2$ and $D3$ are the second, third and fourth columns in Table 9.1. From Table 9.1, we can define the data matrix as:

$$D = \begin{bmatrix} 0\ 0\ 1\ 0\ 8\ 7\ 0\ 0\ 0\ 0\ 1\ 0\ 0\ 0\ 0 \\ 0\ 0\ 1\ 0\ 0\ 0\ 8\ 6\ 1\ 1\ 1\ 0\ 0\ 7\ 6 \\ 3\ 3\ 1\ 3\ 0\ 0\ 0\ 0\ 0\ 1\ 7\ 4\ 5\ 0\ 1 \end{bmatrix} \tag{9.1}$$

The data matrix has one row for each document and one column for each term. Therefore, in this example, its dimension is 3x15. In a SAS data set, this data matrix has 3 rows and 15 columns or 3 observations and 15 variables, where each variable is a term.

If you already know the labels of the documents, you can add an additional column with the labels. For example if we give the label 1 to a document in which the main subject is automobiles and give the label 0 if the main subject of the document is not automobiles. Then the augmented data matrix with an additional column of labels can be used to estimate a logistic regression, which can be used to classify a new document as Auto Related or Not Auto Related.

In the example presented above, the elements of the data matrix are frequencies of occurrences of the terms in the documents. The value of the ij^{th} element of D is the frequency of jth term in the ith document. From the D matrix shown above in Equation 9.1, you can see that the word "health" appears 8 times in Document1.

In practice, adjusted frequencies rather than the raw frequencies are used in a D matrix. The adjusted frequency can be calculated as $tfidf(i,j) = tf(i,j) * \log\left(\dfrac{N}{df(j)}\right)$, where $tf(i,j)$ = frequency of jth term in i^{th} document (same as f_{ij}), $df(j)$ =document frequency (number of documents containing the jth term) and N = Number of documents. $tfidf(i,j)$ is a measure of the relative importance of the jth term in the ith document. We refer to $tfidf(i,j)$ as the adjusted frequency. There are other measures of relative importance of terms in a document.

In the examples presented in this and the next section, raw frequencies (f_{ij}) are used for illustrating the methods of quantification and dimension reduction.

The main tasks of the quantifying textual data are:

1. Constructing the term-document matrix from a collection of documents where the elements of the matrix are the frequency of occurrence of the words in each document. The elements of the term-document matrix can be $tfidf$ as described above, or some other measure. The term-document matrix can be of a very large size such as 500,000x1,000,000 representing 500,000 terms and 1,000,000 documents. The corresponding data matrix has 1,000,000 rows (documents) and 500,000 columns (terms).
    Singular Value Decomposition (SVD) is applied to derive a smaller matrix such as 100x1,000,000, where the rows of the new matrix are called SVD dimensions. These dimensions may be related to meaningful concepts.

After applying the SVD, the data matrix has 1,000,000 rows and 100 columns. The rows are documents (observations) and the columns are SVD dimensions.

The above tasks can be performed by various text mining nodes as described in this chapter.

9.1.2 Dimension Reduction and Latent Semantic Indexing

In contrast to the data matrix used in this example, the real-life data matrices are very large with thousands of documents and tens of thousands of terms. Also, these large data matrices tend to be sparse. *Singular Value Decomposition (SVD)*, which is demonstrated in this section, is used to create smaller and denser matrices from large sparse data matrices by replacing the terms with concepts or topics. The process of creating concepts or topics from the terms is demonstrated in this section. The data matrix with concepts or terms in the columns is much smaller than the original data matrix. In this example, the D matrix (shown earlier in this chapter) which is 3x15 is replaced by a 3x3 data matrix D^*. The rows of D^* are documents, and the columns contain concepts or topics.

Singular Value Decomposition[2] factors the term-document matrix into three matrices in the following way:

$$A = U\Sigma V', \text{ where } A \text{ is a } m \times n \text{ term - document matrix} \tag{9.2}$$

m = number of terms and n = number of documents

U is an $m \times k$ term-concept matrix, where $k \leq r = rank(A)$. The columns of U are eigenvectors of AA' and they are called the Left Singular Vectors. In this example $k = 3$.

Σ is a diagonal matrix of size $k \times k$. The diagonal elements of Σ are singular values which are the square roots of the nonzero eigenvalues of both AA' and $A'A$. The singular values in the diagonal elements of Σ are arranged in a decreasing order (as shown in the example below).

V is a concept-document matrix of size $n \times k$. The columns of V are eigenvectors of $A'A$ and they are called the Right Singular Vectors.

Applying Equation 9.2 to the term-document matrix shown in Table 9.1, we get:

$$A = \begin{bmatrix} 0 & 0 & 3 \\ 0 & 0 & 3 \\ 1 & 1 & 1 \\ 0 & 0 & 3 \\ 8 & 0 & 0 \\ 7 & 0 & 0 \\ 0 & 8 & 0 \\ 0 & 6 & 0 \\ 0 & 1 & 0 \\ 0 & 1 & 1 \\ 1 & 1 & 7 \\ 0 & 0 & 4 \\ 0 & 0 & 5 \\ 0 & 7 & 0 \\ 0 & 6 & 1 \end{bmatrix}, U = \begin{bmatrix} -0.04498 & -0.19748 & -0.18727 \\ -0.04498 & -0.19748 & -0.18727 \\ -0.08883 & -0.10807 & 0.01974 \\ -0.04498 & -0.19748 & -0.18727 \\ -0.02700 & -0.47266 & 0.57880 \\ -0.02363 & -0.41358 & 0.50645 \\ -0.56366 & 0.13471 & 0.07855 \\ -0.42275 & 0.10103 & 0.05891 \\ -0.07046 & 0.01684 & 0.00982 \\ -0.08545 & -0.04899 & -0.05261 \\ -0.17879 & -0.50303 & -0.35481 \\ -0.05998 & -0.26331 & -0.24970 \\ -0.07497 & -0.32914 & -0.31212 \\ -0.49321 & 0.11787 & 0.06873 \\ -0.43774 & 0.03520 & -0.00351 \end{bmatrix}, \Sigma = \begin{bmatrix} 13.86679 & 0.00000 & 0.00000 \\ 0.00000 & 11.10600 & 0.00000 \\ 0.00000 & 0.00000 & 10.41004 \end{bmatrix} \text{ and}$$

$$V = \begin{bmatrix} -0.04680 & -0.65617 & 0.75316 \\ -0.97703 & 0.18701 & 0.10221 \\ -0.20792 & -0.73108 & -0.64985 \end{bmatrix}$$

Table 9.2 shows the columns of the term-concept matrix U.

Table 9.2

Term	Concept 1 (U1)	Concept 2 (U2)	Concept 3 (U3)
bank	-0.04498	-0.19748	-0.18727
deposit	-0.04498	-0.19748	-0.18727
report	-0.08883	-0.10807	0.01974
taxes	-0.04498	-0.19748	-0.18727
health	-0.02700	-0.47266	0.57880
medicine	-0.02363	-0.41358	0.50645
car	-0.56366	0.13471	0.07855
driver	-0.42275	0.10103	0.05891
gasoline	-0.07046	0.01684	0.00982
domestic	-0.08545	-0.04899	-0.05261
foreign	-0.17879	-0.50303	-0.35481
exchange	-0.05998	-0.26331	-0.24970
currency	-0.07497	-0.32914	-0.31212
auto	-0.49321	0.11787	0.06873
dealer	-0.43774	0.03520	-0.00351

The columns in table 9.2 can be considered as *weights* of different terms into various concepts. These weights are analogous to factor loadings in Factor Analysis. If we consider that the vectors U1, U2, and U3 represent different concepts, we will be able to label these concepts based on the weights.

Since the terms "health" and "medicine" have higher weights than other terms in column U3, we may be able to label U3 to represent the concept "health". Since the terms "car," "driver," "gasoline," "auto," and "dealer" have higher weight than other terms in U2, we may be able to label column U2 to represent the concept "auto related". Labeling is not so clear in the case for column U1, but if you compare the weights of the term across different documents, you may get some clues for labeling U1 also. The weight of "bank" in column U1 is higher than in columns U2 and U3. Similarly the terms "deposit," "taxes," "exchange," and "currency" have higher weights in column U1 than in other columns. So we may label column U1 as Economics or Finance.

The columns of the matrix U can be considered as points in an m-dimensional space (m =15 in our example). The values in the columns are the coordinates of the points. By representing the points in a different coordinate system, you may be able to identify the concepts more clearly. Representing the points in a different coordinate system is called *rotation*. To calculate the coordinates of a point represented by a column in the U, you multiply it by a rotation matrix. For example, let the first column of the matrix U be U1. U1 is a point in 15-dimenstional space. To represent this point (U1) in a different coordinate system, you multiply U1 by a rotation matrix R. This multiplication gives new coordinates for the point represented by vector U1. By using the same ration matrix R, new coordinates can be calculated for all the columns of the term-concept matrix U. These new coordinates may show the concepts more clearly.

No rotation is performed in our example, because we were able to identify the concepts from the U matrix directly. In SAS Text Miner the Text Topic node does rotate the matrices generated by the Singular Value Decomposition. In other words, the Text Topic node performs a rotated Singular Value Decomposition.

9.1.2.1 Reducing the Number of Columns in the Data Matrix

Making use of the columns of the U matrix as weights and applying them to the rows of the data matrix D shown in Equation 9.1, we can create a data matrix with fewer columns in the following way. In general the number of columns selected in the U matrix = $k < r = rank(A)$. In our example, $k = 3$, as stated previously.

$$D^* = D \times U \times \Sigma^{-1} \tag{9.3}$$

The original data matrix D has 3 rows and 15 columns, each column representing a term. The new data matrix D^* has the same number of rows but only 3 columns.

In our example:

$$D^* = \begin{bmatrix} -0.04680 & -0.65617 & -0.75316 \\ -0.97703 & 0.18701 & -0.10221 \\ -0.20792 & -0.73108 & 0.64985 \end{bmatrix} \tag{9.4}$$

The new data matrix has 3 rows and 3 columns.

The original data matrix and reduced data matrix is shown in Tables 9.3 and 9.4.

Table 9.3

Document	bank	deposit	report	taxes	health	medicine	car	driver	gasoline	domestic	foreign	exchange	currency	auto	dealer	Target
Document1	0	0	1	0	8	7	0	0	0	0	1	0	0	0	0	H
Document2	0	0	1	0	0	0	8	6	1	1	1	0	0	7	6	A
Document3	3	3	1	3	0	0	0	0	0	1	7	4	5	0	1	E

The data matrix with reduced dimensions appended by a target column is shown in Table 9.4.

Table 9.4

Document	Concept1	Concept2	Concept3	Target
Document1	-0.04680	-0.65617	0.75316	H
Document2	-0.97703	0.18701	0.10221	A
Document3	-0.20792	-0.73108	-0.64985	E

The variables Concept1, Concept2, and Concept 3 for the three documents in the data set shown in Table 9.4 are calculated using the Equation 9.3 and shown in a matrix form in Equation 9.4. I leave it to you to carry out the matrix multiplications shown in Equation 9.3 and arrive at the compressed data matrix shown in Equation 9.4 and hence the values of Concept1, Concept2, and Concept 3 shown in Table 9.4. The variables Concept1, Concept2, and Concept3 shown in Table 9.4 can also be called svd_1, svd_2, and svd_3.

I have added an additional column named Target with known labels of the documents. The label H stands for Health Related, A stands for Auto Related, and E stands for Economics Related.

You can use a data set that looks like Table 9.4 and create a binary target that takes the value if the document is Auto Related and 0 if it is Not Auto Related. You can use such a table for estimating a logistic regression that you can use to predict whether a document is Auto Related or Not Auto Related.

Using Singular Value Decomposition (SVD) to identify patterns in the relationships between the terms and concepts, as we did above, is called Latent Semantic Indexing.

The starting point for the calculation of the Singular Value Decomposition is the term-document matrix such as the one shown in Table 9.1.

9.1.3 Summary of the Steps in Quantifying Textual Information

1. Retrieve the documents from the internet or from a directory on your computer and create a SAS data set. In this data set, the rows represent the documents and the columns contain the text content of the document. Use the %TMFILTER macro or the **Text Import** node.
2. Create terms. Break the stream of characters of the documents[3] into tokens by removing all punctuation marks, reducing words to their stems or roots,[4] and removing stop words (articles) such as a, an, the, at, in, etc. Use the **Text Parsing** node to do these tasks.

3. Reduce the number of terms. Remove redundant or unnecessary terms by using the **Text Filter** node
4. Create a term-document matrix[5] with rows that correspond to terms, columns that correspond to documents (web pages, for example), and whose entities are the adjusted frequencies of occurrences of the terms.
   Create a smaller data matrix by performing a Singular Value Decomposition of the term-document matrix

In the following sections, I show how these steps are implemented in SAS text mining nodes. I use the Federalist Papers to demonstrate this process.

9.2 Retrieving Documents from the World Wide Web

You can retrieve documents from the World Wide Web and create SAS data sets from them using the %TMFILTER macro or the **Text Import** node, which is used only in a client-server environment.

9.2.1 The %TMFILTER Macro

You can retrieve documents from web sites using the %TMFILTER macro, as shown in Display 9.1.

Display 9.1

```
libname tmlib "C:\TextMiner\Public\SASDATA" ;

%tmfilter( dataset=tmlib.text,
    dir=c:\TextMiner\Public\tmfdir,
    destdir=c:\TextMiner\Public\tmfdir_filtered,
    numchars=32000,
    url=http://www.constitution.org/fed/federa01.htm,
    depth=1,
    force=1)
```

The code segment shown in Display 9.1 retrieves Paper 1 of the Federalist Papers from the URL http://www.constitution.org/fed/federa01.htm. When I ran the above macro, six HTML files were created and stored in the directory C:\TextMiner\Public\tmfdir. The file called file1.html contains Federalist Paper 1. A partial view of this paper is shown in Display 9.2.

Display 9.2

The Federalist No. 1

Introduction

Independent Journal
Saturday, October 27, 1787
[Alexander Hamilton]

To the People of the State of New York:

AFTER an unequivocal experience of the inefficacy of the subsisting federal government, you are called upon to deliberate on a new Constitution for the United States of America. The subject speaks its own importance; comprehending in its consequences nothing less than the existence of the UNION, the safety and welfare of the parts of which it is composed, the fate of an empire in many respects the most interesting in the world. It has been frequently remarked that it seems to have been reserved to the people of this country, by their conduct and example, to decide the important question, whether societies of men are really capable or not of establishing good government from reflection and choice, or whether they are forever destined to depend for their political constitutions on accident and force. If there be any truth in the remark, the crisis at which we are arrived may with propriety be regarded as the era in which that decision is to be made; and a wrong election of the part we shall act may, in this view, deserve to be considered as the general misfortune of mankind.

The directory c:\TextMiner\Public\tmfdir_filtered contains text files with the same content as the HTML files. A partial view of the text file created is shown in Display 9.3.

Display 9.3

From the output shown in Display 9.4, you can see that six files were imported and that none of them were truncated.

Display 9.4

Frequency Percent Row Pct Col Pct	Table of TRUNCATED by OMITTED		
		OMITTED	
TRUNCATED		0	Total
0		6	6
		100.00	100.00
		100.00	
		100.00	
Total		6	6
		100.00	100.00

In addition, a SAS data set named TEXT is created and stored in the directory C:\TextMiner\Public\SASDATA. This data set has six rows, each row representing a file that is imported. The first row is relevant to us, since it has Federalist Paper 1. All other files represent other information about the web site itself. Since we are interested in analyzing the Federalist Papers, we can retain only the first row. The data set consists of a variable named Text, which contains the text content of the entire Paper 1.

If you call the macro (shown in Display 9.1) 85 times using a %DO loop and append the SAS data sets created at each iteration , you get a SAS data set with 85 rows, each row representing a document. In this example, each document is a Federalist Paper.

9.3 Creating a SAS Data Set from Text Files

If the text files are previously retrieved and stored in a directory, you can create SAS data sets from them directly. Display 9.5 shows how to create SAS data sets from text files stored in the directory C:\TextMiner\Public\CurrentText\Paper&n, where &n = 1, 2, ..., 85.

Display 9.5

```
libname tmlib "C:\TextMiner\Public\SASDATA" ;

%macro sasdsn(n=,author=);
%tmfilter (dataset=Paper&n,
        dir=c:\TextMiner\Public\CurrentText\Paper&n,
        destdir=c:\TextMiner\Public\filteredText,
        ext=txt,
        language=english ,
        numchars=32000,force=1)
;
```

The arguments in the %TMFILTER macro are:

dataset =	name of the output data set.
dir =	path to the directory that contains the documents to process.
destdir =	path to specify the output in the plain text format.
ext =	extension of the files to be processed.
numchars =	length of the text variable in the output data set.
force =1	keeps the macro from terminating if the directory specified in destdir is not empty.

Display 9.5 (cont'd)

```
data Paper&n ;
 set Paper&n;
 length AUTHOR $ 16;
 length TEXT_A $ 32000;
 length Accessed_A 8 Created_A 8 Extension_A $ 32 Filtered_A $ 48
        Filteredsize_A 8 Language_A $ 7 Modified_A 8
        Name_A $ 11 omitted_A 8  Size_A 8
        Truncated_A 8 URI_A $ 60 ;
 if _N_ = 1 ;
   PAPER = &n;
   AUTHOR = "&author" ;
   TEXT_A = Text ;
   Accessed_A = Accessed;
   Created_A  = Created ;
   Extension_A = Extension ;
   Filtered_A  = Filtered ;
   Filteredsize_A = Filteredsize;
   Language_A = Language;
   Modified_A = Modified ;
   Name_A      = Name ;
   omitted_A  = Omitted ;
   Size_A     = Size ;
   Truncated_A = Truncated ;
   URI_A = URI ;
 drop text Accessed Created Extension  Filtered Filteredsize Language
      Modified Name Omitted  Size Truncated URI ;
 run ;
%if &n eq 1 %then %do;
data tmlib.Federalist;
  set Paper&n;
run;
%end ; %else %do;
proc append base=tmlib.Federalist data=Paper&n FORCE; run;
%end ;
%mend sasdsn ;
```

Using the program shown in Display 9.6, I renamed the variables and also created the target variable.

Display 9.6

```
data tmlib.Federalist2;
  length TEXT $ 32000;
  length Filtered $ 48
        Filteredsize 8 Language $ 7 Modified 8
        Name $ 11 omitted 8  Size 8
        Truncated 8 URI $ 60
        Accessed 8 Created 8 Extension $ 32 ;
  set tmlib.Federalist;
  TEXT = TEXT_A;
  drop TEXT_A;
  URI= URI_A;
  drop URI_A;
  Filtered = Filtered_A ;
  drop Filtered_A ;
  Name =  Name_A ; drop Name_A ;
  Filteredsize = Filteredsize_A ; drop Filteredsize_A;
  Language  = Language_A; drop  Language_A ;
  omitted   = omitted_A ; drop omitted_A;
  Truncated = Truncated_A ; drop Truncated_A;
  Modified = Modified_A; drop Modified_A;
  Size = Size_A; drop Size_A;
  Accessed  = Accessed_A; drop Accessed_A;
  Created = Created_A; drop Created_A;
  Extension= Extension_A; drop Extension_A;
  If author = "HAMILTON" Then TARGET = 1 ; ELSE TARGET = 0 ;
  if author in ('MADISON_HAMILTON', 'JAY') then delete;
run;
```

The %TMFILTER macro can create SAS data sets from Microsoft Word, Microsoft Excel, Adobe Acrobat, and several other types of files.

Display 9.7 shows the contents of the SAS data created by the program shown in Display 9.6.

Display 9.7

#	Variable	Type	Len
14	AUTHOR	Char	16
11	Accessed	Num	8
12	Created	Num	8
13	Extension	Char	32
2	Filtered	Char	48
3	Filteredsize	Num	8
4	Language	Char	7
5	Modified	Num	8
6	Name	Char	11
15	PAPER	Num	8
8	Size	Num	8
16	TARGET	Num	8
1	TEXT	Char	32000
9	Truncated	Num	8
10	URI	Char	60
7	omitted	Num	8

Alphabetic List of Variables and Attributes

In the data set, there is one row per document. The variable PAPER identifies the document, and the variable TEXT contains the entire text content of the document.

9.4 The Text Import Node

The Text Import Node is an interface to the %TMFILTER macro described above.

You can use the **Text Import** node to retrieve files from the Web or from a server directory, and create data sets that can be used by other text mining nodes. The **Text Import** node relies on the SAS Document Conversion Server installed and running on a Windows machine. The machine must be accessible from the SAS Enterprise Miner Server via the host name and port number that were specified at the install time.

If you are not set up in a client-server mode, you can use the %TMFILTER macro as demonstrated in Section 9.2.1.

9.5 Creating a Data Source for Text Mining

Since the steps involved in creating a data source are the same as those demonstrated in Chapter 2, they are not shown here. Display 9.8 shows the variables in the data source.

Display 9.8

Name	Role	Level	Report	Order	Drop	Lower Limit	U
Accessed	Rejected	Interval	No		No	.	
AUTHOR	Rejected	Nominal	No		No	.	
Created	Rejected	Interval	No		No	.	
Extension	Rejected	Nominal	No		No	.	
Filtered	Rejected	Nominal	No		No	.	
Filteredsize	Rejected	Interval	No		No	.	
Language	Rejected	Nominal	No		No	.	
Modified	Rejected	Interval	No		No	.	
Name	Rejected	Nominal	No		No	.	
omitted	Rejected	Interval	No		No	.	
PAPER	ID	Interval	No		No	.	
Size	Rejected	Interval	No		No	.	
TARGET	Target	Interval	No		No	.	
TEXT	Text	Nominal	No		No	.	
Truncated	Rejected	Interval	No		No	.	
URI	Rejected	Nominal	No		No	.	

The roles of the variables TARGET, TEXT, and PAPER are set to Target, Text, and ID respectively. I set the roles of all other variables to Rejected as I do not use them in this chapter.

The variable TARGET takes the value 1 if the author of the paper is Hamilton and 0 otherwise, as defined by the SAS code shown in Display 9.6. The variable TEXT is the entire text of a paper.

9.6 Text Parsing Node

Text parsing is the first step in quantifying textual data contained in a collection of documents.

The **Text Parsing** node tokenizes the text contained in the documents by removing all punctuation marks. The character strings without spaces are called *tokens*, *words* or *terms*. The **Text Parsing** node also does stemming by replacing the words by their base or root. For example, the words "is" and "are" are replaced by the word "be". The **Text Parsing** node also removes stop words, which consist of mostly articles and prepositions. Some examples of stop words are: a, an, the, on, in, above, actually, after, and again.

Display 9.9 shows a flow diagram with the **Text Parsing** node.

Display 9.9

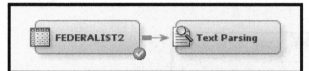

Display 9.10 shows the variables processed by the **Text Parsing** node.

Display 9.10

Name	Use	Report	Role	Level
TEXT	Default	No	Text	Nominal

The **Text Parsing** node processes the variable TEXT included in the analysis and collects the terms from all the documents.

Display 9.11 shows the properties of the **Text Parsing** node.

Display 9.11

Property	Value
General	
Node ID	TextParsing
Imported Data	
Exported Data	
Notes	
Train	
Variables	
Parse	
Parse Variable	
Language	English
Detect	
Different Parts of Speech	Yes
Noun Groups	Yes
Multi-word Terms	SASHELP.ENG_MULTI
Find Entities	None
Custom Entities	
Ignore	
Ignore Parts of Speech	'Aux' 'Conj' 'Det' 'Interj' 'Part' 'Prep' 'Pron'
Ignore Types of Entities	
Ignore Types of Attributes	'Num' 'Punct'
Synonyms	
Stem Terms	Yes
Synonyms	SASHELP.ENGSYNMS
Filter	
Start List	
Stop List	SASHELP.ENGSTOP
Report	
Number of Terms to Display	20000
Status	
Create Time	2/13/13 7:03 AM

You can also provide a start list and stop list of terms. The terms in the start list are included in the collection of terms, and the terms in the stop list are excluded. You can open *SAS Enterprise Miner: Reference Help* to find more details about the properties.

After running the **Text Parsing** node, open the Results to see a number of graphs and a table that shows the terms collected from the documents and their frequencies in the document collection. Display 9.12 shows the ZIPF plot.

Display 9.12

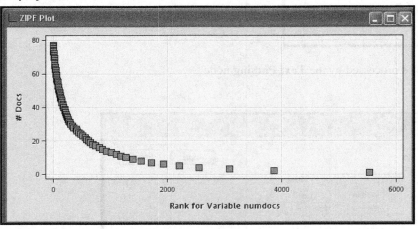

The horizontal axis of the ZIPF plot in Display 9.12 shows the terms by their rank and the vertical axis shows the number of documents in which the term appears. Display 9.13 gives a partial view of the rank of each term and the number of documents where the term appears.

Display 9.13

Term	Role	Attribute	Freq	# Docs	Rank for Variable numdocs	Keep	Parent/Child Status	Parent ID
+ be	...Verb	Alpha	6638	77	1 N	+		135
+ government	...Noun	Alpha	933	77	1 Y	+		75
+ have	...Verb	Alpha	1464	77	1 N	+		110
not	...Adv	Alpha	1133	77	1 N			123
+ state	...Noun	Alpha	1217	77	1 Y	+		218
+ power	...Noun	Alpha	737	76	6 Y	+		223
that	...Adv	Alpha	312	76	6 N			389
other	...Adj	Alpha	384	75	8 N			334
most	...Adv	Alpha	337	74	9 N			102
same	...Adj	Alpha	323	74	9 N			494
+ constitution	...Noun	Alpha	506	73	11 Y	+		80
+ find	...Verb	Alpha	206	73	11 Y	+		3355
+ other	...Noun	Alpha	299	73	11 N	+		488
such	...Adj	Alpha	316	73	11 N			266
any	...Adv	Alpha	209	72	15 N			136
+ great	...Adj	Alpha	340	72	15 Y	+		378
no	...Adv	Alpha	472	72	15 N			492
one	...Num	Alpha	277	72	15 N			244
+ part	...Noun	Alpha	230	72	15 N	+		150

To see term frequencies by document, you need the two data sets &EM_LIB..TEXTPARSING_TERMS and &EM_LIB..TEXTPARSING_TMOUT, where &EM_LIB refers to the directory where these data sets are stored. You can access these data sets via the **SAS Code** node.

Table 9.5 shows selected observations from the data set &EM_LIB..TEXTPARSING_TERMS.

Table 9.5

```
EMWS2.TEXTPARSING_TERMS

                                                                      Parent_
     Obs     Key      Term      Role    Attribute   Freq   numdocs   _ispar    Parent      id

       9    9328    abandon     Noun     Alpha        2       2                    .      9328
      10    2744    abandon     Verb     Alpha        3       3         .        2744     2744
      11    2744    abandon     Verb     Alpha        7       6         +          .      2744
      12    5058    abandoned   Verb     Alpha        3       3         .        2744     2744
      13    5189    abandoning  Verb     Alpha        1       1         .        2744     2744
```

Table 9.6 shows the distinct values of the terms.

Table 9.6

```
DISTINCT TERMS

Obs       Term        Key     Role

  9     abandon      2744     Verb
 10     abandon      9328     Noun
 11     abandoned    5058     Verb
 12     abandoning   5189     Verb
```

Table 9.7 shows the selected observations from &EM_LIB..TEXTPARSING_TMOUT.

Table 9.7

```
EMWS2.TEXTPARSING_TMOUT

Obs     key     document     count

3311    2744        5          1
8146    5058       11          1
8147    5189       11          1
12813   2744       16          1
16395   5058       21          1
16971   5058       22          1
23775   9328       31          1
24908   9328       32          1
48825   2744       64          1
```

Table 9.8 shows a join of Tables 9.6 and 9.7.

Table 9.8

```
TERM_DOCUMENT MATRIX

Obs     key     document     count      Term        Role

 16    2744         5          1      abandon      Verb
 17    2744        16          1      abandon      Verb
 18    9328        31          1      abandon      Noun
 19    9328        32          1      abandon      Noun
 20    2744        64          1      abandon      Verb
 21    5058        11          1      abandoned    Verb
 22    5058        21          1      abandoned    Verb
 23    5058        22          1      abandoned    Verb
 24    5189        11          1      abandoning   Verb
```

You can arrange the data shown in Table 9.8 into a term-document matrix. When the term "abandon" is used as a verb, it is given the key 2744, and when it is used as a noun it is given the key 9328.

Displays 9.14A and 9.14B show the SAS code that you can use to generate Tables 9.5 – 9.8. Display 9.14B shows how to create a term-document matrix using PROC TRANSPOSE. The term-document matrix shows raw frequencies of the terms by document.

Display 9.14A

```
data Textparsing_tmout;
  set &em_lib..Textparsing_tmout;
  rename _Termnum_ = key;
  rename _document_ = document;
  rename _count_ = count;

run;

  proc sql;
  create table terms as
  select distinct term, key, Role
from &em_lib..Textparsing_terms;
quit;

proc print data=&em_lib..Textparsing_terms ;
  VAR KEY TERM ROLE ATTRIBUTE FREQ NUMDOCS _ISPAR PARENT PARENT_ID;
  where term in ('abandon' 'abandoned' 'abandoning');
  title "&EM_LIB..TEXTPARSING_TERMS";
run;

proc print data=terms;
  where term in ('abandon' 'abandoned' 'abandoning');
  title " DISTINCT TERMS";
run;
```

Display 9.14B

```
proc print data=Textparsing_Tmout;
  where key in (9328,2744,5058,5189);
  title "&EM_LIB..TEXTPARSING_TMOUT";
run;
proc sql ;
  create table termdoc as
    select a.* , b.term,b.Role
    from Textparsing_tmout as a left join terms as b
  on a.key = b.key
    order by term , document ;
quit;
proc print data=termdoc ;
  title "TERM DOCUMENT MATRIX ";
  where term in ('abandon', 'abandoned', 'abandoning');
run;
proc sort data=termdoc out=termdocs ;
by key document ;
run;

proc transpose data=termdocs out=termdocmatrix(drop=_name_) prefix=DOC;
  var count ;
  by key ;
  id document ;
run;
```

9.7 Text Filter Node

The **Text Filter** node reduces the number of terms by eliminating unwanted terms and filtering documents using a search query or a where clause. It also adjusts the raw frequencies of terms by applying term weights and frequency weights.

9.7.1 Frequency Weighting

If the frequency of ith term in the jth document is f_{ij}, then you can transform the raw frequencies using a formula such as $g\left(f_{ij}\right) = \log_2\left(f_{ij}+1\right)$. Applying transformations to the raw frequencies is called *frequency weighting* since the transformation implies an implicit weighting. The function $g\left(f_{ij}\right)$ is called the fFquency Weighting Function. The weight calculated using the Frequency Weighting Function is called *frequency weight*.

9.7.2 Term Weighting

Suppose you want to give greater weight to terms that occur infrequently in the document collection. You can apply a weight that is inversely related to the proportion of documents that contain the term. If the proportion of documents that contain the term $t_i = P(t_i)$, then the weight applied is $w_i = \log_2\left(\dfrac{1}{P(t_i)}\right) + 1$, which is

inverse document frequency. w_i is called the *term weight*. Another type of inverse document frequency that is

used as a term weight is $w_i = \dfrac{1}{P(t_i)}$, where $P(t_i)$ is the proportion of documents that contain term t_i.

In the term-document matrix, the raw frequencies are replaced with adjusted frequencies calculated as the product of the term and frequency weights.

9.7.3 Adjusted Frequencies

Adjusted frequencies are obtained by multiplying the term weight with frequency weight.

For example, if we set $g(f_{ij}) = f_{ij}$ and $w_i = \dfrac{1}{P(t_i)}$, then the product of term frequency weight and term

weight is $tfidf_{ij} = f_{ij} \times \dfrac{1}{P(t_i)}$. In general, the adjusted frequency $a_{ij} = w_i \times g(f_{ij})$. In the term-document

matrix and the data matrix created by the **Text Filter** node, the raw frequencies f_{ij} are replaced by the adjusted

frequencies a_{ij}.

For a list of Frequency Weighting methods (frequency weighting functions) and Term Weighting methods (term weight formulae), you can refer online to *SAS Enterprise Miner: Reference Help*. These formulae are reproduced below.

9.7.4 Frequency Weighting Methods

- Binary

$$g(f_{ij}) = 1 \text{ if } j^{th} \text{ term appears in the } i^{th} \text{ document,}$$
$$= 0 \text{ otherwise.}$$

- **Log** (default)

$$g(f_{ij}) = \log_2(f_{ij} + 1)$$

- Raw frequencies

$$g(f_{ij}) = f_{ij}$$

9.7.5 Term Weighting Methods

- Entropy

$$w_i = 1 + \sum_{j=1}^{n} \frac{(f_{ij}/g_i).\log_2(f_{ij}/g_i)}{\log_2(n)} \text{ , where}$$

f_{ij} = The number of times the ith term appears in the jth document

g_i = The number of times the ith term appears in the document collection

$n =$ The number of documents in the collection

w_i = The weight of the ith term

- Inverse Document Frequency (idf)

$$w_i = \log_2\left(\frac{1}{P(t_i)}\right) + 1, \text{ where}$$

$P(t_i) =$ Proportion of documents that contain term t_i

- Mutual Information

$$w_i = \max(C_K)\left[\log\left(\frac{P(t_i, C_k)}{P(t_i)P(C_k)}\right)\right], \text{ where}$$

$P(t_i) =$ The proportion of documents that contain term t_i

$P(C_k) =$ The proportion of documents that belong to category C_k

$P(t_i, C_k) =$ The proportion of documents that contain term t_i and belong to category C_k

If your target is binary, then $k = 2$, $C_1 = '0'$ and $C_2 = '1'$.

- None

$w_i = 1$

Display 9.15 shows a flow diagram with the **Text Filtering** node attached to the **Text Parsing** node.

Display 9.15

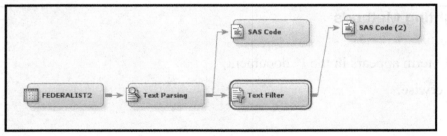

Display 9.16 shows the property settings of the **Text Filter** node.

Display 9.16

Property	Value
General	
Node ID	TextFilter
Imported Data	▭
Exported Data	▭
Notes	▭
Train	
Variables	▭
⊟ Spelling	
├ Check Spelling	No
└ Dictionary	▭
⊟ Weightings	
├ Frequency Weighting	None
└ Term Weight	Inverse Document Frequency
⊟ Term Filters	
├ Minimum Number of Documents	4
├ Maximum Number of Terms	.
└ Import Synonyms	▭
⊟ Document Filters	
├ Search Expression	
└ Subset Documents	▭
⊟ Results	
├ Filter Viewer	▭
├ Spell-Checking Results	▭
└ Exported Synonyms	▭
Report	
Terms to View	All
Number of Terms to Display	20000
Status	
Create Time	2/15/13 6:04 AM

Table 9.9 gives a partial view of the terms retained by the Text Filter node .

Table 9.9

													PARENT_
Obs	KEY	Term	Role	rolestring	Attribute	attrstring	WEIGHT	FREQ	NUMDOCS	KEEP	_ISPAR	PARENT	ID
1	2744	abandon	Verb	Verb	Alpha	Alpha	4.68182	7	6	Y	+	.	2744
2	2744	abandon	Verb	Verb	Alpha	Alpha	0.00000	3	3	Y	.	2744	2744
3	5058	abandoned	Verb	Verb	Alpha	Alpha	0.00000	3	3	Y	.	2744	2744
4	5189	abandoning	Verb	Verb	Alpha	Alpha	0.00000	1	1	Y	.	2744	2744

EMWS6.TEXTFILTER_TERMS

Comparing Table 9.9 with 9.5, you can see that the term "abandon" with the role of a noun is dropped by the **Text Filter** node. From Table 9.10, you can see that the value of the variable KEEP for the noun "abandon" is N. Also the weight assigned to the noun "abandon" is 0, as shown in Table 9.10.

Table 9.10

													PARENT_
Obs	KEY	Term	Role	rolestring	Attribute	attrstring	WEIGHT	FREQ	NUMDOCS	KEEP	_ISPAR	PARENT	ID
1	2744	abandon	Verb	Verb	Alpha	Alpha	4.68182	7	6	Y	+	.	2744
2	2744	abandon	Verb	Verb	Alpha	Alpha	0.00000	3	3	Y	.	2744	2744
3	5058	abandoned	Verb	Verb	Alpha	Alpha	0.00000	3	3	Y	.	2744	2744
4	5189	abandoning	Verb	Verb	Alpha	Alpha	0.00000	1	1	Y	.	2744	2744
5	9328	abandon	Noun	Noun	Alpha	Alpha	0.00000	2	2	N	.	.	9328

EMWS6.TEXTFILTER_TERMS_TMF

Table 9.11 shows the terms retained prior to replacing the terms "abandoned" and "abandoning" by the root verb "abandon".

Table 9.11

TEXT FILTER: DISTINCT TERMS (EMWS6.TEXTFILTER_TERMS)

Obs	Term	KEY	Role
2	abandon	2744	Verb
3	abandoned	5058	Verb
4	abandoning	5189	Verb

Table 9.12 shows the Transaction data set TextFilter_Transaction, created by the **Text Filter** node.

Table 9.12

```
EMWS6.TEXTFILTER_TRANSACTION

 Obs     _TERMNUM_     _DOCUMENT_     _COUNT_

18925      2744             5         4.68182
18926      2744            16         4.68182
18927      2744            64         4.68182
18928      2744            11         9.36365
18929      2744            21         4.68182
18930      2744            22         4.68182
```

Table 9.13 shows the actual terms associated with the term numbers (keys) and the documents in which the terms occur. Since I set the **Frequency Weighting** property to None and the **Term Weight** property to Inverse Document Frequency, the count shown in Table 9.13 is *tfidf* instead of the raw frequency.

Table 9.13

```
TEXT FILTER TERM_DOCUMENT MATRIX


 Obs     key     document     count      Term       Role

  6      2744         5      4.68182    abandon     Verb
  7      2744        11      9.36365    abandon     Verb
  8      2744        16      4.68182    abandon     Verb
  9      2744        21      4.68182    abandon     Verb
 10      2744        22      4.68182    abandon     Verb
 11      2744        64      4.68182    abandon     Verb
```

By transposing the table shown in Table 9.13, you can create a document-term frequency matrix D such as the one shown by Equation 9.1 or a term-document matrix A such as the one shown in Table 9.1 and on the left-hand side of Equation 9.2.

Displays 9.17A – 9.17C show the code used for generating the tables 9.9- 9.13. The code shown in Displays 9.17A – 9.17C is run from the **SAS Code** node attached to the **Text Filter** node, as shown in Display 9.15.

Display 9.17A

```
libname mylib "C:\TheBook\EM12.1\Reports\Chapter9\Text Filter";
run;

proc print data=&EM_LIB..TEXTFILTER_TERMS;
*VAR KEY TERM ROLE ATTRIBUTE FREQ weight NUMDOCS _ISPAR PARENT PARENT_ID;
where term in ('abandon' 'abandoned' 'abandoning');
title "&EM_LIB..TEXTFILTER_TERMS";
run;
proc print data=&EM_LIB..TEXTFILTER_TERMS_TMF;
*VAR KEY TERM ROLE ATTRIBUTE FREQ weight NUMDOCS _ISPAR PARENT PARENT_ID;
where term in ('abandon' 'abandoned' 'abandoning');
title "&EM_LIB..TEXTFILTER_TERMS_TMF";
run;
title;
proc sql;
 create table TFTERMS  as
 select DISTINCT term, key, role from &EM_LIB..TEXTFILTER_TERMS_TMF   ;
quit;

proc sql;
 create table TERMS  as
 select DISTINCT term, key, role from &EM_LIB..TEXTFILTER_TERMS   ;
quit;

proc print data=TERMS;
 where term in ('abandon' 'abandoned' 'abandoning');
 title " TEXT FILTER: DISTINCT TERMS (&EM_LIB..TEXTFILTER_TERMS)";
run;
```

You can view the document data set and the terms used by the **Filter** node by clicking ⬛, located to the right of the **Filter Viewer** property in the Results group.

Display 9.17B

```
data TextFilter_Transaction ;
 set &EM_LIB..TextFilter_Transaction ;
 rename _termnum_ = key;
 rename _document_ = document;
 rename _count_ = count ;
run;


proc print data=&EM_LIB..TextFilter_Transaction  ;
 where _termnum_ in (9328,2744,5058,5189);
 title "&EM_LIB..TEXTFILTER_TRANSACTION";
run;

title;
proc sql ;
 create table TFTERMDOC as
  select a.* , b.term,b.Role
  from TextFilter_Transaction as a left join TFTERMS as b
 on a.key = b.key
 order by term , document ;
quit;
```

Display 9.17C

```
proc print data=TFTERMDOC ;
 title "TEXT FILTER TERM_DOCUMENT MATRIX ";
 where term in ('abandon', 'abandoned', 'abandoning');
run;
title;

proc sort data=TFTERMDOC out=TFTERMDOC2 ;
by key document ;
run;

proc transpose data=TFTERMDOC2 out=TFtermdocmatrix(drop=_name_) prefix=DOC;
 var count ;
 by key  ;
 id document ;
run;
```

9.8 Text Topic Node

The **Text Topic** node creates topics from the terms collected from the document collection done by the **Text Parsing** and **Text Filter** nodes. Different topics are created from different combinations of terms. A Singular Value Decomposition of the term-document matrix is used in creating topics from the terms. Creation of topics from the terms is similar to the derivation of concepts from the term-document matrix using Singular Value Decomposition, demonstrated in Section 9.1.2. Although the procedure used by the **Text Topic** node may be a lot more complex than the simple illustration I presented in section 9.1.2, the illustration gives a general idea of how the **Text Topic** node derives topics from the term-document matrix. I recommend that you review Sections 9.1.1 and 9.1.2, with special attention to Equations 9.1 – 9.4, matrices $A, U, \Sigma, V, D \text{ and } D^*$, and Tables 9.1 – 9.4. In the illustrations in Section 9.1.2, I used the term "concept" instead of "topic." I hope that this switch of terminology does not hamper your understanding of how SVD is used to derive the topics.

Display 9.18 shows the process flow diagram with the **Text Topic** node connected to the **Text Filter** node.

Display 9.18

The property settings of the **Text Topic** node are shown in Display 9.19.

Display 9.19

Property	Value
General	
Node ID	TextTopic
Imported Data	
Exported Data	
Notes	
Train	
Variables	
User Topics	
⊟ Term Topics	
└ Number of Single-term Topics	0
⊟ Learned Topics	
├ Number of Multi-term Topics	25
└ Correlated Topics	No
⊟ Results	
└ Topic Viewer	
Status	
Create Time	2/17/13 9:21 AM

The term-document matrix has 2858 rows (terms) and 77 columns (documents). Selected elements of the -document-term matrix are shown in Table 9.14.

Table 9.14

Selected Elements of the Document-Term Matrix
(texttopic_tmout_normalized)

Obs	_DOCUMENT_	_TERMNUM_	_count_
1	1	3	0.019791
2	5	3	0.017645
3	11	3	0.032581
4	12	3	0.018438
5	14	3	0.017255
6	16	3	0.092146
7	17	3	0.017483
8	18	3	0.017907
9	19	3	0.016800
10	20	3	0.040328
11	21	3	0.019044
12	22	3	0.078707
13	23	3	0.018257
14	24	3	0.056398
15	25	3	0.016915
16	26	3	0.052738
17	29	3	0.078976
18	31	3	0.030899
19	33	3	0.014448
20	34	3	0.014105

Tables 9.14A show s the rows corresponding to the term "abandon" (_termnum_ 2744) in the document-term matrix shown in Table 9.14. The term "abandon" is also present in tables 9.5 – 9.11 and 9.13.

Table 9.14A

The Term 'abandon' in the Document-Term Matrix (texttopic_tmout_normalized)

DOCUMENT	_TERMNUM_	_count_
5	2744	0.038649
11	2744	0.071362
16	2744	0.040366
21	2744	0.041713
22	2744	0.034479
64	2744	0.037896

The **Text Topic** node derives topics from the term-document matrix (texttopic_tmout_normalized shown in Table 9.14) using Singular Value Decomposition as outlined in Section 9.1.2. In the example presented in this section, 25 topics are derived from the 2858 terms. The topics are shown in Table 9.15.

Table 9.15

Topics and Cutoff Values

_displayCat	_topicid	_docCutoff	_termCutoff	_name
Multiple	1	0.533	0.041	+court,+jurisdiction,+tribunal,+court,supreme
Multiple	2	0.479	0.039	+department,executive,legislative department,legislative,+judiciary department
Multiple	3	0.448	0.041	+army,military,+stand army,peace,standing
Multiple	4	0.517	0.040	+state,+power,+law,+clause,+article
Multiple	5	0.479	0.040	+government,+state government,state,people,federal
Multiple	6	0.413	0.036	+jury,+trial,+court,+case,admiralty
Multiple	7	0.424	0.038	+interest,+faction,+majority,+government,+republic
Multiple	8	0.400	0.038	+taxation,+revenue,+tax,+merchant,+state
Multiple	9	0.423	0.037	+representative,people,+state,+number,+house
Multiple	10	0.436	0.036	+bill,+state,+constitution,+right,+clause
Multiple	11	0.426	0.036	+government,+convention,congress,+state,+power
Multiple	12	0.383	0.035	+executive,plurality,+council,responsibility,+punishment
Multiple	13	0.405	0.035	+election,knowledge,+year,+state,+period
Multiple	14	0.367	0.033	+governor,president,+king,york,+state
Multiple	15	0.407	0.034	+state,+government,+majority,federal,+authority
Multiple	16	0.356	0.034	+senate,+treaty,+impeachment,+majority,+make treaty
Multiple	17	0.325	0.034	+state,+war,+republic,+nation,+confederacy
Multiple	18	0.298	0.033	+trade,commerce,+market,navigation,+state
Multiple	19	0.304	0.032	+senate,+vacancy,+appointment,+clause,president
Multiple	20	0.301	0.032	+man,+exclusion,president,+station,+office
Multiple	21	0.331	0.032	+court,+judge,legislative,judicial,judiciary
Multiple	22	0.326	0.032	+election,+senate,+state,national,+elector
Multiple	23	0.333	0.031	+state,+slave,+property,+representation,+inhabitant
Multiple	24	0.292	0.031	militia,+army,military,+state,+government
Multiple	25	0.296	0.031	+state,+convention,+government,+difficulty,+amendment

The topics shown in Table 9.15 are discovered by the **Text Topic** node. Alternatively, you can define the topics, include them in a file, and import the file by opening the User Topics window (click ⌷, located to the right of **User Topics** property in the Properties panel).

The **Text Topic** node creates a data set that you can access from the **SAS Code** node as &em_lib..TextTopic_train. A partial view of this data set is shown in Table 9.16.

Table 9.16

Obs	PAPER	TextTopic_raw3	TextTopic_3	TextTopic_raw10	TextTopic_10	TextTopic_raw14	TextTopic_14	TextTopic_raw18	TextTopic_18	TARGET
1	1	0.189	0	0.220	0	0.056	0	0.106	0	1
2	6	0.256	0	0.155	0	0.175	0	0.343	1	1
3	7	0.206	0	0.253	0	0.174	0	0.307	1	1
4	8	0.770	1	0.178	0	0.123	0	0.246	0	1
5	9	0.179	0	0.202	0	0.104	0	0.126	0	1
6	10	0.178	0	0.256	0	0.070	0	0.180	0	0
7	11	0.269	0	0.154	0	0.139	0	1.009	1	1
8	12	0.242	0	0.202	0	0.112	0	0.478	1	1
9	13	0.164	0	0.126	0	0.092	0	0.212	0	1
10	14	0.253	0	0.206	0	0.097	0	0.220	0	0
11	15	0.293	0	0.263	0	0.139	0	0.204	0	1
12	16	0.290	0	0.223	0	0.118	0	0.124	0	1
13	17	0.189	0	0.194	0	0.096	0	0.161	0	1
14	21	0.210	0	0.251	0	0.136	0	0.181	0	1
15	22	0.348	0	0.421	0	0.233	0	0.338	1	1
16	23	0.386	0	0.221	0	0.113	0	0.160	0	1
17	24	0.758	1	0.307	0	0.140	0	0.195	0	1
18	25	0.681	1	0.259	0	0.151	0	0.163	0	1
19	26	0.858	1	0.318	0	0.196	0	0.133	0	1
20	27	0.159	0	0.198	0	0.097	0	0.094	0	1

Output Data Set Created by the Text Topic node
(Selected Rows and Columns)

The data set name: TextTopic_train

The last column in Table 9.16 shows the variable TARGET, which takes the value 1 if the author of the paper is Hamilton, and 0 otherwise.

The output data set created by the **Text Topic** node contains two sets of variables. One set is represented by TextTopic_Raw$n$ and the other by TextTopic_n. For example, for Topic 1, the variables created are TextTopic_Raw1 and TextTopic_1; for Topic 2, Topic 3, etc., they are TextTopic_Raw2 and TextTopic_2, TextTopic_Raw2 and TextTopic_2, etc.

TextTopic_Raw$n$ is the raw score of a document on Topic n, and TextTopic_n is an indicator variable. The variable is set to 1 if TextTopic_Raw$n$ > the document cut-off value (shown in Table 9.15), and is set to 0 if TextTopic_Raw$n$ < the document cut-off value.

You can use TextTopic_Raw$n$ or TextTopic_n in a Logistic Regression, but you should not use both of them at the same time. If you attach a **Regression** node to the **Text Topic** node, only the raw scores are passed by the **Text Topic** node. If you want both the raw scores and the indicator variables to be passed to the next node, you need to attach a **Metadata** node to the **Text Topic** node and assign the new role of Input to the indicator variable TextTopic_n (n=1 to 25 in this example).

You can get additional information on the topics by opening the Interactive Topic Viewer window; click ⌷ located to the right of the **Topic Viewer** property. The interactive Topic Viewer window opens, as shown in Display 9.20.

Display 9.20

Topic	Category	Term Cutoff
+governor,president,+king,york,+state	Multiple	0.033
+state,+government,+majority,federal,+authority	Multiple	0.034
+senate,+treaty,+impeachment,+majority,+make treaty	Multiple	0.034
+state,+war,+republic,+nation,+confederacy	Multiple	0.034
+trade,commerce,+market,navigation,+state	Multiple	0.033
+senate,+vacancy,+appointment,+clause,president	Multiple	0.032
+man,+exclusion,president,+station,+office	Multiple	0.032
+court,+judge,legislative,judicial,judiciary	Multiple	0.032
+election,+senate,+state,national,+elector	Multiple	0.032

Terms

Topic Weight	+	Term	Role	# Docs	Freq
0.39	+	election	Noun	30	112
0.195	+	senate	Noun	29	173
0.18	+	state	Noun	77	1217
0.172		national	Adj	56	272
0.157	+	elector	Noun	10	24
0.151	+	state legislature	Noun Group	25	59
0.149	+	interest	Noun	50	211
0.141	+	preference	Noun	14	22
0.137	+	legislature	Noun	63	267

Documents

Topic Weight	TEXT	AUTHOR	Accessed	Created	Extension
0.716	WE HAVE seen, that an	HAMILTON	1.646570576E9	1.646570576E9	.txt
0.495	THE natural order of the	HAMILTON	1.646570576E9	1.646570576E9	.txt
0.453	THE more candid	HAMILTON	1.646570576E9	1.646570576E9	.txt
0.384	THE third charge	MADISON	1.646570576E9	1.646570576E9	.txt
0.382	BEFORE we proceed to	HAMILTON	1.646570576E9	1.646570576E9	.txt
0.371	FROM the more general	MADISON	1.646570576E9	1.646570576E9	.txt
0.352	A REVIEW of the	HAMILTON	1.646570576E9	1.646570576E9	.txt
0.334	I SHALL here, perhaps,	MADISON	1.646570576E9	1.646570576E9	.txt
0.317	THE fourth class	MADISON	1.646570576E9	1.646570576E9	.txt

Display 9.20 shows three sections in the Topic Viewer window. Twenty-five topics are listed in the top section. If you scroll down in the top section and click on the 22nd topic, then the middle section shows the weights of different terms in Topic 22 and the bottom section shows the weight of Topic 22 in different documents.

The term "election" has the highest weight (0.39) in Topic 22. Topic 22 has the highest weight (0.716) in paper 60 which was written by Alexander Hamilton.

9.8.1 Developing a Predictive Equation Using the Output Data Set Created by the Text Topic Node

By connecting a **Regression** node to the **Text Topic** node as shown in Display 9.18, you can develop an equation that you can use for predicting who the author of a paper is.

Although I have not done it in this example, in general you should partition the data into Train, Validate, and Test data sets. Since the number of observations is small (77) in the example data set, I used the entire data set for Training only.

In the **Regression** node, I set the **Selection Model** property to Stepwise and **Selection Criterion** to None. With these settings, the model from the final iteration of the Stepwise process is selected. The selected equation is shown in Display 9.21.

Display 9.21

| | | | Standard | Wald | | Standardized | |
Parameter	DF	Estimate	Error	Chi-Square	Pr > ChiSq	Estimate	Exp(Est)
				Analysis of Maximum Likelihood Estimates			
Intercept	1	9.0857	2.7705	10.75	0.0010		999.000
TextTopic_raw10	1	8.1794	3.9221	4.35	0.0370	0.6447	999.000
TextTopic_raw2	1	-15.3612	5.1420	8.92	0.0028	-2.0465	0.000
TextTopic_raw20	1	18.9250	7.6168	6.17	0.0130	1.1384	999.000
TextTopic_raw25	1	-11.3430	4.5910	6.10	0.0135	-0.5844	0.000
TextTopic_raw9	1	-29.1967	8.0413	13.18	0.0003	-2.3848	0.000

From Display 9.21, you can see that those documents that score high on Topics 10 and 20 are more likely to be authored by Alexander Hamilton.

In order to test the equation shown in Display 9.21, you need a Test data set. As mentioned earlier, I used the entire data set for Training only since there are not enough observations.

Note that in the logistic regression shown in Display 9.21, only the raw scores are used. I ran an alternative equation using the indicator variables. To make the indicator variables available, I attached the **Metadata** node to the **TextTopic** node and attached a **Regression** node to the **Metadata** node, as shown in Display 9.22. I set the role of the indicator variables to Input in the **Metadata** node.

Display 9.22

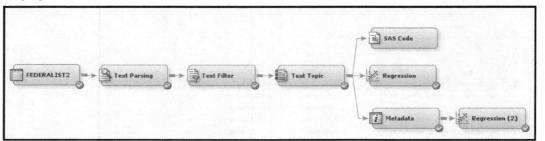

The logistic regression estimated using the indicator variables is shown in Display 9.23.

Display 9.23

| | | | Standard | Wald | | |
Parameter	DF	Estimate	Error	Chi-Square	Pr > ChiSq	Exp(Est)
			Analysis of Maximum Likelihood Estimates			
Intercept	1	-1.8755	0.7567	6.14	0.0132	0.153
TextTopic_11 0	1	1.6800	0.5576	9.08	0.0026	5.365
TextTopic_9 0	1	1.6030	0.5628	8.11	0.0044	4.968

9.9 Text Cluster Node

The **Text Cluster** node divides the documents into non-overlapping clusters. The documents in a cluster are similar with respect to the frequencies of the terms. For developing the clusters, the **Text Cluster** node makes use of the SVD variables created by using the SVD method, as illustrated in Section 9.1.2. The clusters created are useful in classifying documents. The cluster variable, which indicates which cluster an observation belongs to, can be used as categorical input in predictive modeling.

You can develop clusters using a document-term matrix, also known as the data matrix. The rows of a document-term matrix represent the documents and the columns represent terms, and the cells are raw or weighted frequencies of the terms. If there are n terms and m documents, the term-document matrix is of size $n \times m$ and the data matrix (document-term matrix) is of size $m \times n$. The document-term matrix in our example is of size 77×2858 since there are 77 documents and 2858 terms. The document-term matrix is large and sparse as not all the terms appear in every document. The dimensions of the data matrix are reduced by replacing the terms with SVD variables created by means of Singular Value Decomposition of the term-document matrix, as illustrated in Section 9.1.2. Although in the illustration given in Section 9.1.2 many steps such as normalization and rotation of the vectors are skipped, it is still useful to get a general understanding of how Singular Value Decomposition is used for arriving at the special SVD variables. The number of SVD variables (p) is much smaller than the number of terms n. As you can see from the data set exported by the **Cluster** node, the number of SVD terms in this demonstration is 20. That is, $p = 20$. The number of the SVD terms created is determined by the value to which you set the **SVD Resolution** property. You can set the resolution to Low, Medium, or High. Setting the **SVD Resolution** property to Low results in the smallest number of SVD variables, and a value of High results in the largest number of SVD variables.

There are two types of clustering available in the **Text Cluster** node: Hierarchical Clustering andExpectation-Minimization. You can select one of these methods from the Properties panel of the **Text Cluster** node. Following is a brief explanation of the two methods of clustering.

9.9.1 Hierarchical Clustering

If you set the **Cluster Algorithm** property to Hierarchical Clustering, the **Text Cluster** nodes uses Ward's method[6] for creating the clusters. In this method, the observations (documents) are progressively combined into clusters in such a way that an objective function is minimized at each combination step. The error sum of squares for the kth cluster is:

$$E_k = \sum_{i=1}^{m_k} \sum_{j=1}^{p} \left(x_{ijk} - \overline{x}_{jk} \right)^2 \tag{9.5}$$

Where:

E_k = Error sum of squares for the kth cluster

x_{ijk} = Value of the jth variable at the ith observation in the kth cluster

$\overline{x}_{jk}$ = Mean of the jth variable in the kth cluster

m_k = Number of observations (documents) in the kth cluster

p = Number of variables used for clustering

The objective function is the combined Error Sums of Squares given by:

$$E = \sum_{k=1}^{K} E_k \tag{9.6}$$

Where:

K = Number of clusters

At each step, the union of every possible pair of clusters is considered, and the two clusters whose union results in the minimum increase in the error sum of squares (E) are combined.

If the clusters u and v are combined to form a single cluster w, then the increase in error resulting from the union of clusters u and v is given by:

$$\Delta E_{uv} = E_w - \left(E_u + E_v \right) \tag{9.7}$$

Where:

ΔE_{uv} = Increase in the Error sum of squares due to combining clusters u and v

E_w = Error sum of squares for cluster w, which is the union of clusters u and v

E_u = Error sum of squares for cluster u

E_v = Error sum of squares for cluster v

Initially, each observation (document) is regarded as a cluster, and the observations are progressively combined into clusters as described above.

9.9.2 Expectation-Maximization (EM) Clustering

The Expectation-Maximization algorithm assumes a probability density function of the variables in each cluster. If X is a vector of p random variables, then the joint probability distribution of the random vector X in cluster k can be written as $f_k(X \mid \theta_k)$, where θ_k is a set of parameters that characterize the probability distribution. If we assume that the random vector X has a joint normal distribution, the parameters θ_k consist of a mean vector μ_k and a variance-covariance matrix Σ_k, and the joint probability density function of X in cluster k is given by:

$$f_k\left(X \mid \mu_k, \Sigma_k \right) = \frac{1}{\sqrt{(2\pi)^p |\Sigma_k|}} \exp\left[-\frac{1}{2}\left(X - \mu_k \right)' \left(\Sigma \right)^{-1} \left(X - \mu_k \right) \right] \tag{9.8}$$

The probability density function for cluster k given in Equation 9.8 can be evaluated at any observation point by replacing the random vector X with the observed data vector x, as shown in Equation 9.9:

$$f_k\left(x \mid \mu_k, \Sigma_k \right) = \frac{1}{\sqrt{(2\pi)^p |\Sigma_k|}} \exp\left[-\frac{1}{2}\left(x - \mu_k \right)' \left(\Sigma \right)^{-1} \left(x - \mu_k \right) \right] \tag{9.9}$$

We need estimates of the parameters μ_k and Σ_k in order to calculate the probability density function given by Equation 9.8. These parameters are estimated iteratively in EM algorithm. The probability density functions given in Equations 9.8 and 9.9 are conditional probability density functions. The unconditional probability density function evaluated at an observation is:

$$p(x) = \sum_{k=1}^{K} \pi_k f_k(x \mid \mu_k, \Sigma_k) \tag{9.10}$$

where π_k is the probability of cluster k, as defined in the following illustration.

9.9.2.1 An Example of Expectation-Maximization Clustering

In order to demonstrate Expectation-Maximization Clustering, I generated an example data set consisting of 200 observations (documents). I generated the example data set from a mixture of two univariate normal distributions so that the data set contains only one input (x), which can be thought of as a measure of document strength in a given concept. My objective is to create *two clusters* of documents based on the single input. Initially I assigned the documents randomly to two clusters: Cluster A and Cluster B. Starting from these random, initial clusters, the Expectation-Maximization algorithm iteratively produced two distinct clusters. The iterative steps are given below:

Iteration 0 (Initial random clusters)

1. Assign observations randomly to two clusters A and B. These arbitrary clusters are needed to start the EM iteration process.
2. Define two weight variables $w_{ai}(0)$ and $w_{bi}(0)$ as follows:

$w_{ai}(0) = 1$, if the ith observation is assigned to cluster A

$w_{ai}(0) = 0$, otherwise.

$w_{bi}(0) = 1$, if the ith observation is assigned to cluster B

$w_{bi}(0) = 0$ otherwise

(The 0 in the brackets indicates the iteration number.)

3. Calculate the cluster probabilities $\pi_a(0)$ and $\pi_b(0)$ as

$$\pi_a(0) = \frac{\sum_{i=1}^{n} w_{ai}(0)}{n} \text{ and}$$

$$\pi_b(0) = \frac{\sum_{i=1}^{n} w_{bi}(0)}{n},$$

where n is number of observations (200 in this example).

4. Calculate cluster means $\mu_j(0)$ and cluster standard deviations $\sigma_j(0)$, where the subscript j stands for a cluster and 0 in the brackets indicates the iteration number. Since we are considering only two clusters in this example, j takes the values a and b.

$$\mu_a(0) = \frac{\sum_{i=1}^{n} (w_{ai}(0))x_i}{\sum_{i=1}^{n} w_{ai}(0)},$$

where x_i is the value of the input variable for observation (document) i, w_{ai} is the value of the weight variable (as defined in Step 2) for observation i, and $\mu_a(0)$ is the mean of the input variable for cluster A at iteration 0.

Similarly, the mean of the input variable for Cluster B is calculated as:

$$\mu_b(0) = \frac{\sum_{i=1}^{n} (w_{bi}(0))x_i}{\sum_{i=1}^{n} w_{bi}(0)}$$

The standard deviations $\sigma_a(0)$ and $\sigma_b(0)$ for clusters A and B are calculated as

$$\sigma_a^2(0) = \frac{\sum_{i=1}^{n}(w_{ai}(0))(x_i - \mu_a(0))^2}{\sum_{i=1}^{n}w_{ai}(0)} \text{ and } \sigma_b^2(0) = \frac{\sum_{i=1}^{n}(w_{bi}(0))(x_i - \mu_{ba}(0))^2}{\sum_{i=1}^{n}w_{bi}(0)} \text{ respectively.}$$

Iteration 1

1. Calculate the probabilities of cluster membership of the ith observation:

$$P(A|x_i) = P(x_i|A)P(A) = f_A(x_i)\pi_a(0) \text{ and}$$
$$P(B|x_i) = P(x_i|B)P(B) = f_B(x_i)\pi_b(0)$$

where

$$f_A(x_i) = pdf('Normal', x_i, \mu_a(0), \sigma_a(0)) \text{ and}$$
$$f_B(x_i) = pdf('Normal', x_i, \mu_b(0), \sigma_b(0)).$$

2. Calculate the weights for the ith observation:

$$w_{ai}(1) = \frac{P(A|x_i)}{P(A|x_i) + P(B|x_i)} \text{ and}$$

$$w_{bi}(1) = \frac{P(B|x_i)}{P(A|x_i) + P(B|x_i)}.$$

(Since $w_{ai}(1) + w_{bi}(1) = 1$ at each observation, $\sum_{i=1}^{n}w_{ai}(1) + \sum_{i=1}^{n}w_{bi}(1) = n =$ total number of observations in the data set. This is true at each iteration).

3. Calculate the log-likelihood of the ith observation:

$$L_i = \log(P(A|x_i) + P(B|x_i))$$

4. Calculate the log-likelihood for this iteration:

$$LL_i = \sum_{i=1}^{n}L_i$$

5. Calculate cluster probabilities:

$$\pi_a(1) = \frac{\sum_{i=1}^{n}w_{ai}(1)}{n}, \pi_b(1) = \frac{\sum_{i=1}^{n}w_{bi}(1)}{n}$$

6. Calculate cluster means:

$$\mu_a(1) = \frac{\sum_{i=1}^{n}(w_{ai}(1))x_i}{\sum_{i=1}^{n}w_{ai}(1)} \text{ and } \mu_b(1) = \frac{\sum_{i=1}^{n}(w_{bi}(1))x_i}{\sum_{i=1}^{n}w_{bi}(1)}$$

7. Calculate cluster standard deviations:

$$\sigma_a^2(1) = \frac{\sum\limits_{i=1}^{n}(w_{ai}(1))(x_i - \mu_a(1))^2}{\sum\limits_{i=1}^{n} w_{ai}(1)} \quad \text{and} \quad \sigma_b^2(1) = \frac{\sum\limits_{i=1}^{n}(w_{bi}(1))(x_i - \mu_b(1))^2}{\sum\limits_{i=1}^{n} w_{bi}(1)} \quad .$$

Iteration 2

1. Calculate the probabilities of cluster membership of the ith observation:

$$P(A \mid x_i) = P(x_i \mid A)P(A) = f_A(x_i)\pi_a(1) \text{ and}$$
$$P(B \mid x_i) = P(x_i \mid B)P(B) = f_b(x_i)\pi_b(1)$$

where

$$f_A(x_i) = pdf\left('Normal', x_i, \mu_a(1), \sigma_a(1)\right) \text{ and}$$
$$f_B(x_i) = pdf\left('Normal', x_i, \mu_b(1), \sigma_b(1)\right).$$

2. Calculate the weights:

$$w_{ai}(2) = \frac{P(A \mid x_i)}{P(A \mid x_i) + P(B \mid x_i)} \quad \text{and}$$

$$w_{bi}(2) = \frac{P(B \mid x_i)}{P(A \mid x_i) + P(B \mid x_i)} \quad .$$

3. Calculate the log-likelihood of the ith observation:

$$L_i = \log\left(P(A \mid x_i) + P(B \mid x_i)\right)$$

4. Calculate the log-likelihood for this iteration:

$$LL_i = \sum_{i=1}^{n} L_i$$

5. Calculate cluster probabilities:

$$\pi_a(2) = \frac{\sum\limits_{i=1}^{n} w_{ai}(2)}{n} \quad , \pi_b(2) = \frac{\sum\limits_{i=1}^{n} w_{bi}(2)}{n}$$

Since $w_{ai}(2) + w_{bi}(2) = 1$ at each observation, $n = \sum\limits_{i=1}^{n} w_{ai}(2) + \sum\limits_{i=1}^{n} w_{bi}(2)$ at each iteration.

6. Calculate cluster means:

$$\mu_a(2) = \frac{\sum\limits_{i=1}^{n}(w_{ai}(2))x_i}{\sum\limits_{i=1}^{n} w_{ai}(2)} \quad \text{and} \quad \mu_b(2) = \frac{\sum\limits_{i=1}^{n}(w_{bi}(2))x_i}{\sum\limits_{i=1}^{n} w_{bi}(2)}$$

7. Calculate cluster standard deviations:

$$\sigma_a^2(2) = \frac{\sum_{i=1}^{n}(w_{ai}(2))(x_i - \mu_a(2))^2}{\sum_{i=1}^{n}w_{ai}(2)} \quad \text{and} \quad \sigma_b^2(2) = \frac{\sum_{i=1}^{n}(w_{bi}(2))(x_i - \mu_b(2))^2}{\sum_{i=1}^{n}w_{bi}(2)} \quad .$$

You can calculate cluster probabilities, the weights, cluster probabilities and the parameters of the probability distributions in iteration 3, iteration 4 …etc.

In the example shown below, the data set has 200 observations and a single input (x) drawn from a mixture of *two univariate normal distributions. The EM algorithm is applied iteratively, as described above, to create two* clusters. Table 9.18 shows the results of 20 iterations.

Table 9.17

Iteration	Sum of Weights Cluster A	Sum of Weights Cluster B	Cluster Probability P(A)	Cluster Probability P(B)	Mean(x) Cluster A	Std(x) Cluster A	Mean(x) Cluster B	Std(x) Cluster B	Log-Likelihood
1	82.0000	118.000	0.41000	0.59000	0.44561	0.29030	0.42305	0.26267	-24.6028
2	82.0078	117.992	0.41004	0.58996	0.45872	0.30471	0.41394	0.24985	-22.8732
3	81.7817	118.218	0.40891	0.59109	0.48097	0.32520	0.39863	0.22716	-18.6231
4	80.7662	119.234	0.40383	0.59617	0.51127	0.34564	0.37881	0.19606	-11.3025
5	78.7692	121.231	0.39385	0.60615	0.54534	0.35977	0.35885	0.16299	-2.2039
6	76.5177	123.482	0.38259	0.61741	0.57518	0.36557	0.34376	0.13709	4.6692
7	75.4467	124.554	0.37723	0.62277	0.59324	0.36563	0.33481	0.12197	7.6965
8	75.9449	124.055	0.37972	0.62028	0.59976	0.36353	0.32978	0.11380	8.7966
9	77.3480	122.652	0.38674	0.61326	0.59953	0.36119	0.32684	0.10905	9.2741
10	79.0420	120.958	0.39521	0.60479	0.59648	0.35916	0.32501	0.10595	9.5422
11	80.7089	119.291	0.40354	0.59646	0.59265	0.35748	0.32381	0.10372	9.7162
12	82.2212	117.779	0.41111	0.58889	0.58892	0.35610	0.32296	0.10202	9.8353
13	83.5429	116.457	0.41771	0.58229	0.58558	0.35495	0.32234	0.10067	9.9183
14	84.6770	115.323	0.42338	0.57662	0.58271	0.35400	0.32186	0.09957	9.9764
15	85.6408	114.359	0.42820	0.57180	0.58028	0.35320	0.32148	0.09866	10.0171
16	86.4557	113.544	0.43228	0.56772	0.57825	0.35253	0.32117	0.09792	10.0458
17	87.1429	112.857	0.43571	0.56429	0.57654	0.35196	0.32092	0.09730	10.0659
18	87.7214	112.279	0.43861	0.56139	0.57512	0.35149	0.32071	0.09678	10.0800
19	88.2080	111.792	0.44104	0.55896	0.57393	0.35110	0.32054	0.09635	10.0899
20	88.6170	111.383	0.44309	0.55691	0.57294	0.35076	0.32040	0.09599	10.0969

Display 9.24 shows a plot of log-likelihood by iteration.

Display 9.24

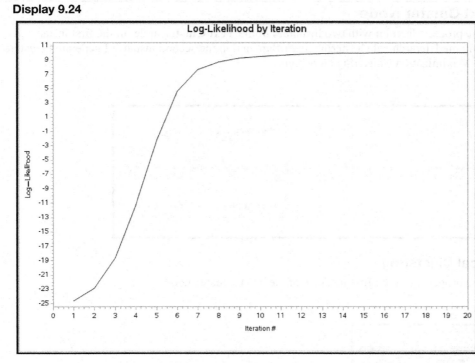

As Table 9.17 and Display 9.24 show, the likelihood increases rapidly from -24.6 in iteration 1 to 7.6965 in iteration 7. After iteration 7, the rate of change in log-likelihood diminishes steadily and log-likelihood becomes increasingly flatter. The difference in the likelihood between iterations 19 and 20 is only 0.007, showing that the solution converged by iteration 19.

Using the weights calculated in the final iteration, I assigned i^{th} observation to Cluster 1 if $w_{ai}(20) > w_{bi}(20)$ and to cluster 2 if $w_{ai}(20) \leq w_{bi}(20)$. The means and the standard deviations of the variable X in the clusters are shown in Table 9.18.

Table 9.18

Analysis Variable : x			
CLUSTER	N Obs	Mean	Std Dev
1	65	0.6684951	0.3664221
2	135	0.3185749	0.0910746

Table 9.18 shows that the mean of the variable x is higher for Cluster 1 than for Cluster 2. Starting with an arbitrary set of clusters, we arrived at two distinct clusters.

In our example, there is only one input, namely, x which is continuous. In calculating the likelihood of each observation, we used normal probability density functions. If you have several continuous variables, you can use a separate normal density function for each variable if they are independent, or you can use a joint normal density function if they are jointly distributed. For ordinal variables, you can use geometric or Poisson distributions and for other categorical variables you can use multinomial distribution. The Expectation-Maximization clustering can be used with variables of different measurement scales by choosing the appropriate probability distributions.

The method outlined can be extended to more than one variable.

For another illustration of the EM algorithm, refer to *Data Mining the Web,*[7] by Markov and Larose.

9.9.3 Using the Text Cluster Node

Display 9.25 shows the process diagram with two instances of the **Text Cluster** node. In the first instance, I generated clusters using the Hierarchical Clustering algorithm, and in the second instance I generated clusters using the Expectation-Maximization Clustering algorithm.

Display 9.25

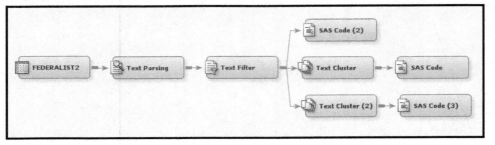

9.9.3.1 Hierarchical Clustering

Display 9.26 shows the properties of the first instance of the **Text Cluster** node.

Display 9.26

Property	Value
General	
Node ID	TextCluster
Imported Data	
Exported Data	
Notes	
Train	
Variables	
Transform	
SVD Resolution	Low
Max SVD Dimensions	100
Cluster	
Exact or Maximum Number	Maximum
Number of Clusters	40
Cluster Algorithm	Hierarchical
Descriptive Terms	15
Status	
Create Time	2/17/13 1:26 PM

With the property settings shown in Display 9.26, the **Text Cluster** node produced five clusters. The cluster frequencies are shown in Display 9.27.

Display 9.27

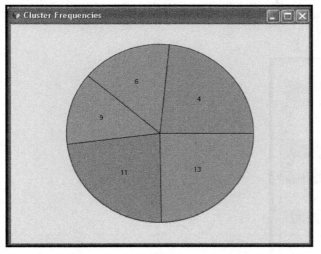

Display 9.28 shows a list of the descriptive terms of each cluster.

Display 9.28

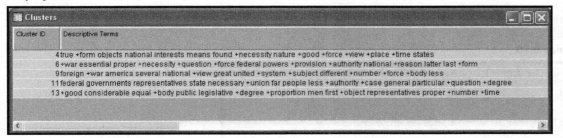

Cluster ID	Descriptive Terms
4	true +form objects national interests means found +necessity nature +good +force +view +place +time states
6	+war essential proper +necessity +question +force federal powers +provision +authority national +reason latter last +form
9	foreign +war america several national +view great united +system +subject different +number +force +body less
11	federal governments representatives state necessary +union far people less +authority +case general particular +question +degree
13	+good considerable equal +body public legislative +degree +proportion men first +object representatives proper +number +time

The target variable in the data set is Target, which takes the value 1 if the author of the paper is Alexander Hamilton and 0 otherwise. Table 9.19 shows the cross tabulation of the clusters and the target variable.

Table 9.19

Table of CLUSTER by TARGET

The FREQ Procedure

Frequency Row Pct	Table of CLUSTER by TARGET		
	TARGET		
CLUSTER(Cluster ID)	0	1	Total
4	6 33.33	12 66.67	18
6	3 25.00	9 75.00	12
9	2 20.00	8 80.00	10
11	8 44.44	10 55.56	18
13	7 36.84	12 63.16	19
Total	26	51	77

Data set: TextCluster_TRAIN

You can see n that Clusters 6 and 9 are dominated by the papers written by Alexander Hamilton.

9.9.3.2 Expectation-Maximization Clustering

Display 9.29 shows the properties settings of the second instance of the **Text Cluster** node.

Display 9.29

Property	Value
General	
Node ID	TextCluster2
Imported Data	
Exported Data	
Notes	
Train	
Variables	
⊟ Transform	
├ SVD Resolution	Low
└ Max SVD Dimensions	100
⊟ Cluster	
├ Exact or Maximum Number	Maximum
├ Number of Clusters	40
├ Cluster Algorithm	Expectation-Maximization
└ Descriptive Terms	15
Status	
Create Time	2/17/13 1:26 PM

Display 9.30 shows the cluster frequencies.

Display 9.30

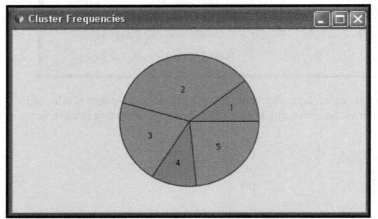

Display 9.31 shows the descriptive terms of the clusters.

Display 9.31

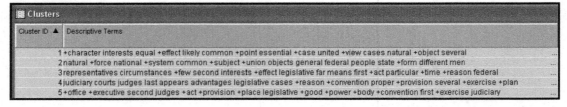

Table 9.20 shows the cross tabulation of the clusters and the target variable.

Table 9.20

```
Expectation-Maximization Clustering
   Table of CLUSTER by TARGET

        The FREQ Procedure
```

Frequency Row Pct	Table of CLUSTER by TARGET			
		TARGET		
	CLUSTER	0	1	Total
	1	4 50.00	4 50.00	8
	2	5 18.52	22 81.48	27
	3	12 75.00	4 25.00	16
	4	3 37.50	5 62.50	8
	5	2 11.11	16 88.89	18
	Total	26	51	77

Data set: TextCluster2_TRAIN

Clusters 2 and 5 are dominated by papers written by Hamilton.

9.10 Exercises

1. Collect URLs by typing the words "Economy," "GDP," "Unemployment Rate," and "European Debt" in Google search. Search for each word separately and collect 5 or 10 links for each word.
2. Repeat the same for the words "health," "vitamins," "Medical," and "health insurance."
3. Use %TMFILTER to download web pages, as shown in Display 9.1. Store the pages as text files in different directories
4. Use %TMFILTER to create SAS data sets, as shown Display 9.5. Include a target variable Target, which takes the value 1 for the documents relating to Economics and 0 for the documents not related to Economics.
5. Create a data source.
6. Create a process flow diagram as shown in Display 9.22.
7. Run the nodes in succession, and finally run the **Regression** node.
   Do the logistic regression results make sense? If not, why not?

Notes

1. The text mining nodes are only available if you license SAS Text Miner.
2. A full Singular Value Decomposition of an $m \times n$ matrix results in a U matrix of size $m \times n$, a Σ matrix of size $n \times m$ and a V matrix of size $n \times m$. Since I have used a reduced Singular Value Decomposition, the dimensions of the matrices are smaller.
3. Character strings without blank spaces are called tokens.
4. The root or stem of the words "is" and "are" is "be".
5. The document-term matrix is a data set that is in a spreadsheet format where the rows represent the documents, columns represent the terms, and the cells show the adjusted frequency of occurrence of the terms in each document.
6. A clear exposition of the Ward's method can be found in Wishart, David. "An Algorithm for Hierarchical Classifications." *Biometrics* Vol. 25, No.1 (March 1969: 165-170.
7. Markov, Zdravko and Daniel T. Larose. 2007. *Data Mining the Web Uncovering Patterns in Web Content, Structure, and Usage*. Wiley-Interscience.

Index

A

B

C

(continued)

(continued)

CPSIA information can be obtained
at www.ICGtesting.com
Printed in the USA
LVOW03s2227240216

476617LV00009B/56/P